SO-ASJ-004

ORGANOMETALLIC REACTIONS

Volume 1

ADVISORY BOARD

G. E. COATES, Professor of Chemistry

Chairman, Department of Chemistry
University of Wyoming
Laramie, Wyoming 82070

H. GILMAN, Professor of Chemistry

Department of Chemistry
Iowa State University
Ames, Iowa 50010

F. HEIN, Professor

Director of Institute of Coordination Compounds of the Academy of Sciences of DDR,
Jena, East Germany

A. N. NESMEYANOV, Professor

Institute for Elementoorganic Compounds
Moscow, U.S.S.R.

G. WITTIG, Professor

Department of Chemistry
University of Heidelberg
Heidelberg, W. Germany

ORGANOMETALLIC REACTIONS

Volume 1

EDITED BY

Ernest I. Becker

Department of Chemistry
University of Massachusetts
Boston, Massachusetts

Minoru Tsutsui

Department of Chemistry
Texas A & M University
College Station, Texas

Wiley-Interscience

A Division of John Wiley & Sons, Inc.

NEW YORK · LONDON · SYDNEY · TORONTO

Copyright © 1970, by John Wiley & Sons, Inc.

All rights reserved. No part of this book may be reproduced
by any means, nor transmitted, nor translated into a machine
language without the written permission of the publisher.

Library of Congress Catalog Card Number: 74-92108

ISBN 0 471 06135 2

Printed in the United States of America

10 9 8 7 6 5 4 3 2 1

Preface

The primary literature on organometallic chemistry has undergone phenomenal growth. The number of papers published from 1955 to 1970 is about equal to all prior literature. Together with this intense activity there has developed a complexity in the literature. Thus specialized texts and teaching texts, a review journal, an advances series, and a research journal have all appeared during this period. The present series also reflects this growth and recognizes that many categories of organometallic compounds now have numerous representatives in the literature.

The purpose of *Organometallic Reactions* is to provide complete chapters on selected categories of organometallic compounds, describing the methods by which they have been synthesized, and the reactions they undergo. The emphasis is on the preparative aspects, although structures of compounds and mechanisms of reactions are briefly discussed and referenced. Tables of all of the compounds prepared in the category under consideration and detailed directions for specific types make these chapters particularly helpful to the preparative chemist. While the specific directions have not been refereed in the same way as are those in *Organic Syntheses* and *Inorganic Syntheses*, the personal experiences of the authors often lend special merit to the procedures and enables the reader to avoid many of the pitfalls frequently encountered in selecting an experimental procedure from the literature.

We acknowledge a debt of gratitude to the contributing authors whose dedication and skill in preparing the manuscripts cannot adequately be rewarded. It has been gratifying to note that virtually all invitations to contribute have been accepted at once. We also owe thanks to the publisher for encouragement and even the "gentle prod" when necessary to see these volumes to their completion.

Ernest I. Becker
Minoru Tsutsui

September 1970
Editors

v

Contents

REDISTRIBUTION REACTIONS OF ORGANOALUMINUM
COMPOUNDS.
    By T. Mole, *C.S.I.R.O. Chemical Research Laboratories,
    Melbourne, Australia*   1

CHEMICAL FIXATION OF MOLECULAR NITROGEN.
    By M. E. Vol'pin and V. B. Shur, *Institute of Organoelemento
    Compounds, Academy of Sciences of U.S.S.R., Moscow,
    U.S.S.R.*   55

REACTIONS OF ORGANOMERCURY COMPOUNDS.
    By L. G. Makarova   119

Author Index   349

Subject Index   371

ORGANOMETALLIC REACTIONS

Volume 1

Redistribution Reactions of Organoaluminum Compounds

T. MOLE

C.S.I.R.O. Chemical Research Laboratories, Melbourne, Australia

I. Introduction 3
  A. Scope of Review 4
  B. Mechanisms of Redistribution 4
  C. Lewis Acidity of Organoaluminum Compounds . . . 5
  D. Self-Association of Organoaluminum Compounds . . . 6
  E. Labilities of Exchangeable Groups 7
  F. Positions of Redistribution Equilibria 8
II. Redistribution between Organoaluminum Compounds . . . 9
  A. Trialkylaluminums 10
    1. Bridge–Terminal Equilibration in Trimethylaluminum . 10
    2. Reactivity of Monomeric Trimethylaluminum toward Exchange 10
    3. Bridge–Terminal Equilibration in Tri-*n*-alkylaluminums 10
    4. Redistribution Equilibria in Mixtures of Trialkylaluminums 11
  B. Alkylaluminum Alkoxides 12
    1. Exchange between Trialkylaluminums and Dialkylaluminum Alkoxides 12
    2. Exchange between Trialkylaluminums and Aluminum Alkoxides 13
  C. Organoaluminum Halides 14
    1. Stability of Halogen-Bridged Dimers 14
    2. Reaction of Triorganoaluminums with Aluminum Halides 15
  D. Unsaturated Organoaluminum Compounds 18
    1. Dialkylarylaluminums 18
    2. Dialkylalkynylaluminums 19
    3. Exchange between Trimethylaluminum and Dimethyl-(phenylethynyl)aluminum 19
  E. Compounds with Mixed Bridges 20

1

F. Organoaluminum Hydrides 22
G. Exchange Reactions of Alkali Metal Tetraorganoaluminates 23
H. Redistribution in the Presence of Lewis Bases 24
III. Exchange between Organoaluminum Compounds and Compounds of Other Metals 26
A. Reaction with Other Organometallic Compounds . . . 26
1. Organic Derivatives of the Electropositive Metals . . 27
2. Organoboron Compounds 27
3. Silicon, Tin, and Lead Compounds 28
B. Reaction with Hydrides Organometallic Hydrides and Organmetallic Compounds 28
C. Reaction with Metal Carboxylates 30
D. Reaction with Alkoxides, Oxides, and Related Compounds 31
1. Zinc 31
2. Boron 31
3. Silicon 31
4. Tin and Lead 32
5. Titanium 32
6. Arsenic and Antimony 32
E. Reactions with Halides 32
1. Beryllium 33
2. Zinc, Cadmium, and Mercury 33
3. Boron 34
4. Gallium, Indium, and Thallium 35
5. Tin 35
6. Lead 37
7. Germanium 38
8. Silicon 38
9. Phosphorus 40
10. Arsenic 40
11. Antimony 40
12. Bismuth 41
13. Transition Metals 41
F. Reaction with Acetylacetonates 43
1. In the Presence of Dipyridyl 43
2. In the Presence of Phosphines 43
3. In the Presence of Other Ligands 45
G. Experimental Methods and Techniques 45
1. Dimethylzinc 46
2. Dimethylcadmium 46
3. Diisobutylcadmium 46
4. Tetraethyltin 47

5. Tetrabutyltin 47
6. Triethyltin Chloride 47
7. Triethylbismuth 47
References   47

I. INTRODUCTION

The development of organometallic chemistry hinges on an under-standing of the main types of reactions by which organometallic synthesis is accomplished. The redistribution or exchange reactions comprise one of these types. In its simplest form the redistribution reaction is defined by Eq. (1):

$$M{-}R + M'{-}R' \longrightarrow M{-}R' + M'{-}R \qquad (1)$$

A group R on metal atom M moves to another metal atom M', while a group R' on M' moves to M. The significance and generality of the reaction appear to have been first realized by Calingaert and Beatty.[1] In recent years Lockhart[2] and Moedritzer[3] have reviewed the reaction generally. Van Wazer and Moedritzer have dealt in detail with the redistribution reactions of the organometallic compounds of silicon and germanium,[4,5] giving particularly valuable treatments of the equilibrium aspects of redistribution, but devoting little attention to the kinetic features of the reaction. Substantial attention is paid to redistribution reactions in two recent reviews of organoaluminum chemistry.[6,7]

The redistribution reactions of organoaluminum compounds are worth particular attention for two reasons. First, kinetic and mechanistic information about redistribution is available more freely for the organoaluminum compounds than for most organometallic compounds; to a large extent this information has been derived from proton magnetic resonance spectroscopic studies on methylaluminum compounds.[8] Second, the use of organoaluminum compounds as polymerization catalysts has led to their ready availability as starting materials for the synthesis of other organometallic compounds by way of redistribution reactions.[6]

In general the metal atoms, M and M', of Eq. (1) are multivalent. Hence in practice redistribution is often not a single reaction, but a sequence of related reactions. For example, reaction of trimethylaluminum with aluminum chloride in ether to produce either dimethylaluminum chloride or methylaluminum dichloride requires two redistributions: either Eq. (2) followed by Eq. (3), or Eq. (2) followed by Eq. (4).[9]

$$Me_3Al \cdot OEt_2 + AlCl_3 \cdot OEt_2 \rightleftharpoons Me_2AlCl \cdot OEt_2 + MeAlCl_2 \cdot OEt_2 \qquad (2)$$

$$Me_3Al \cdot OEt_2 + MeAlCl_2 \cdot OEt_2 \rightleftharpoons 2Me_2AlCl \cdot OEt_2 \qquad (3)$$

$$Me_2AlCl \cdot OEt_2 + AlCl_3 \cdot OEt_2 \rightleftharpoons 2MeAlCl_2 \cdot OEt_2 \qquad (4)$$

Furthermore, the groups R, exchanged in a redistribution, are not necessarily monovalent. The formation of silicone polymers depends upon repeated redistribution reactions in which the exchanging groups are divalent oxygen atoms. Even if R is nominally monovalent (e.g., CH_3—, Cl—, $PhC{\equiv}C$—), it is not necessarily monocoordinate. Herein lies the difficulty in defining the redistribution reaction in rigorous terms.

For example, both dimethylaluminum bromide (1) and dimethylaluminum isopropoxide (2) are dimeric. At 80° in benzene they redistribute their bridging bromo and isopropoxyl groups to give compound (3), which has a mixed bromo–isopropoxyl bridge.[10]

Although this reaction does not fall within the simple definition offered by Eq. (1), it is nevertheless a redistribution reaction within the scope of our present discussion. Self-association, as in compounds (1)–(3), is one of the important features of organoaluminum chemistry. The association of organoaluminum compounds and the stability of the associated molecules toward dissociation frequently determine whether or not redistribution reactions occur.

A. Scope of Review

The purpose of this review is to provide an account of redistribution reactions in which at least one of the two reactants is an organoaluminum compound. Most attention will be paid to those cases in which *both* the reactants are aluminum compounds, so as to put the trivial but fundamental redistribution chemistry of organoaluminum compounds on as firm a base as possible. A less comprehensive account will be given of those reactions in which the second reactant is a derivative of another metal or a metalloid, with the object of indicating without undue detail the scope and limitations of this route to the organometallic compounds of other metals and metalloids.

A relatively superficial treatment will be given to the reactions of organoaluminums with transition metal compounds, since in these cases simple redistribution is the exception rather than the rule. More commonly the over-all result is a reduction of the transition metal compound.

B. Mechanisms of Redistribution

Although redistribution occurs most easily when electronegative groups are exchanged between electropositive metal centers, the reaction does not always proceed merely by ionic dissociation and a fresh regrouping of ions. Such a mechanism is not likely to apply to the reactions of organoaluminum compounds, for these compounds mostly exist as discrete, covalent molecules* and the redistributions are normally carried out in solvents which do not support ionization.

The electronegative groups which are most easily redistributed have lone pairs of electrons, and the metal atoms about which the groups are redistributed are commonly electron-deficient. Thus formation of new covalent bonds is able to proceed in advance of the rupture of the old bonds, providing a low energy path for reaction. If such interactions occur between both groups being exchanged and both metal atoms, then redistribution takes place by a concerted mechanism with a four-centered transition state, as shown in Eq. (5).

$$\text{M—R: + M'—R': } \longrightarrow \left[\begin{array}{c} \text{M} \begin{array}{c} \text{R} \\ \end{array} \text{M'} \leftrightarrow \text{M} \begin{array}{c} \text{R} \\ \end{array} \text{M'} \\ \text{R'} \quad\quad \text{R'} \end{array} \right] \longrightarrow \text{M—R' + M'—R} \tag{5}$$

If, on the other hand, donor–acceptor interactions are not synchronized, then redistribution occurs by at least two steps, and ionic intermediates are involved, as in Eq. (6).

$$\text{M—R: + M'—R' } \longrightarrow \text{[M—R } \rightarrow \text{M'—R'] } \longrightarrow \text{M}^+ + \text{(R—M'—R')}^- \tag{6}$$

If R is an organic group, reaction (6) is recognizable as an electrophilic substitution at a carbon atom. Electrophilic substitution is probably an important redistribution mechanism for organomercury compounds in ionizing solvents.[12] Although there is no significant evidence for redistribution of organoaluminum compounds by such a mechanism, reaction (6) represents a reasonable course for exchange reactions of the complex organoaluminum anions. However, in organoaluminum chemistry we shall generally be concerned with the concerted exchange of groups, as depicted in Eq. (5).

C. Lewis Acidity of Organoaluminum Compounds[13]

The monomeric compounds of tervalent aluminum are strong monobasic Lewis acids. The formation of 1:1 adducts with Lewis bases (e.g., (4)[14] and (5)[15]) makes the aluminum atom four-coordinate and completes a $3s^2, 3p^6$ octet of electrons. Coordination of a second molecule of base requires the occupancy of orbitals of much higher energy and is thus rare in the ground states of organoaluminum molecules.

$$\begin{array}{ccc} \text{Ph}_3\text{Al} \leftarrow\text{OEt}_2 & \text{Ph}_3\text{Al} \leftarrow\text{NMe}_3 & \text{Me}_3\text{N}\rightarrow \text{AlH}_3 \leftarrow\text{NMe}_3 \\ \text{(4)} & \text{(5)} & \text{(6)} \end{array}$$

* See, however, Lehmkuhl and Kobs[11] and Schmidbauer.[11a]

Aluminum hydride forms a bis-trimethylamine adduct (6),[16,17] but many other claims for the existence of pentacoordinate aluminum atoms in organoaluminum compounds are open to doubt.[18]

D. Self-Association of Organoaluminum Compounds[13]

The Lewis acidity of organoaluminum monomers commonly results in self-association, particularly in the absence of solvents which are Lewis bases. Most commonly cyclic dimers are formed, e.g., dimethylaluminum bromide (1)[19] and dimethylaluminum isopropoxide (2). Cyclic trimers are sometimes found (e.g., dimethylaluminum methoxide[15,20] (7)), but steric interactions within a possible trimer are generally so large that the angularly strained dimer occurs instead. Dimethylaluminum phenoxide is particularly interesting in that a dimer (8) and a trimer (9) coexist in slowly attained equilibrium.[21]

$$
\begin{array}{c}
\text{Me}_2\text{Al}-\text{OMe} \\
\diagup \qquad\qquad \diagdown \\
\text{MeO} \qquad\qquad\qquad \text{AlMe}_2 \\
\diagdown \qquad\qquad \diagup \\
\text{Me}_2\text{Al}-\text{OMe}
\end{array}
$$
(7)

$$
\begin{array}{cc}
\begin{array}{c}
\text{OPh} \\
\diagup \quad \diagdown \\
\text{Me}_2\text{Al} \qquad \text{AlMe}_2 \\
\diagdown \quad \diagup \\
\text{OPh}
\end{array}
&
\begin{array}{c}
\text{Me}_2\text{Al}-\text{OPh} \\
\diagup \qquad\qquad \diagdown \\
\text{PhO} \qquad\qquad\qquad \text{AlMe}_2 \\
\diagdown \qquad\qquad \diagup \\
\text{Me}_2\text{Al}-\text{OPh}
\end{array}
\\
(8) & (9)
\end{array}
$$

With more than one highly electronegative substituent on the aluminum atom, coordination numbers greater than four and more complex states of association become possible. Thus aluminum isopropoxide exists both as a trimer and a tetramer, in the latter of which one of the four aluminum atoms is thought to be hexacoordinate.[22]

The Lewis acidities of aluminum compounds are so marked that association may occur even if there are no substituents of high electronegativity on the aluminum atom. The most familiar example of such electron-deficient association is provided by trimethylaluminum dimer (10).[19,23-25]

$$
\begin{array}{c}
\text{H}_3\text{C} \quad \text{CH}_3 \quad \text{CH}_3 \\
\diagdown \diagup \diagdown \diagup \\
\text{Al} \qquad \text{Al} \\
\diagup \diagdown \diagup \diagdown \\
\text{H}_3\text{C} \quad \text{CH}_3 \quad \text{CH}_3
\end{array}
$$
(10)

One sp^3 orbital of the bridging carbon atom overlaps with an sp^3 orbital of each of the two aluminum atoms to form a three-centered bond. Such association is generally not so strong as association through donor–

acceptor bonds. It almost always results in dimers; the dialkylaluminum hydrides, which are hydride-bridged and commonly trimeric,[19,20,26] are notable exceptions.

The association of organoaluminum compounds is highly relevant to any consideration of their redistribution reactions. The four-coordinate aluminum atoms of structures (7)–(10) are crowded, and furthermore are required to use vacant orbitals of relatively high energy in order to function as Lewis acids. Hence the transition states for exchange by concerted mechanisms between these dimeric molecules and other organometallic compounds are difficult to attain. Therefore the dimers are unreactive as compared with monomers.

The greater strength of donor–acceptor bridges as compared with three-centered bridge-bonds is reflected in the stabilities of the bridges toward dissociation in the presence of external bases. Neither dimethylaluminum methoxide nor dimethylaluminum dimethylamide forms complexes with diethyl ether or trimethylamine, while dimethylaluminum thiomethoxide reacts with trimethylamine but not ether.[15] On the other hand, dialkylaluminum halides and trialkylaluminums complex with both these bases.

Since dimethylaluminum chloride and methylaluminum dichloride are chloro-bridged dimers,[19,27,28] it is evident that halogen bridges are preferred to alkyl bridges. The degree of association of triorganoaluminum decreases in the series $Me_3Al > Et_3Al > i\text{-}Bu_3Al$; trimethylaluminum is almost entirely dimeric at ambient temperature, whereas triisobutylaluminum is monomeric.[20,23,29]

Unsaturated organic groups serve as better bridges than saturated organic groups,[7,30,31] and bridge strength decreases in the series: vinyl, alkynyl > aryl > alkyl.[10] Thus dimethylphenylaluminum dimer[30,31] is phenyl bridged, and dimethyl(phenylethynyl)aluminum is a phenylethynyl-bridged dimer.[31–33] Of the known triorganogalliums, only trivinylgallium,[34] dimethyl(phenylethynyl)gallium,[35] dimethyl(octynyl)gallium,[35] and (probably) dimethylvinylgallium[36] are dimeric; trimethylgallium and triphenylgallium are monomeric.[37,38]

From the above information, an order of decreasing bridge-bond strengths may be constructed as follows: —OR, —NR$_2$ > —halogen, —SR, vinyl, alkynyl > aryl > methyl > ethyl > isobutyl. This order is based on equilibrium stabilities, and will not be entirely the same as the order of kinetic labilities.

E. Labilities of Exchangeable Groups

Analogy between the transition state of redistribution reactions (2) and the dimeric associated structures of organoaluminum compounds suggests that the four-centered transition state should be most easily achieved when the groups being redistributed are good bridging groups in organoaluminum dimers. Hence the sequence of bridge strengths quoted above may serve as a reasonable guide to the relative labilities of groups in redistribution reactions. Ruff[39] has found the order of reactivity of mercury compounds toward $AlH_3 \cdot NMe_3$ to be $HgCl_2 > (CH_3C{\equiv}C)_2Hg > (CH_2{=}CH)_2Hg > Ph_2Hg > Bu_2Hg$, and Hartmann[40] has found alkynyltin compounds to be particularly reactive toward redistribution.

If redistribution is looked on as an electrophilic substitution at the group being moved from one metal atom to another, some similarity might be expected between the order of labilities of groups in redistribution reactions and their orders of reactivity in other electrophilic reactions. Thus redistribution labilities might correlate with migratory aptitudes in carbonium-ion rearrangements,[41] and again with the relative ease of transfer of organic groups from the metal atom of an organometallic compound to the carbon atom of a carbonyl group. For example, diethyl(phenylethynyl)aluminum reacts with benzaldehyde to give phenyl(phenylethynyl)carbinol rather than 1-phenylpropanol.[42]

F. Positions of Redistribution Equilibria

The factors which determine the position of equilibrium attained in a series of redistribution reactions have been discussed by Calingaert and Beatty,[1] Lockhart,[2] and Van Wazer and Moedritzer.[3-5] Three broad generalizations may be made:

1. Redistribution proceeds in such a direction as to place the more electronegative group on the more electropositive metal. For example, trimethylaluminum reacts with cadmium acetate to give dimethylcadmium,[43,44] and triphenylaluminum reacts with mercuric chloride to give diphenylmercury.[45]

2. Like groups distribute about metal atoms in a random manner. This generalization was first revealed by Calingaert and Beatty, when they redistributed the alkyl groups in a mixture of equimolar amounts of tetramethyllead and tetraethyllead in the presence of Lewis acid catalyst. Tetramethyllead, trimethylethyllead, dimethyldiethyllead, methyltriethyllead, and tetraethyllead were formed in the mole ratios of 1:4:6:4:1, exactly as expected on a random statistical basis.[46-49] Although such perfectly random distributions have not been found for organoaluminum compounds, random statistics provide a reasonable guide to the product distributions when trimethylaluminum undergoes exchange with triethylaluminum or triphenylaluminum in polar solvents.[50-52]

It is sometimes possible to physically separate a pure compound from a random mixture, and so displace the equilibrium. A mixture of diethylzinc and trimethylaluminum probably redistributes its alkyl groups randomly, but the most volatile component of the equilibrium, dimethylzinc, may nevertheless be distilled from the mixture in good yield.[53]

3. With very few exceptions, unlike groups (in particular groups of differing electronegativities) tend to distribute in such a way as to favor proportionation* products. Thus a 2:1 mixture of triphenylaluminum and aluminum chloride in ether is converted quantitatively to diphenylaluminum chloride, and the same reactants in 1:2 ratio give phenylaluminum dichloride quantitatively.[9] The proportionation reaction is the most

$$2Ph_3Al + AlCl_3 \longrightarrow 3Ph_2AlCl \qquad (7)$$

$$Ph_3Al + 2AlCl_3 \longrightarrow 3PhAlCl_2 \qquad (8)$$

important method of preparing organoaluminum compounds in which the aluminum is bonded to two different groups. Its wide applicability first appears to have been exploited by von Grosse and Mavity.[54]

When a trialkylaluminum is treated with the halide of another less electropositive metal, a balance between factors 1 and 3 determines the over-all stoichiometry of reaction.

For organoaluminum compounds a fourth factor needs to be considered, namely the marked tendency to form those redistribution products which are most strongly associated. For example, a 2:1 mixture of trimethylaluminum with triphenylaluminum produces dimeric dimethylphenylaluminum, because this results in the maximum utilization of phenyl groups in bridging positions.[30,31,50] By contrast in ether, which forms 1:1 adducts with the organoaluminum compounds, the dimethylphenylaluminum is extensively disproportionated to trimethylaluminum and methyldiphenylaluminum.[30,52]

II. REDISTRIBUTION BETWEEN ORGANOALUMINUM COMPOUNDS

With few exceptions, notably triisobutylaluminum, aluminum compounds are associated in hydrocarbon solvents, and this has an important bearing upon the reactivity of the organoaluminum compounds toward redistribution. Both the thermodynamic stabilities of the associated forms relative to one another and the kinetic stabilities of associated forms toward dissociation need to be considered. Some of the most useful kinetic information comes from redistributions which are chemically trivial, such as the interchange of terminal and bridging methyl groups in trimethylaluminum dimer.

* The reverse of disproportionation.

A. Trialkylaluminums

1. Bridge–Terminal Equilibration in Trimethylaluminum

Trimethylaluminum in toluene at $-50°$ shows two proton magnetic resonance peaks in 2:1 intensity ratio, as would be expected for the dimeric structure (10). As the temperature increases to $0°$, the two peaks progressively coalesce to a single peak. This behavior may be traced to an exchange of methyl groups between bridging and terminal positions, and kinetic information about the rate of exchange can be derived from the shape of the coalescing spectrum.[55-59] The exchange is a first-order reaction of trimethylaluminum dimer, and Williams and Brown[59] have argued convincingly that the rate-determining step is dissociation of the dimer to two molecules of monomer. This is followed by rapid reassociation to dimer, the over-all result being an interchange of bridging and terminal methyl groups.

$$
\begin{array}{c}
\text{Me} \\
\diagup \quad \diagdown \\
\text{Me}_2\text{Al} \qquad \text{AlMe}_2 \underset{k_{-1}}{\overset{k_1}{\rightleftharpoons}} 2\text{Me}_3\text{Al} \\
\diagdown \quad \diagup \\
\text{Me}
\end{array} \qquad (9)
$$

Although the dimer dominates the equilibrium, dissociation is fast; $k_1 \sim 10^2 \sec^{-1}$ at $-50°$ in toluene, and $\sim 5 \sec^{-1}$ at $-50°$ in cyclopentane.

2. Reactivity of Monomeric Trimethylaluminum Toward Exchange

Trimethylaluminum is highly reactive toward exchange with monomeric alkyls of the more electropositive metals. The exchange of methyl groups between trimethylaluminum and trimethylgallium or trimethylindium in hydrocarbon solvent is fast even at $-40°$. From proton magnetic resonance observations on the exchange-broadened methyl peaks of trimethylgallium and trimethylindium, Williams and Brown concluded that dissociation of trimethylaluminum dimer is the rate-determining step for exchange.[59,59a] Exchange of methyl groups between trimethylaluminum and dimethylzinc is equally fast and is probably controlled by the same rate-determining step.[8]

On the other hand, exchange between trimethylaluminum and dimethylmercury is much slower,[8] and in this case a dissociation of trimethylaluminum dimer cannot be rate determining.

3. Bridge–Terminal Equilibration in Tri-n-alkylaluminums

In the simpler tri-n-alkylaluminums, the proton magnetic resonance peaks of the α-CH_2 resonances most readily provide information about the dimer–monomer equilibrium.[57,60,61] For triethylaluminum, tripropylaluminum, and tributylaluminum in cyclopentane, the α-CH_2 resonances of the bridging and terminal alkyl groups may be resolved from one another at $-50°$. The temperature at which bridging and terminal peaks coalesce is about $20°$ lower than in the case of trimethylaluminum.[57] Hence it appears that the rate of dissociation of dimer to monomer is greater for the tri-n-alkylaluminums than for trimethylaluminum. Equilibrium (10) also favors the monomer more heavily when R = n-alkyl than when R = methyl.[20,23,29]

$$R_2Al \cdot R_2 \cdot AlR_2 \rightleftharpoons 2R_3Al \tag{10}$$

By contrast, bridge-terminal equilibration in tricyclopropylaluminum dimer is much slower than in trimethyl aluminum dimer; coalescence of the proton magnetic resonance spectrum occurs only at $70°$.[61a]

4. Redistribution Equilibria in Mixtures of Trialkylaluminums

Redistribution equilibria in mixtures of organoaluminum compounds favor those products in which the best bridging groups are used to the fullest extent in bridging. Cryoscopic measurements indicate that monomeric triisobutylaluminum reacts with dimeric trimethylaluminum with the result shown in Eq. (11), so as to allow the methyl groups to serve as

$$4\,i\text{-Bu}_3Al + Al_2Me_6 \longrightarrow 3Me_2\,i\text{-Bu}_4Al_2 \tag{11}$$

bridges to the maximum extent.[20] Proton magnetic resonance spectroscopy confirms this conclusion.[57,62,63] At $-65°C$ a mixture of trimethylaluminum and triisobutylaluminum, mixed in the proportions required by Eq. (11), has only 6% of the methyl groups in the terminal positions of dimers.[62] On the other hand, when the reactants, mixed in the same proportions, attain equilibrium in ether, methyldiisobutylaluminum etherate is not particularly favored. The equilibrium constant, $[Me\,i\text{-Bu}_2Al \cdot OEt_2]^2/[Me_2\,i\text{-BuAl} \cdot OEt_2] \cdot [i\text{-Bu}_3Al \cdot OEt_2] = 7$, is only twice what would be expected for a completely random distribution of methyl and isobutyl groups.[43]

Less preference is shown for a methyl bridging group when the alternative is an ethyl or propyl bridging group.[63] A mixture of triethylaluminum with trimethylaluminum in 2:1 molar ratio has 26% of the methyl groups in terminal positions.[62] There is no direct evidence for the existence of dimers in which two unlike alkyl groups bridge two aluminum atoms, although such dimers are to be expected in a mixture of trialkylaluminums. Hoffmann[56] has postulated the existence of species (11) on the

i-Bu

i-Bu$_2$Al   Al i-Bu$_2$

Me

(11)

basis of the chemical shift of the Me—Al proton resonance in a 5:1 mixture of triisobutylaluminum with trimethylaluminum (see also Yamamoto and Hayamizu[63]).

B. Alkylaluminum Alkoxides

1. Exchange between Trialkylaluminums and Dialkylaluminum Alkoxides[64]

Dissociation of the dimers of dimethylaluminum isopropoxide (12) and dimethylaluminum t-butoxide (13), followed by reassociation to dimers, must result in a substantial proportion of the dimer with a mixed iso-propoxide-t-butoxide bridge (14), characterized by a CH_3—Al proton magnetic resonance chemical shift intermediate between those of (12) and (13).

$$
\begin{array}{ccc}
\underset{\text{(12)}}{\overset{\displaystyle \text{OPr-}i}{Me_2Al \diagup \diagdown AlMe_2}}
& \underset{\text{(13)}}{\overset{\displaystyle \text{OBu-}t}{Me_2Al \diagup \diagdown AlMe_2}}
& \underset{\text{(14)}}{\overset{\displaystyle \text{OPr-}i}{Me_2Al \diagup \diagdown AlMe_2}}
\end{array}
$$

Since no such peak appears during an hour in benzene at 60°C, it may be concluded that the dimeric alkoxides do not readily dissociate.[43] Nevertheless exchange of a methyl group of dimethylaluminum isopropoxide for an ethyl group of triethylaluminum at 40° is complete within a few minutes.[64] This exchange must be occurring without prior dissociation of the dimer of dimethylaluminum isopropoxide.

Triethylaluminum exchanges alkyl groups with dimethylaluminum isopropoxide at least a hundred times faster than trimethylaluminum reacts with diethylaluminum isopropoxide.[64] This difference in reactivity is possibly best understood if the rate-determining step for exchange is reaction of monomeric trialkylaluminum with dimeric dialkylaluminum alkoxide, since triethylaluminum is more extensively dissociated than trimethylaluminum.

Dimeric dialkylaluminum t-butoxides, like dialkylaluminum isopropoxides, exchange their terminal alkyl groups rather than the bridging alkoxyl groups.[64] This confirms the reasonable assumption that terminal groups are in general more exchange-labile than bridging groups in dimeric organoaluminum compounds.

The rate of exchange between trialkylaluminums and dimeric dialkylaluminum alkoxides falls off with increasing size of the alkoxyl group. Reaction of trimethylaluminum with diethylaluminum ethoxide requires less than a minute at 40° in benzene, whereas the corresponding reaction of diethylaluminum isopropoxide requires several hours at 40° and that of diethylaluminum *t*-butoxide much longer still.[64] Furthermore, the diisobutylaluminum alkoxides react more slowly than diethylaluminum alkoxides. For example, exchange between trimethylaluminum and diisobutylaluminum isopropoxide requires several hours at 80° in benzene.[43] From the above evidence it seems that steric effects in the dimeric dialkylaluminum alkoxide are of great importance in determining the rate of exchange.

Because steric interactions are more pronounced in trimeric dialkylaluminum alkoxides than in dimers, greater hindrance to exchange would be expected for the trimers. In fact trimethylaluminum exchanges more slowly with trimeric diethylaluminum methoxide than with dimeric diethylaluminum ethoxide. Many minutes at 40° are required for the methoxide, but one minute suffices for the ethoxide.[64]

Despite the fact that triisobutylaluminum is entirely monomeric, it is far less reactive toward dialkylaluminum alkoxides than is triethylaluminum. Complete exchange between triisobutylaluminum and trimeric dimethylaluminum methoxide requires an hour at 80° in benzene.[43] It appears that isobutyl-bridged transition states for exchange, like isobutyl-bridged structures for organoaluminum dimers, are relatively inaccessible.

2. *Exchange between Trialkylaluminums and Aluminum Alkoxides*

Von Grosse and Mavity[54] described the preparation of dimethylaluminum methoxide, methylaluminum dimethoxide, diethylaluminum ethoxide, and ethylaluminum diethoxide by heating together trialkylaluminum and aluminum alkoxide in the appropriate stoichiometric ratios. The dialkylaluminum alkoxides were well characterized, but the less volatile alkylaluminum dialkoxides were not.

More detailed information on the formation of alkylaluminum alkoxides by the proportionation reaction is available from a proton magnetic resonance study of the reaction of trimethylaluminum with aluminum isopropoxide.[65] Even if the association of the reactants and the products is neglected, three stages in redistribution need to be considered:

$$Me_3Al + Al(OPr\text{-}i)_3 \longrightarrow Me_2AlOPr\text{-}i + MeAl(OPr\text{-}i)_2 \quad (12)$$

$$Me_3Al + MeAl(OPr\text{-}i)_2 \longrightarrow 2Me_2AlOPr\text{-}i \quad (13)$$

$$Me_2AlOPr\text{-}i + Al(OPr\text{-}i)_3 \longrightarrow 2MeAl(OPr\text{-}i)_2 \quad (14)$$

Trimethylaluminum and aluminum isopropoxide, when mixed in 2:1 molar ratio in benzene at 40°, immediately give dimethylaluminum isopropoxide in quantitative yield. Therefore reactions (12) and (13) are fast. If a larger proportion of aluminum isopropoxide is used, a mixture of dimethylaluminum isopropoxide and aluminum isopropoxide results, even after several hours at 85°. Reaction (14) is therefore very slow. By contrast, when trimethylaluminum is reacted with aluminum ethoxide, all stages in redistribution are complete within a few minutes at 40°, although the reaction of the dimethylaluminum alkoxide with aluminum alkoxide is again by far the slowest of the three steps.[65]

In general there should be no difficulty in preparing dialkylaluminum alkoxides by redistribution between trialkylaluminums and aluminum alkoxides in 2:1 molar ratio. Alkylaluminum dialkoxides, on the other hand, can be prepared by redistribution only when the intermediate dialkylaluminum alkoxides are relatively reactive toward exchange. Since this reactivity is limited largely by steric considerations, alkylaluminum di-*n*-alkoxides should be prepared easily by redistribution, whereas alkylaluminum di-*sec*-alkoxides should be preparable only with difficulty. It may not be possible to obtain alkylaluminum di-*t*-alkoxides by proportionation; they are also formed only very slowly by reaction of trialkylaluminums with tertiary alcohols.

The exchange of alkyl groups for alkoxyl groups probably serves as a good model for the exchange of alkyl groups for other good bridging groups (e.g., amino groups, thioalkoxyl groups, and phosphine groups). There is indirect evidence from studies of the polymerization of ethylene that diethylaluminum diethylamide is particularly unreactive toward exchange with trialkylaluminums.[66]

Hydrocarbons represent the best choice of solvent for redistribution reactions between trialkylaluminums and aluminum alkoxides or related compounds. If halogenated solvents are used, there exists the possibility of abstraction of halogen from the solvent. In donor solvents, such as ether, the trialkylaluminum is less reactive toward exchange. For example, there is no reaction between trimethylaluminum and aluminum isopropoxide in ether at room temperature.[64]

C. Organoaluminum Halides

1. Stability of Halogen-Bridged Dimers

The thermodynamic stability of halogen-bridged dimers toward dissociation is less than that of the alkoxyl-bridged dimers but greater than that of the weakly bonded alkyl-bridged dimers. There is some evidence that the halogen-bridged dimers are kinetically as labile as the alkyl-bridged dimers, and, if this is so, the dimer–monomer equilibrium for halogen-bridged dimers must be characterized by higher rates of both association and dissociation.

The chemical shifts of dimethylaluminum bromide (**15**) and dimethylaluminum chloride (**16**) in the 60 Mc/sec proton magnetic resonance spectrum differ by 9 cps at −74° in toluene. The spectrum of a mixture of

(15) and **(16)** shows not only these two peaks but also a third of intermediate chemical shift, which has been assigned to the chloro-bromo-bridged dimer **(17)**. At $-50°$ only a single exchange-averaged peak is observed. The obvious (but not unique) explanation for the spectral coalescence at $-50°$ is that compounds **(15)**, **(16)**, and **(17)** undergo rapid, reversible dissociation to monomers, with a dimer half-life $< 10^{-1}$ sec.[10]

$$
\begin{array}{ccc}
\text{Me}_2\text{Al} \overset{\displaystyle \text{Br}}{\underset{\displaystyle \text{Br}}{\diamond}} \text{AlMe}_2 & \text{Me}_2\text{Al} \overset{\displaystyle \text{Cl}}{\underset{\displaystyle \text{Cl}}{\diamond}} \text{AlMe}_2 & \text{Me}_2\text{Al} \overset{\displaystyle \text{Br}}{\underset{\displaystyle \text{Cl}}{\diamond}} \text{AlMe}_2 \\
\textbf{(15)} & \textbf{(16)} & \textbf{(17)}
\end{array}
$$

$$
\text{Me}_2\text{Al} \overset{\displaystyle \text{Cl}}{\underset{\displaystyle \text{Cl}}{\diamond}} \text{AlMeCl}
$$

(18)

A mixture of methylaluminum sesquichloride **(18)** with dimethylaluminum chloride **(16)** in toluene at $-40°$ shows two proton magnetic resonance peaks, one for Me_2Al terminal groups and the other for MeClAl terminal groups.[10,67,68] Only an average peak is observed for the Me_2Al terminal group, even though this group must be present in both **(16)** and **(18)**, but two peaks are resolved on cooling to $-85°$. This separation may be due to resolution of the two molecular environments of the Me_2Al group, or to resolution of two distinct methyl groups in the terminal Me_2Al moiety in **(18)**, or to a combination of these factors.[10,67] Whatever the detailed explanation, it seems likely that dissociation of the dimers is fast at $-40°$, but slow at $-85°$.

2. Reaction of Triorganoaluminums with Aluminum Halides

If halogen-bridged dimers are labile, there should be no kinetic difficulties in preparing alkylaluminum halides by proportionation reactions. In fact organoaluminum halides or organoaluminum dihalides are obtained quantitatively upon simply mixing the reactants in the appropriate (2:1 or 1:2) molar ratio.[54,69] The products obtained are thermodynamically stable with respect to disproportionation. Reaction may be conducted either without solvent or in hydrocarbons. It also proceeds rapidly and completely in ethereal solvents,[9] but the products are etherates, from which the ether-free compounds are not easily recovered.

Dialkylaluminum fluorides, which are more highly associated than the other dialkylaluminum halides,[70] may be prepared by reaction (15) only if

$$2R_3Al + AlF_3 \longrightarrow 3R_2AlF \tag{15}$$

the aluminum trifluoride is in a finely divided, reactive form.[71,72] They can also be obtained by reaction of dialkylaluminum chlorides with one mole of potassium fluoride.[73] Diethylaluminum azide has been prepared similarly from diethylaluminum chloride and sodium azide.[74]

Proportionation reactions of R_3Al with $AlCl_3$ and $Al(OR')_3$ to give compounds $RAl(OR')Cl$ have been investigated,[75] but there is some doubt about the nature of the bonding in the time-dependent association of these products.[21,76] The formation of alkylaluminum chlorides on treatment of dialkylaluminum alkoxides with aluminum trichloride is the subject of a patent claim.[76a]

More detailed information about the exchange of a methyl group for a halogen group may be obtained from a study of methylaluminum sesquibromide, which, like the other halogen-bridged dimers, is thought to be dissociating reversibly into monomeric units with a frequency greater than 10 times per second at $-50°$ in toluene. Kinetically first-order coalescence of the individual proton magnetic resonance peaks of the terminal Me_2Al and terminal $MeBrAl$ groups occurs only at $0°$ and above. If dissociation of the dibromo-bridged dimer is fast at $-50°$, the Me_2AlBr and $MeAlBr_2$ monomers must nearly all reassociate through dibromo bridges, with only a small proportion of the monomers associating to the labile dimer (19), intermediate in the exchange of methyl groups from one aluminum atom to another.[10]

$$\underset{\textbf{(19)}}{MeBrAl\!\!\diagup^{Br}\!\!\diagdown_{Me}\!\!\diagdown^{\diagup}AlMeBr} \qquad \underset{\textbf{(20)}}{Me_2Al\!\!\diagup^{Br}\!\!\diagdown_{Me}\!\!\diagdown^{\diagup}AlMe_2}$$

The concentration of (19) in methylaluminum sesquibromide must be infinitesimally small, for even in the most favorable case for a methyl–bromo bridge, namely a mixture of trimethylaluminum with dimethylaluminum bromide, proton magnetic resonance spectra at low temperature give no indication of the presence of (20).[10] By contrast, a mixture of trimethylaluminum with dimethylaluminum chloride contains a significant amount of (21a) in toluene,[10,67] but possibly not in cyclopentane.[77]

$$\underset{\textbf{(21)}}{R_2Al\!\!\diagup^{Cl}\!\!\diagdown_{R}\!\!\diagdown^{\diagup}AlR_2} \qquad \underset{\textbf{(22)}}{R_2Al\!\!\diagup^{Cl}\!\!\diagdown_{R}\!\!\diagdown^{\diagup}AlR_2}$$

(a, R = Me; b, R = Et)

No substantial amount of the corresponding ethyl compound (21b) exists in a mixture of triethylaluminum and diethylaluminum chloride.[78]

Exchange of an ethyl group of triethylaluminum for an ethyl or a chloro group of diethylaluminum chloride is slow on the NMR scale at $-20°$, and is thus slower than exchange between the corresponding methyl compounds.[78]

There are numerous references in the literature to organoaluminum sesquihalides, prepared by the action of organic halides on aluminum turnings or powder. The reaction works well for methyl and ethyl halides

$$3RX + 2Al \longrightarrow R_3Al_2X_3 \qquad (16)$$

and for iodobenzene.[29,54,79-81] Bromobenzene and chlorobenzene react only when the aluminum is specially activated.[82] Autocatalytic elimination of hydrogen halide prevents successful application to the higher alkyl halides.[54,79]

There is some question as to whether the sesquihalides exist as such or are merely mixtures of monohalides and dihalides. Equilibrium (17) is

$$2R_2Al \cdot X_2 \cdot AlRX \rightleftharpoons R_2Al \cdot X_2 \cdot AlR_2 + RXAl \cdot X_2 \cdot AlRX \qquad (17)$$

labile and, according to Brandt and Hoffmann,[68] evenly balanced (see, however, Zambelli et al.[83]). Significant amounts of the dimeric disproportionation products must be present, since careful fractionation of alkylaluminum sesquihalides allows separation of dialkylaluminum halide and alkylaluminum dihalide.[54,79,81] For example, methylaluminum sesquichloride, prepared from methyl chloride and aluminum, may be separated into dimethylaluminum chloride and methylaluminum dichloride on distillation at slightly reduced pressure.[54]

Disproportionation of dialkylaluminum halides and alkylaluminum dihalides involves more than just a dissociation of dihalogeno bridges and is more difficult. Trimethylaluminum cannot be separated upon fractionation of methylaluminum sesquichloride or methylaluminum sesquibromide, although methylaluminum dibromide may disproportionate partially to dimethylaluminum bromide and aluminum tribromide on distillation.[54] However, careful fractionation of methylaluminum sesquiiodide at slightly reduced pressure affords a useful laboratory preparation of trimethylaluminum.[29]

The disproportionation of organoaluminum halides may be brought about more effectively by the thermal decomposition of the complex salts which the organoaluminum halides form with alkali metal halides. For example, when methylaluminum sesquichloride is heated with sodium chloride, dimethylaluminum chloride distils leaving behind methylaluminum dichloride as its complex with sodium chloride.[79,81,84,85] Disproportionation is sometimes accomplished more easily and under milder conditions when only $\frac{1}{2}$ mole of alkali metal halide is used for each gram atom of aluminum.[86,87] Disproportionation of ethylaluminum dichloride in the presence of $\frac{1}{2}$ mole of finely powdered KCl (Eq. (18)) is complete

$$2EtAlCl_2 + KCl \longrightarrow Et_2AlCl + KAlCl_4 \qquad (18)$$

within a few hours at 50–70° either in heptane or in the absence of solvent.[86] The $KAlCl_4$ formed may be filtered off, leaving a solution of diethylaluminum chloride. Other bases, including diglyme,[88] are also able to promote partial disproportionation of alkylaluminum dichlorides when used in half-molar amounts.[67,83,87,89,90] Ionic structures may be important under these conditions.[11]

A more deep-seated disproportionation is possible using a full mole of potassium chloride under more drastic conditions. When $KAlEt_2Cl_2$, formed by fusing potassium chloride with diethylaluminum chloride at 80–100°, is heated to 160–200° at 10^{-3} mm pressure, a 2:1 mixture of triethylaluminum and diethylaluminum chloride distills.[91]

In the presence of sodium fluoride extensive disproportionation of alkylaluminum sesquichlorides is possible. A dialkylaluminum chloride reacts with one equivalent of sodium fluoride in hot xylene precipitating sodium chloride and leaving dialkylaluminum fluoride in solution. The dialkylaluminum fluoride forms a complex salt ($NaAlR_2F_2$) with a further equivalent of sodium fluoride, and, when this is heated for several hours to about 200° at low pressure, disproportionation is steadily forced to completion. Trialkylaluminum distills, leaving Na_3AlF_6.[92,92a]

For a fuller discussion of the displacement of disproportionation equilibria by complex formation with alkali metal halides, the reader is referred to reviews by Ziegler,[6] Köster and Binger,[7] and Lehmkuhl.[93]

D. Unsaturated Organoaluminum Compounds

1. Dialkylarylaluminums

Trimethylaluminum and triphenylaluminum in 2:1 molar ratio, alone or in hydrocarbon solvent, lead quantitatively to crystalline dimethylphenylaluminum,[30,50] which is dimeric in benzene, as are the starting materials. The proton magnetic resonance spectrum in toluene at low temperature shows that the phenyl groups serve as bridges in the dimer and that the dimer is not significantly disproportionated.[30,31] Disproportionation is reported to occur on distillation.[7,94] Dimethyl-p-tolylaluminum dimer may be prepared in the same way and has similar properties.[31] Since the methyl group is the most effective of all the alkyl bridging groups, it may be concluded that other aryl-bridged dialkylarylaluminum dimers besides the dimethylarylaluminums will be preparable from trialkylaluminums and triarylaluminums in 2:1 molar ratio and will be reasonably stable (thermodynamically but not kinetically) with respect to dispro-portionation. Two such compounds, diethylphenylaluminum and diiso-butylphenylaluminum, have been prepared as crystalline solids.[50]

No such stability should attach to alkyldiarylaluminums. The stability of dialkylarylaluminums is purely a function of the effectiveness of the aryl bridge, and the monomeric complexes with Lewis bases (e.g., dimethyl-phenylaluminum etherate) are not particularly stable with respect to disproportionation.[50,52]

2. Dialkylalkynylaluminums

What has been said of dialkylarylaluminums probably applies to the dialkylvinylaluminums, which may be prepared by the addition of organo-aluminum compounds to acetylenes.[95] and certainly applies to dialkyl-alkynylaluminums. Alkynyl groups are extremely strong bridging groups generally. It is worth noting that dimethyl(phenylethynyl)gallium and dimethyl(phenylethynyl)indium are dimers and that bis(phenylethynyl)-cadmium and bis(phenylethynyl)zinc are polymers, with the phenylethynyl groups serving as bridges in each case.[35] Dimethyl(phenylethynyl)alu-minum has been prepared by reactions of phenylacetylene with trimethyl-aluminum (further acetylene does not react because of the stability of the phenylethynyl bridge and the unreactivity of dimeric organoaluminum compounds), and dimethyl(α-naphthylethynyl)aluminum and diphenyl-(phenylethynyl)aluminum have been prepared similarly.[32,33] The proton magnetic resonance spectrum of dimethyl(phenylethynyl)aluminum dimer in toluene at low temperature shows that the compound does not dispro-portionate significantly,[31] although disproportionation does occur in ether or tetrahydrofuran.[33]

3. Exchange between Trimethylaluminum and Dimethyl(phenylethynyl)-aluminum

It was earlier postulated that the exchange of alkyl groups between a trialkylaluminum and a dialkylaluminum alkoxide occurs when the mono-meric trialkylaluminum reacts with the dimeric dialkylaluminum alkoxide, but no kinetic proof was offered. The analogous reaction between tri-methylaluminum and dimethyl(phenylethynyl)aluminum has been investi-gated kinetically using proton magnetic resonance spectroscopy.[96] From the widths of the coalescing methyl peaks of the reactants, rate law (19) was deduced:

$$\text{Rate of exchange} = k[Me_6Al_2]^{1/2}[Me_4(PhC\equiv C)_2Al_2] \qquad (19)$$

This is consistent with the following mechanism for exchange:

$$Me_6Al_2 \rightleftharpoons 2Me_3Al \quad \text{fast} \qquad (20)$$

$$Me_3Al + Me_4^*(PhC\equiv C)_2Al_2 \longrightarrow Me_2Me^*Al + MeMe_3^*(PhC\equiv C)_2Al_2 \quad \text{slow} \quad (21)$$

(The asterisks are inserted only to indicate the probable course of the exchange reaction.)

Hence the conclusion is drawn that monomeric trialkylaluminums can undergo exchange with dimeric organoaluminum compounds. Reaction (21) is slow, however, and the analogous reaction of dialkylaluminum alkoxides must be even slower. When both exchanging species readily dissociate (e.g., trialkylaluminums, triarylaluminums, or dialkylaluminum halides), redistribution may occur by a simple series of dissociations and reassociations, avoiding such steps as (21).

E. Compounds with Mixed Bridges

The X-ray crystallographic structure of μ-diphenylamino-μ-methyl-tetramethyldialuminum (22) has been determined by Magnuson and

$$
\begin{array}{c}
\quad NPh_2 \\
\diagup \qquad \diagdown \\
Me_2Al \qquad AlMe_2 \\
\diagdown \qquad \diagup \\
\quad Me
\end{array}
$$

(22)

Stucky.[97] It is the only isolated organoaluminum compound known to associate through unlike bridging groups, but proton magnetic resonance spectroscopy in toluene at low temperature has provided evidence for many other analogous dimeric structures having unlike bridging groups.[10] Table I lists these compounds, the conditions for attainment of equilibrium when the compounds are formed by the proportionation reaction (22), and

$$Me_2Al \cdot X_2 \cdot AlMe_2 + Me_2Al \cdot Y_2 \cdot AlMe_2 \rightleftharpoons 2Me_2Al \cdot XY \cdot AlMe_2 \quad (22)$$

$$K = [Me_2Al \cdot XY \cdot AlMe_2]^2/[Me_2Al \cdot X_2 \cdot AlMe_2] \cdot [Me_2Al \cdot Y_2 \cdot AlMe_2] \quad (23)$$

the approximate values of the equilibrium constant, K, for the proportionation reaction. The values of K range widely on both sides of the value of 4 expected for random pairing of bridging groups. Most interest attaches to the cases in which K has high values and the mixed bridges are thermodynamically stable with respect to disproportionation. The stability of 22 ($10 < K < 100$ in cyclohexane[98]) may be due to the existence of strong steric interactions in the dimethylaluminum diphenylamide dimer (23) with which 22 is in equilibrium. In the absence of such a steric

$$
\begin{array}{c}
\quad NPh_2 \\
\diagup \qquad \diagdown \\
Me_2Al \qquad \qquad AlMe_2 \\
\diagdown \qquad \diagup \\
\quad NPh_2
\end{array}
$$

(23)

effect, mixed bridges between weakly bridging (e.g., alkyl) and strongly bridging (e.g., alkoxyl or amino) groups are probably unstable with respect

TABLE I

Formation and Stability of Compounds $Me_2Al \cdot XY \cdot AlMe_2$
with Unlike Bridging Groups in Toluene[10]

Reproduced by permission of the Editor of the Australian Journal of Chemistry:

Bridging group X	Bridging group Y	Equilibration temperature[a]	K[b]
Methyl	Phenyl	30°	5
Methyl	p-Tolyl	30°	7
Methyl	Phenylethynyl	30°	0.1
Phenyl	Phenylethynyl	30°	1
Chloro	Bromo	30°	1
Bromo	Phenyl	30°	4
Chloro	Phenylethynyl	30°	20
Bromo	Phenylethynyl	30°	60
Bromo	Isopropoxyl	80° 1 day required	100
Phenylethynyl	Isopropoxyl	80° 1 day required	2

[a] Reaction requires only a few minutes except where otherwise noted.
[b] Defined by Eq. (23).

to disproportionation.* The other cases in which K is high are compounds
(24) to **(26)**. Here both X and Y are good bridging groups; one is a

halogen atom, and the other is thought to be a particularly good electron donor. Possibly this combination results in strong mixed bridges
generally.

When both the dimeric reactants $(Me_2AlX)_2$ and $(Me_2AlY)_2$ dissociate
rapidly to monomers, the mixed bridge is formed rapidly. Compounds
with mixed bridges are formed only slowly from dimeric dimethylaluminum
isopropoxide, which is particularly stable with respect to dissociation.

* Careful reaction of trimethylaluminum with a $\frac{1}{2}$ mole of t-butanol gives μ-t-butoxy-
μ-methyltetramethyldialuminum, which disproportionates, but only slowly, to trimethylaluminum and dimethylaluminum t-butoxide.[43]

Possibly reaction occurs by dissociation of both the dimers with symmetrical bridges, followed by association of unlike monomeric units. Alternatively it may occur by reaction of a molecule of monomer with a molecule of the other symmetrical dimer.

F. Organoaluminum Hydrides*

Dialkylaluminum hydrides differ from most of the organoaluminum compounds discussed so far in that they are trimeric,[20,26,100] probably because of the relative freedom from steric interactions across the six-membered hydride-bridged ring. The strength of the hydride bridge is reflected in the recovery of the ether-free hydrides from ethereal solution.[101] The dialkylaluminum hydrides are normally stable with respect to disproportionation, but under some circumstances diethylaluminum hydride throws down a precipitate of aluminum hydride.[101] There is cryoscopic evidence for compounds $R_2AlH \cdot R_3Al$, which presumably have mixed alkyl–hydride bridges.[100]

Dialkylaluminum hydrides may be prepared by the reaction of trialkylaluminum with activated aluminum and hydrogen,[102,103] by hydrogenation of the trialkylaluminum at elevated temperature and pressure,[104] or by elimination of olefin from trialkylaluminum (e.g., from triisobutylaluminum).[100] Redistribution reactions are also useful for the preparation of organoaluminum hydrides. Diphenylaluminum hydride, which has a degree of association of 2.4, may be prepared in 78% yield by reaction of triphenylaluminum with aluminum hydride in 2:1 molar ratio in refluxing benzene and also by reaction of lithium aluminum hydride with triphenylaluminum in benzene.[105]

Treatment of lithium aluminum hydride with trimethylaluminum (or Me_3B or Me_3Ga) in the absence of solvent gives dimethylaluminum hydride,[26] elevated temperatures or long reaction times being required. Equation (24) may not be an adequate description of the redistribution

$$R_3Al + LiAlH_4 \longrightarrow R_2AlH + LiAlH_3R \qquad (24)$$

process, since triethylaluminum reacts with $LiAlH_4$ in ether to give $LiAlH_4$ and $LiAlH_2Et_2$ rather than $LiAlH_3Et$; possibly $LiAlH_3Et$ is more reactive than $LiAlH_4$ toward triethylaluminum.[106] Reaction between $NaAlH_4$ and trialkylaluminum in tetrahydrofuran is complicated by cleavage of the tetrahydrofuran with formation of dialkylaluminum butoxide[106] (see also Miller[107]).

Diethylaluminum chloride reacts with lithium hydride in 1:1 ratio in ether, precipitating lithium chloride and forming diethylaluminum hydride, which may then be distilled. Use of 2 moles of lithium hydride leads

---

* See also Nöth and Wiberg.[99]

to $LiAlH_2Et_2$, which is converted to diethylaluminum hydride on treatment with further diethylaluminum chloride.[101] Preparation of diethylaluminum hydride from diethylaluminum bromide and lithium hydride is easier experimentally.[84] Reaction of sodium hydride with an equimolar amount of diethylaluminum chloride to give diethylaluminum hydride, which begins slowly and may then be violent,[101] is more easily accomplished if a little dialkylaluminum hydride is added initially.[108]

Ethylaluminum dibromide or dichloride reacts smoothly with a slight excess of sodium hydride in benzene in the presence of diethylaluminum hydride, but the ethylaluminum dihydride expected (Eq. (25)) is accompanied by diethylaluminum hydride and $NaAlH_4$.[109]

$$EtAlCl_2 + 2NaH \longrightarrow 2NaCl + EtAlH_2 \qquad (25)$$

The trimethylamine complexes of Me_2AlH, Et_2AlH, $MeAlH_2$, and $EtAlH_2$ have been prepared by redistribution between $Me_3Al \cdot NMe_3$ or $Et_3Al \cdot NMe_3$ and $AlH_3 \cdot NMe_3$. Redistribution between $R_3Al \cdot NMe_3$ and $LiAlH_4$, like reduction of alkylaluminum dichlorides with lithium hydride, was not always successful.[110]

G. Exchange Reactions of Alkali Metal Tetraorganoaluminates*

Lithium tetramethylaluminate is reactive toward exchange, although less so than trimethylaluminum. The mechanism of the trivial exchange between lithium tetramethylaluminate and trimethylaluminum in ether has been studied by observations on coalescing proton magnetic resonance spectra.[112] The rate-determining step is thought to be dissociation of the $LiAlMe_4$ to Li^+ and $AlMe_4^-$. Redistribution between $NaAlEt_4$ and trimethylaluminum gives $NaAlMe_4$, and $NaAlPh_4$ is formed similarly from $NaAlEt_4$ and triphenylaluminum.[113]

The redistribution equilibria attained upon reaction of organoaluminum compounds with alkali metal organoaluminates have been investigated extensively by Ziegler and his collaborators. The results obtained have been reviewed by Lehmkuhl,[93] and so will not be discussed at length here. The positions of equilibrium, and hence the directions in which reactions proceed, must be largely determined by a balance of association energies and the lattice energies of precipitated products. For example reaction (26) (R

$$NaAlEt_3OR + AlEt_3 \longrightarrow NaAlEt_4 + Et_2AlOR \qquad (26)$$

is an alkyl group) proceeds to the right,[93,113] probably with the high association energy of the dialkylaluminum alkoxide providing the driving

---

* Tetraethylaluminates of the alkaline earth metals, $M(AlEt_4)_2$ (M = Ca, Sr, Ba), are also known and have been prepared by reaction of the alkoxides, $M(OR)_2$, with triethylaluminum.[93,111]

force. The corresponding reverse reaction of diethylaluminum phenoxide with $NaAlEt_4$ is complicated by the formation of a $2:1$ complex, $NaAlEt_3$-$(OPh)AlEt_3$, from which triethylaluminum can be distilled leaving $NaAl(OPh)Et_3$.[93]

Redistribution between sodium tetraphenylaluminate and dimethylaluminum chloride to give triphenylaluminum constitutes a final step in the synthesis of triphenylaluminum starting from $NaAlEt_4$ and benzene.[93,114,115] Lithium tetraalkylaluminates may be converted to trialkylaluminums upon treatment with $\frac{1}{3}$ mole of aluminum trichloride in hexane, according to Eq. (27).[101]

$$3LiAlR_4 + AlCl_3 \longrightarrow 3LiCl + 4R_3Al \qquad (27)$$

A freezing-point diagram for mixtures of $NaAlEt_4$ with $NaAlMe_4$ suggests that mixed compounds $NaAlEt_nMe_{4-n}$ ($n = 1, 2, 3$) either are not formed or are not stable.[113] By contrast, when $NaAlH_4$ and $NaAlEt_4$ are fused together in the appropriate stoichiometric ratio for ca. 1 hr at $80°$, crystalline $NaAlHEt_3$ or $NaAlH_2Et_2$ is produced.[116] There is no evidence for the existence of $NaAlH_3Et$,[109,116] and attempts to obtain this compound by reaction of excess sodium hydride with ethylaluminum dichloride resulted in $NaAlH_4$ and $NaAlH_2Et_2$.[109] $LiAlH_4$ and $LiAlH_2Et_2$ resulted similarly from the action of excess lithium hydride on ethylaluminum dibromide.[109] However, $NaAlH_3Bu\text{-}i$ has been prepared by redistribution between $NaAlH_4$ and $NaAlH_2Bu\text{-}i_2$.[106] $NaAlH_2(C{\equiv}CBu)_2$ and $NaAlEt_2(C{\equiv}CPh)_2$ have also been obtained by proportionation.[117]

H. Redistribution in the Presence of Lewis Bases

Complex formation with external Lewis base renders an organoaluminum compound less reactive toward redistribution, just as does self-association. Nevertheless, organoaluminum compounds which dissociate to monomers in ethereal solvents do in general undergo redistribution reactions with one another. For example, redistribution equilibrium is attained within two minutes on mixing trimethylaluminum and aluminum chloride in ether. If the reactants are in the appropriate molar ratio ($2:1$ or $1:2$), dimethylaluminum chloride etherate or methylaluminum dichloride etherate is produced quantitatively.[9] Reaction of trimethylaluminum with triethylaluminum,[51] triisobutylaluminum,[43] or triphenylaluminum[50] is equally fast in ether, but mixtures of products are obtained in all three cases, since the equilibria do not so heavily favor the proportionation products. The dissociation of organoaluminum compounds to monomeric complexes in basic solvents often results in a shift in redistribution equilibrium. For example, dimethylphenylaluminum is stable as its phenyl-bridged dimer in the absence of Lewis bases, but in ether it exists in equilibrium with substantial proportions of its disproportionation products, trimethylaluminum etherate and methyldiphenylaluminum etherate.[50]

There is normally not a particularly pronounced shift in redistribution equilibrium on going from one basic solvent to another. The report that methylaluminum dichloride disproportionates partly in pyridine[9] is almost certainly in error.[67,83] It is, however, worth noting that arylmagnesium halides are much more markedly disproportionated to diarylmagnesiums and magnesium halides in the more strongly basic solvent, tetrahydrofuran, than they are in ether.[118] Dimethylaluminum methoxide, unlike dimethylaluminum isopropoxide, disproportionates extensively in tetrahydrofuran and pyridine.[65]

Redistribution in amines is much slower than redistribution in ether. Exchange between trimethylaluminum and triphenylaluminum at 40° requires about a day in pyridine, lutidine, or triethylamine,[119,120] but is complete within a minute in ether or tetrahydrofuran.[50] Exchange between trimethylaluminum and triisobutylaluminum,[43] triethylaluminum,[51] or aluminum trichloride[43] likewise requires many hours at 40° in pyridine.

Since amines are stronger bases than ether, the retardation of exchange by pyridine suggests that the Lewis base needs to be lost from either one or both of the reacting organoaluminum molecules before exchange can occur. This question has been investigated kinetically for a number of exchange reactions of trimethylaluminum in the presence of polar solvents. Observations have been made on proton magnetic resonance spectra which are collapsing due to fast exchange between the etherates of trimethylaluminum and dimethylethylaluminum in toluene. These show unambiguously that exchange is a second-order reaction in which ether is not lost prior to exchange.[121] The same conclusion was also drawn for exchange between the etherates of dimethylaluminum chloride and dimethylaluminum bromide.[122] On the other hand, exchange between the anisole complex of trimethylaluminum and the anisole complex of dimethylphenylaluminum appears to proceed only after non-rate-determining loss of anisole from one of the reactants.[122] Intermediate kinetic behavior was observed for the fast exchange between the anisole complexes of dimethylaluminum bromide and dimethylaluminum chloride[122] and for the slow exchange between the pyridine complexes of trimethylaluminum and triphenylaluminum.[119]

In summary one may say that the complexes of organoaluminum compounds exchange with one another either without loss of complexing Lewis base or after loss of base from only one of the two reactants. In the latter case activation energy has to be supplied to break a bond prior to exchange, whereas in the former case the transition state is sterically less accessible.

A balance between the donor–acceptor bond strength and steric considerations decides which of the mechanisms dominates, but in any case exchange is slower than it would be in the absence of Lewis base.

III. EXCHANGE BETWEEN ORGANOALUMINUM COMPOUNDS AND COMPOUNDS OF OTHER METALS

The commercially available trialkylaluminums are attractive intermediates for the alkylation of less electropositive metals and metalloids. Although the potential uses have not been fully explored, they are the subject of a large number of patent claims. Jenkner[123] has surveyed many of these claims. As far as possible the present review will be directed to the scientific literature rather than to patent claims (which are often difficult to assess scientifically).

For synthetic purposes one particularly needs to know how many of the three alkyl groups of a trialkylaluminum can be used in alkylation. A knowledge of how self-association and association with solvents influence the occurrence of exchange and the separation of products is also of importance.

The most important of the side reactions which may occur is reduction, particularly by the higher alkylaluminum compounds. This is especially marked upon reaction with compounds of the transition metals, and in practice alkylaluminum compounds (in particular triethylaluminum and triisobutylaluminum) are more often used to reduce transition-metal compounds than to alkylate them.

When alkylaluminum compounds react with compounds of elements which are not readily reduced to lower valence states, reduction is only a problem at high temperature (>150–$200°$) and occurs by elimination of olefin (Eq. (28)), either during redistribution or from the aluminum com-

$$M—C—C—H \longrightarrow M—H + C=C \qquad (28)$$

pounds prior to redistribution. Olefin elimination occurs more easily with triisobutylaluminum than with triethylaluminum.[100,124] Trimethylaluminum[124] and triphenylaluminum,[45] which cannot decompose in this way, nevertheless decompose slowly by other mechanisms above $200°$.

A. Reaction with Other Organometallic Compounds

Problems arising out of the association of reactants and products with each other are minimal when a triorganoaluminum exchanges with an organic derivative (R_nM) or another metal. The redistribution equilibria attained tend to be random, particularly when the groups exchanged are both alkyl groups. Consequently the reaction is of preparative use only when a product has special stability because of association or when it can be physically separated from the relatively random mixture of reactants and products. For example triethylboron may be distilled, leaving behind tribenzylaluminum, when tribenzylboron is redistributed with triethylaluminum.[125] The same method also allows triarylaluminums to be prepared.[53,125]

1. Organic Derivatives of the Electropositive Metals

With the organic compounds of the more electropositive metals, particularly lithium, sodium, potassium, magnesium, and calcium, associated compounds containing both metals (e.g., $LiAlEt_4$, $NaAlPh_4$, $Ca(AlEt_4)_2)$[93,111,126] are formed upon attempted redistribution. Fast redistribution occurs with organic derivatives of the more electropositive metals when such associated compounds are not formed. Thus redistribution of methyl groups between trimethylaluminum and trimethylgallium,[59] trimethylindium[59] or dimethylzinc[43] occurs with a half-life $< 10^{-2}$ sec at room temperature in toluene. Exchange of methyl groups of trimethylaluminum for those of dimethylcadmium is likewise fast on the NMR-time scale,[127] but diphenylmercury is much less reactive toward trimethylaluminum[43] and exchange of methyl groups for phenyl groups requires several minutes at 40°. Redistribution between triethylaluminum and diethylmercury and between triphenylaluminum and diphenylmercury has been demonstrated using isotopic labeling,[128,129] although the temperature (100°) used was probably unnecessarily high.

Upon heating a mixture of trimethylaluminum with diethylzinc or dipropylzinc at 140°, dimethylzinc distills in 50–60% yield.[53] Diethylzinc may similarly be distilled from a redistributing mixture of triethylaluminum and dibutylzinc.[53]

2. Organoboron Compounds

Redistribution of organic groups between organoboron compounds and organoaluminum compounds has received considerable attention.[53,125,130,131] It is much faster than the corresponding reactions between two organoboron compounds, and small amounts of organoaluminum compounds are extremely active catalysts for redistribution of organic groups between organoboron compounds.[125]

Redistribution in mixtures of trimethylaluminum with triphenylboron and tributylboron has been studied,[53] and in each case trimethylboron could be distilled from the redistributing mixture in 50–60% yield on heating at ca. 140°. In similar experiments triethylboron could be distilled from mixtures of triethylaluminum with triphenylboron, tributylboron, or triisobutylboron.[53]

Since organoboron compounds do not associate, exchange between organoboron and organoaluminum compounds tends to proceed in such a direction as to put the best bridging groups on aluminum. Thus the equilibrium established starting from a mixture of a trialkylaluminum and dialkyl(1-alkenyl)borane strongly favors the dimeric dialkyl(1-alkenyl)-aluminum.[7] The relatively high steric requirements in trialkylaluminum dimers may also influence the position of equilibrium. For example, redistribution between equimolecular amounts of triethylaluminum and tributylboron leads to a statistical distribution of alkyl groups, but in the products derived from a mixture of trimethylaluminum and tributylboron there is a marked preference for the methyl groups to occupy both the terminal and the bridging positions in the organoaluminum dimers.[125]

Unusual cyclic boron compounds may be prepared by hydroboration of dienes and trienes and by pyrolysis of simpler organoboron-compounds.[132–136] Redistribution between these compounds and trialkylaluminums leads to the corresponding cyclic aluminum compounds.[131,134,137]

3. Silicon, Tin, and Lead Compounds

Since the Lewis acidity of an alkylmetallic compound and the basicity of the exchanging groups are of importance in determining how easily redistribution reactions occur, the tetraalkyl derivatives of the group IVB elements (particularly silicon and germanium) are relatively unreactive toward redistribution. Exchange of alkyl groups between tributylaluminum and tetraethyltin requires several days at 140°.[138] Isotopic labeling has been used to investigate exchange between triphenylaluminum and the tetraphenyl derivatives of silicon, tin, and lead.[129] Tetraphenylsilane and tetraphenyltin do not exchange significantly during 5 hr at 100°, but tetraphenyllead reacts appreciably (it is still less reactive than diphenylmercury). There is appreciable exchange of ethyl groups between tetraethyllead and triethylaluminum during 5 hr at 100°.[128]

B. Reaction with Hydrides

The hydrides of lithium, beryllium, magnesium, zinc, and cadmium have been prepared by reaction of the corresponding metal alkyls (MeLi, Me_2Be, Et_2Mg, Me_2Zn, and Me_2Cd) with lithium aluminum hydride in ether.[139] By contrast metallic mercury and hydrogen were formed from dimethylmercury with lithium aluminum hydride.[139] Methylberyllium hydride has been prepared by treatment of dimethylaluminum hydride with excess dimethylberyllium in ether.[139]

Exchange between diethylaluminum hydride and diethylzinc or diethylcadmium in the absence of solvent at 25–80° results in the formation of triethylaluminum in good yield, deposition of zinc or cadmium, and evolution of hydrogen.[140] Diethylaluminum hydride reacts with diethylmercury similarly except that both hydrogen and ethane are evolved, but the corresponding reaction with diethylmagnesium results in magnesium hydride and triethylaluminum.[140]

Organomercury compounds react with $LiAlH_4$, AlH_3NMe_3, or AlH_2NMe_2, evolving hydrogen, depositing mercury, and forming the corresponding organoaluminum compounds.[39] For example:

$$LiAlH_4 + 2Hg(CH{=}CH_2)_2 \longrightarrow LiAl(CH{=}CH_2)_4 + 2Hg + 2H_2 \qquad (29)$$

$$AlH_2{\cdot}NMe_2 + HgPh_2 \longrightarrow Ph_2Al{\cdot}NMe_2 + Hg + H_2 \qquad (30)$$

The reactivity of the aluminum compounds decreases in the series $LiAlH_4 > AlH_3{\cdot}NMe_3 > AlH_2{\cdot}NMe_2$.[39] A variety of organoaluminum compounds, including $LiAl(CH{=}CH_2)_4$,[39,141] trivinylaluminum,[141] tri-(perfluorovinyl)aluminum,[142] and the trimethylamine complexes of the dialkylaluminum hydrides and alkylaluminum dihydrides,[110] have been prepared in this way.

Neumann and his co-workers have studied redistribution between trialkylaluminums and alkyltin hydrides in cyclohexane.[138,143] The alkyltin hydrides may be prepared by reaction of dialkylaluminum hydrides with alkyltin halides, dialkylamides, or alkoxides; exclusive exchange of a hydride group for a halogen, dialkylamino, or alkoxyl group occurs.[144-146] A few hours at 70° are required for complete reaction between the trialkylaluminum and alkyltin hydride. The hydride groups of the organotin compound are exchanged rather than the alkyl groups, both because the hydride group serves as an effective transient bridge and because the dialkylaluminum hydrides formed are stabilized by self-association. The rate-law for exchange between tributylaluminum and tributyltin hydride is of first order in the concentration of tributyltin hydride but of one-half order in the concentration of dimeric tributylaluminum.[138] The monomer of tributylaluminum must therefore be involved in the rate-determining step, which is thought to be reaction (31). The observed over-all activation energy of 19.2 kcal/mole suggests an activation energy of about 10–12 kcal/mole for reaction (31) itself.

$$Bu_3Al + Bu_3SnH \longrightarrow \left[Bu_2Al \underset{H}{\overset{Bu}{\diamondsuit}} SnBu_3 \right] \longrightarrow Bu_2AlH + Bu_4Sn \qquad (31)$$

Alkyltin hydrides are much more reactive than tetraalkyltins towards trialkylaluminums.[138] Exchange of a hydride group for a hydride group is even more facile.[147] Diisobutylaluminum hydride reacts with triethyltin deuteride, as expressed by Eq. (32).

$$i\text{-Bu}_2\text{AlH} + \text{Et}_3\text{SnD} \longrightarrow \left[i\text{-Bu}_2\text{Al} \underset{D}{\overset{H}{\diagup\diagdown}} \text{SnEt}_3 \right] \longrightarrow i\text{-Bu}_2\text{AlD} + \text{Et}_3\text{SnH} \quad (32)$$

However the analogous reactions of diethylaluminum deuteride with triethylsilane or triethylgermane did not occur even at 60°.[147] Triethylsilane and triethylgermane are also unreactive toward triethylaluminum.[138]

C. Reaction with Metal Carboxylates

Alkylation of metal carboxylates by means of trialkylaluminums is the subject of a patent claim.[44] The method works quite well despite the possibility of alkylation of the carboxylate groups and of proton abstraction from the carboxylate group. For example, dimethylzinc, diisobutylzinc, dimethylcadmium, and diisobutylcadmium have been prepared in 40–60% yield by adding the appropriate trialkylaluminum (one $R_3\text{Al}$ per Zn(OAc)_2 or Cd(OAc)_2) to the metal acetate in the absence of solvent, heating to 100–150°C, and distilling out the dialkylzinc or dialkylcadmium at atmospheric or moderately reduced pressure.[43] Reaction of mercuric acetate with tripropylaluminum in hexane gives an 80% yield of dipropylmercury after hydrolysis.[148] No more than two of the three alkyl groups on the aluminum atom are used in these alkylations. The main advantage of the method is the ready availability of the metal acetates in anhydrous form.

Tetraethyllead may be produced in good yield by the action of either triethylaluminum or sodium tetraethylaluminate on lead(II) acetate or lead tetraacetate in toluene at temperatures of 0–100°, the tetraethyllead being isolated after hydrolysis of the products.[148–149a] Reaction of triethylaluminum with lead tetracarboxylates may be carried out in either hydrocarbons or glycol ethers at ca. 100°. Generally only about 50% of the available lead is converted to tetraethyllead,[150] although higher yields are attainable, particularly when excess triethylaluminum is used.[151] The low yields are probably associated with reduction of Pb^{IV} to Pb^{II} and subsequent precipitation of lead metal (see later). By contrast, reaction of triphenylaluminum with lead tetraacetate in toluene gives tetraphenyllead in excellent yield.[151]

Alkylation of organotin carboxylates by trialkylaluminums has allowed the preparation of tetraalkyltins with mixed alkyl groups. For example, diethyldibenzyltin and ethyltributyltin have been prepared by the action of triethylaluminum on dibenzyltin diacetate and tributyltin acetate, respectively.[152]

D. Reaction with Alkoxides, Oxides, and Related Compounds

1. Zinc

When trimethylaluminum is reacted with zinc ethoxide in the 2:3 ratio required by Eq. (33) and heated to 120° in the absence of solvent, dimethyl-

$$2Me_3Al + 3Zn(OEt)_2 \longrightarrow 3Me_2Zn + 2Al(OEt)_3 \qquad (33)$$

zinc may be distilled from the mixture in 60% yield.[43] It appears that only two of the three methyl groups of trimethylaluminum are effectively used in alkylation. Here, as in the alkylation of metal carboxylates, heat is needed to drive the reaction to completion. This is necessary because the alkylaluminum alkoxides are much poorer alkylating agents than trialkylaluminums. In view of the earlier discussion of the reaction of trimethylaluminum with aluminum alkoxides, it is clear that the shorter chain n-alkoxides represent the best choice of substrates for alkylation.

2. Boron

Trialkylborates are reported to be alkylated in excellent yields by trialkylaluminums or alkylaluminum halides, apparently with all of the alkyl groups being transferred from aluminum.[153–155] The compounds R_2BOR' and $RB(OR')_2$, intermediate in the complete alkylation of the trialkylborate, are volatile and monomeric, and hence may be distilled out from the reaction mixture if desired.[123]

Trialkylboroxines and trialkoxyboroxines are readily alkylated by trialkylaluminums and alkylaluminum halides.[155–158] The boroxines required may be formed *in situ* from B_2O_3.[158] For example, reaction of triethylaluminum with a mixture of trimethylborate and B_2O_3 in diethylene glycol dimethyl ether at 75° produces triethylboron in ca. 80% yield,[159] and reaction of tributylboroxine with methylaluminum sesquiiodide or trimethylaluminum leads to dimethylbutylborane.[155] B_2O_3 and $Na_2B_4O_7$ are less suitable reactants, but have given high yields with dialkylaluminum halides.[155,160] B_2O_3 has given variable results in reaction with trialkylaluminums.[123,156,158,160a]

Triethylboron has also been obtained in high yields by the action of triethylaluminum on B-trichloroborazine or B-triphenylborazine.[161]

3. Silicon

Tetraalkylsilicates react with trialkylaluminums at elevated temperatures in inert solvents or in the absence of solvents to yield mixtures of alkylsilicon alkoxides. The trialkylaluminums may be used as their etherates, the ether being evolved as dialkylaluminum alkoxides are formed. For example, a mixture of equimolar amounts of triethylaluminum etherate and tetrabutylsilicate produced Et_3SiOBu (20%), $Et_2Si(OBu)_2$ (45%), and $EtSi(OBu)_3$ (35%) on heating to 190–210°, with only two of the ethyl groups of the triethylaluminum being used.[123,162,163] Alkylation appears to become increasingly difficult as the number of alkoxyl groups on the silicon atom decreases. Polyorganosiloxanes can be alkylated by trialkylaluminums, but hexaorganodisiloxanes are cleaved only at elevated temperatures with reduction to a trialkylsilane resulting rather than alkylation to a tetraalkylsilane.[164]

Silica in suitable form (e.g., chromatographic silica gel) gives tetraalkylsilanes when treated with the lower alkylaluminum halides or their complexes with alkali metal halides at high temperatures (200°C).[165]

4. Tin and Lead

The reaction of tin tetraethoxide with triethylaluminum in a mole ratio of 1:1.4 in hot benzene gives tetraethyltin in 73% yield.[166] Moderate yields of tetraethyllead are obtainable by the action of triethylaluminum, sodium tetraethylaluminate, or lithium tetraethylaluminate on lead(II) oxide or sulfide[150,167,168]; in every case less than 50% of the lead is converted to tetraethyllead. Organotin oxides, R_2SnO and $(R_3Sn)_2O$, and tin(IV) sulfide have also been alkylated by trialkylaluminums to give tetraalkyltins, which in some cases have two different alkyl groups ($R_2R_2'Sn$ and $R_3R'Sn$).[152] Alkylation of trimethyltin dimethylamide by triethylaluminum allows the preparation of dimethylaluminum dimethylamide.[169]

5. Titanium

An 80% yield of bis(cyclooctatetraene)titanium[170] has been obtained by the action of excess triethylaluminum and excess cyclooctatetraene on titanium tetrabutoxide at 80°. Use of lesser amounts of cyclooctatetraene led to tris(cyclooctatetraene)dititanium.[170,171]

6. Arsenic and Antimony

Reaction of excess trialkylaluminum with finely powdered As_2O_3 or Sb_2O_3 at 60°, followed by distillation at reduced pressure, allows the isolation of trialkylarsines or trialkylstibines in yields of up to 90%.[172,173]

E. Reactions with Halides

This is the most important method of obtaining organic derivatives of metals and metalloids from organoaluminum compounds. It has been the subject of a large number of patent claims, which cannot be surveyed fully here. The reader is referred to articles by Jenkner[123] for a more comprehensive review of the patent claims, particularly those of Jenkner himself.

The commonly available organoaluminum alkylating agents are the triorganoaluminums themselves, organoaluminum halides, organoaluminum alkoxides, and alkali metal tetraorganoaluminates. Generally the triorganoaluminums and their alkali metal derivatives represent the alkylating agents of choice.

Diorganoaluminum alkoxides are weaker (slower) alkylating agents but have the advantage that, being self-associated, they are not strong Lewis acids. For this reason they have been used in the reduction and alkylation of transition metal compounds.

Organoaluminum halides are poorer alkylating agents than the triorganoaluminums. Furthermore the higher Lewis acidity of the halide is disadvantageous in many cases. Accordingly the halides are generally a poorer choice of alkylating agent than the trialkylaluminums, and are only preferred when alkylation is facile and when the alkylaluminum halides are more easily available.

Sundermeyer and Verbeek[174] report a process whereby the sodium chloride complex of methylaluminum dichloride is formed by the action of methyl chloride on aluminum metal in a molten mixture of sodium chloride and aluminum chloride. The aluminum is previously generated *in situ* by electrolysis of the molten salts. Reaction of metal halides with the $NaAlMeCl_3$, also carried out in the molten salts, affords good yields of various methylmetallic compounds, including tetramethylsilane, trimethylboron, dimethylmercury, and tetramethyltin.

Alkylation of chlorides by trialkylaluminums produces alkylaluminum chlorides as intermediates. So, just as in the case of alkylaluminum halide starting materials, the problems of Lewis acidity are present. The organoaluminum halides complex with the other organometallic halides formed, preventing reaction from going to completion and hindering the separation of products.

1. Beryllium

Ethylberyllium chloride is claimed to be formed upon reaction of triethylaluminum with beryllium chloride.[175] It is accompanied by diethylaluminum chloride which may be distilled off. Some information is also available on the reaction of bis(dimethylamino)beryllium with trimethylaluminum and methylaluminum hydrides.[176]

2. Zinc, Cadmium, and Mercury

The highly exothermic reaction between triethylaluminum and zinc chloride in the 2:3 molar ratio required by Eq. (34) gives only poor yields of diethylzinc on distillation.

$$2Et_3Al + 3ZnCl_2 \longrightarrow 3Et_2Zn + 2AlCl_3 \qquad (34)$$

An equilibrium is attained in which ethylzinc and ethylaluminum chlorides dominate[177] (see also Ziegler[178]). Eisch[179] prepared dimethylzinc by the action of trimethylaluminum on zinc chloride in the ratio required by Eq. (35). Only one of the three alkyl groups of trimethylaluminum is

$$2Me_3Al + ZnCl_2 \longrightarrow Me_2Zn + 2Me_2AlCl \tag{35}$$

used in the alkylation. The same stoichiometry is required for the alkylation of cadmium chloride and of zinc chloride by other trialkylaluminums.[123,179,180]

The reaction of trialkylaluminums with mercuric bromide or mercuric chloride in 1:1 molar ratio in ether leads to moderate-to-good yields of dialkylmercury compounds, accompanied by alkylmercuric halides.[148,181] Reaction proceeds quantitatively in the opposite direction when aluminum bromide reacts with methylpentafluorophenylmercury.[182,183] Pentafluorophenylaluminum dibromide or bis(pentafluorophenyl)aluminum bromide is produced, depending on the ratio in which the reactants are used (Eqs. (36) and (37)).

$$C_6F_5HgMe + AlBr_3 \longrightarrow C_6F_5AlBr_2 + MeHgBr \tag{36}$$

$$C_6F_5HgMe + C_6F_5AlBr_2 \longrightarrow (C_6F_5)_2AlBr + MeHgBr \tag{37}$$

These cases are exceptional and proceed in the directions shown, first, because of the high electronegativity of the pentafluorophenyl group and, second, because of the insolubility of methylmercuric bromide.

Dialkylmercury compounds are claimed to be formed almost quantitatively, according to Eq. (38), from trialkylaluminums and mercuric chloride

$$2R_3Al + 2NaCl + 3HgCl_2 \longrightarrow 3R_2Hg + 2NaAlCl_4 \tag{38}$$

in the presence of sodium chloride.[123] Diethylmercury may also be obtained in good yield by the alkylation of mercurous chloride according to Eq. (39).[123]

$$2Et_3Al + 2NaCl + 3Hg_2Cl_2 \longrightarrow 3Et_2Hg + 3Hg + 2NaAlCl_4 \tag{39}$$

3. Boron

Trialkylborons can be distilled from the almost quantitative reaction of trialkylaluminum etherate with boron trifluoride etherate in 1:1 molar ratio. This represents an extremely good preparation of trialkylborons.[153,181] The reaction is facile in the earlier stages, but heating to up to 200° may be required to complete alkylation. Alkylation of powdered potassium tetrafluoroborate by trialkylaluminums in either the trialkylboron or an inert hydrocarbon as solvent represents an equally good preparation of trialkylborons.[153] In this case there is the added advantage that it is not necessary to separate the product from ether. Once again temperatures of 200° are necessary to complete the evolution of product.

The alkylation of boron trichloride by trialkylaluminums in the presence of sodium chloride is reported to proceed at lower temperatures than the alkylation of boron trifluoride etherate, and this should be advantageous in the preparation of more thermally unstable trialkylborons.[123] Alkylboron chlorides have also been prepared by reaction of trialkylaluminums with boron trichloride,[123,177] and methylboron dichloride and ethylboron dichloride may be prepared from boron trifluoride with boiling alkylaluminum dichloride, as shown in Eq. (40).[184]

$$RAlCl_2 + BF_3 \longrightarrow RBCl_2 + AlF_3 \tag{40}$$

4. Gallium, Indium, and Thallium

Only one of the three alkyl groups of a trialkylaluminum can be effectively used in alkylation of gallium trichloride or indium trichloride.[178,185*] Typically 3 moles of trialkylaluminum are added to each mole of gallium or indium trichloride in pentane. After one hour at reflux, the pentane is distilled off, potassium chloride is added, and the trialkylgallium or trialkylindium is distilled (50–90% yield). If gallium tribromide is used as starting material, addition of potassium chloride is not necessary.[179]

Treatment of thallium(III) chloride with excess trialkylaluminum and then hydrolysis is reported to produce dialkylthallium chloride.[148] In some cases thallium(I) chloride is formed instead.

5. Tin†

The difficulties arising out of association of alkylation products with organoaluminum halides are illustrated by the reaction of triethylaluminum with tin tetrachloride without solvent and in a molar ratio of 4:3, as required by Eq. (41). Only up to 10% tetraethyltin is obtained, along with

$$4Et_3Al + 3SnCl_4 \longrightarrow 3Et_4Sn + 4AlCl_3 \tag{41}$$

74% triethyltin chloride and some diethyltin dichloride.[166] Crystalline complexes of composition $R_3SnCl \cdot AlCl_3$ and $R_2SnCl_2 \cdot AlCl_3$ are known.[186] Formation of such complexes in the redistribution reactions may be avoided by the addition of sodium chloride, an ether, or an amine, so as to

---

* See also Zakharin and Okhlobystin.[177]
† See also Luijten and van der Kerk.[185a]

complex the aluminum chloride and allow reaction to proceed to completion. Particularly good results were obtained by Neumann,[166] when he added tin tetrachloride to cooled trialkylaluminum (3:4 molar ratio), followed by excess diethyl ether or dibutyl ether, then heated the reaction mixture briefly. The method of separating the tetraalkyltin so formed depended on its physical properties and the boiling point of the ether. Tetraethyltin, prepared in the presence of diethyl ether, was distilled at reduced pressure, while tetrabutyltin was washed free of aluminum compounds and separated from dibutyl ether by distillation at reduced pressure. On the other hand tetraoctyltin, also prepared in the presence of dibutyl ether, formed a separate phase at the end of the exchange reaction. Lower ratios of trialkylaluminum to tin tetrahalide allowed the preparation of various alkyltin halides. Neumann obtained ethyltin trichloride in good yield by reaction of diethylaluminum ethoxide with two moles of tin tetrachloride.[166]

The use of alkali metal halides as complexing agents during the alkylation of tin tetrahalides by trialkylaluminums and alkylaluminum halides has been investigated by Zakharkin[187] and by van Egmont.[187a] Tetraethyltin, tetraisobutyltin, triethyltin chloride, diethyltin dichloride, diethyltin dibromide, and diisobutyltin dichloride were prepared by this method.[187]

Complete alkylation of tin tetrachloride by a trialkylaluminum can be accomplished without a complexing agent if an excess of trialkylaluminum is used (only approximately two of the three alkyl groups being transferred to tin). Reaction of tin tetrachloride with a moderate excess of triethylaluminum in refluxing hexane gives a 71% yield of tetraethyltin after hydrolysis,[148] and an even better yield (85%) is obtained by reacting the tin tetrachloride with a moderate excess of lithium tetraethylaluminate in refluxing benzene.[188] The action of phenylaluminum sesquichloride on tin tetrachloride has given tetraphenyltin and phenyltin trichloride in high yields,[82] and an 86% yield of trimethylethyltin has been obtained from trimethyltin chloride with diethylaluminum chloride.[189] Johnson has described the partial alkylation of tin tetrachloride and alkyltin chlorides to trialkyltin chlorides by alkylaluminum sesquichlorides, as well as the conversion of the trialkyltin chlorides to tetraalkyltins under the influence of a mole of trialkylaluminum.[189a] Partial alkylation of tin tetrachloride by triisobutylaluminum, followed by hydrolysis to give diisobutyltin oxide, has also been reported.[177]

Heterocyclic aluminum compounds, prepared by reaction of diisobutylaluminum hydride with dienes or from the corresponding organoboron compounds, may be reacted with stannic chloride or organotin chlorides to produce heterocyclic tin compounds. For example, the action of 3,3-dimethylpenta-1,4-diene on diisobutylaluminum hydride (reaction (42)) produces (27), which with stannic chloride gives the cyclic tin compound (28), and with dibutyltin dichloride gives (29).[190]

$$i\text{-Bu}_2\text{AlH} + \text{CH}=\text{CH}-\text{C(CH}_3)_2-\text{CH}=\text{CH}_2 \longrightarrow (27) + 4(\text{CH}_3)_2\text{C}=\text{CH}_2 \quad (42)$$

$$(\text{CH}_3)_2\text{C}\underset{\text{CH}_2-\text{CH}_2}{\overset{\text{CH}_2-\text{CH}_2}{<}}\text{Al}\cdot(\text{CH}_2)_2\cdot\text{C(CH}_3)_2\cdot(\text{CH}_2)_2\cdot\text{Al}\underset{\text{CH}_2-\text{CH}_2}{\overset{\text{CH}_2-\text{CH}_2}{>}}\text{C(CH}_3)_2$$

$$(27)$$

$$(\text{CH}_3)_2\text{C}\underset{\text{CH}_2-\text{CH}_2}{\overset{\text{CH}_2-\text{CH}_2}{<}}\text{Sn}\underset{\text{CH}_2-\text{CH}_2}{\overset{\text{CH}_2-\text{CH}_2}{>}}\text{C(CH}_3)_2$$

$$(28)$$

$$(\text{C}_4\text{H}_9)_2\text{Sn}\underset{\text{CH}_2-\text{CH}_2}{\overset{\text{CH}_2-\text{CH}_2}{<}}\text{C(CH}_3)_2$$

$$(29)$$

6. Lead*

Gilman and Apperson[191] found extensive formation of triphenyllead chloride and diphenyllead dichloride upon treatment of tetraphenyllead with aluminum chloride. Similarly tetraethyllead is dealkylated by aluminum trichloride and even ethylaluminum dichloride.[191] It is therefore clear that the alkylation of trialkyllead chlorides by alkylaluminum dichlorides and even dialkylaluminum chlorides (reactions (43) and (44)) will be carried to completion only with difficulty.

$$\text{RAlCl}_2 + \text{R}_3\text{PbCl} \rightleftharpoons \text{R}_4\text{Pb} + \text{AlCl}_3 \qquad (43)$$

$$\text{R}_2\text{AlCl} + \text{R}_3\text{PbCl} \rightleftharpoons \text{R}_4\text{Pb} + \text{RAlCl}_2 \qquad (44)$$

A further difficulty in the preparation of tetraalkylleads by alkylation of lead(II) halides is that only a 50% conversion of lead can be anticipated, the rest of the lead being precipitated (see Eq. (45)).[151,167]

$$6\text{Pb}^{\text{II}}\text{Cl}_2 + 4\text{R}_3\text{Al} \longrightarrow 3\text{R}_4\text{Pb}^{\text{IV}} + 3\text{Pb} + 4\text{AlCl}_3 \qquad (45)$$

This problem cannot be avoided by use of lead(IV) halides, since reduction to lead(II) occurs at an intermediate stage in alkylation (probably $\text{RPbCl}_3 \rightarrow \text{RCl} + \text{PbCl}_2$), and reformation of Pb^{IV} and precipitation of lead then follows. Frey and Cook obtained only 50% yields of tetraethyllead on alkylation of K_2PbCl_6 with triethylaluminum ($2\text{Et}_3\text{Al}$ per K_2PbCl_6) in toluene or diglyme.[151] Reduction to lead(II) and precipitation of lead is not such a serious problem with the lead tetracarboxylates.

* See also Willemsens and van der Kerk.[190a]

The lead which is precipitated in the reaction of a lead(II) halide with a trialkylaluminum is reactive toward alkyl halides. Thus better yields of tetraethyllead are obtained by the action of triethylaluminum and ethyl iodide upon lead(II) chloride,[192] and the simultaneous reaction of sodium tetraethylaluminate and ethyl chloride with activated lead metal in ether at 80° produces tetraethyllead in 80–90% yield.[193]

7. Germanium*

The alkylation of germanium tetrachloride by trialkylaluminums results in reasonable yields of tetraalkylgermanes. For example, reaction of germanium tetrachloride with a small excess of triisobutylaluminum or triethylaluminum at 120–123° for 6 hr gave tetraisobutylgermane or tetraethylgermane in 73% yield upon hydrolysis.[177] In another experiment under comparable conditions with a larger excess of triethylaluminum in the presence of sodium chloride, tetraethylgermane was obtained in 57% yield, accompanied by hexaethyldigermane (15%)[194] (sodium chloride is unlikely to account for the difference in the two tetraethylgermane preparations). Tetraisobutylgermane, accompanied by isobutylpolygermanes, was prepared from triisobutylaluminum in an experiment similarly carried out with added sodium chloride.[194]

When germanium tetrachloride and trimethylaluminum (1:2 molar ratio) were heated at 130° with sodium chloride, tetramethylgermane was distilled out in 73% yield, accompanied by unchanged germanium tetrachloride.[194] From this it seems that the slowest stage in reaction is alkylation of the parent germanium tetrachloride. Sodium chloride accelerates the reaction but does not raise the yield of tetramethylgermane.

Tetraethylgermane has been obtained in high yield from ethylaluminum sesquibromide and germanium tetrachloride at 130° in the presence of sodium chloride.[187] Methylation of germanium tetrachloride by methylaluminum sesquichloride in diglyme solvent was accomplished in reasonable yield only when sodium chloride was present; unfortunately some cleavage of the solvent to β-chloroethyl methyl ether also occurred.[194]

8. Silicon†

Redistribution reactions of organoaluminum compounds with silicon halides are slow, and redistribution between organosilicon compounds is even slower, so that alkylation of a silicon tetrahalide generally gives a complex mixture of products, R_nSiX_{4-n} ($n = 1$ to 4), rather than a single product.

* See also Glockling and Hooton.[193a]
† For further discussion see Voorhoeve.[194a]

Alkylation of silicon tetrachloride by methylaluminum sesquichloride requires high temperature (250°). As with germanium tetrachloride, the alkylation becomes easier as the number of chlorine atoms on the central atom decreases.[195,196] Even when silicon tetrachloride is in excess, tetramethylsilane is the dominant product in the early stages of reaction, but eventually an equilibrium is attained in which the methylsilicon chlorides dominate.[195] Reaction between trimethylaluminum and trimethylfluorosilane is facile. A complex, Me_3Si—$F \rightarrow AlEt_3$ (cf. $Et_3AlF \cdot AlEt_3^-$ ion), forms, which decomposes at 30–35°, trimethylethylsilane and diethylaluminum chloride being separated on distillation.[197] Trimethylethylsilane is also formed readily when trimethylbromosilane is treated with ethylaluminum dichloride in the absence of solvent, or when trimethylsilyl isothiocyanate is treated with ethylaluminum dichloride or diethylaluminum chloride.[189] Reaction of triethylfluorosilane with dialkylaluminum hydrides provides a route to triethylsilane.[71]

Dimethylaluminum iodide reacts with a slight excess of triphenylsilyl azide under mild conditions (0° in pentane) to give dimethylaluminum azide quantitatively (reaction (46)), but the action of a 33% excess of dimethylaluminum iodide on trimethylsilyl azide under slightly more drastic conditions leads to tetramethylsilane and methylaluminum iodide azide (reaction (47)).[198] Excess trimethylsilyl azide converts the latter product to methylaluminum diazide.

$$Me_2AlI + Ph_3SiN_3 \longrightarrow Me_2AlN_3 + Ph_3SiI \tag{46}$$

$$Me_2AlI + Me_3SiN_3 \longrightarrow MeAlIN_3 + Me_4Si \tag{47}$$

The difficulty in obtaining high yields in the alkylation of silicon tetrachloride[123] is probably due to the high temperature required for alkylation, particularly by alkylaluminum dichlorides. The final stage in alkylation (reaction (48)) is reversible.[199]

$$RAlCl_2 + R_3SiCl \rightleftharpoons AlCl_3 + R_4Si \tag{48}$$

(dearylation of arylsilicon halides by aluminum chloride is also known[200]).

More complete utilization of the alkyl groups of a trialkylaluminum is possible when silicon tetrafluoride is alkylated, but high temperature is again required and a mixture of products (R_4Si, R_3SiF, R_2SiF_2, and $RSiF_3$) results.[123] Reaction of triethylaluminum etherate with Na_2SiF_6 at 200° likewise produces a mixture of products.[123]

Reaction of heterocyclic organoaluminum compounds with alkylchlorosilanes or alkylfluorosilanes provides a route to the corresponding heterocyclic silicon compounds.[131] The partial alkylation of alkylsilicon chlorides and silicon tetrachloride by sodium tetraethylaluminate is covered in a patent claim.[201]

9. Phosphorus

Alkylation of phosphorus trichloride by trialkylaluminums[181,202] or lithium tetraethylaluminate[188] gives only poor yields of trialkylphosphines, first because the alkylphosphorus(III) chlorides are Lewis bases and so do not readily alkylate, and second because the phosphorus(III) compounds are rendered even less reactive when they form complexes with aluminum compounds. By contrast, triphenylphosphine is obtained in good yield on reaction of phosphorus trichloride with phenylaluminum sesquichloride.[82] Alkyldichlorophosphines have been prepared in reasonable yield by reacting trialkylaluminums, in some cases as their etherates, with excess phosphorus trichloride; the product was distilled from the reaction accompanied by excess PCl_3.[202]

Trialkylphosphine oxides cannot be prepared adequately by reaction of trialkylaluminums with $POCl_3$. The corresponding reaction of trimethylaluminum or triphenylaluminum with $PSCl_3$ in a 1:1 molar ratio in refluxing hexane gives Me_3PS or Ph_3PS in high yields.[203] The preparation of compounds R_2PSCl and $RPSCl_2$ was also investigated. Alkylation by a deficit of alkylaluminum dichloride led to $RPSCl_2$, but the use of a low ratio of dialkylaluminum chloride or trialkylaluminum to $PSCl_3$ resulted in a mixture of products in which R_2PSCl dominated rather than $RPSCl_2$. By contrast, $PSBr_3$ could be alkylated to $EtPSBr_2$ by use of triethylaluminum in 3:1 mole ratio, and $EtPS(NMe_2)_2$ could be obtained in 80% yield by reaction of one mole of $PS(NMe_2)_2Cl$ with $\frac{1}{3}$ mole of triethylaluminum.[203]

10. Arsenic

Trialkylarsines are obtained in good yields by reaction of trialkylaluminums either with arsenic trifluoride or with arsenic trichloride in the presence of sodium chloride. Lithium tetraethylaluminate has also been used to prepare triethylarsine from arsenic trichloride in 31% yield.[188] A mixture of diisobutylchloroarsine and isobutyldichloroarsine was obtained from arsenic trichloride and triisobutylaluminum in ether.[181]

11. Antimony

Trialkylstibines have been obtained in 60–80% yields by addition of a slight excess of trialkylaluminum to a suspension of finely powdered antimony trifluoride in ether and subsequently distilling.[123,148,181] Triethylstibine has also been prepared in 65% yield from antimony trifluoride and a moderate excess of lithium tetraethylaluminate.[188]

The alkylation of antimony trichloride by trialkylaluminums, particularly in the presence of sodium chloride, and the conversion of antimony pentachloride to trialkylantimony dichlorides have also been claimed,[123] but a complex situation (comparable with the alkylation of lead(II) chloride) is revealed by a recent study of the reactions of antimony compounds with triethylaluminum and ethylaluminum chlorides in hexane.[204] Antimony trichloride reacts with excess triethylaluminum to give hexane-soluble triethylstibine in only low yield, accompanied by antimony metal (corresponding to $Sb^{III} \rightarrow Sb^0 + Sb^V$), and a large amount of a hexane-insoluble tetraethylstibonium salt, $[Et_4Sb]^+[Et_5Al_2Cl_2]^-$. Reaction with excess diethylaluminum chloride proceeds similarly to give $[Et_4Sb]^+[AlCl_4]^-$. Tetraethylstibonium derivatives could also be obtained upon reaction of triethylantimony dichloride with triethylaluminum or ethylaluminum chlorides. Both $SbEt_3$ and $SbEt_5$ complex with ethylaluminum compounds.[204,205].

12. Bismuth

Trialkylbismuths are obtained in excellent yields by reaction of slight excess of trialkylaluminum with bismuth trichloride in ether.[148] An equally good yield is obtained by the action of a moderate excess of lithium tetraethylaluminate on bismuth trichloride.[188*]

13. Transition Metals

Most of the reactions of organoaluminum compounds with transition metal halides are not simple redistribution reactions, but relatively uncomplicated redistributions have been observed for the halides and cyclopentadienyl halides of titanium and zirconium. Reaction of $(Cp_2ZrCl)_2O$ with trimethylaluminum ($2Me_3Al$ per $(Cp_2ZrCl)_2O$) in benzene at 65–70° gives a 53% yield of $Cp_2ZrClMe$,[206] and even better results may be obtained at lower temperatures in methylene dichloride solvent.[207] Treatment of Cp_2ZrCl_2 with excess triethylaluminum[208] results in ethane evolution and the formation of a compound, which is possibly (**30**).

$$Et_2Al \diagdown^{Cl}_{Cl} (Cp_2Zr)-CH_2-CH_2-(Cp_2Zr) \diagup^{Cl}_{Cl} AlEt_2$$

(**30**)

Three distinct phases in the reaction of Cp_2TiCl_2 with diethylaluminum chloride have been recognized: first, formation of a complex between the reactants; second, exchange of an ethyl group for a chloro group to give complexed $Cp_2TiEtCl$; and third, reduction to $Cp_2Ti^{III}Cl$.[209] The $Cp_2Ti^{III}Cl$

* See also Jenkner.[123]

is also complexed with organoaluminum halide, e.g., with Et_2AlCl in (31).[210-212] Upon prolonged treatment of Cp_2TiCl_2 with trimethyl-

$$Cp_2Ti \underset{Cl}{\overset{Cl}{<}} \underset{Et}{\overset{Et}{>}} Al$$

(31)

aluminum, methane is evolved and an unusual compound is formed which is thought to have $TiCH_2Al$ groupings.[213]

Titanium tetrachloride reacts with dimethylaluminum chloride in 1:1 ratio to give methyltitanium trichloride, which may be separated by distillation if the methylaluminum dichloride also produced is complexed with sodium chloride.[214,215] Methyltitanium trichloride is reasonably stable in the absence of aluminum compounds. The most important gaseous product of its decomposition is methane rather than ethane.[216] Dimethyltitanium dichloride, obtained by reaction of methyltitanium trichloride with trimethylaluminum at $-20°$, ethyltitanium trichloride, and isobutyltitanium trichloride are much less stable than methyltitanium trichloride.[214]

When phosphine complexes of ruthenium(II) dichloride and rhodium(I) chloride are reacted with trimethylaluminum, the corresponding Me_2Ru^{II} and $MeRh^I$ complexes are formed, but reaction with triethylaluminum or triisopropylaluminum gives the corresponding Ru^{II} and Rh^I hydrido complexes.[217,218]

The reduction of transition metal compounds by organoaluminum compounds is not well understood, even though it has been the subject of many studies.[214,216,219-221] It seems likely that an alkyl- or hydridoderivative of the transition metal is first formed, and that this subsequently decomposes to give a compound of the metal in a lower valence state. One of the surprising features of the reduction is that often, and particularly in the cases of methylmetallic compounds, the alkyl group (R) obtained from the organoaluminum compound ends up largely as the corresponding alkane (RH), rather than as dimerization product (R—R) plus equivalent amounts of RH and alkene (R minus H). Formation of RH is not entirely due to abstraction of a hydrogen atom from solvent.

Reduction of transition metal halides by organoaluminum compounds is useful preparatively. For example, reduction of $CrCl_3$, $MoCl_5$, or WCl_6 by triethylaluminum in benzene or ether in the presence of carbon monoxide under pressure gives $Cr(CO)_6$, $Mo(CO)_6$, or $W(CO)_6$, respectively, in good yields.[222,223] $Mn_2(CO)_{10}$ has been prepared similarly from manganese(II) acetate.[222] Reduction by triethylaluminum in the presence of olefins, acetylenes, or arenes has been used in the preparation of complexes of transition metals in lower valence states (e.g., hexamethylbenzene$_2$Cr^I ion,[224] mesitylene$_2$Cr^I ion,[225] and bis(cyclooctatetraene)titanium[170]).

F. Reaction with Acetylacetonates

1. In the Presence of Dipyridyl

Compounds with transition metal–carbon sigma bonds are often stable only when the transition metal carries appropriate stabilizing ligands (phosphines, 2,2′-dipyridyl, etc.). Even in the presence of these ligands, the alkylation of transition-metal halides by trialkylaluminum compounds may not be effective, because the alkylaluminum compounds may complex with the stabilizing ligands, thus allowing decomposition of the desired alkylmetallic compounds.

The reaction of transition metal acetylacetonates with alkylaluminum alkoxides avoids these difficulties, since the starting alkoxide and the acetylacetonate–alkoxide products are both self-associated. This method has been successfully exploited in recent years.[226] Diethyl(2,2′-dipyridyl)-nickel and other dialkyl(2,2′-dipyridyl)nickels have been prepared by reaction of nickel(II) acetylacetonate with dialkylaluminum alkoxides in the presence of dipyridyl in ether.[227-229]

Diethylbis(2,2′-dipyridyl)iron(II)[229,230] has been obtained by reaction of diethylaluminum ethoxide with dipyridyl and iron (III) acetylacetonate at $-20°$ to $+10°$ in ether, and a compound thought to be diethylbis-(2′,2′-dipyridyl)cobalt(II) has been prepared similarly from cobalt(III) acetylacetonate.[231]

2. In the Presence of Phosphines

Quite different results are obtained on reacting iron(III) and cobalt(III) acetylacetonates with diethylaluminum ethoxide in the presence of phosphine ligands, presumably because the alkylmetallic compounds are now less stable. In the presence of 1,2-bis(diphenylphosphino)ethane, the (ethylene)iron(0) compound (32) and the hydridocobalt(I) compound (33) are obtained.[232]

$$\left[\begin{array}{c} CH_2—Ph_2P \\ \\ CH_2—Ph_2P \end{array}\right]_2 \searrow \atop \nearrow Fe \longleftarrow \begin{array}{c} CH_2 \\ \| \\ CH_2 \end{array} \qquad \left[\begin{array}{c} CH_2—Ph_2P \\ \\ CH_2—Ph_2P \end{array}\right]_2 \searrow \atop \nearrow Co—H$$

$$(32) \qquad\qquad (33)$$

In the presence of triphenylphosphine reduction of cobalt(III) acetylacetate by diethylaluminum ethoxide occurs similarly, although there is some doubt about the structures of the products. Under a nitrogen atmosphere a compound results which was originally formulated as (nitrogen)tris(triphenylphosphine)cobalt(0),[233] but which now appears to be the corresponding cobalt(I) hydride, $H\text{—}Co(PPh_3)_2 \leftarrow N\equiv N$.[234,235]* The nitrogen may be reversibly displaced by hydrogen to give $H_3Co(PPh_3)_3$.[236-238] Reduction of the cobalt(III) acetylacetonate by diethylaluminum ethoxide under an atmosphere of argon results in an ethylene complex, formulated as (ethylene)tris(triphenylphosphine)cobalt(0) (possibly this is also a Co^I hydride), from which the ethylene may be reversibly displaced by nitrogen to give the nitrogen derivative.[236] The hydrido(nitrogen)tris(triphenylphosphine)cobalt has also been formed by reduction of cobalt(III) acetylacetonate with triisobutylaluminum in toluene under nitrogen.[239]

Reflecting the greater stability of methylmetallic compounds, the reaction of dimethylaluminum methoxide with cobalt(III) acetylacetonate gives methyltris(triphenylphosphine)cobalt(I).[233]

Reaction of cobalt(III) acetylacetonate with diethylaluminum ethoxide at room temperature in benzene in the presence of tetrakis(diphenylphosphinomethyl)methane is unusual in that it produces a mixed hydridoethylcobalt(III) compound (34).[240]

$$Et_2HCo \xleftarrow{\begin{array}{c} Ph_2P\text{—}CH_2 \\ Ph_2P\text{—}CH_2 \\ Ph_2P\text{—}CH_2 \end{array}} C\text{—}CH_2\cdot PPh_2$$

(34)

Rhodium(III) acetylacetonate is reduced by triethylaluminum in the presence of triphenylphosphine to give hydridotetrakis(triphenylphosphine)rhodium(I) [241] (cf. $(PPh_3)_3RhH$ from $(PPh_3)_3RhCl$ and triisopropylaluminum[218]). The corresponding reaction of ruthenium(III) acetylacetonate or ruthenium trichloride gives a compound which is probably dihydridotetrakis(triphenylphosphine)ruthenium(II), from which one molecule of triphenylphosphine may be reversibly displaced by nitrogen.[241] The resulting compound, $(Ph_3P)_3Ru^{II}(N_2)H_2$, may also be prepared by reaction of $(Ph_3P)_3Ru^{II}HCl$ with triethylaluminum in ether under nitrogen.[241a]

Treatment of nickel(II) acetylacetonate with triethylaluminum and tertiary phosphines or phosphites in benzene at 20° produces (ethylene)-nickel(0) compounds, $(R_3P)_2Ni\cdot C_2H_4$ (R = ethyl, cyclohexyl, phenyl, o-phenylphenoxy, etc.) in good yields, which are improved still further in the presence of added ethylene.[242] A compound thought to be (cyclohexyl$_3$P)$_2$Ni(N$_2$), from which nitrogen may be reversibly displaced, is formed from nickel(II) acetylacetonate, trimethylaluminum, and tricyclohexylphosphine under nitrogen in toluene.[243]

* See, however, the views of Allen and Bottomley.[235a]

3. In the Presence of Other Ligands

By contrast, when nickel(II) acetylacetonate (1 mole) reacts with trimethylaluminum (2 moles) and tri(o-phenylphenyl)phosphite (2 moles), the dimethylnickel(II) compound (35) is formed in yields of at least 30%.[228]

(35)

Reaction of diethylaluminum ethoxide with nickel acetylacetonate in the presence of 1,5-cyclooctadiene in benzene at 0–5° gives bis(1,5-cyclooctadiene)nickel(0).[244] In the alternative presence of hexaphenylethane, bistritylnickel(II) results.[245]

G. Experimental Methods and Techniques

Almost all organoaluminum compounds are sensitive to air, moisture, and hydroxylic solvents. The trialkylaluminums and alkylaluminum halides, which are most commonly used, are dangerously reactive toward air. Experiments with these materials should not be undertaken without careful attention to experimental technique and the safety of the experimenter in the event of breakage of apparatus or other accident. Protective clothing (face shields and leather gloves) and a well organized supply of inert blanketing gas (nitrogen; sometimes argon is preferred) are essential. The organoaluminum compounds may be handled with less risk in inert (hydrocarbon) solvents, particularly upon being deliberately destroyed. Redistribution reactions between organoaluminum compounds and halides, alkoxides, carboxylates, etc. of other elements are generally highly exothermic and precautions need to be taken accordingly.

Most experiments on a small to medium scale can be carried out in ordinary laboratory glassware. Silicone grease is attacked by organoaluminum compounds. Serum caps, lubricated syringes, and stainless steel hypodermic tubing (ca. 0.05-in. diameter with sharpened tips) allow easy transfer of reactive materials from vessel to vessel. A low (5 psi) positive pressure of nitrogen or argon should be used for filtration or transfer from vessel to vessel. Glove box techniques are invaluable for flexible handling of organometallic compounds at the beginning and end of experiments.

Almost all the simpler alkylaluminum compounds may be obtained from the commercially available trialkylaluminum compounds by redistribution reactions, and triphenylaluminum may be obtained by metalation of benzene starting with triethylaluminum. Other aryl and unsaturated organoaluminum compounds are most commonly prepared by reaction of organolithium or Grignard reagents with aluminum chloride or by reaction of aluminum with organomercurials, but we shall not concern ourselves with these reactions in this review.

The method by which products are isolated after being formed by a redistribution reaction depends upon the properties of the products. If they are sensitive to air or moisture, as are the trialkylaluminums, they are most commonly distilled from the unhydrolyzed reaction products. Occasionally crystallization is useful. If the desired products are not sensitive to water, the simplest procedure is generally to remove the aluminum compounds by hydrolysis and then to separate the desired products by conventional methods (distillation, etc.).

1. Dimethylzinc[43]

a. From Diethylzinc.* Trimethylaluminum (1.44 g) was added by syringe through a serum cap to diethylzinc (2.46 g) in a nitrogen-flushed distillation apparatus. The temperature was slowly raised to 140°, while dimethylzinc (1.43 g, 75% yield), bp 46°, distilled. The apparatus was dismantled under a nitrogen atmosphere, and the dimethylzinc was transferred by syringe to glass ampoules, which were cooled and sealed.

b. From Zinc Ethoxide. Trimethylaluminum (10.1 g) was added slowly by syringe to a suspension of zinc ethoxide (31.1 g) in toluene (15 ml) in a nitrogen-flushed distillation apparatus. The reactants were then heated to 120°. Dimethylzinc (11.6 g., 61% yield) distilled.

2. Dimethylcadmium[43]

Trimethylaluminum (4.32 g) was slowly added by syringe to anhydrous cadmium acetate (13.6 g) in a slowly stirred, externally cooled distillation apparatus. As the temperature was slowly raised to 180°, dimethylcadmium (5.0 g, 60% yield), bp 104–106°, distilled, and was later purified by redistillation from a small quantity of cadmium acetate.

3. Diisobutylcadmium[43]

In a similar manner triisobutylaluminum (11.9 g) was reacted with cadmium acetate (12.9 g). The reactants were heated to 90° for 10 min, then the pressure was lowered so that diisobutylcadmium (6.7 g, 54% yield), bp 70°/8 mm, distilled.

---

* Trimethylaluminum and diethylzinc are available commercially at moderate prices.

4. Tetraethyltin[166]

Tin tetrachloride (78.2 g) was added at 0° to stirred triethylaluminum (46.0 g) in a three-necked flask fitted with a reflux condenser. Diethyl ether (45 ml) was then added, and reaction was completed by heating at 80° for 30 min. Tetraethyltin (68.1 g, 97% yield), bp 62°/10 mm, was distilled directly from the flask.

5. Tetrabutyltin[166]

Similarly tin tetrachloride (39.1 g) and dibutyl ether (30 g) were reacted with tributylaluminum (39.6 g). When reaction was completed, the products were hydrolyzed by water and dilute hydrochloric acid. The organic layer was distilled to give dibutyl ether and then tetrabutyltin (46.7 g, 90% yield), bp 149°/12 mm.

6. Triethyltin Chloride[166]

Similarly tin tetrachloride (52 g) and dibutyl ether (25 ml) were reacted with triethylaluminum (22.8 g). After 1 hr at 80° gas chromatography showed the presence of small amounts of tetraethyltin and diethyltin dichloride. After a further hour at 100°, fractionation gave triethyltin chloride (45.1 g, 94% yield), bp 90–93°/10 mm.

7. Triethylbismuth[148]

Triethylaluminum (5.0 g), dissolved in ether (20 ml), was added slowly to freshly sublimed bismuth trichloride (13 g) in ether (20 ml). The products were hydrolyzed with water under a nitrogen atmosphere, and the ether layer was dried over calcium chloride and distilled under nitrogen to give triethylbismuth (10.5 g, 86% yield), bp 104–105°/76 mm.

REFERENCES

References are given to English translations of foreign language journals, whenever these are freely available.

1. G. Calingaert and H. A. Beatty, *Organic Chemistry, An Advanced Treatise*, 2nd ed. Vol. 2, H. Gilman, Ed., Wiley, New York, 1943, p. 1806.
2. J. C. Lockhart, *Chem. Rev.*, **65**, 131 (1965).
3. K. Moedritzer, *Advances in Organometallic Chemistry*, Vol. 6, F. G. A. Stone and R. West, Eds., Academic Press, New York, 1968, p. 171.
4. K. Moedritzer, *Organometal. Chem. Rev.*, **1**, 179 (1966).
5. J. R. Van Wazer and K. Moedritzer, *Angew. Chem. Internat. Ed.*, **5**, 341 (1966).
6. K. Ziegler, *Organometallic Chemistry*, H. H. Zeiss, Ed., Reinhold Publ., New York, 1960, p. 194.
7. R. Köster and P. Binger, *Advances in Inorganic Chemistry and Radiochemistry*, Vol. 7, H. J. Emeleus and A. G. Sharpe, Eds., Academic Press, New York, 1965, p. 263.

8. N. S. Ham and T. Mole, *Progress in Nuclear Magnetic Resonance Spectroscopy*, Vol. 4, J. W. Emsley, J. Feeney, and L. H. Sutcliffe, Eds., Pergamon Press, Oxford, 1969, in the press.

9. T. Mole, *Australian J. Chem.*, **17**, 1050 (1964).

10. E. A. Jeffery, T. Mole, and J. K. Saunders, *Australian J. Chem.*, **21**, 649 (1968).

11. H. Lehmkuhl and H. D. Kobs, *Ann. Chem*, **719**, 11 (1968).

11a. H. Schmidbauer, W. Wolfsberger and K. Schwirten, *Chem. Ber.*, **102**, 556 (1969).

12. C. K. Ingold, *Helv. Chim. Acta*, **47**, 1191 (1964).

13. G. E. Coates and K. Wade, *Organometallic Compounds*, 3rd ed., Vol. 1, Methuen, London, 1967, p. 295 ff.

14. S. Hilpert and G. Grüttner, *Chem. Ber.*, **45**, 2828 (1912).

15. N. Davidson and H. C. Brown, *J. Am. Chem. Soc.*, **64**, 316 (1942).

16. E. Wiberg, H. Graf, M. Schmidt, and R. Uson, *Z. Naturforsch.*, **7b**, 578 (1952).

17. C. W. Heitsch, C. E. Nordman, and R. W. Parry, *Inorg. Chem.*, **2**, 508 (1963).

18. Ref. 13, p. 306.

19. L. O. Brockway and N. R. Davidson, *J. Am. Chem. Soc.*, **63**, 3287 (1941).

20. E. G. Hoffmann, *Ann. Chem.*, **629**, 104 (1960).

21. E. A. Jeffery and T. Mole, *Australian J. Chem.*, **21**, 2683 (1968).

22. V. J. Shiner, D. Whittaker, and V. P. Fernandez, *J. Am. Chem. Soc.*, **85**, 2318 (1963).

23. A. W. Laubengayer and W. F. Gilliam, *J. Am. Chem. Soc.*, **63**, 477 (1941).

24. P. H. Lewis and R. E. Rundle, *J. Chem. Phys.*, **21**, 986 (1953).

25. R. G. Vranka and E. L. Amma, *J. Am. Chem. Soc.*, **89**, 3121 (1967).

26. T. Wartik and H. I. Schlesinger, *J. Am. Chem. Soc.*, **75**, 835 (1957).

27. O. Yamamoto, *Bull. Chem. Soc. Japan*, **35**, 619 (1962).

28. G. Allegra, G. Perego, and A. Immirzi, *Makromol. Chem.*, **61**, 69 (1963).

29. K. S. Pitzer and H. S. Gutowski, *J. Am. Chem. Soc.*, **68**, 2204 (1946).

30. E. A. Jeffery, T. Mole, and J. K. Saunders, *Chem. Commun.*, **1967**, 696.

31. E. A. Jeffery, T. Mole, and J. K. Saunders, *Australian J. Chem.*, **21**, 137 (1968).

32. T. Mole and J. R. Surtees, *Chem. Ind.*, **1963**, 1727.

33. T. Mole and J. R. Surtees, *Australian J. Chem.*, **17**, 1229 (1965).

34. J. P. Oliver and L. G. Stevens, *J. Inorg. Nucl. Chem.*, **24**, 953 (1962).

35. E. A. Jeffery and T. Mole, *J. Organometal. Chem.*, **11**, 393 (1968).

36. H. D. Visser and J. P. Oliver, *J. Am. Chem. Soc.*, **90**, 3579 (1968).

37. W. Strohmeier, K. Hümphner, K. Miltenberger, and F. Seifert, *Ber. Bunseges. Phys. Chem.*, **63**, 537 (1959).

38. N. Muller and A. L. Ottermat, *Inorg. Chem.*, **4**, 296 (1967).

39. J. K. Ruff, *J. Am. Chem. Soc.*, **83**, 1798 (1961).

40. H. Hartmann, *Ann. Chem.*, **714**, 1 (1968).

41. G. W. Wheland, *Advanced Organic Chemistry*, 2nd ed., Wiley, New York, 1960, p. 573 ff.

42. L. I. Zakharkin, V. V. Gavrilenko, and L. L. Ivanov, *J. Gen. Chem. U.S.S.R. (English transl.)*, **37**, 937 (1967).

43. T. Mole, unpublished results.

44. S. M. Blitzer and T. H. Pearson, U.S. Patent 2,969,381; *Chem. Abstr.* **55**, 9282 (1961).

45. T. Mole, *Australian J. Chem.*, **16**, 794 (1963).

46. G. Calingaert and H. A. Beatty, *J. Am. Chem. Soc.*, **61**, 2748 (1939).

47. G. Calingaert, H. A. Beatty, and H. R. Neal, *J. Am. Chem. Soc.*, **61**, 2755 (1939).

48. G. Calingaert and H. Soroos, *J. Am. Chem. Soc.*, **61**, 2758 (1939).
49. G. Calingaert, H. A. Beatty, and H. Soroos, *J. Am. Chem. Soc.*, **62**, 1099 (1940).
50. T. Mole and J. R. Surtees, *Australian J. Chem.*, **17**, 310 (1964).
51. T. Mole, *Australian J. Chem.*, **18**, 1183 (1965).
52. T. Mole, *Exchange Reactions*, Internat. Atomic Energy Agency, Vienna (1965) p. 327.
53. L. I. Zakharkin and O. Y. Okhlobystin, *J. Gen. Chem. U.S.S.R.* (*English transl.*), **30**, 2109 (1960).
54. A. von Grosse and J. M. Mavity, *J. Org. Chem.*, **5**, 106 (1940).
55. N. Muller and D. E. Pritchard, *J. Am. Chem. Soc.*, **82**, 248 (1960).
56. E. G. Hoffman, *Trans. Faraday Soc.*, **58**, 642 (1962).
57. K. C. Ramey, J. F. O'Brien, I. Hasegawa, and A. E. Borchert, *J. Phys. Chem.*, **69**, 3418 (1965).
58. M. P. Groenewege, J. Smidt, and H. De Vries, *J. Am. Chem. Soc.*, **82**, 4425 (1960).
59. K. C. Williams and T. L. Brown, *J. Am. Chem. Soc.*, **88**, 5460 (1966).
59a. E. A. Jeffery and T. Mole, *Australian J. Chem.*, **22**, 1129 (1969).
60. O. Yamamoto, *Bull. Chem. Soc. Japan*, **37**, 1125 (1964).
61. J. Smidt, M. P. Groenewege, and H. De Vries, *Rec. Trav. Chim.*, **81**, 729 (1962).
61a. D. A. Sanders and J. P. Oliver, *J. Am. Chem. Soc.*, **90**, 5910 (1968).
62. E. G. Hoffmann, *Bull. Soc. Chim. France*, **1963**, 1467.
63. O. Yamamoto and K. Hayamizu, *J. Phys. Chem.*, **72**, 822 (1968).
64. T. Mole, *Australian J. Chem.*, **19**, 381 (1966).
65. T. Mole, *Australian J. Chem.*, **19**, 373 (1966).
66. K. Ziegler and W. R. Kroll, *Ann. Chem.*, **629**, 167 (1960).
67. A. Zambelli, A. L. Segre, A. Marinangelli, and G. Gatti, *Chim. Ind.*, **48**, 1 (1966).
68. J. Brandt and E. G. Hoffmann, *Brennstoff-Chem.*, **45**, 200 (1962).
69. H. Reinheckel and K. Haage, *J. Prakt. Chem.*, [4] **33**, 70 (1966).
70. G. F. Lengnick and A. W. Laubengayer, *Inorg. Chem.*, **5**, 503 (1966).
71. H. Jenkner, *Z. Naturforsch.*, **12b**, 809 (1957).
72. Kali-Chemie, German Patent 1,009,630; *Chem. Abstr.*, **53**, 21667 (1959).
73. K. Ziegler and R. Köster, *Ann. Chem.*, **608**, 1 (1957).
74. M. I. Prince and K. Weiss, *J. Organometal. Chem.*, **5**, 584 (1966).
75. Badische-Anilin & Soda Fabrik, German Patent 1,090,209; *Chem. Abstr.*, **56**, 3513 (1962).
76. H. Scherer and G. Seydel, *Angew. Chem.*, **75**, 846 (1963).
76a. A. J. Lundeen and D. M. Coyne, U.S. Patent 3,290,349; *Chem. Abstr.*, **66**, 95172 (1967).
77. Y. Sakurada, M. L. Huggins, and W. R. Anderson, Jr., *J. Phys. Chem.*, **68**, 1934 (1964).
78. K. Hatada and H. Yuki, *Tetrahedron Letters*, **1967**, 5227.
79. C. J. Marsel, E. O. Kalil, A. Reidlinger, and L. Kramer, *Metal Organic Compounds*, ACS Advances in Chemistry Series No. 23 (1959) p. 172.
80. V. Grignard and R. Jenkins, *Compt. Rend.*, **179**, 89 (1924).
81. V. Hnizda and C. A. Kraus, *J. Am. Chem. Soc.*, **60**, 2276 (1938).
82. D. Wittenberg, *Ann. Chem.*, **654**, 23 (1962).
83. A. Zambelli, G. Gatti, A. Marinangelli, F. Cabassi, and I. Pasquon, *Chim. Ind.* **48**, 333 (1966).

84. L. I. Zakharkin and I. M. Khorlina, *Bull. Acad. Sci. U.S.S.R. (English transl.)*, **1960**, 133.
85. L. I. Zakharkin and I. M. Khorlina, *J. Gen. Chem. U.S.S.R. (English transl.)*, **30**, 1905 (1960).
86. H. Martin, R. Reinäcker, and K. Ziegler, *Brennstoff-Chem.*, **47**, 33 (1966).
87. A. Zambelli, J. Di Pietro, and G. Gatti, *J. Polymer. Sci. A* **1**, 403 (1963).
88. H. Lehmkuhl and R. Schäfer, *Ann. Chem.*, **705**, 23 (1967).
89. G. Natta, A. Zambelli, I. Pasquon, G. Gatti, and D. De Luca, *Makromol. Chem.*, **70**, 206 (1964).
90. R. Tarao and S. Takeda, *Bull. Chem. Soc. Japan*, **40**, 650 (1967).
91. R. Köster and W. R. Kroll, *Ann. Chem.*, **629**, 50 (1960).
92. K. Ziegler, British Patent 779,873; *Chem. Abstr.*, **51**, 17981 (1957).
92a. K. Ziegler, German Patent 934,649; *Chem. Abstr.*, **52**, 18216 (1958).
93. H. Lehmkuhl, *Angew. Chem. Intern. Ed.*, **3**, 107 (1964).
94. H. Lehmkuhl and R. Schäfer, *Ann. Chem.*, **705**, 32 (1967).
95. G. Wilke and H. Müller, *Ann. Chem.*, **629**, 222 (1960).
96. N. S. Ham, E. A. Jeffery, and T. Mole, *Australian J. Chem.*, **21**, 2687 (1968).
97. V. R. Magnuson and G. D. Stucky, *J. Am. Chem. Soc.*, **90**, 3269 (1968).
98. M. Kawai, T. Ogawa, and K. Hirota, *Bull. Chem. Soc. Japan*, **37**, 1302 (1964).
99. H. Nöth and E. Wiberg, *Fortsch. Chem. Forsch.*, **8**, 396 (1967).
100. K. Ziegler, W. R. Kroll, W. Larbig, and O. W. Steudel, *Ann. Chem.*, **629**, 53 (1960).
101. K. Ziegler, H. G. Gellert, H. Martin, K. Nagel, and J. Schneider, *Ann. Chem.*, **589**, 91 (1954).
102. K. Ziegler, H. G. Gellert, K. Zosel, H. Lehmkuhl, and W. Pfohl, *Angew. Chem.*, **67**, 424 (1955).
103. K. Ziegler, H. G. Gellert, H. Lehmkuhl, W. Pfohl, and K. Zosel, *Ann. Chem.*, **629**, 1 (1960).
104. H. E. Podall, H. E. Petree, and J. R. Zietz, *J. Org. Chem.*, **24**, 1222 (1959).
105. J. R. Surtees, *Chem. Ind.*, **1964**, 1260.
106. L. I. Zakharkin and V. V. Gavrilenko, *Bull. Acad. Sci. U.S.S.R. (English transl.)*, **1965**, 624.
107. D. B. Miller, *J. Organometal. Chem.* **14**, 253 (1968).
108. L. I. Zakharkin and I. M. Khorlina, *Bull. Acad. Sci. U.S.S.R. (English transl.)*, **1961**, 1768.
109. L. I. Zakharkin, V. V. Gavrilenko, and I. M. Khorlina, *Bull. Acad. Sci. U.S.S.R. (English transl.)*, **1962**, 405.
110. F. M. Peters, B. Bartocha, and A. J. Bilbo, *Can. J. Chem.*, **41**, 1051 (1963).
111. H. Lehmkuhl and W. Eisenbach, *Ann. Chem.*, **705**, 42 (1967).
112. K. C. Williams and T. L. Brown, *J. Am. Chem. Soc.*, **88**, 4134 (1966).
113. K. Ziegler, *Bull. Soc. Chim. France*, **1963**, 1456.
114. K. Ziegler, German Patent 1,161,895; *Chem. Abstr.*, **60**, 12050 (1964).
115. L. I. Zakharkin and V. V. Gavrilenko, *Bull. Acad. Sci. U.S.S.R. (English transl.)*, **1963**, 1737.
116. P. Kobetz, W. E. Becker, R. C. Pinkerton, and J. B. Honeycutt, *Inorg. Chem.*, **2**, 859 (1963).
117. L. I. Zakharkin, V. V. Gavrilenko, and L. L. Ivanov, *J. Gen. Chem. U.S.S.R. (English transl.)*, **35**, 1677 (1965).
118. D. F. Evans and V. Fazakerly, *Chem. Commun.*, **1968**, 974.
119. E. A. Jeffery and T. Mole, *Australian J. Chem.*, **21**, 1497 (1968).

120. E. A. Jeffery, unpublished results.
121. N. S. Ham, E. A. Jeffery, T. Mole, and J. K. Saunders, *Australian J. Chem.*, **20**, 2641 (1967).
122. N. S. Ham, E. A. Jeffery, T. Mole, and J. K. Saunders, *Australian J. Chem.*, **21**, 659 (1968).
123. H. Jenkner, *Chem. Zeitung*, **86**, 527 and 563 (1962).
124. K. Ziegler, K. Nagel, and W. Pfohl, *Ann. Chem.*, **629**, 210 (1960).
125. R. Köster and G. Bruno, *Ann. Chem.*, **629**, 89 (1960).
126. Ref. 13, p. 313 ff.
127. C. R. McCoy and A. L. Allred, *J. Am. Chem. Soc.*, **84**, 912 (1962).
128. L. M. Nazarova, *J. Gen. Chem. U.S.S.R.* (*English transl.*), **29**, 2636 (1959).
129. L. M. Nazarova, *J. Gen. Chem. U.S.S.R.* (*English transl.*), **31**, 1034 (1961).
130. L. I. Zakharkin and O. Y. Okhlobystin, *Bull. Acad. Sci. U.S.S.R.* (*English transl.*), **1959**, 172.
131. Ref. 7, p. 300 ff.
132. R. Köster, *Advances in Organometallic Chemistry*, Vol. 2, F. G. A. Stone and R. West, Eds., Academic Press, New York, 1964, p. 257.
133. R. Köster, *Progress in Boron Chemistry*, Vol. 1, H. Steinberg and A. L. McCloskey, Eds., Macmillan, New York, 1964, p. 289.
134. R. Köster and G. Benedikt, *Angew. Chem. Intern. Ed.*, **1**, 507 (1962).
135. R. Köster, *Angew. Chem. Intern. Ed.*, **3**, 174 (1964).
136. R. Köster, G. Griasnow, W. Larbig, and P. Binger, *Ann. Chem.*, **672**, 1 (1964).
137. Studiengesellschaft-Kohle, French Patent 1,360,431; *Chem. Abstr.*, **61**, 9526 (1964).
138. B. Schneider and W. P. Neumann, *Ann. Chem.*, **707**, 7 (1967).
139. G. D. Barbaras, G. Dillard, A. E. Finhold, T. Wartik, K. E. Wilzbach, and H. I. Schlesinger, *J. Am. Chem. Soc.*, **73**, 4585 (1957).
140. L. I. Zakharkin and I. M. Khorlina, *J. Gen. Chem. U.S.S.R.* (*English transl.*), **32**, 2740 (1962).
141. B. Bartocha, A. J. Bilbo, D. E. Bublitz, and M. Y. Gray, *Angew. Chem.*, **72**, 36 (1960).
142. B. Bartocha and A. J. Bilbo, *J. Am. Chem. Soc.*, **83**, 2202 (1961).
143. W. P. Neumann, H. Niermann, and B. Schneider, *Angew. Chem. Intern. Ed.*, **2**, 547 (1963).
144. W. P. Neumann and H. Niermann, *Ann. Chem.*, **653**, 164 (1962).
145. M. R. Kula, J. Lorberth, and E. Amberger, *Chem. Ber.*, **97**, 2087 (1964).
146. Studiengesellschaft-Kohle, British Patent 951,150; *Chem. Abstr.*, **60**, 13271 (1964).
147. W. P. Neumann and R. Sommer, *Angew. Chem. Intern. Ed.*, **2**, 547 (1963).
148. L. I. Zakharkin and O. Y. Okhlobystin, *Bull. Acad. Sci. U.S.S.R.* (*English transl.*), **1959**, 1853.
149. S. M. Blitzer and T. H. Pearson, U.S. Patent 2,859,231; *Chem. Abstr.*, **53**, 9150 (1959).
149a. W. Dahlig, S. Pasynkiewicz and K. Wazynski, *Przemysl. Chem.*, **39**, 436 (1960); *Chem. Abstr.*, **55**, 15335 (1961).
150. T. H. Pearson, S. M. Blitzer, D. R. Carley, W. T. McKay, R. L. Ray, L. L. Sims, and J. R. Zietz, *Metal Organic Compounds*, *ACS Advances in Chemistry Series No. 23* (1959), p. 299
151. F. W. Frey and S. E. Cook, *J. Am. Chem. Soc.*, **82**, 530 (1960).
152. J. R. Mangham, U.S. Patent 3095,433; *Chem. Abstr.*, **59**, 6440 (1963).

153. R. Köster, *Ann. Chem.*, **618**, 31 (1958).
154. R. Köster, *Angew. Chem.*, **70**, 371 (1958).
155. J. Iyoda and J. Shiihara, *Bull. Chem. Soc. Japan*, **32**, 304 (1959).
156. J. C. Perrine and R. N. Neller, *J. Am. Chem. Soc.*, **80**, 1823 (1958).
157. L. I. Zakharkin and O. Y. Okhlobystin, *Bull. Acad. Sci. U.S.S.R.* (*English transl.*), **1959**, 1100.
158. E. C. Ashby, *J. Am. Chem. Soc.*, **81**, 4791 (1959).
159. Kali-Chemie, German Patent 1,067,814; *Chem. Abstr.*, **55**, 13291 (1961); see also ref. 123.
160. Kali-Chemie, German Patents 1,121,048 and 1,127,356; *Chem. Abstr.*, **57**, 15152 and 11232 (1962).
160a. Societa Edison, Netherlands Patent Appl. 645,120; *Chem. Abstr.*, **64**, 754 (1966).
161. W. G. Woods and A. L. McCloskey, *Inorg. Chem.*, **2**, 861 (1963).
162. Kali-Chemie, German Patent 974,764; *Chem. Abstr.*, **56**, 4795 (1962).
163. Kali-Chemie, German Patent 974,338; *Chem. Abstr.*, **56**, 1481 (1962).
164. H. Jenkner, *Z. Naturforsch.*, **14b**, 133 (1959).
165. R. C. Anderson and G. J. Sleddon, *Chem. Ind.*, **1960**, 1335.
166. W. P. Neumann, *Ann. Chem.*, **653**, 157 (1962).
167. S. M. Blitzer and T. H. Pearson, British Patents 2,859,225, 2,859,227 and 2,859,228; *Chem. Abstr.*, **53**, 9149–9150 (1959).
168. S. M. Blitzer and T. H. Pearson, U.S. Patent 2,989,558; *Chem. Abstr.*, **55**, 23345 (1961).
169. T. A. George and M. F. Lappert, *Chem. Commun.*, **1966**, 463.
170. H. Breil and G. Wilke, *Angew. Chem. Intern. Ed.*, **5**, 898 (1966).
171. H. Dietrich and H. Dierks, *Angew. Chem. Intern. Ed.*, **5**, 899 (1966).
172. W. Stamm and A. Breindel, *Angew. Chem. Intern. Ed.*, **3**, 66 (1964).
173. Stauffer Chemical Co., French Patent 1,338,836; *Chem. Abstr.*, **60**, 3015 (1964).
174. W. Sundermeyer and W. Verbeek, *Angew. Chem. Intern. Ed.*, **5**, 1 (1966).
175. Montecatini, Italian Patent 645,593; *Chem. Abstr.*, **60**, 3004 (1964).
176. F. M. Peters and N. R. Fetter, *J. Organometal. Chem.*, **4**, 181 (1965).
177. L. I. Zakharkin and O. Y. Okhlobystin, *J. Gen. Chem. U.S.S.R.* (*English transl.*), **31**, 3417 (1961).
178. K. Ziegler, British Patent 836,734; *Chem. Abstr.*, **55**, 3435 (1961).
179. J. J. Eisch, *J. Am. Chem. Soc.*, **84**, 3605 (1962).
180. E. Huether, U.S. Patent 3,124,604; *Chem. Abstr.*, **60**, 15909 (1964).
181. L. I. Zakharkin and O. Y. Okhlobystin, *Proc. Acad. Sci. U.S.S.R.* (*English transl.*), **116**, 857 (1957).
182. R. D. Chambers and D. Cunningham, *Tetrahedron Letters*, **1965**, 2389.
182a. R. D. Chambers and J. A. Cunningham *J. Chem. Soc.* (*c*), **1967**, 2185.
183. R. D. Chambers and T. Chivers, *Organometal. Chem. Rev.*, **1**, 279(1966).
184. B. Lengyel and B. Csakvari, *Z. Anorg. Allgem. Chem.*, **322**, 103 (1963).
185. Siemens-Schuckertwerke, German Patent 1,158,977; *Chem. Abstr.*, **60**, 6867 (1964).
185a. J. G. A. Luijten and J. G. M. van der Kerk, *Organometallic Chemistry of the Group 4 Elements*, A. G. MacDiarmid, Ed., Dekker, New York, 1968, p. 112.
186. W. P. Neumann, R. Schick, and R Köster, *Angew. Chem. Intern. Ed.*, **3**, 385 (1964).
187. L. I. Zakharkin, O. Y. Okhlobystin, and B. N. Strunin, *J. Pract. Chem. U.S.S.R.* (*English transl.*), **36**, 1969 (1963).

187a. J. G. van Egmont, M. J. Janssen, J. G. A. Luijten and G. J. M. van der Kerk, *J. Applied Chem.*, **12**, 17 (1962).

188. R. S. Dickson and B. O. West, *Australian J. Chem.*, **15**, 710 (1962).

189. J. R. Horder and M. F. Lappert, *J. Chem. Soc. A* **1968**, 1167.

189a. W. K. Johnson, *J. Org. Chem.*, **25**, 2253 (1960).

190. R. Polster, *Ann. Chem.*, **654**, 20 (1962).

190a. L. C. Willemsens and G. J. M. van der Kerk, *Organometallic Chemistry of the Group 4 Elements*, A. G. MacDiarmid, Ed. Dekker, New York, 1968, p. 202.

191. H. Gilman and L. D. Apperson, *J. Org. Chem.*, **4**, 162 (1939).

192. E. H. Dobratz, U.S. Patent 2,816,123; *Chem. Abstr.*, **52**, 7344 (1958).

193. F. W. Frey, P. Kobetz, G. C. Robinson, and T. O. Sistrunk, *J. Org. Chem.*, **26**, 2950 (1961).

193a. F. Glockling and K. A. Hooton, *Organometallic Compounds of the Group 4 Elements*, A. G. MacDiarmid, Ed., Dekker, New York, 1968, p. 22.

194. F. Glockling and J. R. C. Light, *J. Chem. Soc. A*, **1967**, 623.

194a. R. J. H. Voorhoeve, *Organohalosilanes: Precursors to Silicones*, Elsevier, Amsterdam, 1967.

195. B. Lengyel and T. Szekely, *Z. Anorg. Allgem. Chem.*, **287**, 273 (1956),

196. B. Lengyel, T. Szekely, S. Jenei, and G. Garze, *Z. Anorg. Allgem. Chem.*, **323**, 65 (1963).

197. H. Schmidbauer and H. F. Klein, *Angew. Chem. Intern. Ed.*, **5**, 726 (1966).

198. von N. Wiberg, W. C. Joo, and H. Henke, *Inorg. Nucl. Chem. Letters*, **3**, 267 (1967).

199. G. A. Russell, *J. Am. Chem. Soc.*, **81**, 4815 (1959).

200. A. Y. Jakubovich and G. V. Motserev, *Zh. Obshch. Khimii*, **23**, 771 (1953).

201. Ethyl Corp., French Patent 1,429,549; *Chem. Abstr.*, **65**, 10621 (1966).

202. O. Y. Okhlobystin and L. I. Zakharkin, *Bull. Acad. Sci. U.S.S.R. (English transl.)*, **1958**, 977.

203. L. Maier, *Helv. Chim. Acta*, **47**, 27 (1964).

204. Y. Takashi and I. Aishima, *J. Organometal. Chem.*, **8**, 209 (1967).

205. Y. Takashi, *J. Organometal. Chem.*, **8**, 225 (1967).

206. J. R. Surtees, *Chem. Commun.*, **1965**, 567.

207. W. Adcock, unpublished results.

208. H. Sinn and E. Kolk, *J. Organometal. Chem.*, **6**, 373 (1966); see also H. Sinn and G. Oppermann, *Angew. Chem. Intern. Ed.*, **5**, 962 (1966).

209. W. P. Long and D. S. Breslow, *J. Am. Chem. Soc.*, **82**, 1953 (1960).

210. G. Natta, P. Corradini, and I. W. Bassi, *J. Am. Chem. Soc.*, **80**, 755 (1958).

211. G. Natta, P. Pino, G. Mazzanti, and U. Giannini, *J. Am. Chem. Soc.*, **79**, 2975 (1957).

212. G. Natta, P. Pino, G. Mazzanti, and U. Giannini, *J. Inorg. Nucl. Chem.*, **8**, 612 (1958).

213. Dow Chemical Co., Belgian Patent 635,987; *Chem. Abstr.*, **62**, 4052 (1965).

214. C. Beermann and H. Bestian, *Angew. Chem.*, **71**, 618 (1959).

215. C. Beermann, German Patent 1,100,022; *Chem. Abstr.*, **55**, 25757 (1961).

216. H. De Vries, *Rec. Trav. Chim.*, **80**, 866 (1961).

217. K. C. Dewhurst, W. Keim, and C. A. Reilly, *155th Am. Chem. Soc. Meeting* Abstract M81 (1968).

218. W. Keim, *J. Organometal. Chem.*, **8**, P25 (1967).

219. M. I. Prince and K. Weiss, *J. Organometal. Chem.*, **2**, 166, 251 (1964).

220. C. Eden and H. Feilchenfield, *Tetrahedron*, **18**, 233 (1962).

221. D. F. Herman and W. K. Nelson, *J. Am. Chem. Soc.*, **75**, 3882 (1953).
222. H. Podall, J. H. Dunn, and H. Shapiro, *J. Am. Chem. Soc.*, **82**, 1325 (1960).
223. L. I. Zakharkin, V. V. Gavrilenko, and O. Y. Okhlobystin, *Bull. Acad. Sci. U.S.S.R. (English transl.)*, **1958**, 92.
224. G. Wilke and M. Kröner, *Angew. Chem.*, **71**, 574 (1959).
225. M. Tsutsui and G. Chang, *Can. J. Chem.*, **41**, 1255 (1963).
226. Studiengesellschaft-Kohle, French Patent 1,320,729; *Chem. Abstr.*, **59**, 14026 (1963).
227. T. Saito, Y. Uchida, A. Misono, A. Yamamoto, K. Morifuji, and S. Ikeda, *J. Am. Chem. Soc.*, **88**, 5198 (1966).
228. G. Wilke and G. Herrmann, *Angew. Chem. Intern. Ed.*, **5**, 581 (1966).
229. A. Yamamoto, K. Morifuji, S. Ikeda, T. Saito, Y. Uchida, and A. Misono, *J. Am. Chem. Soc.*, **87**, 4652 (1965).
230. A. Yamamoto, K. Morifuji, S. Ikeda, T. Saito, Y. Uchida, and A. Misono, *J. Am. Chem. Soc.*, **90**, 1878 (1968).
231. T. Saito, Y. Uchida, A. Misono, A. Yamamoto, K. Morifuji, and S. Ikeda, *J. Organometal. Chem.*, **6**, 572 (1966).
232. G. Hata, H. Kondo, and A. Miyake, *J. Am. Chem. Soc.*, **90**, 2278 (1968).
233. A. Yamamoto, S. Kitazume, L. S. Pu, and S. Ikeda, *Chem. Commun.*, **1967**, 79.
234. J. H. Enemark, B. R. Davis, J. A. McGinnety, and J. A. Ibers, *Chem. Commun.*, **1968**, 96.
235. A. Misono, Y. Uchida, M. Hidai, and M. Araki, *Chem. Commun.*, **1968**, 1044.
235a. A. D. Allen and F. Bottomley, *Accounts Chem. Research*, **1**, 360 (1968).
236. A. Yamamoto, L. S. Pu, S. Kitazume, and S. Ikeda, *J. Am. Chem. Soc.*, **89**, 3071 (1967).
237. A. Sacco and M. Rossi, *Chem. Commun.*, **1967**, 316.
238. A. Sacco and M. Rossi, *Inorg. Chim. Acta*, **2**, 127 (1968).
239. A. Misono, Y. Uchida, and T. Saito, *Bull. Chem. Soc. Japan*, **40**, 700 (1967).
240. J. Ellermann and W. H. Gruber, *Angew. Chem. Intern. Ed.*, **7**, 129 (1968).
241. A. Yamamoto, S. Kitazume, and S. Ikeda, *J. Am. Chem. Soc.*, **90**, 1089 (1968).
241a. W. H. Knoth, *J. Am. Chem. Soc.*, **90**, 7172 (1968).
242. G. Wilke and G. Herrmann, *Angew. Chem. Intern. Ed.*, **1**, 549 (1962).
243. P. W. Jolly and K. Jonas, *Angew. Chem. Intern. Ed.*, **7**, 731 (1968).
244. Studiengesellschaft-Kohle, German Patent 1.191,375; *Chem. Abstr.*, **63**, 7045 (1965).
245. G. Wilke and H. Schott, *Angew. Chem. Intern. Ed.*, **5**, 583 (1966).

Chemical Fixation of Molecular Nitrogen

M. E. Vol'pin and V. B. Shur

Institute of Organoelement Compounds, Academy of Sciences of U.S.S.R., Moscow, U.S.S.R.

I.	Activation of Molecular Nitrogen by Transition Metal Compounds .	57
	A. Reaction of Nitrogen Reduction under Mild Conditions .	57
	B. The Mechanism of Reaction	70
	C. The Catalytic Nitrogen Fixation	73
	D. Reaction of Nitrogen with Hydrogen	77
	E. Amines Formation in Molecular Nitrogen Reactions . .	81
II.	Molecular Nitrogen Complexes	87
III.	Conclusion	100
IV.	Experimental Procedures	101
	1. Nitrogen Reaction with the System $Cp_2TiCl_2 + C_2H_5MgBr$ in Ether at Atmospheric Pressure . . .	101
	2. Nitrogen Reaction with the System $Cp_2TiCl_2 + C_2H_5MgBr$ in Ether under Pressure	101
	3. Nitrogen Reaction with the System: Transition Metal Compound $+ Mg + MgI_2$ under Pressure	101
	4. Nitrogen Reaction with Hydrogen in the Presence of a System $Ti(OC_2H_5)_4 + Al(i\text{-}Bu)_3$ in Toluene	102
	5. Nitrogen Reaction with the System $Cp_2TiCl_2 + C_6H_5Li$ in Ether	102
	6. The Synthesis of Nitrogenpentaammineruthenium(II) Complexes $[(NH_3)_5RuN_2]X_2$	103
V.	Supplement	104
	References	113

Molecular nitrogen occupies a special place in the series of compounds with multiple bonds because of its exceptional inertness.

Unlike highly reactive olefins and acetylenes, nitrogen reacts with most elements and compounds only at high temperatures and pressures and often only in the presence of catalysts.

The difficulties encountered in solving the problem of the introduction of nitrogen into reactions under mild conditions are very great and are of a thermodynamic and kinetic character.

The nitrogen molecule is a thermodynamically very stable compound. The dissociation energy of N_2 is high—225 kcal/mole. As a result, many reactions, for example, bromination of molecular nitrogen, its oxidation by molecular oxygen, and hydration, are strongly endothermic. High positive values of ΔF_{295}° are characteristic of these reactions and at moderate temperatures and pressures the equilibrium positions for these reactions are far toward the starting compounds. Thus, in some cases the thermodynamics of reactions makes them practically unrealizable at moderate temperatures and pressures.

However, the thermodynamic hindrances narrowing the range of possible reactions are not the sole determining factors. Many reactions involving molecular nitrogen are thermodynamically advantageous at moderate temperatures, but the rate of these processes under such conditions is usually very low. For example, the reaction of N_2 with H_2 thermodynamically favors NH_3 at room temperature. However, the reaction rate under these conditions is almost zero. According to Frankenburg's data[1] the activation energy for the first step of the hydrogenation of nitrogen to ammonia ($N_2 + H_2 \rightarrow NH{=}NH$) is very high and amounts to ~ 100 kcal/mole.

The high activation energies for reactions involving molecular nitrogen and consequently the slow rates of their occurrence at low temperatures are mostly due to the difficulty of breaking the first of the three bonds. The energy of this bond is 125 kcal/mole and this greatly exceeds the respective $C{=}N$ and $C{=}C$ bond energies (63 and 53 kcal/mole).

Bond energies, kcal/mole

C—C 80 $\}\Delta = 65$	C—N 73 $\}\Delta = 74$	N—N 37 $\}\Delta = 63$
C=C 145$\{$	C=N 147$\{$	N=N 100$\{$
C≡C 198$\}\Delta = 53$	C≡N 210$\}\Delta = 63$	N≡N 225$\}\Delta = 125$

Further transition from a double $N{=}N$ bond to a single one would require 63 kcal/mole and breaking the last bond would take only ~ 37 kcal/mole.

Thus, to solve the problem of nitrogen fixation under mild conditions it is necessary to find out the way to activate the molecule. From the very beginning of the development of the chemistry of nitrogen all attempts at activation of this inert molecule followed two directions which were different in principle.

One of these directions ("physical activation") involves the use of different kinds of violent forces such as electric discharge, radiation, and so on. In this case the N_2 molecule is activated as a result of different processes of excitation, ionization, or dissociation, and thus becomes capable of reacting with many inorganic and organic compounds. The recent development of this trend resulted in the working out of such promising methods of nitrogen fixation as low-temperature plasma techniques (see, for example, ref. 2).

Another direction is the chemical activation of nitrogen, the principal idea of which consists of the use of catalysts. About 60 years ago Haber succeeded in showing that some transition metals (Os, Fe, Mo, W, and so on) can catalyze the nitrogen hydrogenation reaction by molecular hydrogen to form NH_3. This discovery had great meaning because it gave the possibility of industrial production of nitrogen-containing compounds from nitrogen in the air. In fact, further work by Haber and other investigators permitted the use of this interesting reaction on a large scale as an economical industrial process which is today the principal source of fixed nitrogen. Nevertheless, in spite of the large amount of work which was done to improve the first Haber catalysts, the conditions of the catalytic synthesis of ammonia remain rather severe (temperature of 400–500°C, pressure of 100–1000 atm).

I. ACTIVATION OF MOLECULAR NITROGEN BY TRANSITION METAL COMPOUNDS.

A. Reaction of Nitrogen Reduction under Mild Conditions

At present there is a possibility of activating molecular nitrogen under milder conditions. This is connected with intensive development in recent years of the chemistry of transition-metal coordination compounds. Discovery of the ferrocenelike compounds and other compounds with unusual types of bonds and intensive development of carbonyls and other complexes resulted in obtaining new types of complex compounds sometimes displaying unusual chemical properties and capable of catalyzing a variety of organic reactions (oxidation, hydrogenation, cyclization, polymerization, and so on).

In many cases such complex catalysts are more effective than classic catalysts based on metals or their oxides.

Very important for understanding the catalytic conversion of unsaturated compounds was the discovery and further investigation of π-complexes of olefins and acetylenes. Such complexes appeared to be formed as intermediates in different reactions of unsaturated compounds catalyzed by the transition-metal derivatives.

Formation of a π-complex leads to weakening of the π-bond in the olefin molecule due to the decrease of the original bonding and the increase in internuclear distance. As a result the double bond appears to be partially broken and the olefin becomes, so to speak, ready for further addition reactions. On the whole the activation energy of addition decreases.

Thus complex formation of an unsaturated compound is a rather universal and highly effective way of multiple-bond activation in addition reactions.

It would be natural to attempt to use this way for activation of the multiple bond of molecular nitrogen. Some years ago it was suggested[3,4,11] that molecular nitrogen, in spite of its inertness, could form similar complexes with compounds of transition metals and thus become activated for further reactions. Along with the complexes of π-type (1), nitrogen can also form linear complexes (2) (similar to CO complexes), and various dinuclear complexes, for example 3 and 4.

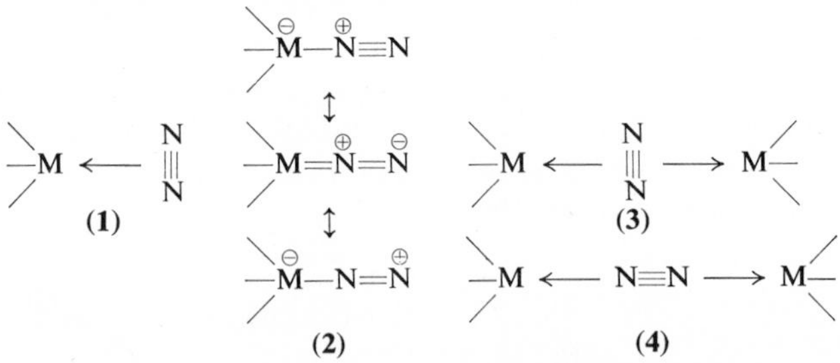

It has been shown above that the high energies of activation in molecular nitrogen reactions are connected with the difficult cleavage of the first of the three N_2 bonds. It might have been expected that formation of the complex of N_2 with a transition-metal compound will lead to a considerable weakening of this bond as a result of which nitrogen will be able to react readily.

On the basis of these concepts an investigation of molecular nitrogen reactions in the presence of transition-metal compounds was undertaken.[3-13] At first an attempt was made to reduce nitrogen under conditions of its possible complexing with transition-metal compounds.

$$L_nM + N_2 \longrightarrow L_nM \cdot N_2 \xrightarrow{[e]} 2[N^{3-}]$$

To realize this reaction it is necessary that the reaction system should display both reducing properties and a capability of forming complexes with N_2.

From this point of view, the systems which are formed by the reaction on transition-metal salts and complexes by carbanion donors (Li-, Mg-, or

Al-organic compounds) or hydride ion donors ($LiAlH_4$, $NaBH_4$, etc) are especially promising.

Such systems are of a pronounced reducing character and have an almost universal capability of complex formation with very different unsaturated compounds (carbon monoxide, olefins, acetylenes, aromatic hydrocarbons, etc.). This makes them capable of catalyzing various reactions involving multiple bonds, such as polymerization, cyclization, hydrogenation, isomerization, etc.

It may be expected that nitrogen, similar to CO and other unsaturated compounds, will also be capable of forming complexes with derivatives of transition metals in such systems, thus becoming activated for the next reduction reaction.

Indeed, it has been found that a number of systems obtained by interaction of various transition-metal derivatives with Li-, Mg-, and Al-organic compounds can react with molecular nitrogen at room temperature.[3-7,10,11,13] As a result, the N_2 molecule is reduced to form products of the nitride type giving ammonia after hydrolysis.

A necessary condition for occurrence of a reaction of such type is the simultaneous presence of both components in the reaction mixture, namely, of transition-metal derivatives and organometallic compounds. In the absence of even one of them nitrogen is not reduced and ammonia is not formed.

To prove that the source of the ammonia formed is N_2 rather than possible impurities in the reagents used, special experiments were made using molecular nitrogen labeled with [15]N.[6] For this purpose labeled nitrogen was introduced at room temperature into the reaction with one of the active systems ($CrCl_3$ + C_2H_5MgBr in ether) and ammonia obtained after hydrolysis was investigated for ^{15}N by means of the mass spectroscopy technique. It was found that the NH_3 formed contained practically the same amount of ^{15}N as the nitrogen introduced in the reaction.

TABLE I
Nitrogen Fixing Systems

Transition metal compound	Reducer
1. Metal halogenides[3,8,15]	RLi, RMgX, R_3Al,[3,4,5,7] $LiAlH_4$,[3] Li,
2. Alkoxy derivatives[7,28]	Mg,[8] K[28], Al[61,62]
3. Acetylacetonates[7]	naphthalene-lithium,[15]
4. Phosphinic complexes[10]	naphthalene-sodium,[28, 60]
5. π-Cyclopentadienyl complexes[4,5,8,12,13,28]	etc.

TABLE II

Nitrogen Fixation by the System MCl_n + C_2H_5MgBr in Ether[a] (Temperature 20°C, Pressure of N_2 150 atm)[3]

Salt(MCl_n):	$TiCl_4$	$MnCl_2$	$CrCl_3$	$MoCl_5$	WCl_6	$FeCl_3$	$CoCl_2$	$NiCl_2$	$PdCl_2$	$CuCl$
Yield of NH_3 in moles per mole of salt[b]	0.12	0.02	0.17	0.08	0.15	0.09	0	0	0	0

[a] Molar ratio of C_2H_5MgBr:MCl_n—9:1.
[b] Time of reaction 10–11 hr.

TABLE III

Nitrogen Fixation by Systems Based on Transition Metal Acetylacetonates and Organometallic Compounds[a] (Temperature 20°C, Pressure of N_2 70–100 atm)[7]

Transition metal compound	Yield of NH_3 in moles per mole of transition metal compound[b]		
	C_2H_5MgBr in ether	n-C_4H_9Li in n-heptane	$(i$-$C_4H_9)_3Al$ in toluene
$VO(acac)_2$	0.12	0.32	0.07
$Cr(acac)_3$	0.03	0.01	0.14
$MoO_2(acac)_2$	0.14	0.07	0.09
$Mn(acac)_3$	0.01	0.01	0.09

[a] The molar ratio of C_2H_5MgBr and n-C_4H_9Li to a transition metal compound is 9:1, for $(i$-$C_4H_9)_3Al$, 6:1.
[b] Time of reaction 9–10 hr.

The reaction found is of a general nature (Table I). Besides the simple halogenides[3,11] (Table II) and alkoxy derivatives[7] very different complex compounds of transition metals such as acetylacetonates[7,11] (Table III), phosphine complexes,[10] cyclopentadienyl complexes (Table IV),[4,5,11,13] etc., enter this reaction.

Most active are derivatives of Ti, V, Cr, Mo, W, and Fe, that is, of metals on the left side of the transition period. Less active or quite inactive are the Co, Ni, Pd, and Pt compounds, that is, metals which occupy the

TABLE IV

Nitrogen Fixation by Systems Based on Titanium Derivatives and Organometallic Compounds[a] (Temperature 20°C, Pressure of N_2 100–150 atm)[4,5,11,13]

Titanium compound	Yield of NH_3 in moles per mole of titanium compound[b]		
	C_2H_5MgBr in ether	n-C_4H_9Li in n-heptane	$(i$-$C_4H_9)_3Al$ in toluene
$TiCl_4$	0.11	0.29	0.32
$CpTiCl_3$	0.23	0.07	0.15
Cp_2TiCl_2	0.90	0.50	0

[a] The molar ratio of C_2H_5MgBr and n-C_4H_9Li to a titanium compound is everywhere 9:1, for $(i$-$C_4H_9)_3Al$ it is 6:1.
[b] Time of reaction 10–11 hr.

right part of the transition period. In many cases the Zr, Mn, and Cu compounds also have little activity.

The presence of a ligand at the transition metal atom often leads to a considerable increase of the activity in the reaction of nitrogen reduction. For instance, in passing from $TiCl_4$ to Cp_2TiCl_2 in mixtures with Li-and Mg-organic compounds (Table IV) the yield of NH_3 greatly increases and in the case of the system $Cp_2TiCl_2 + C_2H_5MgBr$ in ether it amounts to 0.9–1.0 moles per mole of Cp_2TiCl_2. In this case large amounts of NH_3 (0.7 and more moles per mole of titanium compound) are also formed upon simply passing N_2 through the reaction mixture at room temperature and atmospheric pressure.

As the organometallic component lithium and magnesium compounds as well as aluminum-organic compounds can be used. Thus typical Ziegler-Natta catalysts for stereospecific polymerization of olefins are also capable of reacting with N_2 at room temperature. It explains the inhibiting action of nitrogen on the polymerization that was observed in some investigations.[14]

The nitrogen-fixing activity of the systems depends on the nature of the organometallic compound and in particular on the character of the radical R, bound to the metal atom.[4]

It has been found that organometallic compounds containing very different aliphatic and aromatic radicals can be used in reactions with nitrogen (Table V). However, these compounds differ in their activity. Sometimes these differences can be very marked. For instance, in passing

TABLE V

Effect of the Grignard Reagent Character on Nitrogen Fixation by the System $CrCl_3 + RMgX$ in Ether (Temperature 20°C, Pressure of N_2 150 atm)[4]

$RMgX^a$	The NH_3 yield in moles per mole of $CrCl_3^b$
CH_3MgJ	0.04
C_2H_5MgBr	0.17
$n\text{-}C_3H_7MgBr$	0.30
$n\text{-}C_4H_9MgBr$	0.27
$i\text{-}C_3H_7MgBr$	0.26
$i\text{-}C_4H_9MgBr$	0.19
C_6H_5MgBr	0.04
$C_6H_5CH_2MgBr$	0.02
$CH_2{=}CH{-}CH_2MgBr$	0.02

a Molar ratio $RMgX:CrCl_3$, 9:1.
b Time of reaction: 10–11 hr.

TABLE VI

The Effect of the Ratio of Reagents on Nitrogen Fixation (Temperature 20°C, Pressure of N_2 150 atm)[a]

RMgBr(RLi):MX$_n$ in moles per mole	The NH_3[a] yield in moles per mole of a transition metal compound (MX$_n$)[b]			
	Cp$_2$TiCl$_2$ + C$_2$H$_5$MgBr in ether	CrCl$_3$ + C$_2$H$_5$MgBr in ether	Cp$_2$TiCl$_2$ + PhLi[a] in ether	Cp$_2$TiPh$_2$ + PhLi[a] in ether
1:1	0.05	0.01	—	0.10
2:1	0.24	0.03	0.12	—
3:1	0.40	0.05	0.45	0.57
5:1	—	—	0.80	—
9:1	0.90	0.17	—	—
16:1	0.91	0.17	—	—
100:1	—	—	0.70	—

[a] For systems Cp$_2$TiCl$_2$ + PhLi and Cp$_2$TiPh$_2$ + PhLi the yield values given are the summary of quantities of NH_3 and $PhNH_2$ (see page 82).

[b] Time of reaction 10–11 hr.

from CH_3MgI to C_2H_5MgBr in mixtures with $CrCl_3$ in ether the amount of ammonia formed increases more than fourfold. Using n-C_3H_7MgBr the yield of ammonia increases still more. But further transition to n-C_4H_9MgBr has no effect on the reaction.

The molar ratio of an organometallic compound to the transition metal derivative is very important.[4] It was found that maximum activity in reaction with N_2 could be obtained only by using sufficiently high excesses of organometallic compounds over transition-metal compounds in a ratio of 5–9:1 (Table VI). Further increase of this ratio has no great effect on the amount of the ammonia formed but its decrease leads to a markedly lower effectiveness of the nitrogen-fixing system. This phenomenon is of a quite general character and is observed for many of the systems studied.

The nature of the solvent also plays an important part in the reaction.[3,4] It was found that diethyl ether is the best solvent when organomagnesium compounds are used. A reaction carried out in more solvating solvents such as tetrahydrofuran (THF) or dimethoxyethane (DME) results as a rule in a marked deceleration of nitrogen fixation and a lower yield of ammonia (Table VII). A similar phenomenon is observed on substitution of di-n-propyl and di-n-butyl ether for diethyl ether. In case of Li-organic compounds good results are obtained using hydrocarbon solvents (n-heptane, toluene, etc.) as well as diethyl ether. Organoaluminum compounds could be used as usual in toluene, n-heptane, and similar solvents.

The effect of addition of some unsaturated compounds capable of competing with nitrogen for a place in the coordination sphere of a transition metal atom was studied.[4] Such complexing agents as CO and acetylenes (e.g., PhC≡CPh) appeared to suppress the nitrogen fixation completely

TABLE VII

The Effect of a Solvent on Nitrogen-Fixing Activity (Temperature 20°C, Time of Reaction 10–11 hr)[3,4]

System	Pressure of N_2 in atm	The NH_3 yield in moles per mole of a transition metal compound		
		Ether	THF	DME
Cp_2TiCl_2 + C_2H_5MgBr	150	0.90	0.40	—
$CrCl_3$ + C_2H_5MgBr	150	0.17	0.03	—
$CrCl_3$ + C_6H_5MgBr	150	0.04	0.01	—
$CrCl_3$ + Mg + I_2	100	0.65	0.01	—
$CrCl_3$ + $LiAlH_4$	1	0.02	0	0
$CrCl_3$ + $LiAlH_4$	150	0.07	0.04	0.03

TABLE VIII

The Effect of Inhibitors on Nitrogen Fixation by System $CrCl_3$ + C_2H_5MgBr[a]
in Ether[b] (Temperature 20°C, Pressure of N_2 150 atm)[4]

Inhibitor	The quantity of inhibitor in moles per mole of $CrCl_3$	The NH_3 yield in moles per mole of $CrCl_3$
Absent	—	0.17
CO	6	0
O_2	3	0.01
$PhC{\equiv}CPh$	3	0.01
$PhC{\equiv}CPh$	6	0
$C_4H_9CH{=}CH_2$	6	0.07
$PhCH{=}CHPh$	3	0.15
C_6H_6	2	0.14

[a] Molar ratio $C_2H_5MgBr:CrCl_3$, 9:1.
[b] Time of reaction 10–11 hr.

(Table VIII). A similar effect, though somewhat weaker, is shown by some olefins. Thus introduction of 1-hexene into the reaction reduces by half the yield of NH_3. Small amounts of oxygen do not affect the reaction, though excesses practically stop nitrogen fixation.

Organometallic compounds used in the reaction play an important function in that they reduce the initial derivative of a transition metal to a low valent state capable of nitrogen complex formation and can also participate in reduction of the coordinated N_2 molecule.

An attempt was made on this basis to replace organometallic compounds in nitrogen-fixing systems by nonorganic reducing agents. It has been found that in a number of cases nonorganic reducing agents such as alkali or alkaline earth metals (Li, Mg),[8] hydrides (LiAlH$_4$),[3] and others could actually be used instead of Li-, Mg, and Al-organic compounds. The systems obtained often proved to be even more active in reactions with N_2 than systems involving organometallic compounds.

First of all, this pertains to systems with metallic Mg as a reducing agent in the presence of an activator (I_2 or MgI_2).[8] In this case the yields of ammonia for all transition-metal compounds studied appeared to be much higher and at proper conditions they could reach up to 1–2 moles per mole of salt (Table IX). The most active are titanium compounds (TiCl$_4$, Cp$_2$TiCl$_2$) and also the metal halides of VIa group metals (CrCl$_3$, MoCl$_5$, WCl$_6$). FeCl$_3$ and CoCl$_2$ are less active and NiCl$_2$ and CuCl are quite inactive. Thus, the same regularity is observed both for systems involving Mg

TABLE IX

Nitrogen Fixation by the System: Transition Metal Compound + Mg + I_2 (or MgI_2) (Temperature 20°, Pressure of N_2 90–100 atm)[8]

Transition metal compound	The NH_3 yield in moles per mole of a transition metal compound	
	Mg + I_2	Mg + MgI_2
$TiCl_4$	1.3	1.25
Cp_2TiCl_2	1.14	1.0
$CrCl_3$	0.65	0.44
$MoCl_5$	0.62	—
WCl_6	0.59	—
$FeCl_3$	0.09	—
$CoCl_2$	0.13	—
$NiCl_2$	0	—
$CuCl$	0	—

as a reducer and for systems using Li-, Mg-, and Al-organic compounds. In both cases the transition-metal compounds of the left part of the transition period appeared to be most active in N_2 reduction.

Other metals such as lithium in the presence of an activator, such as lithium iodide can also be used along with Mg as a reducing agent. An example of this is the system Cp_2TiCl_2 + Li + LiI reacting with nitrogen at room temperature with formation of ammonia in an amount of 0.33 moles per mole of Cp_2TiCl_2 upon hydrolysis.

Under certain conditions lithium and magnesium metals are known to react with nitrogen, yielding corresponding nitrides, though under the conditions employed for nitrogen fixation neither Li nor Mg nor mixtures with LiI or MgI_2 as activators react with N_2 to any noticeable degree. Intensive nitrogen fixation was observed only in the presence of appropriate transition-metal halides.

The nitrogen-fixing activity of systems involving inorganic reducing agents is very sensitive to factors such as the nature of the solvent, the ratio of reagents, and so on. The regularities observed are, in general, similar to those of systems involving Li-, Mg-, and Al-organic compounds. Diethyl ether, or its mixtures with aromatic hydrocarbons, is the most convenient solvent if Mg and Li, as well as $LiAlH_4$, are used as reducing agents. The substitution of THF or DME for ether results in a markedly lower activity of the system (Table VII).

Henrici-Olivé and Olivé have shown that a solution of naphthalene-lithium in THF[15] can also be used as a reducing agent in nitrogen-fixing systems. This compound reduces the halides of transition metals yielding solutions containing arenic hydrido-σ-naphthylic complexes of the anion type (5). Depending on the excess of the reducing agent these complexes can carry up to six negative charges.[16]

$$\left[\begin{array}{c} (\pi\text{-}C_{10}H_8)_{2-n} \\ \vdots \\ (\sigma\text{-}C_{10}H_7)_n \cdots\cdots M \cdots\cdots S_m \\ \vdots \\ H_n \end{array} \right]^{x-} \quad Li_x^+$$

(5)

where $0 \leq n \leq 2$, $x \leq 6$, and S_m-solvent

It was found that the solutions obtained on interaction of VCl_3, $CrCl_3$, and $TiCl_4$ with naphthalene-lithium in THF can readily reduce nitrogen at room temperature with formation of products yielding ammonia after hydrolysis. The nitrogen-fixing activity of these systems as well as of the systems described above is, to a considerable degree, a function of the amount of the reducing agent. At a sufficiently high excess of naphthalene-lithium the ammonia yield can attain in a number of cases (VCl_3, nitrogen pressure: 120 atm) up to 2 moles per mole of the transition metal compound. Such systems also show considerable activity at atmospheric pressure (VCl_3, NH_3 yield: 0.9 moles per mole of VCl_3).

Van Tamelen, Boche, and Greeley[60] have also employed the naphthalene anion radicals as reducing agent in the nitrogen-fixing system.

They have observed the nitrogen absorption at room temperature and atmospheric pressure at the nitrogen reaction with a mixture of titanium tetraisopropoxide and an excess of sodium naphthalene in ethereal solvent. During the absorption process, which is continued from 30 up to 60 min, nitrogen is reduced and ammonia is formed upon hydrolysis of the reaction products. The ammonia yield depends on the concentrations and ratios of reagents, the solvent nature, etc., and under optimal conditions is as high as 1.3 moles per mole of the initial titanium compound.

It is of interest that considerable amounts of ammonia are formed in this reaction with air passing through the reaction mixture. In such a case, however, due to the sodium naphthalene oxidation as well as that of the active N_2-fixing species the ammonia yields turn out to be 2–3 times less than with pure nitrogen.

As in all nitrogen-fixing systems described earlier, the hydrolysis of the reaction mixture, along with the ammonia formation, leads to destruction of the system due to the decomposition of the active N_2-fixing species and of the reducing agent. Van Tamelen, Boche, and Greeley[60] have found that introduction of a new portion of reducing agent into the hydrolyzed reaction mixture after the ammonia has been removed leads to partial recovery of nitrogen-fixing activity of the system. Thus with the operations of nitrogen fixation, hydrolysis, ammonia removal, and introduction of additional amount of reducing agent having been repeated five-fold, they managed to obtain ammonia in as much as 3.4 moles per mole of initial titanium tetraisopropoxide.

All the systems mentioned above are not catalytic. In studying the kinetics of nitrogen fixation[4] it was found that the curves describing the NH_3 yield as a function of time are of a similar character for various nitrogen-fixing systems involving Li-, Mg- and Al-organic compounds as well as inorganic reducing agents.

In all cases observed the increase in ammonia yield with time occurs only up to a certain limit. After this the yield of ammonia practically stops changing though an excess of reducing agent is still present in the mixture. The maximum yield of NH_3 depends in general upon the type of system used and in a number of cases it can attain 1–2 moles per mole of the transition-metal compound.

For a number of systems kinetic curves are of more complicated, S-shaped character. In this case a marked induction period is observed along with the saturation region (Figs. 1 and 2).

The system of $Cp_2TiCl_2(Cp_2TiPh_2)$ + PhLi in ether is a typical example. Investigation of this system by the ESR method as well as simultaneous

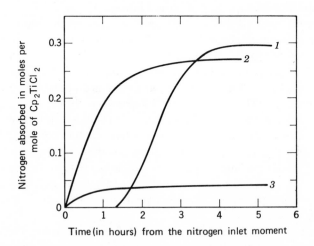

Fig. 1. Plot of the nitrogen absorption by the system Cp_2TiCl_2 + PhLi in ether versus time (temperature 20°C, molar ratio of PhLi : Cp_2TiCl_2 = 5 : 1, initial pressure of N_2 30 atm). (*1*) Nitrogen was introduced at the moment of mixing of Cp_2TiCl_2 with PhLi; (*2*) Nitrogen was introduced 80 min after mixing of Cp_2TiCl_2 with PhLi; (*3*) Nitrogen was introduced 4 hr after mixing of Cp_2TiCl_2 with PhLi.

kinetic study of nitrogen fixation by the manometric method have shown that intensive reduction of the initial titanium compound starts at once after the mixing of Cp_2TiCl_2 (or Cp_2TiPh_2) with PhLi. [13,36] However, the nitrogen absorption is observed only in approximately 80 min after the mixing of reagents (Fig. 1, curve 1). A similar phenomenon is also characteristic for the system $Cp_2TiCl_2 + Mg + MgI_2$ (Fig. 2).

ESR studies of the system $Cp_2TiCl_2 + Mg + MgI_2$ have shown that the immediate reduction of the initial Cp_2TiCl_2 takes place with the appearance of $Ti^{(III)}$ compounds. But in 50–60 min the ESR spectrum shows a sharp change. The absorption of nitrogen begins approximately at the same time.

Thus the induction period in nitrogen fixation is obviously connected with formation of an active transition metal compound responsible for nitrogen fixation in the reaction mixture. Indeed, if a mixture of Cp_2TiCl_2 and PhLi is kept for 80 min in an argon atmosphere and N_2 is then introduced, the nitrogen absorption begins at once without any induction period (Fig. 1, curve 2).

Nitrogen fixation in the presence of transition-metal compounds was also observed in aqueous solutions, though to a very small extent. In 1964 Haight and Scott[17] reported that a prolonged (up to two weeks) cathodic reduction of nitrogen, or reduction by action of $SnCl_2$ in the presence of aqueous solutions of molybdate and tungstate ions yielded a small amount of ammonia (from 0.001 to 0.0001 of a mole).*

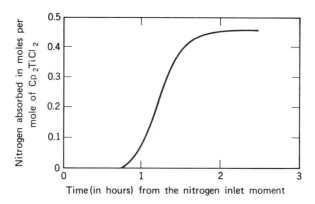

Fig. 2. Plot of the nitrogen absorption by the system $Cp_2TiCl_2 + Mg + MgI_2$ in ether–benzene mixture versus time (temperature 20°C, molar ratio of $Cp_2TiCl_2:Mg: MgI_2 = 1:8:3$).

*Recently Shilov and Shilova[113] reinvestigated these data[17,18] and failed to get any labeled ammonia employing N_2^{15} in the reactions.

Iatsimirskii and Pavlova[18] studied the reduction of various compounds of transition metals by zinc or zinc amalgam in acidic water solutions in the presence of nitrogen. They also observed the formation of small amounts of ammonia when using solutions of $TiCl_4$, $VOSO_4$, $NbCl_5$, $K_2Cr_2O_7$, Na_2MoO_4, Na_2WO_4, $KReO_4$, and $TaCl_5$.*

B. The Mechanism of Reaction

The information available at present is not yet sufficient for judging the detailed mechanism of nitrogen fixation by the systems described above. There is yet another complication of the problem. As a result of the variety of nitrogen-fixing systems the paths of activation of the nitrogen molecule and its subsequent reduction may be different depending upon the initial reactants and the reaction conditions.

Nevertheless it can be supposed on the basis of data obtained that some main features are common for all types of nitrogen-fixing systems no matter what the character of the transition-metal compound and of the reducing agent.

One can offer the following general scheme of the mechanism of nitrogen fixation:

$$L_nM \xrightarrow[(1)]{[e]} L_mM \xrightarrow[(2)]{N_2} \left\{ \begin{matrix} L_mM \cdot N_2 \\ \text{or} \\ L_mM \cdot N_2 \cdot ML_m \end{matrix} \right\} \xrightarrow[(3)]{[e]} [N^{3-}] \xrightarrow[(4)]{H_2O} NH_3$$
$$\quad (A) \qquad\quad (B) \qquad\qquad\qquad (C) \qquad\qquad\qquad (D)$$

Stage 1

At the first stage the initial salt or transition-metal complex (A) is reduced to form a low-valent compound (B), capable of forming a complex with N_2. When Li-, Mg- and Al-organic compounds are used in a system as reducing agents, this process occurs through intermediate formation of unstable organometallic compounds of transition metals

$$\overset{\diagdown}{\underset{\diagup}{}}M{-}Cl + RLi \xrightarrow{-LiCl} \overset{\diagdown}{\underset{\diagup}{}}M{-}R \xrightarrow{-[R]} \overset{\diagdown}{\underset{\diagup}{}}M$$

Both the kinetic data and investigation of the reaction by the ESR method point to the necessity of a preliminary reduction of the initial derivative of the transition metal (see above).

In a number of cases the compound B, which is responsible for the nitrogen-fixing activity, could be an anion complex. Similar complexes are usually formed in reactions of transition-metal derivatives with excess organometallic compounds[19-22,26] and some other strong reducing agents such as naphthalene lithium. In the last case Henrici-Olivé and Olivé[15]

*See footnote on page 69.

observed direct participation of anion complexes in nitrogen fixation (see above).

Depending on the type of the system used the active compound (B) can be considerably different in its stability. In a number of cases it proves to be unstable and can rather quickly decompose in further reactions. For instance, if a mixture of Cp_2TiCl_2 and PhLi is initially kept under Ar for four hours, then the amount of absorbed nitrogen appeared to be almost eight times less (Fig. 1, curve 3). A similar phenomenon is also characteristic of the system $Cp_2TiCl_2 + Mg + MgI_2$.

However in some cases this active compound is quite stable and can be preserved in a solution for a long time without further changes. For instance, Maskill and Pratt[23] showed that the nitrogen-fixing activity of the system $Cp_2TiCl_2 + C_2H_5MgBr$ hardly changes if the mixture of these two components is kept under Ar for six hours before nitrogen introduction. The authors believe that the nitrogen-fixing activity of the system $Cp_2TiCl_2 + C_2H_5MgBr$ is due to a dicyclopentadienic complex of zero-valent titanium $(C_5H_6)_2Ti$ formed as a result of reduction. This compound, as a dimer, reacts with nitrogen, reducing it to form a corresponding nitride derivative of titanium.

$$2[Ti^0(C_5H_6)_2] \rightleftharpoons [Ti_2^0(C_5H_6)_4] \xrightarrow{N_2} [(C_5H_6)_2TiN]_{1 \text{ or } 2}$$

Stage 2

At stage 2 the complex of N_2 with an active compound of a transition metal is formed.

This complex, as mentioned above, (p. 58), can be of a symmetrical structure (1), linear structure (2), or can be binuclear (3). A nitrogen can be displaced out of the complex by stronger complexing agents such as CO, acetylenes, etc. Indeed, when a reaction is carried out in the presence of CO, tolane, or some olefins, nitrogen fixation is completely inhibited (Table VIII).

The possibility of formation of nitrogen complexes in reactions of N_2 with similar systems has been confirmed by a number of investigations.[48,49,51,56] Thus Yamamoto, Kitazume, Pu, and Ikeda[51] have found that the reaction of N_2 with a mixture of $Co(acac)_3$ and AlR_3 in the presence of PPh_3 results in the formation of a nitrogen complex of composition $(PPh_3)_3CoH(N_2)$.[89] The nitrogen molecule in this complex is actually displaced by the action of CO, ethylene, and some other compounds (see below).

Stage 3

As a result of complex formation the nitrogen molecule becomes activated and then reduced. In principle, reduction of N_2 can be realized in two ways. The first way is by an electron transfer from a reducing agent. Both the reduced form of a transition-metal compound activating nitrogen and the excess of lithium, magnesium, RLi, RMgX, etc., can serve as sources of electrons. The electron transfer to the nitrogen molecule can take place via preliminary formation of anion complexes of transition metals.

Another way, which cannot be excluded yet, is a hydrogen transfer to nitrogen due to dehydrogenation of the organometallic compound, lithium aluminum hydride, solvent, etc.[3,4] In this case corresponding transition-metal hydrides can be formed as intermediates.

Such a method of nitrogen reduction has been proposed by Brintzinger for an explanation of the nitrogen fixation by the system Cp_2TiCl_2 + C_2H_5MgCl.[24,25] It has been shown using the ESR method that the reaction between Cp_2TiCl_2 and C_2H_5MgCl yields a hydride derivative of titanium (III) as a final product to which structure 6 was initially ascribed.

$$2Cp_2TiCl_2 + 2C_2H_5MgCl \longrightarrow 2Cp_2TiCl(C_2H_5) \xrightarrow{-2[C_2H_5]}$$

$$[Cp_2TiCl]_2 \xrightarrow{2C_2H_5MgCl} 2Cp_2TiC_2H_5 \xrightarrow{-2C_2H_4} [Cp_2TiH]_2$$

$$(6)$$

Later on it appeared[26] that the product structure was that of an anion dihydride complex $[Cp_2TiH_2]^{\ominus}$. According to Brintzinger the nitrogen-fixing activity of the system Cp_2TiCl_2 + C_2H_5MgCl is due to the formation of a hydride compound of titanium, and in the course of the reaction one molecule of nitrogen is inserted into the Ti—H bond. Thus in the proposed mechanism nitrogen reduction is realized in part by hydrogen-atom transfer from the β-position of the ethyl radical of C_2H_5MgCl. However, Brintzinger's mechanism does not agree with the results of Nechiporenko, Tabrina, Shilova, and Shilov,[27] who have studied this reaction using C_2D_5MgBr and showed the absence of N—D bonds in the nitrogen-fixation products. Apparently additional investigations using different methods are necessary for final elucidation of the nitrogen-reduction mechanism in this system.

Van Tamelen, Boche, Ela, and Fechter[28] have recently obtained results showing that nitrogen reduction by nitrogen-fixing systems can, in a number of cases, occur with participation of hydrogen atoms of the solvent. It was found that small amounts of free ammonia are carried away from the reaction mixture with a nitrogen current during prolonged (up to several weeks) bubbling of nitrogen through the mixture of $(RO)_2TiCl_2$ (where R = $tert$-C_4H_9, CH_3, C_6H_{11}, Ph) and metallic potassium in diglyme at room temperature and atmospheric pressure. A similar phenomenon has also been observed for the system Cp_2TiCl_2 + naphthalene sodium in THF or diglyme. The ammonia formation before hydrolysis shows that a hydrogen transfer to the N_2 molecule is taking place in the process of nitrogen fixation. The authors believe that in this case the ether–like solvents are a source of hydrogen in the nitrogen reduction. Indeed, the yield of free NH_3 is not practically affected by the nature of R in the initial $(RO)_2TiCl_2$, but it sharply decreases when the diglyme or THF are substituted by xylene as a solvent.

Stage 4

In the reduction process a nitrogen molecule is split and the compounds (D) of the reduced nitrogen (conventionally marked as $[N^{3-}]$) are formed. Their character depends on the nature of a nitrogen-fixing system.

In most cases studied these compounds appear to be of nitride type in which nitrogen can be bound to a transition metal atom as well as to a reducing agent metal. The hydrolysis of such compounds yields the ammonia. For instance, the product of nitrogen reaction with the system $Cp_2TiCl_2 + Mg + MgI_2$ is a cyclopentadienylic derivative of titanium, in which a nitrogen atom is bound to a titanium atom and, apparently, with magnesium.

In the cases when the nitrogen reduction is partially due to a hydrogen transfer from RLi, RMgx, solvents, etc., the product of nitrogen fixation must contain nitrogen–hydrogen bonds besides metal–nitrogen bonds also. Finally, in the limiting case described by van Tamelen and co-workers,[28] the final product of reaction can be free ammonia even before hydrolysis.

C. The Catalytic Nitrogen Fixation

As has been mentioned, all the systems considered above are non-catalytic. All attempts to get any catalytic nitrogen fixation by means of increasing the amounts of reducing agents, the reaction times, etc., failed. In all cases the yield of ammonia did not exceed 1–2 moles per mole of the transition-metal compound.

The noncatalytic character of these reactions may well be due to the active transition metal compound (B), is in some cases unstable and rather rapidly destroyed by side reactions (cf. Section I-B). Another more general reason consists in the formation of a strong nitridelike bond between the transition metal and the nitrogen atoms as a result of the molecular nitrogen reduction which may prevent regeneration of the active transition-metal compound (B) and cause the reaction to be noncatalytic.

In some cases the nitride bonds are known to be capable of cleavage under the action of acids. It has enabled us to suggest that a nitrogen-fixing system might become catalytic when the nitrogen reaction is carried out in the presence of acid. However, since the protic acids destroy nitrogen-fixing systems the possibility of catalytic nitrogen fixation was investigated by means of introducing aprotic acids in the system.

The nitrogen reactions with the systems consisting of aluminum, aluminum halides, and the transition-metal compounds such as $TiCl_4$, $Ti(OR)_4$, $ZrCl_4$, $CrCl_3$, $MoCl_5$, WCl_6, $MnCl_2$, $FeBr_3$, $FeCl_3$, $NiBr_2$, $CoCl_2$, etc., have been studied.[61,62] These systems are of considerable reductive power and include the aprotic acid. The systems of such kind are widely used for the synthesis of arene[63] and carbonyl[64] complexes of transition metals along with polymerization and cyclization of olefins. It has been found[61,62,116] that by means of these systems the catalytic nitrogen fixation can be realized too.

Thus nitrogen is capable of reaction with a mixture of $TiCl_4$ ($TiBr_4$), Al, $AlBr_3$ at a temperature of 50° and more. As a result of the reaction, nitrogen is reduced to form products with metal–nitrogen bonds, the latter hydrolyzing to yield ammonia.* The reaction can be carried out in the presence of solvent (e.g., benzene) as well as without any solvent added, that is, in molten $AlBr_3$.

In this case the nitrogen reaction turned out to be of catalytic character. With increasing amounts of Al and $AlBr_3$, the ammonia yield steadily increases and can be raised up to 200 and more moles per mole of initial $TiCl_4$ (cf. Table X).

TABLE X

Nitrogen Fixation by the System $TiCl_4$ + Al + $AlBr_3$[a]

Molar ratio $TiCl_4 : Al : AlBr_3$	C_6H_6, moles per mole of $TiCl_4$	Reaction time, hr	Ammonia yield, moles per mole of $TiCl_4$
1:2:1	9	8	0.80
1:6:1	9	8	0.84
1:6:2	9	8	1.25
1:12:0	9	8	0.03
1:12:16	9	8	4.2
1:12:16	0	30	6.44
1:12:16	18	30	6.65
1:12:33	0	30	10.7
1:150:200	0	30	95
1:600:1000	0	30	284

[a] Temperature 130°C; nitrogen pressure ~ 100 atm.

*A little amount of hydrazine is also formed in this reaction.[166] The hydrazine formation in course of nitrogen fixation was also observed by Shilov et al.[115] and van Jamelen et al.[117]

It is worth noting that without any $TiCl_4$ added, neither aluminum, nor its mixture with $AlBr_3$, reacts with nitrogen. The titanium derivative is thus a catalyst of nitrogen reduction by aluminum.

Of great importance in this reaction is aluminum bromide. The nitrogen-fixing activity of this system has been found to be considerably influenced by the quantity of $AlBr_3$, sharply increasing with increasing quantity of $AlBr_3$.

The time-dependence of the ammonia yield at different $AlBr_3 : TiCl_4$ ratios employed (see Fig. 3) has pointed out that first of all the reaction rate is affected by the variation of aluminum bromide. However, this is not the sole effect of aluminum bromide on the reaction. The $AlBr_3$ content in the mixture turned out to determine also the consumption of aluminum in the reaction, i.e., the maximum ammonia yield. The complete conversion of aluminum was found to be attainable, using the sufficient excess of aluminum bromide only (Fig. 3, curve 1). With insufficient amounts of $AlBr_3$, the complete introduction of aluminum failed in the reaction and nitrogen fixation stopped despite remaining considerable unreacted aluminum (Fig. 3, curve 3).

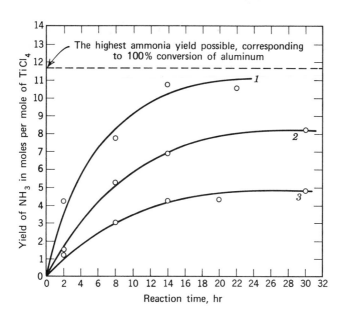

Fig. 3. The time-dependence of the ammonia yield in the nitrogen reaction with the system $TiCl_4 + Al + AlBr_3$ (temperature 130°C, nitrogen pressure ~ 100 atm). Molar ratio $TiCl_4 : Al : AlBr_3 = (1)$ 1:12:33; (2) 1:12:22; (3) 1:12:16.

The data obtained are in agreement with the supposition that catalyst regeneration in the nitrogen-fixation process is realized with the participation of aluminum bromide which splits the Ti—N bonds formed. Indeed, in this case aluminum bromide should be gradually consumed and, when it has disappeared from the reaction mixture, the nitrogen fixation stops in spite of an excess of aluminum present.

It is worth noting, however, that the $AlBr_3$ role probably does not consist in its possible participation in a catalyst regeneration only. As a matter of fact, it has been found that without any $AlBr_3$ added the system not only becomes noncatalytic, but even practically loses its nitrogen-fixing activity. Aluminum bromide seems to be necessary also for the realization of some other reaction stages.

In accordance with the data reported, the following possible scheme of the catalytic nitrogen fixation can be offered.

1. $TiX_4 \xrightarrow{\text{Al + AlBr}_3} TiX_2 \cdot nAlBr_3$, where X = Cl or Br

2. $TiX_2 \cdot nAlBr_3 + N_2 \longrightarrow N_2 \cdot TiX_2 \cdot nAlBr_3$

3 a. $N_2 \cdot TiX_2 \cdot nAlBr_3 \xrightarrow{\text{Al + AlBr}_3} X_2Ti-N\begin{smallmatrix}Al \\ Al\end{smallmatrix}$

b. $N_2 \cdot TiX_2 \cdot nAlBr_3 \xrightarrow{\text{TiX}_2 n AlBr_3} X_2Ti-N\begin{smallmatrix}TiX_2 \\ TiX_2\end{smallmatrix}$

4. $X_2Ti-N\diagup + AlBr_3 \longrightarrow N-(Al\diagup)_3 + TiX_3$

5. $TiX_3 \xrightarrow{\text{Al + AlBr}_3} TiX_2 \cdot nAlBr_3$

In the first stage the initial $TiCl_4(TiBr_4)$ is reduced to a divalent titanium compound, the latter in the form of its adduct with aluminum bromide ($TiX_2 \cdot nAlBr_3$). This adduct is probably a catalyst for nitrogen fixation. With the reaction being carried out in benzene a corresponding arenic complex of divalent titanium can work similarly as catalyst. Natta and coworkers[65] as well as Martin and Vohwinkel[66] have reported that on heating of TiX_4 with a mixture of Al and AlX_3 (X = Cl) in benzene a stable complex $C_6H_6 \cdot TiX_2 \cdot 2AlX_3$ was formed. It has been found[62] that in the presence of $AlBr_3$ this complex is indeed capable of catalyzing the nitrogen reduction with aluminum. Thus on heating (130°) the mixture of Al, $AlBr_3$, and catalytic amounts of $C_6H_6 \cdot TiCl_2 \cdot 2AlCl_3$ (in molar ratio of $Al:AlBr_3: C_6H_6 \cdot TiCl_2 \cdot 2AlCl_3 = 150:250:1$) with nitrogen the latter is reduced to form 115 moles of ammonia per mole of titanium complex upon hydrolysis. This corresponds to nearly 80% conversion of aluminum.

In stage 2 of the foregoing scheme the nitrogen complex with divalent titanium is formed. As a result of the complex formation, the nitrogen molecule becomes activated for further reduction. The nitrogen reduction can be brought about at the expense of electrons of divalent titanium and aluminum.* This should lead to the nitride–like products containing Ti—N and Al—N bonds (stage 3a).

Another possibility consists in the nitrogen reduction by means of divalent titanium only, without any aluminum metal participation (stage 3b). Thus it has been found[61,62] that the complex $C_6H_6 \cdot TiCl_2 \cdot 2AlCl_3$ reduces nitrogen at 130°C to form a compound, the latter nearly corresponding to $C_6H_6 (TiCl_2 \cdot 2AlCl_3)_3N$ in its chemical composition. This compound is hydrolyzed to give a stoichiometric yield of ammonia.

On the further stages 4 and 5 the action of $AlBr_3$ and Al leads to the cleavage of Ti—N bonds and the reduction of titanium into the divalent state, that is, the catalyst regeneration takes place. As a result of these reactions the compound with Al—N bonds is formed, yielding ammonia upon hydrolysis.

The data obtained make it possible to construct the other catalytic nitrogen-fixing systems. Thus aluminum can be replaced by $LiAlH_4$ in these reactions.[62,116] In the presence of aluminum halides and $TiCl_4$ as a catalyst, nitrogen was reduced by $LiAlH_4$ to form ~ 100 moles of NH_3 per mole of $TiCl_4$ (temperature 70°C).**

D. Reaction of Nitrogen with Hydrogen

The accomplishment of reaction between nitrogen and hydrogen under mild conditions is one of the interesting problems in the chemistry of molecular nitrogen. It was mentioned above that this reaction is thermodynamically advantageous just at low temperatures. But under such conditions the reaction is very slow even if the highly active Haber catalysts are used.

In search of more effective activating systems for this reaction the complex catalysts for homogeneous hydrogenation of unsaturated compounds represent considerable interest.

The hydrogenation of acetylenes and olefins over complex catalysts is

* The reaction products of aluminum with aluminum bromide (e.g., Al_4Br_6, Al_2^0,[67] etc.) rather than aluminum as such may be the real reducing agents in this reaction.

** Van Jamelen et al.[118] recently reported the electrochemical system in which $Ti(O-i-Pz)_y$ in the presence of $Ae(O-i-Pz)_3$ and naphthalene catalytically effect reduction of nitrogen to the ammonia (yield of $NH_3 \sim 6$, 2 moles per mole of titanium compound, reaction time–11 days, pressure of N_2–7 atm).

known to consist of three main steps: activation of H_2 by its formation of a hydride with a transition metal derivative, activation of an unsaturated compound as a result of π-complex formation and, at the last, insertion of a coordinated molecule of an unsaturated compound into the transition metal–hydride bond:

$$L_nM + H_2 \longrightarrow L_nM\!-\!H \xrightarrow{CH_2=CH_2} L_nM \begin{smallmatrix} CH_2=CH_2 \\[2pt] \diagup \\[-2pt] \diagdown \\[2pt] H \end{smallmatrix} \longrightarrow$$

$$L_nM\!-\!CH_2CH_3 \xrightarrow{H_2} L_nM\!-\!H + CH_3CH_3$$

In order to realize the reaction between N_2 and H_2 in a similar way it is necessary to have systems capable of activating both molecular nitrogen and molecular hydrogen.

The above systems obtained in reactions of transition-metal derivatives with Li-, Mg-, and Al-organic compounds and which appeared to be active in nitrogen-reduction reactions just meet these requirements. The organometallic derivatives of transition metals formed in these systems readily react with hydrogen yielding corresponding hydrides that proved to be in a number of cases catalytically active in the hydrogenation of unsaturated compounds.[29]

However, in studying the effect of H_2 on nitrogen fixation by such systems it was found that, as a rule, the introduction of hydrogen either has no effect on the yield of ammonia or even inhibits the reaction.[4,7,111] A similar phenomenon was observed in using systems involving acetylacetonate,[7] cyclopentadienyl and other complexes of transition metals in mixtures with organometallic compounds.

The extent of inhibition of nitrogen fixation by molecular hydrogen depends to a considerable degree on the nature of the system and also on the ratio of N_2 to H_2 in the gaseous mixture. For instance, in the system $Cp_2TiCl_2 + C_2H_5MgBr$ in ether a considerable decrease in ammonia yield is observed in the presence of even a small amount of hydrogen (Fig. 4). With further increase in the partial pressure of H_2 the inhibiting effect of the latter increases and when the ratio of N_2 to H_2 is $1:3$ in a gaseous mixture the nitrogen-fixing activity of the system practically tends to zero.

In other cases inhibition is of a less marked character. The system $CpTi(OC_2H_5)_3 + Al(i\text{-}C_4H_9)_3$ in toluene can serve as an example. In the rather wide range of ratios $N_2:H_2$ the hydrogen practically does not change the efficiency of nitrogen fixation by this system (Fig. 5). But at higher H_2 pressures the inhibition is also observed.

The inhibiting effect of H_2 observed in nitrogen fixation may be accounted for by the fact that hydrogen, like acetylenes, carbon monoxide,

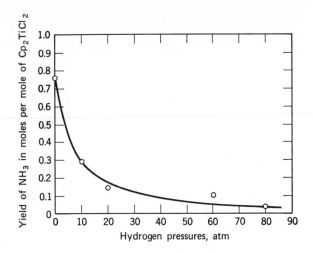

Fig. 4. The effect of H_2 on nitrogen fixation by the system $Cp_2TiCl_2 + C_2H_5MgBr$ in ether (temperature 20°C, molar ratio of $C_2H_5MgBr:Cp_2TiCl_2 = 9:1$, time of reaction 7 hr, pressure of N_2 30 atm).

and olefins, is capable of displacing a nitrogen molecule out of the co-ordinate sphere of the transition metal atom thus hindering N_2 activation for subsequent reduction.

This supposition finds its support in the data of Sacco and Rossi[55] as well as those of Yamamoto, Pu, Kitazume, and Ikeda,[52] who have found

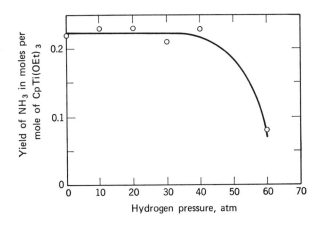

Fig. 5. The effect of H_2 on nitrogen fixation by the system $CpTi(OEt)_3 + (i\text{-}C_4H_9)_3Al$ in toluene (temperature 20°C, molar ratio of $(i\text{-}C_4H_9)_3Al:CpTi(OEt)_3 = 6:1$, time of reaction 7 hr, pressure of N_2 30 atm).

that the reaction of stable nitrogen complex of Co with hydrogen does not result in nitrogen hydrogenation but in reversible displacement of N_2 out of the complex.

$$(PPh_3)_3Co(N_2)H + H_2 \rightleftharpoons (PPh_3)_3CoH_3 + N_2$$

These data show that N_2 and H_2 can actually compete for a site in the coordination sphere of a transition-metal atom. The result of this competition is a function of the relative stabilities of the respective nitrogen and hydride complexes and also of the partial pressures of N_2 and H_2. Thus the inhibiting effect of hydrogen is a serious obstacle for the hydrogenation of nitrogen under mild conditions.

Nevertheless such action of hydrogen on nitrogen fixation is not general. At present a number of systems are known for which an increase in the yield of ammonia in the presence of hydrogen is observed.[7,9] One of these is the system formed by the reaction of $Ti(OC_2H_5)_4$ with excess $Al(i\text{-}C_4H_9)_3$ in toluene.[7] This system is homogeneous and is capable of reacting with nitrogen at room temperature yielding 0.33 moles of ammonia per mole of $Ti(OC_2H_5)_4$ after hydrolysis. In studying the effect of hydrogen on this reaction it was found that introduction of H_2 results in a noticeable increase in the yield of NH_3 over a wide range of ratios $N_2:H_2$ (Fig. 6). At the same time a maximum yield of NH_3 was observed at a ratio of $N_2:H_2$ approximately equal to 3:1. Under these conditions the amount of ammonia yielded is about 0.6 moles per mole of $Ti(OC_2H_5)_4$, which is approximately twice that in the absence of H_2. It would be reasonable to suppose that

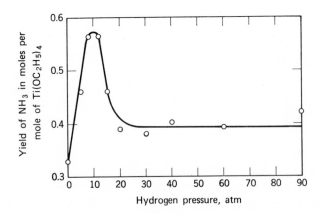

Fig. 6. The effect of H_2 on nitrogen fixation by the system $Ti(OC_2H_5)_4$ + $(i\text{-}C_4H_9)_3Al$ in toluene (temperature 20°C, molar ratio of $(i\text{-}C_4H_9)_3Al:Ti(OC_2H_5)_4$ = 6:1, time of reaction 10 hr, pressure of N_2 30 atm).

such an increase in the yield of NH_3 is due to hydrogenation of nitrogen with molecular hydrogen.

Recently Shvetsov, Korabliova, Nechiporenko, Ilatovskaya, and Khidekel[30] have obtained direct proof for such a suggestion. The authors have studied the reaction of $Ti(OC_2H_5)_4$ and $Al(i-C_4H_9)_3$ with a mixture of N_2 and D_2. After the end of the reaction the products were hydrolyzed by small amounts of water, and the ammonia obtained was analyzed for deuterium. For this purpose the aqueous solution of ammonia obtained after hydrolysis was treated with $LiAlH_4$ and the isotope composition of the hydrogen liberated was then investigated. It was found that besides H_2, the hydrogen contained a noticeable amount of HD. In control experiments the hydrogen formed after treatment with $LiAlH_4$ contained practically no HD, if only D_2 in the absence of nitrogen was introduced into the reaction with a mixture of $Ti(OC_2H_5)_4$ and $Al(i-C_4H_9)_3$. The data obtained indicate that in this case the formation of ammonia is at least partially due to the reaction of nitrogen with hydrogen that seems to occur via the insertion of a nitrogen molecule into the Ti-H bond. Thus the possibility of hydrogenation of nitrogen by molecular hydrogen over complex homogeneous catalysts has been shown. A patent[31] by Glemser should be mentioned in this connection. It deals with the possibility of the formation of NH_3 from N_2 and H_2 under mild conditions in the presence of $H_2Mo_5O_7(OH)_3$ that can be prepared on reduction of MoO_3. However, Shilov et al.[105] failed to reproduce these results.

Another case is known in which the introduction of hydrogen into the reaction leads to an increase in the yield of ammonia.[9] It has been found that individual Cp_2Ti as well as dimethyl-titanocene $Cp_2Ti(CH_3)_2$ and diphenyltitanocene Cp_2TiPh_2 in the course of their thermal decomposition at 90°C to Cp_2Ti in organic solvents react with nitrogen yielding compounds of the nitride type.* Hydrolysis of these nitride compounds leads to ammonia. The amount of ammonia produced after hydrolysis appeared to increase threefold when this reaction was carried out in the presence of a mixture of nitrogen with hydrogen. It might be that the increase in the yield of NH_3 observed is also due to the reaction of N_2 with H_2.

E. Amines Formation in Molecular Nitrogen Reactions

The capacity of N_2 for forming complexes with derivatives of transition metals gives possibilities for the synthesis of nitrogen-containing organic compounds directly from molecular nitrogen.

*This reaction proceeds only by heating. At room temperature Cp_2Ti does not reduce nitrogen but such reaction takes place if excess of RMgX, RLi, etc. is added to Cp_2Ti.

It is known that carbon monoxide, olefins, and acetylenes reacting with organometallic compounds of transition metals are capable of insertion into the transition metal–carbon bond. The corresponding complexes of unsaturated compounds with a transition-metal atom are intermediates in these reactions.

$$L_mM{-}R{-}\begin{cases} \xrightarrow{CO} L_mM\!\!\begin{array}{c}{\diagup}C{=}O\\{\diagdown}R\end{array} \longrightarrow L_mM{-}\overset{\displaystyle O}{\overset{\|}{C}}{-}R \\[2em] \xrightarrow{CH_2=CH_2} L_mM\!\!\begin{array}{c}{\diagup}CH_2{=}CH_2\\{\diagdown}R\end{array} \longrightarrow L_mM{-}CH_2CH_2R \end{cases}$$

One can expect that nitrogen would also enter a similar reaction of insertion ultimately to yield organic nitrogen-containing compounds. Indeed it appeared that noticeable amounts of amines[12,112] were obtained in reactions of N_2 with some systems involving derivatives of transition metals and organometallic compounds (i.e., under conditions of the transition metal–carbon bond formation).

The mixture of titanocene dichloride Cp_2TiCl_2 with excess of C_6H_5Li in ether is one of such systems. This system reacts with N_2 at a noticeable rate at room temperature (see Table VI and Fig. 1). After hydrolysis the products of the reaction contain a considerable amount of aniline along with ammonia. A small amount of o-aminodiphenyl is also formed.

The $C_6H_5NH_2$ and NH_2 yields are 0.15 and 0.65 moles per mole of Cp_2TiCl_2, respectively. Approximately the same amount of aniline is obtained in the reaction of N_2 with a mixture of $Cp_2Ti(C_6H_5)_2$ and C_6H_5Li in ether.*

It seems most probable that aniline formation in these reactions proceeds via nitrogen insertion into the $Ti{-}C_6H_5$ bond of some low-valent titanium derivative arising due to the reduction of the initial Cp_2TiCl_2 or $Cp_2Ti(C_6H_5)_2$ with excess of C_6H_5Li.[36]

$$L_mTi{-}C_6H_5 \xrightarrow{N_2} L_mTi\!\!\begin{array}{c}{\diagup}N{\equiv}N\\{\diagdown}C_6H_5\end{array} \longrightarrow \underset{(7)}{L_mTi{-}N{=}N{-}C_6H_5} \xrightarrow{[e]}$$

$$[C_6H_5N^{2-}] \xrightarrow{H_2O} C_6H_5NH_2$$

*Phenyl lithium itself, without any added titanium compound, does not react with nitrogen under these conditions, neither do Cp_2TiCl_2 and $Cp_2Ti(C_6H_5)_2$ react with nitrogen without the excess of PhLi.

A phenylazo derivative of titanium (**7**) obtained as a result of nitrogen insertion is subjected to further reductive splitting with production of *N*-metal substituted aniline (conventionally marked on the scheme as $[C_6H_5N^{2-}]$), hydrolysis of which yields free aniline.

The formation of a titanium phenylazoderivative as an intermediate product is postulated in the proposed scheme of the aniline production mechanism. Some time ago compounds of a similar type were obtained by the reaction of aromatic diazonium salts with anionic[32,33] or hydride[34,35] complexes of transition metals

$$ArN_2^{\oplus}BF_4^{\ominus} + [CpMo(CO)_3]^{\ominus}Na^{\oplus} \xrightarrow{\ -CO\ } Ar{-}N{=}N{-}Mo(CO)_2Cp$$

$$ArN_2^{\oplus}BF_4^{\ominus} + (Et_3P)_2PtH(Cl) \longrightarrow Ar{-}N{=}N{-}PtCl(PEt_3)_2$$

In the presence of reducing agents the arylazo group of these compounds is actually subjected to a reductive splitting with formation of aromatic amines.

$$Ar{-}N\{N{-}PtCl(PEt_3)_2 \xrightarrow{\ H_2/Pt\ } ArNH_2$$

However, there exists another possibility of aniline formation in the nitrogen fixation reaction mentioned above.

It has been shown recently[36] that in the reaction of Cp_2TiPh_2 or Cp_2TiCl_2 with excess phenyllithium, triphenylene is one of the products. It points to the possible intermediate formation of dehydrobenzene in this reaction.

Dehydrobenzene is known to readily add various nucleophilic reagents, particularly amides of alkaline metals. Therefore, the products of nitrogen fixation containing metal–nitrogen bonds (and serving as a source of ammonia), could add to dehydrobenzene yielding aniline after hydrolysis.

$$\text{[benzyne]} + \text{Li-NM}_2 \longrightarrow \text{[o-C}_6\text{H}_4(\text{NM}_2)(\text{Li})] \xrightarrow{\text{H}_2\text{O}} \text{[C}_6\text{H}_5\text{NH}_2]$$

where M = Li or Ti

The nitrogen reaction with systems obtained by the interaction of Cp_2TiCl_2 with p-, m-, and o-tolyllithium in ether was studied in order to examine the possibility of such a mechanism for the formation of aniline. Dehydrotoluenes must be obtained in this case leading to mixtures of isomeric toluidines if Li-NM$_2$ is added to them.

$$L_mTi\!-\!\!\bigcirc\!\!-\!CH_3 \xrightarrow{-L_mTiH} \text{[dehydrotoluene]} \xrightarrow[\text{2. H}_2\text{O}]{\text{1. LiNM}_2} \text{[}p\text{-toluidine]} + \text{[}m\text{-toluidine]}$$

$$L_mTi\!-\!\!\bigcirc(CH_3) \xrightarrow{-L_mTiH} \text{[dehydrotoluene]} \xrightarrow[\text{2. H}_2\text{O}]{\text{1. LiNM}_2} \text{[}o\text{-toluidine]} + \text{[}m\text{-toluidine]}$$

$$L_mTi\!-\!\!\bigcirc(CH_3) \xrightarrow{-L_mTiH} \text{[dehydrotoluene]} + \text{[dehydrotoluene]} \xrightarrow[\text{2. H}_2\text{O}]{\text{1. LiNM}_2}$$

$$\text{[}p\text{-toluidine]} + \text{[}m\text{-toluidine]} + \text{[}o\text{-toluidine]}$$

Thus, if in the reaction of N$_2$ with the system Cp_2TiCl_2 + ArLi amines are formed on the dehydrobenzene mechanism an equimolecular mixture of p- and m-toluidines will be obtained from p-tolyllithium and a mixture of o- and m-toliudines from o-tolyllithium, whereas m-tolyllithium would yield a mixture of all three isomeric toluidines in a molar ratio of $p:m:o = 1:2:1$. In reality, if polarization of the triple bond of dehydrotoluenes under the effect of the electron-donating CH$_3$ group is taken into account

TABLE XI

Nitrogen Fixation by the Systems Cp_2TiCl_2 + p-, m- and o-$CH_3C_6H_4Li$ in Ether (Temperature 20°C, Pressure of N_2 100 atm)

ArLi[a]	The yield of nitrogen fixation products in moles per mole of Cp_2TiCl_2[b]	
	NH_3	$CH_3C_6H_4NH_2$
p-$CH_3C_6H_4Li$	0.25	0.07
m-$CH_3C_6H_4Li$	0.22	0.06
o-$CH_3C_6H_4Li$	0.41	0.06

[a] Molar ratio ArLi:Cp_2TiCl_2, 5:1.
[b] Time of reaction 10–11 hr.

the ratio of the isomeric toluidines in the case of the dehydrobenzene mechanism should differ from those given, tending to some increase in the m-isomer in all cases.

It has been found that the systems obtained from Cp_2TiCl_2 and p-tolyllithium as well as from m- and o-tolyllithium react with nitrogen yielding toluidines (Table XI). Product analysis (Table XII) showed that practically only p-toluidine is produced in the reaction of N_2 with a mixture of Cp_2TiCl_2 and p-tolyllithium and almost pure m-toluidine is obtained from a mixture of Cp_2TiCl_2 with m-tolyllithium (Table XII).

Thus the data obtained contradict the dehydrobenzene mechanism of amine formation in this reaction and are in agreement with the mechanism of the nitrogen insertion into the titanium–carbon bond.

Unusual results were obtained in studying the reaction of nitrogen with a mixture of Cp_2TiCl_2 and o-tolyllithium. In this case all the three isomeric toluidines—o-toluidine, m-toluidine, and a small amount of p-toluidine—are formed. The content of isomeric toluidines in the mixture changes from experiment to experiment and usually consists of 40–77% for o-,

TABLE XII

Toluidine Product Composition in the Reaction of Nitrogen with Systems: Cp_2TiCl_2 + p- and m-$CH_3C_6H_4Li$ in Ether

ArLi	Ratio of toluidines, %		
	p-	m-	o-
p-$CH_3C_6H_4Li$	96	4	—
m-$CH_3C_6H_4Li$	3	97	—

52–21% for *m*-, and 8–2% for *p*-toluidine. The results obtained could be obviously accounted for by the possibility of isomerization of the tolyl-titanium bonds, proceeding via compounds of the dehydrobenzene type.

This process would proceed most readily for a compound with an *o*-tolyltitanium bond which is less stable owing to steric hindrance and can isomerize into a more stable *m*-tolyltitanium bond.

The formation of dehydrobenzene also provides an explanation for the presence of small amounts of *o*-aminodiphenyl in products of the reaction of Cp_2TiCl_2 and C_6H_5Li with nitrogen. *o*-Lithiumdiphenyl that can be formed by the addition of excess of PhLi to dehydrobenzene, is apparently the source of this product.

The possibility of the production of aliphatic amine in the reaction of N_2 with Cp_2TiCl_2 and aliphatic Li- and Mg-organic compounds has been also studied. It turned out that in such cases aliphatic amines either are not formed at all, or are formed in very low yields (< 0.002 moles per mole of Cp_2TiCl_2).

Thus in passing from C_6H_5Li to $CH_3C_6H_4Li$ and further to aliphatic organo-Li compounds the yield of amines sharply decreases. The stability of the Ti—R bond decreases in the same sequence. It is consistent with the mechanism of amine formation via nitrogen insertion into the titanium–carbon bond. Indeed, if the Ti—R bond is unstable, and thus its lifetime in the solution is very short, the probability of insertion of nitrogen into this bond should also become lower.

II. MOLECULAR NITROGEN COMPLEXES

An intermediate formation of nitrogen complexes with transition-metal compounds was suggested for all reactions discussed above. Up to most recent times these complexes were hypothetical and it was only quite recently that the first representatives of this very interesting class of compounds were obtained.

A simple qualitative consideration of nitrogen complexes suggests[13] that they must show some characteristic peculiarities making them different from similar complexes of acetylenes and olefins and relating them with complexes of carbon monoxide. It is known that in the case of CO the most stable transition-metal complexes are those in which the metal is in a low valence state, that is, in the zero-valent state or even in a negatively charged state. This makes carbon monoxide differ from many other unsaturated compounds.

It can be explained by the fact that the highest occupied electron level of CO is located much lower (-14 eV) than, for example, the respective levels in a molecule of acetylene (-11.4 eV). For this reason carbon monoxide is a very weak base, forming only unstable complexes even with strong Lewis acids. For the same reason complexes in which CO is bound to a metal atom only by means of the donor–acceptor bond are unstable. The stability of a complex can considerably increase if, along with the donor–acceptor interaction, an effective back-donation is realized, that is, transfer of electrons from the metal to vacant CO orbitals. However, the later interaction will be considerable only in the event the positive charge on the metal is not too high and will be most effective if the charge is zero or even somewhat negative.

The upper occupied level of the nitrogen molecule is located still lower (-15.6 eV) than that of CO, whereas the lower vacant levels are rather near to each other ($E_{N_2} = +7.5$ eV $E_{CO} = +7.0$ eV). It follows that N_2 as a base is weaker than CO, and the role of back-donation in the nitrogen complexes must be even greater than in metal carbonyls. From this point of view the most stable complexes with nitrogen would be formed first by transition-metal compounds which have only a low positive, zero, or even negative charge on the metal.

First information about the possible existence of nitrogen complexes was obtained in 1964 by Eischens and Jacknow[37] in studying N_2 chemisorption on transition metal surfaces. The authors studied the chemisorption of nitrogen on Ni at $-100°C$ and $+30°C$ using infrared spectroscopy. At these temperatures nitrogen forms a weak absorption compound with surface atoms of nickel. It turned out that chemisorption of nitrogen is accompanied by appearance in the infrared spectrum of an intense absorption band at 2202 cm^{-1} ascribed to stretching vibration of the N≡N bond and indicating the formation of a surface complex of nitrogen with nickel atoms. The authors believe that the nitrogen surface complex has a linear structure (2), since, in the case of a symmetrical complex (1), the intensity of the absorption band of N≡N bond must be low. In the presence of CO, O_2, and NO the intensity of the band at 2202 cm^{-1} decreases considerably, it can be explained by postulating a competing displacement of N_2 from the nickel surface.

An attempt was made to observe spectroscopically the hydrogenation of the nitrogen surface complex. It turned out that on adding H_2, the 2202 cm^{-1} band intensity decreases and a new absorption band appears at 2254 cm^{-1}. The authors believe that the decrease in the 2202 cm^{-1} band intensity is partially due to the hydrogenation of nitrogen on the surface, and the appearance of the new band at 2254 cm^{-1} is accounted for by

formation of a $Ni\diagdown\begin{smallmatrix}H\\\\N{\equiv}N\end{smallmatrix}$ compound that could be an intermediate in hydrogenation of the N_2 surface complex.

$$\overset{|}{\underset{|}{Ni}} + N_2 \longrightarrow \overset{|}{\underset{|}{Ni}}{-}N{\equiv}N \xrightarrow{\ H_2\ } \overset{|}{\underset{|}{Ni}}\diagup\diagdown\begin{smallmatrix}H\\\\N{\equiv}N\end{smallmatrix} \longrightarrow$$

$$\overset{|}{\underset{|}{Ni}}{-}N{=}NH \xrightarrow{\ H_2\ } \cdots \longrightarrow \overset{|}{\underset{|}{Ni}}{-}NH_3$$

The formation of similar nitrogen complexes was recently observed in spectroscopic studies of nitrogen chemisorption on the surface of metallic cobalt.[38]

The results of Eischens and Jacknow indicating the possible formation of nitrogen complexes by N_2 chemisorption on the transition metal surfaces are of great importance for understanding the mechanism of the catalytic synthesis of ammonia according to Haber-Bosch, in which nitrogen chemisorption is one of the most important stages of the process.

One of the outstanding achievements of the last few years in the chemistry of coordinated compounds is the preparation and isolation of molecular nitrogen complexes with transition-metal compounds.

The first complex (8) of this type was obtained in 1965 by Allen and Senoff[39] upon reduction of $RuCl_3$ by hydrazine in aqueous solution. Other derivatives of ruthenium[40,41] — $(NH_4)_2RuCl_6$, $K_2[Ru(H_2O)Cl_5]$, $[Ru(NH_3)_5H_2O](CH_3SO_3)_3$, and $(PR_3)_3RuCl_3$ may be used in this reaction instead of $RuCl_3$.

$$RuCl_3 + N_2H_4 \longrightarrow [Ru(NH_3)_5N_2]X_2$$
$$(8)$$

The nitrogen–ruthenium complex (8) is of a cation character and proved to be a sufficiently stable compound to be isolated in a pure state and to be identified. It was obtained as corresponding halides ($X = Cl^-$, Br^-, and I^-), forofluoride ($X = BF_4^-$) and hexafluorophosphate ($X = PF_6^-$). All these compounds are diamagnetic salts, moderately stable toward air and decomposing gradually in water. In the absence of water they are stable for several weeks, though they decompose on heating, yielding nitrogen. A nitrogen molecule is bound with ruthenium relatively weakly and can be displaced by compounds such as ammonia, pyridine, etc.

$$[Ru(NH_3)_5N_2]X_2 + NH_3 \longrightarrow [Ru(NH_3)_6]X_2 + N_2$$

A complex decomposition with liberation of nitrogen is also observed on heating with hydrohalic acid solutions.

An intensive absorption band in the region of 2105–2167 cm^{-1}, corresponding to the stretching vibrations of the coordinate $N\equiv N$ bond is observed in the infrared spectrum 8 (the vibration frequency for gaseous nitrogen in Raman spectrum is 2331 cm^{-1}). Moreover there is a band in the region of 508–474 cm^{-1}, corresponding to stretching vibrations of the Ru—N bond in the Ru—N_2 fragment. All these data undoubtedly point to coordination of the nitrogen molecule with the ruthenium atom.

The reaction of nitrogen complex 8 with reducing agents ($NaBH_4$, Zn, the Devarda's alloy) has been studied.[39,49] It was found that, under the action of an alkali solution of $NaBH_4$, the coordinated nitrogen molecule is reduced forming ammonia in amount of 1 mole per mole of initial complex 8*.

Bottomley and Nyborg[42] have published the first data on X-ray structural investigation of a nitrogen–ruthenium complex. In spite of a number of experimental difficulties the authors conclude that the structure of the complex is octahedral and the nitrogen molecule is bound with the Ru atom linearly (Ru—N—N), that is, in the same way as carbon monoxide in metal carbonyls. The distances Ru–N and N–N are correspondingly 2.11 and 1.12 Å.

The method of Allen and Senoff (reaction of metal salt with hydrazine) was later applied by Borod'ko, Bukreev, Kozub, Khidekel, and Shilov[43,49] for synthesis of osmium complexes of molecular nitrogen. It was found that reduction of K_2OsCl_6, $OsCl_4$, OsO_4, and $OsOHCl_3$ by hydrazine-hydrate in aqueous or tetrahydrofuran solutions yields dark-colored osmium compounds exhibiting three absorption bands at 2010, 2095, and 2168 cm^{-1} in the infrared spectrum. All three bands are displaced by 60–68 cm^{-1} when the reduction is carried out with labeled hydrazine hydrate, $^{15}N_2H_4 \cdot H_2O$. The authors suggest that the three infrared bands

---

*Recently these data[39,40] have been reinvestigated by Shilov et al.[79] and Chatt et al.[80] (see "Supplement," page 105).

correspond to three states of molecular nitrogen coordinated with Os (II).*
The compounds obtained are stable in aqueous solutions and to air. They
are much more stable toward heating than the respective ruthenium complexes: no marked decomposition is observed for a long time even at
200°. At higher temperatures the bands at 2095 and 2168 cm^{-1} gradually
disappear from the infrared spectra of products but the band at 2010 cm^{-1}
remains. It proves that stability of the complex characterized by the band
at 2010 cm^{-1} is highest. The compounds obtained decompose in sulfuric
acid yielding nitrogen.

Formation of the osmium complex of molecular nitrogen via reduction
of osmium compounds by hydrazine was also observed independently by
Allen and Stevens.[44] They succeeded, however, in isolating a pure nitrogen
complex (9) and in identifying it.

$$(NH_4)_2OsCl_6 + N_2H_4 \longrightarrow [Os(NH_3)_5N_2]X_2$$
$$(9)$$

where X = Cl$^-$, Br$^-$, I$^-$, ClO$_4^-$, BF$_4^-$, and BPh$_4^-$.

It turned out to be similar in its composition to a corresponding
ruthenium complex (8) and is apparently one of the osmium–nitrogen
complexes observed by Borod'ko and co-workers by means of infrared
spectroscopy (see above).

The osmium complex (9) is a saltlike, diamagnetic compound, markedly
more stable than the corresponding ruthenium nitrogen complex. The
chloride is the most stable compound, remaining unchanged in air for a
long time and decomposing with liberation of nitrogen only on strong
heating in vacuum. The bromide is of the same stability, but other salts
are somewhat less stable. A characteristic feature of the osmium nitrogen
complex is its extraordinary stability toward hydrohalic acids. Unlike the
ruthenium nitrogen complex the osmium complex does not decompose in
hydrohalic acid solutions even on boiling.

There is an intense absorption band in the infrared spectrum 9 corresponding to the stretching vibration of the coordinate N≡N bond. But it
is markedly shifted to the long wavelength region (2064–2010 cm^{-1}) as compared to bands characteristic of the respective ruthenium complex. On the
other hand the absorption bands corresponding to stretching vibrations of
the Os—N bond in Os—N$_2$ appears to be shifted respectively to the region of
shorter waves (546–520 cm^{-1}). It indicates a greater stability of the Os—N$_2$
bond as compared to the Ru—N$_2$ bond in accordance with the chemical
data mentioned above.

*According to the new data[100,105] these three infra-red bands correspond actually
to two nitrogen complexes, i.e. mononitrogen complex (9) [Os(NH$_3$)$_5$N$_2$]X$_2$ ($\nu_{N \equiv N}$
=2010 cm^{-1}) and *Bis*-nitrogen complex *cis*-[Os(NH$_3$)$_4$(N$_2$)$_2$]X$_2$ ($\nu_{N \equiv N}$=2095 and
2168 cm^{-1}, see Supplement, page 111).

The nitrogen complex formation in the reactions of metal salts with hydrazine seems to proceed by the intrasphere dehydrogenation of the hydrazine molecule coordinated to the transition-metal atom. It has been found by Chatt and Fergusson[68] that nitrogen complexes could be also obtained by dehydrogenation of the ammonia molecule. The reduction of ruthenium(III) chloride with zinc in aqueous ammonia solution has been investigated. It was found that N_2 and H_2 were formed during the reduction. The reaction product, $[Ru(NH_3)_6]Cl_2$, was contaminated with a small amount of nitrogen–ruthenium complex $[Ru(NH_3)_5N_2]Cl_2$ obtained earlier by Allen and Senoff[39,49] by means of reaction of ruthenium(III) chloride with N_2H_4. This is evidenced by (1) evolution of nitrogen when the complex is degraded with concentrated hydrochloric acid and (2) by the presence of a band at 2120 cm^{-1} in the infrared spectrum of the product corresponding to the stretching vibrations of the N≡N bond characteristic of $[Ru(NH_3)_5N_2]Cl_2$.

The nitrogen complex yield in this reaction depends on the strength of the ammonia solution, increasing from 1–3% up to 15% by passing from aqueous to liquid ammonia.

Chatt and Fergusson suggest that the ammonia which is dehydrogenated with the ruthenium ion participation during the reaction is a source of coordinated nitrogen in the complex $[Ru(NH_3)_5N_2]Cl_2$. Indeed the band at 2120 cm^{-1} in infrared of the reaction product was found unaffected when the reaction was carried out *in vacuo* as well as under argon with atmospheric nitrogen thoroughly excluded.

Collman and Kang offered another method for obtaining nitrogen complexes, based on reaction of transition metal compounds with azides.[45,46] By means of this method they succeeded in obtaining iridium complexes of molecular nitrogen. In reaction of the triphenylphosphine carbonyl complex of monovalent iridium (10) with azides of organic acids they isolated a complex (11), in which N_2 was bound to the Ir atom by a coordinated bond.

where R = Ph, NH_2, C_4H_3O.

An intense band at 2095 cm^{-1} corresponding to the stretching vibration of the triple N≡N bond coordinated with metal is in the infrared spectrum of this complex. The authors ascribe to complex **11** a linear structure.

Complex **11** decomposes slowly at room temperature and rapidly at 110°C with the result that molecular nitrogen is liberated. By treatment of the complex with carbon monoxide, $P(C_6H_5)_3$, and acetylenes the displacement of N_2 from the coordination sphere takes place. A similar process also occurs under the action of HCl and CCl_4.

No nitrogen displacement is observed in the reaction of complex **11** with diethyl maleate, but a new complex **12** is formed.

(12)

The new complex (**12**) is even more stable than complex **11**. Its fast decomposition with liberation of N_2 was observed only at 165°C. The infrared spectrum of this complex exhibits an intense band characteristic of stretching vibrations of the N≡N bond but it is rather displaced to the short wavelength region (2190 cm^{-1}).

Ukhin, Shvetsov, and Khidekel[47] applied a similar method for synthesis of the rhodium–nitrogen complex. They obtained a complex of the composition $(PPh_3)_2Rh(N_2)Cl$ in reaction of $(PPh_3)_2Rh(CO)Cl$ with an excess of butyric acid azide. Its infrared spectrum resembles the spectrum of the similar iridium complex (**11**). An intensive band at 2152 cm^{-1} corresponding to a stretching vibration of the N≡N bond was observed in this case. The rhodium complex is stable in inert atmosphere at low temperature and is fairly stable at room temperature. On dissolving in some organic solvents (benzene, toluene, chloroform) the complex decomposes, liberating nitrogen.

Allen and co-workers[40] found that the azide method can also be used for synthesis of the ruthenium–nitrogen complex (**8**) mentioned above.

$$[Ru(NH_3)_5H_2O]X_3 + NaN_3 \longrightarrow [Ru(NH_3)_5N_2]X_2$$

All nitrogen complexes of Ru^{II}, Os^{II}, Ir^{I}, and Rh^{I} described above were obtained by an indirect method. Shilov, Shilova, and Borod'ko[48] were the first to succeed in observing by means of infrared spectra the formation of similar complexes in the direct reaction of N_2 with transition-metal compounds. They have studied the reduction of $RuCl_3$ or $RuCl_3(OH)$ by zinc amalgam in a THF solution in the presence of N_2 at room temperature and observed the band at 2140 cm^{-1} in the infrared spectrum of products. This band shifted to 2070 cm^{-1} on use of a heavy isotope—$^{15}N_2$ instead of the usual $^{14}N_2$. The same band (2140 cm^{-1}) appeared on reduction of $RuCl_3OH$ by hydrazine in a THF solution in the absence of N_2. These results point to the presence in the reaction mixture of Ru compounds containing molecular nitrogen in the coordination sphere of the metal atom. The formation of nitrogen complexes was also observed in reaction of ruthenium salts with organomagnesium compounds (C_2H_5MgBr, C_6H_5MgBr) in the presence of N_2.[49]

The reaction with asymmetrically labeled nitrogen ($^{14}N\equiv N^{15}$) has been studied[49] in order to find out the structure of the ruthenium complex obtained. In this case there are two possibilities of the nitrogen coordination with ruthenium for linear configuration of the Ru$\cdots N_2$ fragment:

$$Ru—^{15}N\equiv^{14}N \quad \text{and} \quad Ru—^{14}N\equiv^{15}N$$

This must result in splitting of the infrared absorption band corresponding to the stretching vibration of the $N\equiv N$ bond. It was found that the outline of the band (at 2115 cm^{-1}) for the complex with $^{14}N\equiv^{15}N$ actually corresponds to superposition of two bands with splitting close to 7 cm^{-1}. This is in agreement with the linear configuration of the Ru$\cdots N_2$ fragment in the complex.

The authors succeeded in isolating the complex formed and in investigating its properties.[49] It turned out that the composition of the complex obtained from $RuCl_3(H_2O)_n$ and Zn/Hg in THF resembles that of $RuCl_2N_2(H_2O)_2\cdot THF$. The nitrogen molecule is strongly bound to a ruthenium atom and, one difference from the complex $[Ru(NH_3)_5N_2]X_2$, is that nitrogen is not displaced under the action of bases such as ammonia, pyridine, and water. Instead, the displacement of other ligands coordinated with ruthenium is observed. It gives the possibility of preparing the Allen-Senoff complex (8) from the $RuCl_2N_2(H_2O)\cdot THF$. Thus $RuCl_2N_2(H_2O)$ $\cdot THF$ converts on reacting with ammonia to a complex which, after treating with KI, becomes identical to the complex $[Ru(NH_3)_5N_2]I_2$ obtained by Allen and Senoff from $RuCl_3$ and hydrazine in the absence of nitrogen.

Harrison and Taube[59] observed the formation of $[Ru(NH_3)_5N_2]X_2$ on direct interaction of nitrogen with the aquapentaammineruthenium complex (13).

$$[Ru(NH_3)_5H_2O]X_2 + N_2 \xrightarrow[H_2O]{20°} [Ru(NH_3)_5N_2]X_2$$
(13)

It was found that when nitrogen at room temperature and atmospheric pressure was passed through the $[Ru(NH_3)_5H_2O]X_2$ solution there appears in the spectrum a peak at 221 mμ characteristic of the Allen-Senoff complex. When the solution formed is treated by Ce^{IV} salt the nitrogen is liberated. Similar liberation of nitrogen is observed also in the oxidation of $[Ru(NH_3)_5N_2]X_2$ by Ce^{IV} salt. Harrison and Taube succeeded in isolating this complex from the solution as the borofluroide salt. It is identical to a borofluoride derivative of the Allen-Senoff complex.

With the investigation continued, Harrison, Weissberger, and Taube[69] have discovered that in this reaction along with Allen-Senoff's complex another complex, $[(NH_3)_5RuN_2Ru(NH_3)_5]X_4$, is formed. In the latter the nitrogen molecule is bound to both ruthenium atoms. This compound, which exemplifies the first binuclear nitrogen complex, is rather stable and is isolated in the individual state as fluoroborate ($X = BF_4^-$). The elemental analyses and the volume of nitrogen evolved as a result of the complex oxidation with Ce^{IV} salt are in rather good agreement with the binuclear structure of the compound.

In contrast to the majority of other nitrogen complexes known, a characteristic of this one is the absence of a sharp and strong band in the range of 2000–2300 cm^{-1} in the infrared. Instead, a broad and weak band is observed at about 2060 cm^{-1}. The latter band can be ascribed either to N—N bond stretching in the complex, or to some other bond vibrations. At any rate, however, the absence of any strong absorption in the region of coordinated N≡N bond stretching point out that the nitrogen molecule in binuclear ion is probably bound symmetrically.

The water solutions of the binuclear complex show characteristic absorption at 262 mμ in the ultraviolet region.

The binuclear complex formation in this reaction seems to be due to initial formation of Allen-Senoff's complex and its further reaction with aquapentammineruthenium complex (13) still present.

$$[(NH_3)_5RuN_2]X_2 + [(NH_3)_5RuH_2O]X_2 \rightleftharpoons [(NH_3)_5RuN_2Ru(NH_3)_5]X_4 + H_2O$$
$$\textbf{(8)} \qquad\qquad\qquad \textbf{(13)}$$

It has indeed been found that on mixing of individual complexes $[(NH_3)_5RuN_2]X_2$ and $[(NH_3)_5RuH_2O]X_2$ in aqueous solution the intensity of a band at 221 mμ, characteristic of the Allen-Senoff ion, decreases slowly. At the same time a band appears at 262 mμ, which is characteristic of the binuclear ion. This reaction turned out to be reversible. When dissolved in water, the binuclear complex dissociates and initial complexes 8 and 13 again appear in solution. The dissociation equilibrium is considerably displaced toward the binuclear ion and the complexes $[(NH_3)_5RuN_2]X_2$ and $[(NH_3)_5RuH_2O]X_2$ are thus present in significant amounts only in highly diluted water solutions. By this reason in the nitrogen reaction with $[(NH_3)_5RuH_2O]X_2$ considerable amounts of the

Allen-Senoff complex can be obtained only with rather dilute (e.g., $10^{-3}\ M$) solutions employed. In more concentrated solutions (e.g., $\sim 0.1\ M$) the main reaction product is the binuclear nitrogen complex.

Allen and Bottomley[70] have investigated the possibility of ruthenium nitrogen complex formation under the conditions of the direct reaction of atmospheric nitrogen with $[Ru(NH_3)_5H_2O]X_2$.

With air passed at room temperature and atmospheric pressure through the aqueous solution of $[Ru(NH_3)_5H_2O]X_2$, which had been obtained by reduction of a $0.1M$ solution of $[Ru(NH_3)_5Cl]Cl_2$ in $0.1M\ H_2SO_4$ with amalgamated Zn, they have observed the appearance of a strong band at 263 mμ and of a weak band at 221 mμ. On treatment of the solution with $NaBF_4$ a product was obtained which turned out to be a binuclear ruthenium complex $[(NH_3)_5RuN_2Ru(NH_3)_5]X_4$ contaminated with small amounts of $[Ru(NH_3)_5N_2]X_2$ ($X = BF_4^-$) according to the infrared and ultraviolet evidence. When heated with concentrated aqueous ammonia, the binuclear nitrogen complex was converted into the Allen-Senoff ion, the latter being isolated as the fluoroborate and identified by means of its infrared and ultraviolet spectra.

It is worth noting that along with nitrogen complex formation the initial Ru^{II} compound was partially oxidized with oxygen to Ru^{III} and thus removed from the nitrogen reaction, when air was passed through the $[Ru(NH_3)_5H_2O]^{2+}$ solution. For this reason authors were forced repeatedly to reduce the Ru^{III} compound back to the divalent state in an argon atmosphere in order to obtain the nitrogen complex in sufficient quantity.

Nevertheless, Allen and Bottomley's findings are of great importance. These results as well as those by Van Tamelen et al.[60] indicate that under some conditions the molecular nitrogen is capable of more or less effective competition with oxygen in the coordination with transition metal ion. This circumstance opens new possibilities to get direct utilization of atmospheric nitrogen.

Interesting nitrogen complexes have been obtained in investigating nitrogen reactions with cobalt compounds. Yamamoto, Kitazume, Pu, and Ikeda[51] have studied the reaction of nitrogen with a mixture of cobalt acetylacetonate and diethylaluminum ethoxide in the presence of triphenylphosphine. They found that the nitrogen complex (14) of composition $[Co(PPh_3)_3N_2]$ is formed in this reaction.*

$$Co(acac)_3 + 3AlEt_2OEt + 3PPh_3 + N_2 \xrightarrow[\text{or toluene}]{\text{ether}} [(PPh_3)_3Co \cdot N_2]$$

$$(14)$$

*Recent investigations[89] have shown that this complex 14 is actually of structure $(PPh_3)_3CoH(N_2)$, being thus identical with complex 15 (See Supplement, page 107).

Accordingly, the hydride and ethylene complexes, formed under the action of H_2 and C_2H_4 on nitrogen complex 14 are of structure $(PPh_3)_3CoH_3$ and $(PPh_3)_3CoH(C_2H_4)$

Complex **14** is a crystalline compound moderately stable in the absence of air and decomposing above 80°C with liberation of N_2. The infrared spectrum of the complex exhibits an intense band at 2088 cm^{-1} (stretching vibration of the coordinate N≡N band).

Under the action of hydrogen and ethylene the nitrogen molecule is displaced from the coordination sphere of the Co atom and the corresponding dihydride and ethylene complexes are formed.[52]

$$[Co(PPh_3)_3 \cdot N_2] + H_2 \rightleftharpoons [(PPh_3)_3CoH_2] + N_2$$

$$[Co(PPh_3)_3 \cdot N_2] + C_2H_4 \rightleftharpoons [(PPh_3)_3Co \cdot C_2H_4] + N_2$$

Nitrogen displacement is also observed under the action of ammonia, carbon monoxide, and carbon dioxide.[52] With H_2, C_2H_4, and NH_3 this process is reversible: in the presence of N_2 the nitrogen complex (**14**) is again formed. But carbon monoxide and carbon dioxide displace nitrogen irreversibly.

The mechanism of formation of the nitrogen complex (**14**) has been studied.[53] It was shown that the reaction of $Co(acac)_3$ with $AlEt_2(OEt)$ and PPh_3 in the absence of N_2 yields an ethylene complex $[(PPh_3)_3CoC_2H_4]$ which is converted to the nitrogen complex (**14**) under the action of N_2. An unstable σ-ethyl derivative of cobalt, $[C_2H_5Co(PPh_3)_3]$, seems to be an intermediate in this reaction. Use of dimethylaluminum ethoxide $Me_2Al(OEt)$ results in the formation of $[CH_3Co(PPh_3)_3]$, which is rather stable and can be isolated. Under the action of N_2 it also yields a nitrogen complex (**14**). Thus in this case the nitrogen complex is formed as a result of replacement of the methyl group by a nitrogen molecule.

$$Co(acac)_3 + AlEt_2OEt + PPh_3 \longrightarrow [(Ph_3P)_3Co—C_2H_5] \longrightarrow [(PPh_3)_3Co \cdot C_2H_4]$$
$$\downarrow N_2$$

$$Co(acac)_3 + AlMe_2OEt + PPh_3 \longrightarrow [(PPh_3)_3Co—CH_3] \xrightarrow{N_2} [(PPh_3)_3Co \cdot N_2]$$

Misono, Uchida, Saito, and Song[54] have obtained the nitrogen complex (**14**) in a some similar way. At first they prepared the corresponding dihydride complex $[H_2Co(PPh_3)_3]$ by reaction of cobalt acetylacetonate with triisobutylaluminum and PPh_3 in the presence of H_2. By further action of N_2 the authors obtained the same nitrogen complex (**14**) as that described by Yamamoto and collaborators

$$Co(acac)_3 + (i\text{-}C_4H_9)_3Al + 3PPh_3 \xrightarrow[10°C]{H_2} [H_2Co(PPh_3)_3] \xrightarrow{N_2} [(PPh_3)_3CoN_2]$$

Sacco and Rossi[55] have described a cobalt triphenylphosphine complex with molecular nitrogen $(PPh_3)CoH(N_2)$ (**15**) of a similar nature having a Co—H bond. It was obtained by reaction of N_2 with the cobalt trihydridephosphine complex $(PPh_3)_3CoH_3$. The reaction occurs at room

$$(PPh_3)_3CoH_3 + N_2 \rightleftharpoons (PPh_3)_3CoH(N_2) + H_2$$
$$\textbf{(15)}$$

temperature and atmospheric pressure and is reversible; a trihydride complex is again formed in the presence of H_2. The nitrogen complex (15) is a crystalline compound moderately stable in air and soluble in nonpolar solvents. An intense absorption band corresponding to the stretching vibration of the $N\equiv N$ bond (2080–2084 cm^{-1}) is observed in the infrared spectrum but the Co—H bond is not seen in the spectrum.

Nevertheless the authors give convincing proofs in favor of the hydride character of the nitrogen complex obtained. Its thermal decomposition should be expected to yield 1.5 moles of gas containing 66.6% of N_2 and 33.3% of H_2. Treatment with $Ph_2PCH_2CH_2PPh_2$ results in formation of $HCo(Ph_2PCH_2CH_2PPh_2)_2$ and one mole of N_2. Moreover, complex 15 is diamagnetic in accordance with its pentacoordinate nature. Similar nitrogen complexes were obtained by Sacco and Rossi using other phosphines, such as $PEtPh_2$ and PEt_2Ph instead of PPh_3. According to Misono, Uchida, and Saito,[56] some of the complex (15) is formed under conditions of the preparation of $[(PPh_3)_3CoN_2]$, that is, in reacting N_2 with a mixture of $Co(acac)_3$, triisobutylaluminum and PPh_3. The authors believe that this is confirmed by the presence of about 10% of hydrogen in the nitrogen formed on thermal decomposition of the nitrogen complex.*

$$Co(acac)_3 + (i\text{-}C_4H_9)_3Al + 3PPh_3 + N_2 \xrightarrow[-50-15°C]{toluene}$$
$$[(PPh_3)_3CoN_2] + (PPh_3)_3CoH(N_2)$$

Enemark, Davis, McGinnety, and Ibers[71] have investigated the reaction products of nitrogen with a mixture of $Co(acac)_3$, PPh_3, and aluminumtriethyl. They have found the nitrogen cobalt complex with three triphenylphosphine ligands to be formed also under these conditions. Careful investigation of this substance has shown that the complex obtained is a mixture of two compounds. These compounds seem to be a mixture of two modifications of the same nitrogen complex, which differ only in the amount of ether of crystallization, which also was involved in the product composition.

X-ray study of one of these modifications with one ether molecule per mole of the complex showed this compound to be monomeric, the nitrogen molecule being bound to the cobalt atom practically linearly (Co—N—N). The Co—N and N—N distances are 1.80 and 1.16 Å, respectively.

The arrangement of three triphenylphosphine and a nitrogen ligand at cobalt atom is best fitted to a trigonal-bipyramidal structure of the complex. In the case of such a structure one of the tops of the bipyramid remains vacant and, in the authors' opinion, can be occupied by a hydride ligand.

*Apparently the authors dealt with the single nitrogen complex $(PPh_3)_3CoH(N_2)$ (See the footnote on page 95).

Thus the nitrogen complex obtained is treated as five-coordinate with the structure of the Sacco-Rossi complex, i.e., $(PPh_3)_3CoH(N_2)$.

Summarizing the data considered and having in mind the complications with the Co—H bond identification a question arises: Are not the authors dealing with the same nitrogen complex of hydride type in all cases?

The interaction of hydrogen with triphenylphosphine-nitrogen cobalt complexes was mentioned above to lead to evolution of molecular nitrogen from the coordination sphere of the transition-metal atom. It has been shown by Parshall[72] that this process can be accompanied with the interesting exchange reaction between H_2 and hydrogen atoms of phenyl rings in Ph_3P groups.

Parshall has studied the reaction of the complex $(PPh_3)_3CoH(N_2)$ with D_2 and found that the aromatic nuclei of the Ph_3P groups become deuterated, with enrichment of the gaseous phase with light hydrogen. The isotope exchange takes place in benzene at 25°, involving, besides the Co—H bond, as many as 18 C—H bonds of the phenyl groups. With PMR spectroscopy it was found that deuterium is localized in the o-positions of the phenyl rings practically exclusively.

In Parshall's opinion, the results observed can be accounted for with the capability of the o-hydrogen atoms to migrate reversibly to the coordinatively unsaturated cobalt atom, the latter arising as a result of N_2 elimination from the starting complex molecule.

Yamamoto, Kitazume, and Ikeda[73] have investigated the possibility of formation of the ruthenium and rhodium molecular nitrogen complexes with triphenylphosphine ligands.

With $RuCl_3$ or $Ru(acac)_3$ being reduced at room temperature by means of Et_3Al solution in THF or benzene in the presence of Ph_3P, they obtained a diamagnetic complex, the structure of which was not established exactly. The analytic and infrared data as well as chemical properties of the compound isolated agree best with the dihydride structure $H_2Ru(PPh_3)_4$ or with σ-phenylmonohydride structure $HRu(PPh_3)_3(PPh_2C_6H_4)$.

With nitrogen bubbling through the benzene solution of this complex, a band at 2143 cm^{-1} appears in the infrared spectrum. This band is shifted up to 2110 cm^{-1} when nitrogen containing $^{29}N_2$ is employed, indicating that the ruthenium complex with coordinated nitrogen is present in solution.

The nitrogen complex formation in this reaction seems to involve a preliminary dissociation step of the initial ruthenium compound, with concomitant elimination of the triphenylphosphine ligand. As a result of the dissociation, one of the coordination sites of the ruthenium atom becomes vacant and then is occupied by the nitrogen molecule. In favor of this mechanism is evidence that introduction of an excess of tripehylphosphine in the reaction mixture prevents the nitrogen complex formation.

The molecular nitrogen–ruthenium complex thus obtained is far less stable than the corresponding cobalt–nitrogen complexes (14, 15) reported earlier. The low stability of the ruthenium complex prevents the isolation of this compound and determination of its structure. In this complex nitrogen is so weakly bonded with the ruthenium atom that it can be removed from the coordination sphere by a simple blowing off the solution with argon at room temperature, that is, by lowering the partial nitrogen pressure. It is of interest that this process is completely reversible: in a nitrogen atmosphere the nitrogen complex is regenerated again and this cycle can be repeated many times. Similar reversible nitrogen displacement from the complex is observed also under the action of H_2 and NH_3.

Synthesis of molecular nitrogen rhodium complex with triphenylphosphine ligands has also been attempted.[73] For that purpose the reaction of $RhCl_3$ as well as of $Rh(acac)_3$ with Et_3Al in the presence of Ph_3P was investigated. This reaction led to a diamagnetic product, to which the structure $HRh(PPh_3)_4$ was ascribed on the grounds of analytic and infrared data as well as of its chemical properties. This compound turned out, however, to be unreactive with N_2 at room temperature and atmospheric pressure, in contrast to the above-mentioned hydridephosphine ruthenium complex.

An unusual nitrogen complex has been described by Johnson and Beveridge.[57] They found that a solution of N,N-disalicylaldehyde-1,3-propanediiminomanganese (16) in benzene reversibly absorbs molecular nitrogen at room temperature to form an adduct in a 1:1 ratio. Similar adducts are also formed with O_2 and CO. A characteristic feature of the nitrogen complex is the absence of an absorption band in the range of 2000–2200 cm^{-1} corresponding to the stretching vibration of the coordinate nitrogen triple bond. Instead, the infrared spectrum exhibits absorption bands at 1295, 820, 620, and 502 cm^{-1} besides the bands characteristic of the initial compound (16). Similar anomalies in spectra are observed for the adduct with CO. Johnson and Beveridge consider that the nitrogen adduct, like the adducts with CO and O_2, are similar to a symmetrical oxygen complex obtained by Vaska.[58,59] It is worth noting, however, that there was not works, confirming the formation of such unusual nitrogen complex.

Thus, molecular nitrogen complexes ("nitrogenyls"[43]) with Mn, Co, Ru, Rh, Ir, and Os compounds have been described up to now. Apparently the number of such complexes will continue to grow rapidly. It should be noted that all the nitrogen complexes obtained are transition-metal derivatives mainly of the right side of transition periods. These complexes are highly stable and the nitrogen they contain seems to be insufficiently reactive.

On the other hand, most active in nitrogen reduction to ammonia are the transition-metal compounds belonging to the left part of transition period (Ti, V, Cr, Mo, W, Fe). However, stable nitrogen complexes have not yet been obtained with these compounds. Apparently in these cases the nitrogen complexes are less stable and more reactive.

III. CONCLUSION

Thus, investigations of recent years have shown that the nitrogen molecule considered before as highly inert is capable of entering under mild conditions in various reactions in the presence of transition-metal compounds. Nitrogen may be reduced by various reducing agents with formation of compounds of the nitride type yielding ammonia after hydrolysis. It may also react with hydrogen inserting into the metal hydride bond and finally to insert into the transition metal–carbon bond forming amines. Apparently the nitrogen complexes with transition metal compounds are the intermediate compounds in all these reactions. In some cases these complexes were obtained and isolated.

Comparison of these results with data on biological nitrogen fixation and ammonia synthesis over metal catalysts permits one to establish a number of general features inherent to all these processes.[13]

In all cases the nitrogen-fixing system involves a transition metal compound (or the metal as such) and a reducing agent. The function of the transition metal seems to be formation of a complex with N_2 which leads to nitrogen activation for the subsequent reduction. Thus, nitrogen complexing is a fairly universal way of activation of this inert molecule.

It may be supposed that further success in this field will permit the creation of methods of practical importance of nitrogen fixation under mild conditions.

IV. EXPERIMENTAL PROCEDURES

1. Nitrogen Reaction with the System Cp_2TiCl_2 + C_2H_5MgBr in Ether at Atmospheric Pressure

An ether solution (45 ml) of ethylmagnesium bromide (containing 9.45×10^{-2} mole of C_2H_5MgBr) is placed under nitrogen in a 100-ml flask equipped with a stirrer providing for gas circulation, dropping funnel, and an efficient reflux condenser connected with an effluent bubbler containing 30 ml of 20% H_2SO_4. The flask is cooled to $-75°C$ and 2.6 g (1.05×10^{-2} mole) of Cp_2TiCl_2 are added. A flow of oxygen-free nitrogen is bubbled through the mixture with efficient stirring and the vessel is allowed to warm to room temperature. At this temperature nitrogen is bubbled through the mixture for 7–8 hr. Ether is added periodically to keep the volume constant.

The nitrogen flow is stopped and the reaction mixture is carefully decomposed by a small amount of methanol and then by 20% H_2SO_4. The contents of the flask and the bubbler are washed with distilled water and then evaporated to a volume of ~ 60 ml. The remainder is made strongly alkaline with 40% KOH, and ammonia together with water is distilled into a flask containing $0.1N$ HCl. Excess acid is titrated with $0.1N$ NaOH. The ammonia yield is ~ 0.7 mole per mole of Cp_2TiCl_2.

2. Nitrogen Reaction with the System Cp_2TiCl_2 + C_2H_5MgBr in Ether under Pressure

An ether solution (30 ml) of ethylmagnesium bromide (containing 6.3×10^{-2} mole of C_2H_5MgBr) and a sealed, thin-walled, glass ampoule containing 1.74 g (7.02×10^{-3} mole) of Cp_2TiCl_2 are introduced under Ar or N_2 into a 50-ml, stainless steel autoclave. The autoclave is filled with nitrogen to a pressure of 150 atm. Under such conditions the ampoule bursts. The autoclave is shaken at room temperature for 10–11 hr. Nitrogen is then vented, passing through the washer with 20% H_2SO_4, and the mixture is carefully decomposed first by methanol and then by 20% H_2SO_4. The gases liberated are passed through the same sulfuric acid washer. The autoclave and the washer contents are then washed with distilled water into the flask and evaporated to a volume of ~ 60 ml. The ammonia is determined using the method described above. The ammonia yield is 0.9–1.0 mole per mole of Cp_2TiCl_2.

3. Nitrogen Reaction with the System: Transition Metal Compound + Mg + MgI_2 under Pressure

In a 100-ml, three-necked flask, equipped with a stirrer and a reflux condenser (with calcium chloride drying tube), is placed 36 ml of absolute benzene, 24 ml of absolute ether, and 3.6 g of Mg. Iodine (10 g) is added in portions with stirring. After addition of the last portion of I_2 the mixture is stirred until it becomes decolorized. The MgI_2 solution obtained is decanted from the excess of Mg. The concentration of MgI_2 solution is ~ 0.66 mole/liter. Magnesium shaving (1.36 g), 30 ml of the MgI_2 solution obtained, and 4×10^{-3} mole of anhydrous salt of the transition metal complex sealed in a thin-walled glass ampoule are introduced into a 50-ml, stainless steel autoclave under Ar or N_2. The autoclave is filled with nitrogen to a pressure of 80–100 atm and is shaken for 7 hr at room temperature. Further treatment of the reaction mixture and quantitative determination of NH_3 is made using the method described above.

The reaction using a Li + LiI mixture as a reducer is carried out similarly.

4. Nitrogen Reaction with Hydrogen in the Presence of a System $Ti(OC_2H_5)_4$ + Al(i-Bu)$_3$ in Toluene

A solution of 3.56 g (1.8 $\times$ 10^{-2} mole) (i-C_4H_9)$_3$Al in 30 ml of absolute toluene and a sealed, thin-walled, glass ampoule containing ~ 0.68 g (3×10^{-3} mole) of $Ti(OC_2H_5)_4$ are introduced into a 50-ml, stainless steel autoclave under Ar or N_2. The autoclave is filled by a mixture of nitrogen and hydrogen. The N_2 pressure is 30 atm; the H_2 pressure is 10 atm. The autoclave is shaken at room temperature for 9–10 hr. Treatment of the reaction mixture and quantitative determination of the NH_3 obtained is carried out using the method described above. The yield of ammonia is ~ 0.55 mole per mole of $Ti(OC_2H_5)_4$.

5. Nitrogen Reaction with the System Cp_2TiCl_2 + C_6H_5Li in Ether

Preparation of Aniline. A solution of 30 ml of phenyllithium in ether (3.6 $\times$ 10^{-2} mole of PhLi) and 1.74 g (7.02 $\times$ 10^{-3} mole) of Cp_2TiCl_2 in a sealed, thin-walled, glass ampoule are introduced into a 50-ml, stainless steel autoclave under argon or nitrogen. The autoclave is filled with nitrogen up to 100 atm pressure and then shaken at room temperature for 10–11 hr. Nitrogen is vented from the autoclave and the reaction mixture is decomposed with methanol and then with 20% H_2SO_4. The acid solution is washed with distilled water into a flask and is evaporated to ~ 60 ml. The remainder is basified and aniline and ammonia are steam-distilled into a flask containing 0.1N HCl. Excess acid is titrated and the combined NH_3 and $PhNH_2$ are thus determined. Acid is then added to the solution and $PhNH_2$ is determined separately by titrating with 0.1N NaNO$_2$.

The NH_3 yield is ~ 0.65 moles per mole of Cp_2TiCl_2. The $PhNH_2$ yield is ~ 0.15 moles per mole of Cp_2TiCl_2.

To isolate aniline and o-aminodiphenyl (also formed in small amounts), the reaction mixture is hydrolyzed by water after the experiment. Aniline and o-aminodiphenyl are extracted with ether and then extracted from the ether by 5–10% H_2SO_4. The acid solution is basified and the amines again taken up in ether. The ether solution is dried over KOH and then evaporated. The residue, which is a mixture of amines, is separated by means of chromatography on alumina. The products are eluted with benzene collecting o-aminodiphenyl and then aniline. The benzene solutions of aniline and o-aminodiphenyl obtained by elution are evaporated and the amines are converted to respective benzoyl derivatives using the Schotten-Baumann method.

Benzanilide (mp 165°C from alcohol) and o-benzoylaminodiphenyl (mp 87°C from aqueous alcohol) are obtained as a result of benzoylation.

6. The Synthesis of Nitrogenpentaammineruthenium(II) Complexes [(NH₃)₅RuN₂]X₂[40]

1. Ruthenium trichloride (1.0 g) is dissolved in water (12 ml), and hydrazine hydrate (10 ml) is added carefully over a period of 5 min. The initial vigorous, exothermic reaction is allowed to subside; the mixture is stirred overnight and filtered several times by gravity, leaving a black residue. The appropriate anion (saturated aqueous solution) is added to the filtrate to precipitate the desired salt. The iodide, bromide, tetrafluoroborate, and hexafluorophosphate salts precipitate immediately; the chloride salt is obtained by allowing the solution to evaporate at room temperature. The product (0.5 g) is collected by filtration, washed with water, alcohol, and ether, and air dried.

Similarly, when hydrazine hydrate (10 ml) is added to ammonium hexachlororuthenium(IV) (1.0 g), potassiumpentachloroaquoruthenium(III) (1.0 g), or aquopentaammineruthenium(III) methanesulfonate (1.0 g) dissolved in the minimum quantity of water, approximately 40% yields of nitrogenpentaammineruthenium(II) salts are obtained. Reaction times are 1 hr for $[Ru(NH_3)_5H_2O](CH_3SO_3)_3$ and 12 hr for $K_2[RuCl_5H_2O]$ and $(NH_4)_2[RuCl_6]$.

2. Aquopentaammineruthenium(III) methanesulfonate (1.0 g) is dissolved in water (15 ml), and sodium azide (1.0 g) is added. The solution is adjusted to approximately pH 7 with methanesulfonic acid and gently warmed for 20 min. The mixture is filtered and cooled, and the product precipitated by addition of the appropriate anion (saturated aqueous solution), yield 0.80 g (83%).

Caution: under certain circumstances it is possible to obtain explosive products from this reaction.[40]

V. SUPPLEMENT*

Since the review was written a number of new papers have appeared in literature dealing with the nitrogen fixation.

Henrici-Olivé and Olivé[74] investigated (by ESR method) the reduction of Cp_2TiCl_2 with excess of lithiumnaphthalene in THF. They found that this system reacts with nitrogen at room temperature and atmospheric pressure to yield ammonia upon hydrolysis as much as ~ 0.96 moles per mole of Cp_2TiCl_2. With the reaction carried out in Ar and at molar ratio $NpLi : Cp_2TiCl_2 = 3\text{--}4$, a paramagnetic titanium derivative was found to form in solution, indicating itself by characteristic ESR signal. SFS analysis of this signal led the authors to the conclusion that the paramagnetic entity most likely has a structure of the anionic hydride titanium complex

$$\left[Cp_2Ti \underset{H}{\overset{H}{\diamond}} TiCp_2 \right]^{x-} x\,Li^+,$$

where x probably equals unity. It is interesting that the ESR signal disappears under the action of N_2. With the reduction carried out in a nitrogen atmosphere from the beginning, this signal does not appear at all. The paramagnetic hydride complex is thus sensitive to nitrogen, but its actual role in the nitrogen fixation is not clear at the present. Similar results have been obtained with the THF solution of NpNa as reducing agent.

Van Tamelen and Akermark[75] reported the electrolytic reduction of nitrogen under mild conditions. The electrolysis was carried out in solution of $Ti(OPr\text{-}i)_4$ and $AlCl_3$ (molar ratio 1:1.5) in 1,2-dimethoxyethane. As a result of prolonged (2 days) reduction of nitrogen ~ 0.1 moles of ammonia per mole of the starting titanium compound have been obtained (after hydrolysis).

The investigation of stable transition metal–nitrogen complexes has been intensively developed. Bottomley and Nyborg[76] reported the detailed X-ray investigation of the complex $[Ru(NH_3)_5N_2]Cl_2$. The authors confirm the conclusion they made previously[42] that fragment Ru—N—N in the complex is of linear structure. These data are in good agreement with the X-ray measurements on $[Ru(NH_3)_5N_2]\,(BF_4)_2$ carried out by Hodgson and Ibers[76].

To elucidate the structure of the binuclear ruthenium complex of nitrogen[69] Chatt, Nikolsky, Richards, and Sanders[77] investigated the Raman spectra of solid $[(NH_3)_5RuN_2Ru(NH_3)_5](BF_4)_4$. The strong absorption at $2100 \pm 2\,cm^{-1}$ has been found in the spectra of the N_2 metal complex.

*Up to April 1969.

This band is shifted to 2030 ± 2 cm^{-1} in the binuclear complex containing $^{15}N_2$. Relatively high frequency of the Raman band, assigned to the coordinated $N\equiv N$ bond stretching, along with some other evidence lead to the conclusion that the binuclear nitrogen complex is of linear structure $[(NH_3)_5Ru^{II}-N\equiv N-Ru^{II}(NH_3)_5]^{4+}$. The $N\equiv N$ bond stretching in the Raman spectrum of mononuclear ruthenium complex of nitrogen $[Ru(NH_3)_5N_2](BF_4)_2$ appears at 2133 ± 1 cm^{-1}.

Itzkovitch and Page[78] have studied the kinetics of the nitrogen reaction with aqueous solutions of pentaammineruthenium(II) complex $[Ru^{II}(NH_3)_5X]$, obtained by means of electrolytic reduction of $[Ru^{III}(NH_3)_5Cl]Cl_2$. As it was mentioned above (see page 94) this reaction leads to mono- and binuclear ruthenium(II) complexes of nitrogen:

$$[Ru^{II}(NH_3)_5X] + N_2 \xrightarrow{k_1} [Ru^{II}(NH_3)_5N_2]^{2+}$$

$$[Ru^{II}(NH_3)_5X] + [Ru^{II}(NH_3)_5N_2]^{2+} \xrightarrow{k_2} [(NH_3)_5Ru^{II}N_2Ru^{II}(NH_3)_5]^{4+}$$

It has been found that the rate constants for the reactions of $[Ru^{II}(NH_3)_5X]$ with $N_2(k_1)$ and of $[Ru^{II}(NH_3)_5X]$ with $[Ru^{II}(NH_3)_5N_2]^{2+}(k_2)$ are close in value. It means the reactivity of coordinated nitrogen in the $[Ru(NH_3)_5N_2]^{2+}$ ion is practically the same as that of free nitrogen. Although this conclusion is quite correct only for the reaction of $[Ru^{II}(NH_3)_5N_2]^{2+}$ with $[Ru^{II}(NH_3)_5X]$, there are additional indications of little reactivity of coordinated N_2 in $[Ru^{II}(NH_3)_5N_2]^{2+}$ ion and also in other stable nitrogen complexes described recently.

Borod'ko, Shilova, and Shilov[79] as well as Chatt, Richards, Fergusson, and Love[80] reinvestigated the data of Allen et al.[39,40] that the nitrogen ligand in the complex $[Ru(NH_3)_5N_2]X_2$ could be reduced by $NaBH_4$ to give ammonia (cf. pp. 89). With $[Ru(NH_3)_5{}^{15}N_2]X_2$ employed in the reaction, both groups of investigators have shown that the ammonia formed contains practically no ^{15}N. Thus, coordinated nitrogen in the complex $[Ru(NH_3)_5N_2]X_2$ is not reduced by means of $NaBH_4$. According to Chatt et al.[80] the discrepancy with Allen's results[39,40] can be due to the fact that samples of $[Ru(NH_3)_5N_2]X_2$ prepared by Allen's procedure[39,40] (cf above pp. 88) contain up to 50% of hydrazine complex such as $[Ru(NH_3)_5N_2H_4]X_2$. This latter may be a source of additional ammonia in reaction of the complex with $NaBH_4$.

Chatt et al.[77,80] have studied the possibility of nitrogen ligand reduction in some other complexes such as $[Os(NH_3)_5N_2]^{2+}$, $(Ph_3P)_3CoH(N_2)$, $(Ph_3P)_2IrCl(N_2)$, and $[(NH_3)_5RuN_2Ru(NH_3)_5]^{4+}$. However, in spite of using various reducing agents, ammonia formation from nitrogen could never be observed. Polarographic reduction of nitrogen in complexes

$[Os(NH_3)_5N_2]^{2+}$,[81] $[Ru(NH_3)_5N_2]^{2+}$,[78] and $[(NH_3)_5RuN_2Ru(NH_3)_5]^{4+}$ [78] in aqueous solutions also failed.

Shilova, Shilov, and Vostroknutova[82] investigated the kinetics of thermal decomposition of $[RuN_2Cl_2(H_2O)_2(THF)]$ in solution and found a unimolecular mechanism of nitrogen evolution:

$$Ru^{II}L_5N_2 \rightleftharpoons Ru^{II}L_5 + N_2$$

$$Ru^{II}L_5 + L' \longrightarrow Ru^{II}L_5L'$$

where $L = Cl$, H_2O, THF; $L' =$ probably H_2O. When the decomposition was carried out under nitrogen pressure, the rate of reaction decreased and in the presence of $^{15}N_2$ an exchange of coordinated nitrogen was observed to give the corresponding "labeled" complex $RuL_5{}^{15}N_2$.

New data on iridium complexes of nitrogen[83-87] have been published.

Collman, Kubota, Vastine, Sun, and Kang[83] have investigated the kinetics of formation of $(Ph_3P)_2Ir(N_2)Cl$ from Vaska's complex and organic azides. They have found that this reaction proceeds via a bimolecular mechanism. The interaction of organic azides with the analogs of Vaska's complex such as $(R_3P)_2Ir(CO)X$, where $X = Br$, I, N_3, and $R_3P = PhEt_2P$, Ph_2CH_3P is of similar character. However, nitrogen complexes formed are too unstable to be isolated in this case.

The reactions of $(Ph_3P)_2Ir(N_2)Cl$ with Ph_3P, CO, and $RC{\equiv}CR$, proceeding with nitrogen liberation, were found to yield the corresponding phosphine, carbonyl, and acetylene complexes of iridium.[83-87]

$$(Ph_3P)_2Ir(N_2)Cl \begin{cases} \xrightarrow{Ph_3P} (Ph_3P)_3IrCl + N_2 \\ \xrightarrow{CO} (Ph_3P)_2Ir(CO)Cl + N_2 \\ \xrightarrow{RC{\equiv}CR} (Ph_3P)_2Ir(R_2C_2)Cl + N_2 \end{cases}$$

It is of interest that the coordinatively unsaturated $(PPh_3)_2Ir(N_2)Cl$ does not react with H_2 at pressures up to 4 atm, unlike its carbonyl analog $(Ph_3P)_2Ir(CO)Cl$ which is known to be capable of facile and reversible addition of hydrogen. It is explained by the assumptions that N_2 is a stronger π-acid and a weaker σ-donor as compared with CO. Both infrared[83] and electrochemical data[78] on nitrogen complexes also are in agreement with assumptions[13,40] (cf. p. 87) about the importance of π-back bonding for the nitrogen coordination with transition metal atom.

Kinetic investigations[83,84] of formation of complexes $(R_3P)_2Rh(N_2)X$ from $(R_3P)_2Rh(CO)X$ ($R_3P = Ph_3P$, $PhEt_2P$; $X = Cl$, Br, I) and organic azides have also shown the bimolecular mechanism of these reactions. The rhodium complexes of nitrogen $(R_3P)_2Rh(N_2)X$ are considerably less stable than those of iridium $(R_3P)_2Ir(N_2)X$.

Sacco and Rossi[88] have reported results of a detailed study concerning the synthesis and properties of nitrogen complexes of cobalt $(PR_3)_3CoH(N_2)$, where $R_3P = PPh_3$, $PEtPh_2$, and PEt_2Ph. The data obtained give additional confirmation to the hydride character of the compounds.

The reaction of $(PPh_3)_3CoH(N_2)$ with I_2 yields $CoI(PPh_3)_3$, stoichiometric amounts of N_2 and H_2 being formed:

$$(PPh_3)_3CoH(N_2) + 0.5I_2 \longrightarrow N_2 + O, 5H_2 + (PPh_3)_3CoI$$

Approximately stoichiometric amounts of N_2 and H_2 are also obtained in the reactions of the complex with NO, O_2 and HCl:

$$(PPh_3)_3CoH(N_2) + NO \longrightarrow N_2 + 0.5H_2 + (PPh_3)_3Co(NO)$$

$$(PPh_3)_3CoH(N_2) + HCl \xrightarrow{10°} N_2 + H_2 + (PPh_3)_3CoCl$$

The reaction with CCl_4 leads to N_2, $CHCl_3$, and $(PPh_3)_3CoCl$:

$$(PPh_3)_3CoH(N_2) + CCl_4 \longrightarrow N_2 + CHCl_3 + (PPh_3)_3CoCl$$

Thermal decomposition of the complex is more complicated. In this case some amount of benzene together with N_2 and H_2 are formed on account of competitive cleavage of Ph—P bond in the phosphine ligand.

$$(PPh_3)_3CoH(N_2) \overbrace{\begin{array}{l} \xrightarrow{110-150°} N_2 + 0.5\,H_2 + [(PPh_3)_3Co] \\ \xrightarrow{110-150°} N_2 + PhH + (PPh_3)_2CoPPh_2 \end{array}}$$

A remarkable fact was observed by Sacco and Rossi in the reaction of the nitrogen cobalt complex with $Ph_2PCH_2CH_2PPh_2$:

$$(PPh_3)_3CoH(N_2) + 2Ph_2PCH_2CH_2PPh_2 \longrightarrow$$
$$N_2 + CoH(Ph_2PCH_2CH_2PPh_2)_2 + 3Ph_3P$$

It was found that a mixture of $CoD(Ph_2PCH_2CH_2PPh_2)_2$ and $CoH(Ph_2PCH_2CH_2PPh_2)_2$ is formed on using of $(Ph_3P)_3CoD(N_2)$ in this reaction. The same mixture is obtained when $[P(C_6D_5)_3]_3CoH(N_2)$ reacts with $Ph_2PCH_2CH_2PPh_2$. These results indicate that the exchange occurs between hydrogen bound with the metal and that of phenyl rings in triphenylphosphine ligands.

Misono, Uchida, Hidai, and Araki[89] have shown by NMR methods that the cobalt nitrogen complex (14), arising in reactions of N_2 with $Co(acac)_3$, R_3Al, and Ph_3P and previously formulated as $[(Ph_3P)_3CoN_2]$, in fact contains Co—H bond being thus identical with the Sacco–Rossi complex, $(Ph_3P)_3CoH(N_2)$, and is also verified by the chemical properties of this compound. Under the action of CO the complex liberates N_2 to give a mixture of $(Ph_3P)_3CoH(CO)$, $(Ph_3P)_2CoH(CO)_2$, and $[Co(CO)_3PPh_3]_2$.[90]

$$(Ph_3P)_3CoH(N_2) + CO \xrightarrow{20°}$$
$$(Ph_3P)_3CoH(CO) + (Ph_3P)_2CoH(CO)_2 + [Co(CO)_3PPh_3]_2 + N_2$$

The presence of the Co—H bond in the first two reaction products was shown by infrared.

Also of hydride nature are the analogs of $(Ph_3P)_3CoH(N_2)$ such as $[(p\text{-}CH_3C_6H_4)_3P]_3CoH(N_2)$, $(Ph_2EtP)_3CoH(N_2)$, $(n\text{-}Bu_3P)_3CoH(N_2)$ and $(Et_3P)_3CoH(N_2)$ which are formed in the nitrogen reactions with a mixture of $Co(acac)_3$, $i\text{-}Bu_3Al$ and the corresponding phosphine.[106] Like $(Ph_3P)_3CoH(N_2)$ these compounds exhibit the characteristic NMR signal of hydride hydrogen at $\tau = 29\text{-}31$ (a quartet with relative intensities $1:3:3:1$, $I_{P-H} = 50$ cps). In the infrared spectra of the complexes, however, the Co—H bond was absent. There were absorption bands due to only $N\equiv N$ bond stretching at $2000\text{-}2100$ cm^{-1}.

Lorberth, Nöth, and Rinze[107] obtained a number of nitrogen cobalt complexes with various phosphine ligands using $n\text{-}Bu_2AlH$ as reducing agent.

$$Co(acac)_{2 \text{ or } 3} + n\text{-}Bu_2AlH + R_3P \xrightarrow{\text{Ar or } H_2} (R_3P)_3CoH_3$$

$$(R_3P)_3CoH_3 + N_2 \longrightarrow (R_3P)_3CoH(N_2) + H_2$$

where $R_3P = Ph_3P$, $(p\text{-}CH_3C_6H_4)_3P$, $(p\text{-}ClC_6H_4)_3P$, $(p\text{-}FC_6H_4)_3P$.

The absence of the Co—H bond absorption in the infrared spectra of the nitrogen complexes $(R_3P)_3CoH(N_2)$ and the presence of these bands in the spectra of $(R_3P)_3CoH_3$ may be accounted for by an enhanced hydride character of the hydrogen atom in $(R_3P)_3CoH(N_2)$ relative to $(R_3P)_3CoH_3$. The latter is supported by a comparison of NMR spectra of these compounds.

Lee and Cubberly[108] isolated the complex $(Ph_3P)_3CoH(N_2)$ as well as its tris-octyl-and tris-butylphosphine analogs in the reaction of nitrogen with a mixture of cobalt chloride, $NaBH_4$, and suitable phosphine in diglyme.

$$CoCl_2 + NaBH_4 + PR_3 + N_2 \longrightarrow (R_3P)_3CoH(N_2)$$

The authors report that the nitrogen complexes of iron and nickel can be also obtained similarly, but structures of these compounds were not proven. In agreement with the data of Sacco and Rossi[88] the reaction of $(Ph_3P)_3CoH(N_2)$ with CCl_4 was found[108] to proceed with nitrogen evolution and formation of $CHCl_3$. Other organopolyhalogenides such as $BrCCl_3$, $PhCCl_3$, $PhCHBr_2$, 2,5-dichlorobenzotrichloride and CCl_3CCl_3 reacted similarly yielding $CHCl_3$, $PhCHCl_2$, $PhCH_2Br$, 2,5-dichlorobenzylidene chloride and Cl_3CCHCl_2, respectively. Nitrogen complexes of iron and nickel behave similarly in these reactions.

The interaction of $(Ph_3P)_3CoH(N_2)$ with carbon dioxide,[90,91] formic acid,[91] aldehydes,[90] and nitriles[109] has been investigated.

Pu, Yamamoto, and Ikeda[91] have found that, in the reaction of $(Ph_3P)_3CoH(N_2)$ with CO_2 in the presence of Ph_3P, nitrogen was liberated to give the formate complex of cobalt:

$$(Ph_3P)_3CoH(N_2) + CO_2 \xrightarrow[Ph_3P]{20°} HCOOCo(Ph_3P)_3 + N_2$$

A similar complex was formed in the reaction of $(Ph_3P)_3CoH(N_2)$ with formic acid:

$$(Ph_3P)_3CoH(N_2) + HCOOH \longrightarrow HCOOCo(PPh_3)_3 + N_2 + H_2$$

Without addition of Ph_3P the reaction of $(Ph_3P)_3CoH(N_2)$ with CO_2 is more complicated. In this case along with formate complex an indeterminate amount of $[Co(CO)(PPh_3)_3]_x$ (x is probably 2) has been isolated. The same compound was obtained in the reaction of $(Ph_3P)_3CoH(N_2)$ with $PhCH_2N{=}C{=}O$.

The formation of the cobalt formate complex in the reaction $(Ph_3P)_3CoH(N_2)$ with CO_2 was also observed independently by Misono, Uchida, Hidai, and Kuse.[90] The authors studied this reaction in the absence of Ph_3P and also isolated a small amount of $[Co(CO)(PPh_3)_3]_x$.

The nitrogen complex of cobalt $(Ph_3P)_3CoH(N_2)$ reacts with aldehydes leading to their decarbonylation.[90] Thus in the reaction of $(Ph_3P)_3CoH(N_2)$ with formaldehyde $(Ph_3P)_3CoH(CO)$ has been isolated.

It has been found[109] that the reaction of $(Ph_3P)_3CoH(N_2)$ with organonitriles (CH_3CN, C_2H_5CN, $PhCN$, $CH_2{=}CHCN$) leads to nitrogen evolution and formation of the corresponding nitrile cobalt complexes. With CH_3CN and C_2H_5CN the reaction is reversible: $(Ph_3P)_3CoH(N_2)$ is regenerated under the action of N_2 on the nitrile complex.

$$(Ph_3P)_3CoH(N_2) + RCN \rightleftharpoons (Ph_3P)_3CoH(RCN) + N_2$$

where $R = CH_3$, C_2H_5. However, in the case of $PhCN$ and $CH_2{=}CHCN$ nitrogen displacement from the complex is irreversible.

A series of papers dealing with the application of nitrogen complexes as homogeneous catalysts of organic reactions has been published. Pu, Yamamoto, and Ikeda[92] found that the nitrogen cobalt complex $(Ph_3P)_3CoH(N_2)$ catalyzed the hydrogenation, dimerization, and isomerization of olefins, oxidation of Ph_3P to Ph_3PO, and reduction of N_2O.

According to Collman et al.[83-85] the iridium nitrogen complex $(PPh_3)_2Ir(N_2)Cl$ may be used as catalyst in the oxidation of some olefins with oxygen as well as isocyanate formation from organic azides and CO.

A number of new nitrogen complexes has been obtained. Campbell, Dias, Green, Saito, and Swanwick[93] have shown by infrared that a nitrogen complex of iron is formed in the reaction of N_2 with $Fe(acac)_3$, $i\text{-}Bu_3Al$, and Et_3P. However, the structure of this complex is unknown. A similar nitrogen complex seems to be formed in the reactions of N_2 with a mixture of tris-allyliron or bis-cyclooctatetraene complex of iron (O) and Et_3P. An attempt to obtain the rhodium nitrogen complex from tris-allylrhodium, Et_3P, and N_2 failed.

Sacco and Aresta[94] observed the formation of nitrogen complexes of iron in the reaction of N_2 with $(PR_3)_3FeH_2$:

$$FeCl_2 \cdot 2H_2O + NaBH_4 + PR_3 \xrightarrow[\text{EtOH}]{\text{Ar or } H_2} (PR_3)_3FeH_2$$

$$(PR_3)_3FeH_2 + N_2 \xrightarrow[\text{1 atm}]{20°C} (PR_3)_3FeH_2(N_2)$$

where PR_3 = $PEtPh_2$, $PBuPh_2$. The complex $(PEtPh_2)_3FeH_2(N_2)$ is diamagnetic. It is decomposed with formation of N_2 and H_2 on heating in vacuum (80°) as well as under the action of I_2 and HCl. Carbon monoxide also displaces N_2, but Fe—H bonds remain in this case:

$$(PEtPh_2)_3FeH_2(N_2) + CO \xrightarrow[\text{1 atm}]{20°C} N_2 + (PEtPh_2)_3FeH_2(CO)$$

In sunlight the solid complex undergoes reversible decomposition with liberation of H_2:

$$(PEtPh_2)_3FeH_2(N_2) \rightleftharpoons H_2 + FeH(C_6H_4PEtPh)N_2(PEtPh_2)_2$$

In the reaction with $AlEt_3$ no reduction of the coordinated N_2 takes place, that is, the nitrogen ligand in the iron complex is also of small reactivity. It is interesting that under the action of $AlEt_3$ the phosphine rather than the nitrogen ligand was displaced. As a result the new five-coordinate nitrogen complex of iron $(PEtPh_2)_2FeH_2(N_2)$ was formed.

$$(PEtPh_2)_3FeH_2(N_2) + AlEt_3 \longrightarrow (PEtPh_2)_2FeH_2(N_2) + Et_3Al \cdot PEtPh_2$$

Knotes[95] has obtained the ruthenium complex of nitrogen $(Ph_3P)_3RuH_2(N_2)$ similar to the iron complexes $(R_3P)_3FeH_2(N_2)$, described by Sacco and Aresta.[94] The ruthenium complex is decomposed under

$$(Ph_3P)_3RuHCl + Et_3Al + N_2 \xrightarrow{Et_2O} (Ph_3P)_3RuH_2(N_2)$$

the action of HCl to give N_2 and H_2. The nitrogen ligand in the complex was displaced by Ph_3P, NH_3, and H_2. In the last two cases $(Ph_3P)_3Ru(H_2)NH_3$ and $(Ph_3P)_3RuH_4$ were formed, respectively, the reactions being reversible. Like $(Ph_3P)_3CoH(N_2)$, ruthenium complex $(Ph_3P)_3RuH_2(N_2)$ exchanges the hydrogen orthoatoms of the phenyl rings in the Ph_3P ligands by deuterium under the action of D_2. The ruthenium nitrogen complex which was obtained previously by Yamamoto et al.[73] (cf p. 98) by the reaction of N_2 with benzene solution of $(Ph_3P)_4RuH_2$, may also be formulated as $(Ph_3P)_3RuH_2(N_2)$.

Chatt, Leigh, and Mingos[96] have found that the reaction product of $OsCl_3(n\text{-}Bu_2PPh)_3$ with hydrazine hydrate is actually a mixture of 90% $OsCl_3(n\text{-}Bu_2PPh)_3$ and 10% nitrogen osmium complex rather than the single complex $OsHCl_2(n\text{-}Bu_2PPh)_3$ as it was previously reported.[97,98] It is supposed that the structure of the nitrogen osmium complex is $OsCl_2(N_2)(n\text{-}Bu_2PPh)_3$.

Jolly and Jonas[99] have obtained the first nitrogen complex of nickel:

$$2Ni(acac)_2 + 4R_3P + 4Al(CH_3)_3 + N_2 \longrightarrow$$
$$N_2[Ni(PR_3)_2]_2 + 4(CH_3)_2Al(acac) + 4[CH_3]$$

where $R = C_6H_{11}$. The nitrogen ligand in this binuclear complex is displaced under the action of olefins, phosphines, and CO. In benzene solution equilibrium exists between mono- and binuclear nitrogen complexes of nickel:

$$N_2[Ni(PR_3)_2]_2 \rightleftharpoons N_2Ni(PR_3)_2 + Ni(PR_3)_2$$

The structure assumed for the binuclear nitrogen complex is $(PR_3)_2Ni\text{---}N\equiv N\text{---}Ni(PR_3)_2$.

Das, Pratt, Smith, Swinden, and Woolcock[81] have investigated the possibility of obtaining binuclear nitrogen complexes in which the N_2 molecule was bound to two different metal atoms. On mixing of $[Os(NH_3)_5N_2]X_2$ and $AgClO_4$ in aqueous solution at low temperature they observed, in the infrared, a pronounced shift of the $N\equiv N$ bond stretching frequency into short wave region (~ 185 cm^{-1}). The shift could be reversed by treatment of the silver adduct with aqueous NaCl. It is supposed that the adduct has the structure of asymmetric binuclear complex $[(NH_3)_5OsN_2Ag]^{3+}$. Similar results have been observed on mixing of $[Os(NH_3)_5N_2]X_2$ with a solution of $CuCl_2$.

First representatives of the so-called bis-nitrogen complexes have been reported.

Scheidegger, Armor, and Taube[100] have obtained the bis-nitrogen complex of osmium $[Os(NH_3)_4(N_2)_2]Cl_2$ by diazotization of coordinated ammonia in $[Os(NH_3)_5N_2]Cl_2$.

$$[Os(NH_3)_5N_2]Cl_2 + HNO_2 \xrightarrow{20°} [Os(NH_3)_4(N_2)_2]Cl_2$$

In the infrared of the compound obtained there are two bands (at 2120 and 2175 cm^{-1}) in the region of the $N\equiv N$ bond stretching. It indicates to a *cis*-configuration of the nitrogen ligands in the complex. The bis-nitrogen complex of osmium is less stable than the corresponding mononitrogen complex: its decomposition in solutions occurs already at 50°. It is interesting that $[Ru(NH_3)_5N_2]^{3+}$ ion does not form a bis-nitrogen complex when treated with HNO_2. In this case the nitrogen is liberated to give $[Ru(NH_3)_5NO]^{2+}$.

Nevertheless Kane-Maguire, Sheridan, Basolo, and Pearson[101] managed to isolate a bis-nitrogen complex of ruthenium(II), but with another ligand surrounding; namely cis-$[Ru^{II}(en)_2(N_2)_2]^{2+}$. To obtain this compound the authors started from the azide-nitrogen complex of ruthenium cis-$[Ru^{II}(en)_2N_3(N_2)]^{1+}$, arising in the thermal decomposition of cis-$[Ru^{III}(en)_2(N_3)_2]$.

$$cis\text{-}[Ru^{III}(en)_2(N_3)_2]PF_6 \xrightarrow{65°} cis\text{-}[Ru^{II}(en)_2N_3(N_2)]PF_6$$

Upon reaction of the azide–nitrogen complex with HNO_2 followed by the treatment with $NaBPh_4$ the bis-nitrogen ruthenium complex cis-$[Ru^{II}(en)_2(N_2)_2](BPh_4)_2$ could be isolated in the mixture with a new mononitrogen complex cis-$[Ru^{II}(en)_2H_2O(N_2)]BPh_4)_2(\nu_{N\equiv N} = 2130\ cm^{-1})$

$$cis\text{-}[Ru^{II}(en)_2N_3(N_2)]^+ \xrightarrow[0°]{HNO_2} cis\text{-}[Ru^{II}(en)_2(N_2)_2]^{2+} + N_2O$$

The bis-nitrogen ruthenium(II) complex is still less stable than the bis-nitrogen complex of osmium(II). It is rapidly decomposed at room temperature. In the infrared spectrum of $[Ru(en)_2(N_2)_2](BPh_4)_2$ there are also two bands in the region of $N\equiv N$ stretching (at 2220 and 2190 cm^{-1}). It is in agreement with the cis-configuration of the complex. The mononitrogen complex cis-$[Ru(en)_2H_2O(N_2)]^{2+}$ is much more stable. Unlike $[Ru(NH_3)_5H_2O]^{2+}$ it is incapable of reacting with N_2.

Chernikov, Kuz'min, and Borod'ko[102] investigated the nitrogen chemisorption on Ni with IR spectroscopy. They confirmed the supposition of Eischens and Jacknow[37] that the surface nitrogen complex formed is linear. The temperature dependence of the equilibrium constant of the reaction:

$$Ni^{surf.} + N_2 \rightleftharpoons Ni^{surf.}\cdot N_2$$

allowed them to evaluate the bond energy of Ni with N_2. The bond energy depends on degree of surface covering (θ) and equals -9.2 ± 1.1 kcal/mole ($\theta \approx 0.05$) and -5.2 ± 1 kcal/mole ($\theta \approx 0.2$).

Shustorovich[110] has theoretically discussed the possibility of a cyclic dimer of molecular nitrogen (nitrogen analog of cyclobutadiene) to be stabilized by complex formation with transition metal compounds.

A number of reviews on nitrogen fixation have been published.[103–105]

Thus the investigation of nitrogen fixation is being intensely pursued. There are two closely related lines of research. One of them is the study of the nitrogen reactions accompanied by reduction of nitrogen into ammonia derivatives or into nitrogen-containing organic compounds. The second line concerns the study of the complexes of molecular nitrogen and to the attempts of the nitrogen fixation via stable nitrogen complexes.

The variety of nitrogen complexes is rapidly expanding, so that today it includes derivatives of almost all of the VIII group elements except palladium and platinum. There are numerous mononitrogen complexes as well as first representatives of binuclear and bisnitrogen complexes. All these complexes seem to be linear. In some cases it was shown by direct X-ray measurements. Characteristic of the known nitrogen complexes is the presence of the sharp and strong band in infrared or Raman spectra at 2000–2300 cm^{-1} assigned to the coordinated $N \equiv N$ bond stretch. In the series of isostructural complexes of metals belonging to the same subgroup of periodic table the strength of the $M—N_2$ bond apparently increases downward. Bisnitrogen complexes are less stable than corresponding mononitrogen complexes. All the facts combined indicate the particular importance of the π-back bonding in the formation of nitrogen complexes. This is supported in particular by the observation that all the nitrogen complexes known by now are the derivatives of transition metals, which are either in their lower valent states or surrounded by ligands with well-pronounced donor properties (ammonia, diamines, etc.).

One of the most intriguing features of the nitrogen complexes is the small reactivity of coordinated nitrogen. As it has been mentioned above, all the attempts failed to reduce the nitrogen ligand in the complexes or to involve it into some other reactions. This correlates with a slight effectivity of the derivatives of VIII group elements in nitrogen fixation to form ammonia or nitrogen-containing organic compounds. Thus there is the gap between elements belonging to left and right parts of the transition series. Derivatives of VIII group transition metals are capable of formation of the nitrogen complexes, but not active in its fixation to yield ammonia. On the other hand the most active in the nitrogen reduction to ammonia are the compounds of IV–VI group transition metals. However, their stable complexes are still unknown. In these cases nitrogen complexes formed are apparently less stable, but more reactive instead.*

REFERENCES

1. W. G. Frankenburg, *Catalysis*, **3**, 186 (1955).
2. L. S. Polak, *Kinetics and Thermodynamics of Chemical Reaction in Low-Temperature Plasma*, Moscow, 1965.
3. M. E. Vol'pin and V. B. Shur, *Dokl. Akad. Nauk SSSR*, **156**, 1102 (1964).
4. M. E. Vol'pin and V. B. Shur, *Vestn. Akad. Nauk SSSR*, **1965** (No. 1), 51.
5. M. E. Vol'pin, V. B. Shur, and M. A. Ilatavskaya, *Izvest. Akad. Nauk SSSR, Ser. Khim.*, **1964**, 1728.

*Recently van Tamelen et al.[114] and Shilov et al.[115] have observed the formation of nitrogen titanium complexes. The nitrogen molecule in these complexes is really capable of reduction.

6. M. E. Vol'pin, V. B. Shur, and L. P. Bichin, *Izvest. Akad. Nauk SSSR, Ser. Khim.*, **1965**, 720.
7. M. E. Vol'pin, M. A. Ilatovskaya, E. I. Larikov, M. L. Khidekel', Yu. A. Shetsov, and V. B. Shur, *Dokl. Akad. Nauk SSSR*, **164**, 331 (1965).
8. M. E. Vol'pin, A. A. Belii, and V. B. Shur, *Izvest. Akad. Nauk SSSR, Ser. Khim.*, **1965**, 2225.
9. M. E. Vol'pin, V. B. Shur, V. N. Latyaeva, L. I. Vyshinskaya, and L. A. Shulgaitser, *Izvest. Akad. Nauk SSSR, Ser. Khim.*, **1966**, 385.
10. M. E. Vol'pin, N. K. Chapovskaya, and V. B. Shur, *Izvest. Akad. Nauk SSSR, Ser. Khim.*, **1966**, 1083.
11. M. E. Vol'pin and V. B. Shur, *Nature*, **209**, 1236 (1966).
12. M. E. Vol'pin and V. B. Shur, *Izvest. Akad. Nauk SSSR, Ser. Khim*, **1966**, 1873.
13. M. E. Vol'pin and V. B. Shur, *Zh. Vses. Khim. Obshchestva im. D. I. Mendeleeva*, **12** (No. 1), 31 (1967).
14. A. Shindler, Symposium on Macromolecular Chemistry July 1-6, 1963, Section IA (Part I).
15. G. Henrici-Olivé and S. Olivé, *Angew. Chem. (Intern. Ed. English)*, **6** (No. 10), 873-874 (1967).
16. G. Henrici-Olivé and S. Olivé, *Angew, Chem. (Intern. Ed. English)*, **6** (No. 10), 873 (1967).
17. G. P. Haight and R. Scott, *J. Am. Chem. Soc.*, **86**, 743 (1964).
18. K. B. Iatsimirskii and V. K. Pavlova, *Dokl. Akad. Nauk SSSR*, **165**, 130 (1965).
19. F. Hein and R. Weiss, *Z. Anorg. Allgem. Chem.*, **295**, 145 (1958).
20. F. Hein and D. Tille, *Z. Anorg. Allgem. Chem.*, **329**, 72 (1964).
21. E. Kurras and J. Otto, *J. Organomet. Chem.*, **4** (No. 2), 114 (1965).
22. B. Sarry, *Angew. Chem. Intern. Ed. English*, **6**, 571 (1967).
23. R. Maskill and J. M. Pratt, *Chem. Commun.*, **1967** (No. 18), 950; *J. Chem. Soc. (A)*, **N8**, 1914 (1968).
24. H. Brintzinger, *J. Am. Chem. Soc.*, **88**, 4305 (1966).
25. H. Brintzinger, *J. Am. Chem. Soc.*, **88**, 4307 (1966).
26. H. Brintzinger, *J. Am. Chem. Soc.*, **89**, 6871 (1967).
27. G. N. Nechiporenko, G. M. Tabrina, A. K. Shilova, and A. E. Shilov, *Dokl. Akad. Nauk SSSR*, **164**, 1062 (1965).
28. E. E. van Tamelen, G. Boche, S. W. Ela, and R. B. Fechter, *J. Am. Chem. Soc.*, **89**, 5707 (1967).
29. M. F. Sloan, A. S. Matlack, and D. S. Breslow, *J. Am. Chem. Soc.*, **85**, 4014 (1963).
30. Yu. A. Shvetsov, L. G. Korabliova, G. N. Nechiporenko, M. A. Ilatovskaya, and M. L. Khidekel', *Dokl. Akad. Nauk SSSR*, **178**, 400 (1968).
31. O. Glemser, Ger., **956**, 674 (1957).
32. R. B. King and M. B. Bisnette, *J. Am. Chem. Soc.*, **86**, 5694 (1964).
33. R. B. King and M. B. Bisnette, *Inorg. Chem.*, **5** (No. 2), 300 (1966).
34. G. W. Parshall, *J. Am. Chem. Soc.*, **87**, 2133 (1965).
35. G. W. Parshall, *J. Am. Chem. Soc.*, **89**, 1822 (1967).
36. V. N. Latyaeva, L. I. Vyshinskaya, V. B. Shur, L. A. Fyodorov, and M. E. Vol'pin, *Dokl. Akad. Nauk SSSR*, **179**, 875 (1968); *J. Organomet. Chem.*, **16**, 103 (1969).
37. R. P. Eischens and J. Jacknow, *Proc. III Intern. Congr. Catalysis, Amsterdam*, **1**, 627 (1964).

38. Yu. G. Borod'ko, A. E. Shilov, and A. A. Shteinman, *Dokl. Akad. Nauk SSSR*, **168**, 581 (1966).
39. A. D. Allen and C. V. Senoff, *Chem. Commun.*, **1965** (No. 24), 621.
40. A. D. Allen, F. Bottomley, R. O. Harris, V. P. Reinsalu, and C. V. Senoff, *J. Am. Chem. Soc.*, **89**, 5595 (1967).
41. J. Chatt, G. J. Leigh, and R. J. Paske, Chem. Soc. Autumn Meeting, Sussex, Sept. 1966.
42. F. Bottomley and S. C. Nyburg, *Chem. Commun.*, **1966**, 897.
43. Yu. G. Borod'ko, V. S. Bukreev, G. I. Kozub, M. L. Khidekel', and A. E. Shilov, *Zh. Strukt. Khim.*, **8** (No. 3), 542 (1967).
44. A. D. Allen and J. R. Stevens, *Chem. Commun.*, **1967**, 1147.
45. J. P. Collman and J. W. Kang, *J. Am. Chem. Soc.*, **88**, 3459 (1966).
46. J. P. Collman, M. Kubota, Tui-Yuan Sun, and F. Vastine, *J. Am. Chem. Soc.*, **89**, 169 (1967).
47. L. Yu. Ukhin, Yu. A. Shvetsov, and M. L. Khidekel', *Izvest. Akad. Nauk SSSR, Ser. Khim.*, **1967** (No. 4), 957.
48. A. E. Shilov, A. K. Shilova, and Yu. G. Borod'ko, *Kinetika i Kataliz*, **7**, 768 (1966).
49. Yu. G. Borod'ko, A. K. Shilova, and A. E. Shilov, *Dokl. Akad. Nauk SSSR*, **176** (No. 6), 1297 (1967).
50. D. F. Harrison and H. Taube, *J. Am. Chem. Soc.*, **89**, 5706 (1967).
51. A. Yamamoto, S. Kitazume, L. S. Pu, and S. Ikeda, *Chem. Commun.*. **1967**, 79.
52. A. Yamamoto, L. S. Pu, S. Kitazume, and S. Ikeda, *J. Am. Chem. Soc.*, **89**, 3071 (1967).
53. A. Yamamoto, S. Kitazume, L. S. Pu, and S. Ikeda, *X. International Conference on Coordination Chemistry*, Tokyo, 1967. (Theses of reports.)
54. A. Misono, Y. Uchida, T. Saito, and K. M. Song, *Chem. Commun.*, **1967**, 419.
55. A. Sacco and M. Rossi, *Chem. Commun.*, **1967**, 316.
56. A. Misono, Y. Uchida, and T. Saito, *Bull. Chem. Soc. Japan*, **40** (No. 3), 700 (1967).
57. G. L. Johnson and W. D. Beveridge, *Inorg. Nucl. Chem. Letters*, **3** (No. 8), 323 (1967).
58. L. Vaska, *Science*, **140**, 809 (1963).
59. J. A. Ibers and S. T. LaPlaca, *Science*, **145**, 920 (1964).
60. E. E. van Tamelen, G. Boche, and R. G. Greeley, *J. Am. Chem. Soc.*, **90**, 1677 (1968).
61. M. E. Vol'pin, M. A. Ilatovskaya, L. V. Kosyakova, and V. B. Shur, *Dokl. Akad. Nauk SSSR*, **180**, 103 (1968).
62. M. E. Vol'pin, M. A. Ilatovskaya, L. V. Kosyakova, and V. B. Shur, *Chem. Commun.*, **1968**, 1074.
63. E. O. Fischer and W. Hafner, *Z. Naturforsch.*, **10b**, 665 (1955).
64. E. O. Fischer, W. Hafner, and K. Ofele, *Chem. Ber.*, **92**, 3040 (1959).
65. G. Natta, G. Mazzanti, and G. Pregaglia, *Gazz. Chim. Ital.*, **89** (No. 9), 2065 (1959).
66. H. Martin and F. Vohwinkel, *Chem. Ber.*, **94**, 2416 (1961).
67. J. Thonstad, *Can. J. Chem.*, **42** (No. 12), 2739 (1964).
68. J. Chatt and J. E. Fergusson, *Chem. Commun.*, **1968** (No. 3), 126.
69. D. F. Harrison, E. Weissberger, and H. Taube, *Science*. **159**, 320 (1968).
70. A. D. Allen and F. Bottomley, *Can. J. Chem.*, **46**, 469 (1968).

71. J. H. Enemark, R. R. Davis, J. A. McGinnety, and J. A. Ibers, *Chem. Commun.*, **1968** (No. 2), 96.
72. G. W. Parshall, *J. Am. Chem. Soc.*, **90**, 1669 (1968).
73. A. Yamamoto, S. Kitazume, and S. Ikeda, *J. Am. Chem. Soc.*, **90**, 1089 (1968).
74. G. Henrici-Olivé and S. Olivé, *Angew Chem.* (*Intern. Ed.*), **7** (No. 5), 386 (1968).
75. E. E. van Tamelen and B. Akermark, *J. Am. Chem. Soc.*, **90**, 4492 (1968).
76. F. Bottomley and S. C. Nyborg, *Acta Crystallogr., Sect. B*, **24**, pt. 10, 1289 (1968).
77. J. Chatt, A. B. Nikolsky, R. L. Richards, and J. R. Sanders, *Chem. Commun.*, No. 4, 154 (1969).
78. I. J. Itzkovitch and J. A. Page, *Can. J. Chem.*, **46**, N 16, 2743 (1968).
79. Yu. G. Borod'ko, A. K. Shilova, and A. E. Shilov, Chemical Society Symposium on N_2, O_2, and Hydrogen as ligands for metals, Leeds, April, 1968; *Zhur. Fiz. Khim.*, in press.
80. J. Chatt, R. L. Richards, J. E. Fergusson, and J. L. Love, *Chem. Commun.*, N 23, 1522 (1968).
81. P. K. Das, J. M. Pratt, R. G. Smith, G. Swinden, and W. J. V. Woolcock, *Chem. Commun.*, N 23, 1539 (1968).
82. A. K. Shilova, A. E. Shilov, and Z. N. Vostroknutova, *Kinet. Kataliz*, **9**, 924 (1968).
83. J. P. Collman, M. Kubota, F. D. Vastine, J. Y. Sun, and J. W. Kang, *J. Am. Chem. Soc.*, **90**, 5430 (1968).
84. J. P. Collman, *Accounts Chem. Research*, **1** (No. 5), 136 (1968).
85. J. P. Collman, M. Kubota, and J. W. Hosking, *J. Am. Chem. Soc.*, **89**, 4809 (1967).
86. J. P. Collman and J. W. Kang, *J. Am. Chem. Soc.*, **89** (No. 4), 844 (1967).
87. J. P. Collman, J. W. Kang, W. F. Little, and M. F. Sullivan, *Inorg. Chem.*, **7**, (No. 7), 1298 (1968).
88. A. Sacco and M. Rossi, *Inorg. Chim. Acta*, **2** (No. 2), 127 (1968).
89. A. Misono, Y. Uchida, M. Hidai, and M. Araki, *Chem. Commun.*, No. 17, 1044 (1968).
90. A. Misono, Y. Uchida, M. Hidai, and T. Kuse, *Chem. Commun.*, No. 16, 981 (1968).
91. L. S. Pu, A. Yamamoto, and S. Ikeda, *J. Am. Chem. Soc.*, **90**, 3896 (1968).
92. L. S. Pu, A. Yamamoto, and S. Ikeda, *J. Am. Chem. Soc.*, **90**, 7170 (1968).
93. C. H. Campbell, A. R. Dias, M. L. H. Green, T. Saito, and M. G. Swanwick, *J. Organomet. Chem.*, **14** (No. 2), 349 (1968).
94. A. Sacco and M. Aresta, *Chem. Commun.*, No. 20, 1223 (1968).
95. W. H. Knotes, *J. Am. Chem. Soc.*, **90**, 7172 (1968).
96. J. Chatt, G. I. Leigh, and D. M. P. Mingos, *Chem. Ind.*, No. 4, 109 (1969).
97. J. Chatt, G. I. Leigh, D. M. P. Mingos, and R. J. Paske, *Chem. Ind.*, No. 31, 1324 (1967).
98. J. Chatt, G. I. Leigh, and R. J. Paske, *Chem. Commun.*, No. 14, 671 (1967).
99. P. W. Jolly and K. Jonas, *Angew. Chem.* (*Intern. Ed.*), **7**, 731 (1968).
100. H. A. Scheidegger, J. N. Armor, and H. Taube, *J. Am. Chem. Soc.*, **90**, 3263 (1968).
101. L. A. P. Kane-Maguire, P. S. Sheridan, F. Basolo, and R. G. Pearson, *J. Am. Chem. Soc.*, **90** (No. 19), 5295 (1968).
102. S. S. Chernikov, S. G. Kuz'min, and Yu. G. Borod'ko, *Zhur. Fiz. Khim.*, No. 8, 2038 (1968).

103. R. Murray and D. C. Smith, *Coord. Chem. Revs.*, 3, (No. 4), 429 (1968).
104. S. Tyrlik and S. Malinowski, *Wiadomości Chemiczne*, 23, 95 (1969).
105. Yu. G. Borod'ko and A. E. Shilov, *Uspekhi Khim.*, 38 (No. 5), 761 (1969).
106. A. Misono, Y. Uchida, T. Saito, M. Hidai, and M. Araki, *Inorg. Chem.*, 8 (No. 1), 168 (1969).
107. J. Lorberth, H. Nöth, and P. V. Rinze, *J. Organomet. Chem.*, 16 (No. 1), p. 1 (1969).
108. J. B. Lee and B. Cubberly; *Tetrahedron Letters*, No. 13, 1061 (1969).
109. A. Misono, Y. Uchida, M. Hidai, and T. Kuse, *Chem. Commun.*, No. 5D, 208 (1969).
110. E. M. Shustorovich, *Zhur. Strukt. Khim.*, 10 (No. 1), 159 (1969).
111. M. L. Khidekel' and Yu. B. Grebenshchikov, *Izvest. Akad. Nauk SSSR, Ser. Khim.*, 747 (1965).
112. M. E. Vol'pin, V. B. Shur, R. V. Kudryavtsev and L. A. Prodayko, *Chem. Commun.*, 1038 (1968).
113. A. E. Shilov and A. K. Shilova, *Kinetika i Kataliz*, 10, 1163 (1969).
114. E. E. van Tamelen, R. B. Fechter, S. W. Schneller, G. Boche, R. H. Greeley and B. Akermark, *J. Am. Chem. Soc.* 91 (N6), 1551 (1969).
115. A. E. Shilov, A. K. Shilova and E. F. Kvashina, *Kinetika i Kataliz*, 10, 1402 (1969).
116. M. E. Vol'pin, M. A. Ylatovskaya, and V. B. Shur, *Kinetika i Kataliz*, 11, 333 (1970).
117. E. E. van Tamelen, R. B. Fechter and S. W. Schneller, *J. Am. Chem. Soc.* 91, 7196 (1969).
118. E. E. van Tamelen, D. A. Seeley, *J. Am. Chem, Soc.* 91. 5194 (1969).

Reaction of Organomercury Compounds, Part 1

L. G. MAKAROVA

Institute of Organoelemento Compounds,
Academy of Science, Moscow, U.S.S.R.

Part 1, Chapters 1–11

Chapter 1. Reactions with Halides and Other Salts as Well as with Alkyl and Aryl Halides of Some Elements

I. Introduction 129
II. Reaction Mechanism 130
III. Reaction Conditions 135
IV. Application of the Reactions 136
  A. Reactions with Salts of Group III Elements 136
    1. Boron Halides 136
    2. Indium Halides 137
    3. Thallium Salts 137
  B. Reactions with Salts and Other Compounds of Group IV Elements . 139
    1. Silicon Halides and Alkyl-silicon Halides 139
    2. Tetrachlorogermanium and Alkylgermanium Halides 139
    3. Tin(IV) Chloride 140
    4. Lead Tetraacylate 140
    5. Titanium(IV) Chloride and $(C_5H_5)_2TiCl_2$ 141
  C. Reactions with Halides, Alkylhalides, and Other Salts of Group V Elements 141
    1. Phosphorus Halides and Alkylhalides 141
    2. Halides and Alkylhalides of Arsenic(III) 142
    3. Antimony Trichloride 144
    4. Bismuth Halides and Aryl Halides 144
    5. Vanadium and Vanadyl Halides 144
  D. Reactions with Alkyl Halides, Halides, and Some Other Inorganic Compounds of Group VI Elements 145
    1. Halides and Inorganic Compounds of Sulfur 145
    2. Selenium Halides, Alkyl Halides, and Dioxide 145
    3. Tellurium Halides and Aryl Halides 146
    4. Uranium Tetrafluoride 147
  E. Reactions with Halides and Alkyl Halides of Group VII Elements . 147
    1. Iodine Chlorides and $RICl_2$ 147
    2. $RAtCl_2$ 148
  F. Reactions with Halides of Group VIII Elements 148

V. Preparative Syntheses 148
  A. Preparation of Organoboron Derivatives 148
    1. Preparation of Di-β-chlorovinylboron Chloride from Boron
       Trichloride and *trans–trans*-Di-β-chlorovinylmercury. 148
    2. Preparation of ArBBr$_2$ Type Compounds 149
    3. Preparation of Phenylboron Dibromide C$_6$H$_5$BBr$_2$ 149
    4. Preparation of Diphenylboron Bromide, (C$_6$H$_5$)$_2$BBr 149
    5. Preparation of Diphenylboron Chloride 149
  B. Preparation of Organothallium Derivatives 149
    1. Preparation of Phenylthallium Diisobutyrate 149
    2. Preparation of Diphenylthallium Isobutyrate 150
  C. Preparation of Organosilicon Compounds 150
    1. Reaction of Triethylsilane Iodide with Bis(carbomethoxymethyl)-
       mercury 150
  D. Preparation of Organolead Compounds 150
    1. Preparation of Di-β-chlorovinyl Diacetate 150
    2. Preparation of Di-*o*-anisyllead Diacetate,
       (*o*-CH$_3$OC$_6$H$_4$)Pb(OOCCH$_3$)$_2$ 151
    3. Preparation of α-Naphthyllead Triisobutyrate. 151
    4. Preparation of Di-α-naphthyllead Diacetate 151
  E. Preparation of Organophosphorus Compounds 151
    1. Preparation of *n*-Butylphosphine Dichloride 151
    2. Preparation of Vinylphosphine Dichloride. 152
    3. Preparation of *m*-Chlorophenylphosphine Dichloride 152
    4. Preparation of Diphenylphosphine Chloride 153
    5. Preparation of *p*-Tolyl-phenylphosphine Chloride 153
  F. Preparation of Organoarsenic Compounds 153
    1. Preparation of Ethylarsine Dichloride 153
    2. Preparation of CF$_2$=CFAsCl$_2$ 153
    3. Preparation of α-Naphthylarsine Dichloride 153
    4. Preparation of Phenyl-*p*-tolylarsine Chloride 154
    5. Preparation of Diphenylarsine Chloride 154
  G. Preparation of Organoselenium Compounds 154
    1. Preparation of Diethylselenium 154
  H. Preparation of Organic Derivatives of Group VII Elements . . . 154
    1. Preparation of Diphenyliodonium Chloride 154
    2. Preparation of Diphenyliodonium Chloride 173
  Table I 155
  References 173

Chapter 2. Methylene Insertion and Addition Reactions by Organomercury Compounds

I. Introduction 179
II. Reaction Mechanisms 179
III. Reaction Conditions 180
IV. Application of the Reaction 180
  A. Introduction of Methylene (Halo- and Dihalomethylene) in Com-
    pounds Containing a Mobile Atom of H 180

1. Methylene Insertion into a C—H Bond 180
2. Methylene Insertion into a Si—H Bond 180
3. Methylene Insertion into a Ge—H Bond 181
4. Reaction with Trialkyltin(IV) Hydride 181
5. Methylene Insertion into O—H Bonds of Carboxylic Acids and
   Alcohols and the HCl Bond 181
B. Mono- and Dihalomethylene Addition to a C=C Bond. Synthesis of
   Mono- and Gem-Dihaloderivatives of Cyclopropane 181
   1. Action of $C_6H_5HgCHal_3$ and $C_6H_5HgCHHal_2$ 181
   2. Reactions of $C_6H_5HgCHal_3$ + NaI 184
   3. Reactions of Halomethylmercury Compounds 185
C. Preparation of 1-Haloolefins from Aldehydes and Ketones 185
D. Reactions of $C_6H_5HgCRHalBr$ with $P(OAlk)_3$ (R = H or Hal) . . 185
E. Miscellaneous Reactions 185
V. Preparative Syntheses 186
A. Preparation of gem-Dihalocyclopropane 186
B. Preparation of Hexachlorocyclopropane 187
C. Preparation of 1,1-Dichloro-2-vinylcyclopropane 187
D. Preparation of 7,7-Dichloronorcarane 188
E. Reactions with trans-Crotonic Acid 188
References 188

Chapter 3. Exchange Reactions with Metals and Metal Alloys

I. Introduction 189
II. Reaction Mechanism and Stereochemistry 189
III. Reaction Conditions 193
IV. Application of the Reaction 194
A. Reactions of Organomercury Compounds with Alkali Metals . . . 194
   1. Metallic Lithium 193
   2. Sodium 195
   3. Potassium 196
B. Reactions with Metals of Group II 197
   1. Beryllium 197
   2. Magnesium 197
   3. Zinc 197
   4. Cadmium 198
C. Reactions with Metals of Group III 198
   1. Aluminum 198
   2. Gallium 199
   3. Indium 199
D. Reactions with Metals and Metal Alloys of Group IV 200
   1. Tin and Tin Alloys 200
E. Reactions with Metals and Metal Alloys of Group V 201
F. Reactions with Elements of Group VIII 201
G. Reactions with Actinides 201
V. Preparative Syntheses 201
A. Preparation of Soluble Lithium Alkyls 201
B. Preparation of Vinylalkali—Metal Compounds 202
C. Preparation of p-Tolyllithium Solution 204

D. Preparation of Ethylsodium. 204
E. Preparation of Benzylsodium 204
F. Preparation of Diphenylberyllium 204
G. Preparation of Diethylmagnesium 204
H. Preparation of the Etherate of Diphenylmagnesium 205
I. Preparation of Dimethylzinc 205
J. Preparation of Di-p-chlorophenylzinc 205
K. Preparation of Trimethylaluminum 205
L. Preparation of Triphenylaluminum 205
M. Preparation of Triethylgallium 206
N. Preparation of Triphenylgallium 206
O. Reaction of *trans*-β-Chlorovinylmercury(II) Chloride with a Tin and
    Sodium Alloy. 207
P. Preparation of Tetra-p-tolyltin 207
Q. Preparation of Triphenyltin Chloride 207
Table II   208
References. 224

Chapter 4. Reactions with Salts-Reducing Agents, and Low-Valence Organometallic Compounds

I. Introduction 227
II. Application of the Reaction. 227
A. Synthesis of Organogermanium Compounds 227
1. Reactions with Germanium(II) Halides 227
B. Synthesis of organotin Compounds 228
1. Reactions with Tin(II) Salts 228
2. Reactions with Organotin Compounds 231
III. Preparative Syntheses 231
A. Preparation of Organogermanium Compounds 231
1. Reaction of Di-p-tolylmercury with Germanium(II) Iodide . . . 231
B. Preparation of Di-*cis*-β-chlorovinyltin(IV) Chloride 232
C. Preparation of Diethyltin(IV) Chloride 232
D. Preparation of Di-p-chlorophenyltin(IV) Chloride 232
E. Preparation of Diphenyldiethyltin 239
F. Preparation of Triethyltin by the Reaction of Hexaethylditin with
    Phenylmercury(II) Chloride 239
G. Reaction of Hexaethylditin with the Methyl Ester of Mercury-
    bisacetic Acid 239
Table III   233
References 239

Chapter 5. Reactions with Hydrides and Alkyl and Arylhydrides of Elements and Their Metallic Derivatives

I. Introduction 241
II. Application of the Reaction. 241
A. Reactions with Group III Hydrides 241
1. Synthesis of Organoaluminum Compounds 241

B. Reactions with Hydrides (and Metallic Derivatives) of Group IV
   Elements   241
   1. Syntheses of Organosilicon Compounds   241
   2. Syntheses of Organogermanium Compounds   242
   3. Syntheses of Organotin Compounds   242
III. Preparative Syntheses   242
A. Reactions of Dibutylgermane with the Methyl Ester of Mercurybisacetic Acid   242
   1. 1:1 Reagent Ratio   242
   2. 1:2 Reagent Ratio   246
B. Reaction of the Methyl Ester of Dibutylgermaniumacetic Acid with
   the Methyl Ester of Mercurybisacetic Acid   246
C. Reaction of Triethyltin(IV) Hydride with the Propyl Ester of Mercurybisacetic Acid   246
Table IV   243
References   246

Chapter 6. Reactions with Elements of the Sulfur Group

I. Application of the Reaction   248
II. Preparative Syntheses   248
A. Preparation of Di-o-tolylsulfide   248
B. Preparation of Di-o-tolylselenide   250
C. Preparation of Di-o-tolyltelluride   250
Table V   249
References   250

Chapter 7. Reactions with Reducing Agents—Hydrogen Donors

I. Introduction   251
II. Application of the Reaction   251
A. Reaction with Molecular Hydrogen   251
B. Reactions with Alcohols and Other Compounds which are Subject to
   Dehydration   252
C. Reduction by Hydrides and Hydride Derivatives of Group III and IV
   Elements   253
D. Substitution of Hydrogen for Mercury by Action of Symmetrizators .   254
E. Other Hydrogen for Mercury Substitutions   256
F. Reactions of Organomercury Compounds in the Presence of Substances
   Susceptible to Dehydrogenation by Light Initiation and Other Initiating
   Agents Causing Radical Decomposition   267
Table VI   257
References   267

Chapter 8. Coupling Reactions

I. Introduction   271
II. Application of the Reaction   271
References   274

Chapter 9. Mercury-to-Acid Exchange Reactions

I. Introduction 275
    A. Substitution Reactions of Hydrogen for Mercury (Protodemercuration
       Reaction) 275
    B. Demercuration Reaction 275
    C. Elimination Reaction, Deoxydemercuration 275
II. Hydrogen-for-Mercury Substitution (Protodemercuration) 276
    A. Reaction Mechanism 276
    B. Effect of Acid Type 278
    C. Effect of the Type of Organomercury Compound 279
       1. The Kharasch Series 281
III. Demercuration Reactions 284
    A. Mechanism of the Demercuration Reaction 284
    B. Effect of Organomercury Salt Type and Other Factors 284
IV. Deoxymercuration (Elimination) Reaction 286
    A. General Survey 286
    B. Mechanism of the Deoxymercuration Reaction 290
    C. Effect of Organomercury Compound Type 292
    D. Miscellaneous Observations 293
    References 294

Chapter 10. Reactions with Alkyl and Acyl Halides

I. Introduction 301
II. Reactions with Alkyl Halides 301
    A. Reactions Forming a New C—C Bond 301
    B. Formation of Nonsaturated Compounds 304
    C. Other Reactions with Alkyl Halides 305
III. Reactions with Acid Halides 307
    Table VII 311
    References 312

Chapter 11. Reactions with Halogens

I. Introduction 325
II. Reaction Stereochemistry, Kinetics and Mechanism 325
    A. Reaction Stereochemistry 325
       1. Compounds of Mercury at Olefinic Carbon Atoms 325
       2. Compounds of Mercury at a Saturated Carbon Atom 327
    B. Kinetics and Mechanism of the Reaction 331
III. Reaction Conditions 335
    A. Effect of Halogen Type 335
    B. Related Reactions 336
    C. Effect of Organomercury Compound Type 337
    D. Secondary Reactions 340
IV. Miscellaneous Observations 341
V. Preparative Syntheses 342
    1. 2-Iodo-n-butane from Di-sec-butylmercury 342
    2. trans-β-Chlorovinylmercury Chloride 342

3. *cis*-α-Bromostilbene 342
4. *o*-Iodophenol 342
5. Cleavage of Bis-(perchlorovinyl)mercury with Bromine 343
6. Bromination of Mercury Acetylides 343
7. Preparation of Bromo-vinylacetylene 343
8. Preparation of 2-Bromoheptafluoropropane 344
9. Preparation of 2-Chlorodiphenylene 344
10. Preparation of Dimethyl Ester of Exo-*cis*-4-oxy-brom-3,6-endoxohexa-
    hydrophthalic Acid 344
11. Preparation of 2,4,5-Trijodo-3-nitrothiophene 344
12. Preparation of 2,2″–Dibromo-o-terphenyl 345
13. Preparation of Bromoferrocene 345
References 345

Part 2 (To appear in Volume II)

**Chapter 12. Radical Exchange Reactions between Organomercury Compounds
and Organic Compounds of Other Elements**

I. Introduction
II. Reaction Mechanism
III. Application of the Reaction
   A. Exchange with Compounds of Other Elements
   B. Radical Redistribution Reactions between Organomercury
      Compounds
   References

**Chapter 13. Reactions with Nitrogen-Containing Compounds: Nitric Oxides,
Nitric Oxide Derivatives, Diazocompounds and Carbodiimides**

I. Reactions with Nitric Oxides
   A. Formation of Diazo Compounds by the Reaction of Nitric Oxides with
      Organomercury Compounds
II. Reactions with Diazo Compounds
III. Reactions of Phenyl(bromodichloromethyl)mercury with carbodiimides
   References

Chapter 14. Reactions with Ketenes and Their Derivatives

I. Introduction
II. Formation of Ketones
   1. Introduction and Application of the Reaction
   2. Reaction Mechanism
   3. Reaction Conditions
III. Formation of Unsymmetrical Organomercury Compounds

IV. Reactions of Ketene with β-Mercurated Ethers
   A. Application and Conditions of the Reaction
   B. Reaction Mechanism
 V. Reactions with α-Mercurated Oxocompounds
   A. Application and Conditions of the Reaction
   B. Reaction Mechanism
VI. Reactions of α-Mercurated Carbonyl Compounds with Ketene Dimer
   A. Application and Conditions of the Reaction
   B. Reaction Mechanism
   References

Chapter 15. Reactions $RHgX \rightarrow R_2Hg$ and Reverse Reactions

  I. Introduction
 II. Mechanism of Symmetrization and Reverse Reactions
III. Symmetrization Agents
   A. Reducing Symmetrizators
     1. Sodium Stannite
     2. Metals and Metal–Base Alloys
     3. Hydrazine, its Derivatives, Hydroxylamine
     4. Electrolytic Symmetrization
     5. Alcoholates of Alkali Metals
     6. Ferrous Oxide Hydrate and its Salts
     7. Sodium Hydrosulfite
     8. Salts of Divalent Chromium
   B. Symmetrizators Entraining Hg^{2+} in the Form of Un-ionized Compounds
     1. Alkali Metal Iodides
     2. Potassium (Sodium) Cyanide
     3. Potassium (Sodium) Thiocyanate
     4. Sodium Thiosulfate
     5. Ammonia
     6. Pyridine
     7. Tertiary Phosphines
     8. Triethanolamine
     9. Iron Pentacarbonyl and its Derivatives
    10. Alkali Sulfides
    11. Potassium (Sodium) Hydroxide
    12. Calcium Chloride
   C. Symmetrizators Producing Completely Substituted Unsymmetrical Organomercury Compounds which then Disproportionate
     1. Unsaturated Compounds
     2. n-Butyllithium
     3. Diphenylmercury
     4. Disproportionation of Unsymmetrical Compounds
   D. Symmetrizator Acting as an Adsorbent
     1. Aluminum Oxide
   E. Thermal Symmetrization
IV. Symmetrization Reaction, Application of the Reaction
   A. Aliphatic Mercury Compounds
     1. Mercury at a Saturated Carbon Atom

2. Olefinic Derivatives with Mercury at an Unsaturated Carbon Atom which are not Addition Products of Mercury Salts to Multiple Bonds
3. Addition Products of Mercury Salts to Unsaturated Compounds
B. Mercury Derivatives of Aralkyls
C. Mercury Derivatives of Alicyclic Compounds
D. Mercury Derivatives of Aromatic Compounds
   1. Mercurated Hydrocarbons, their Halogen- and Nitro Derivatives
   2. Mercurated Phenols
   3. Mercurated Aromatic Amines
   4. Mercurated Aromatic Ketones
   5. Mercury Derivatives of Aromatic Acids, their Esters and Halogen- and Nitro-substituted Compounds
E. Mercury Derivatives of Heterocyclic Compounds
   1. Mercury Derivatives of Superaromatic Heterocyclic Compounds
   2. Mercury Derivatives of Other Heterocyclic Compounds
F. Mercury Derivatives of Metallocenes
G. Cases when Symmetrization Reactions Fail to Occur
V. Reverse Symmetrization Reaction
VI. Preparative Syntheses
A. Symmetrization of Alkylmercury Salts by Electrolysis
   1. Electrolysis of Propylmercury Sulfate
   2. Electrolysis of Ethylmercury Succinate
B. Preparation of 1,6-Dimercuracyclodecane
C. Preparation of Bistrifluoromethylmercury
D. Preparation of Diallylmercury
E. Symmetrization of Acetic Acid Salts of 3-Oxy-, 3-Methoxy-, and 3-Ethoxy-2,2,3-Trimethylbutylmercury
F. Preparation of Bis-(perchlorovinyl)mercury
G. Preparation of cis,cis-Di-β-chlorovinylmercury
H. Preparation of Bis-carbomethoxymercury
I. Preparation of Diacetonylmercury
J. Preparation of Dibenzylmercury
K. Symmetrization of Diastereomer of 2-Chloromercurycamphane by Hydrazine Hydrate
L. Preparation of Di-p-bromophenylmercury
M. Preparation of Di-p-Aminophenylmercury
N. Preparation of Di-m-oxyphenylmercury
O Preparation of Mercury-bis-p-benzoic Acid Ethyl Ester by Copper without Ammonia
P. Preparation of Mercury-bis-p-Benzoic Acid Methyl Ester by Copper and Ammonia
Q. Preparation of Di-o-phenethylmercury
R. Preparation of Bispentafluorophenylmercury
S. Preparation of 2,2'-Difurylmercury
T. Preparation of Mercury-bis(2,4-Diphenyl-selenophene-5)
U. Preparation of Di-3-pyridylmercury
V. Preparation of Diferrocenylmercury
W. Preparation of Ethylmercury Phosphate
Y. Preparation of Ethylmercury Nitrate
References

Chapter 16. Decomposition of Organomercury Compounds by Oxygen [and Other Oxidants], Light and Thermal Treatment

  I. Introduction
 II. Decomposition by Oxygen and Other Oxidants
     A. Oxidation by Oxygen
        1. Introduction
        2. Reaction Mechanism
        3. Application of the Reaction
     B. Oxidation by Acylperoxides
        1. Application of the Reaction
        2. Reaction Mechanism
     C. Oxidation by *tert*-Butyl Hydroperoxide
     D. Oxidation by Divalent Mercury Salts
        1. Application of the Reaction
        2. Reaction Mechanism
     E. Oxidation by Other Oxidants
     F. Autooxidation of Organomercury Salts
        1. Application of the Reaction
        2. Reaction Mechanism
     G. *trans–cis* Isomerization of Organomercury Compounds of the Ethylene Series by Peroxides
III. Light-Induced Reactions of Organomercury Compounds
     A. Introduction
     B. Photolysis in Solution
        1. Aliphatic Solvents Free of Halides
        2. Aromatic Compounds Used as Solvents
        3. Aliphatic Halogen-Containing Solvents
     C. Photolysis in the Gas Phase
     D. *cis–trans* Isomerization of Organomercury Compounds of the Ethylene Series by Ultraviolet Light
 IV. Thermal Decomposition of Organomercury Compounds
     A. Introduction
     B. Pyrolysis in the Absence of Solvent
     C. Pyrolysis in Solution
     D. Thermal *cis–trans* Isomerizations of Unsaturated Organomercury Compounds
     References

Chapter 17. Reaction with Alkalis

     References

Chapter 1

Reactions with Halides and Other Salts as Well as with Alkyl, (Aryl)Halides of Some Elements

I. INTRODUCTION

Completely substituted organomercury compounds as well as organomercury salts react with halides and other salts of various elements, leading to their organic compounds. These reactions are of great importance as preparative-type reactions and are depicted by the following equations:

$$R_2Hg + M^nX_n \longrightarrow RM^nX_{n-1} + RHgX$$
$$R_2Hg + 2M^nX_n \longrightarrow 2RM^nX_{n-1} + HgX_2$$
$$RHgX + M^nX_n \longrightarrow RM^nX_{n-1} + HgX_2$$

In exchange reactions of this type, fewer halides react with organomercury compounds than with organomagnesium (or compounds of lithium, zinc, aluminum, etc.) since the latter are more reactive. However, organomercury compounds have an advantage in the above reactions, since the possibility exists of organic substituent exchange with the halides, for example, the exchange of substituents containing functional groups which are reactive to Grignard reagents (or lithium, zinc, and other compounds). In addition, some polymercury compounds may also participate in the reaction. Furthermore, organomercury compounds possessing no reducing properties may be involved in the reaction with those halides (ICl_3, $RICl_2$) which can be reduced by organomagnesium compounds.

Organomercury compounds may react with halides (salts and other derivatives) of B, Tl, In, Si, Ge, Sn, Pb, Ti, P, As, Sb, Bi, V, S, Se, Te, I, and At. In the cases of B, Tl, Pb, P, As, and partially Si, the above reactions are important synthetic methods for the preparation of their organic derivatives.

Mechanistically interesting among these reactions are the exchange reactions of organomercury compounds with Hg* Hal_2 (where Hg* is a radioactive isotope) which have, however, no preparative significance. The above-mentioned R_2Hg and RHgCl react with $R'ICl_2$ as well ($R' = $ Ar or β-ClCH=CH).

No exchange with organomercury compounds proceeds with halides of the first group elements, alkali earth metals, Sc, Ga, and the following transition metal groups: Cr to Ni; Y to Pd; La to Pt, and actinides. As a rule, iron group metal halides as well as those of Cu and Mn initiate the decomposition of the organomercury compound into radicals, and give no organometallic compounds. Special cases of the reactions with $FeCl_3$ and cyclopentadienylcarbonyl iron derivatives will be discussed below.

Also considered in this chapter are exchange reactions of the halogens in alkyl- and arylhalides of various elements with an organic substituent of the organomercury compound.

II. REACTION MECHANISM

Nesmeyanov and Borisov first investigated the stereochemistry and mechanism of radical exchange of organomercury compounds with metal salts.

These studies involved the exchange reactions of stereoisomeric β-chlorovinyl (and other alkenyl) mercury derivatives with salts of Tl, Sn, B, Pb, etc. (and also reverse reactions), which showed that the above reactions occurred with retention of configuration. An example is the reaction of *cis,cis*-di-β-chlorovinylmercury with $TlCl_3$ giving only *cis,cis*-di-β-chlorovinylthallium[1]:

$$\left(\begin{array}{c} \text{H} \quad\quad \text{H} \\ \diagdown\!\!\diagup \\ \text{C}\!=\!\text{C} \\ \diagup\!\!\diagdown \\ \text{Cl} \end{array}\right)_2 \text{Hg} + \text{TlCl}_3 \longrightarrow \left(\begin{array}{c} \text{H} \quad\quad \text{H} \\ \diagdown\!\!\diagup \\ \text{C}\!=\!\text{C} \\ \diagup\!\!\diagdown \\ \text{Cl} \end{array}\right)_2 \text{TlCl} + \text{HgCl}_2$$

The reaction of *trans* isomers with BCl_3 was analogous,[2] as represented by the following reaction:

$$\left(\begin{array}{c} \text{H} \\ \diagdown \\ \text{C}\!=\!\text{C} \\ \diagup \quad\quad \diagdown \\ \text{Cl} \quad\quad \text{H} \end{array}\right) \text{HgCl} + \text{BCl}_3 \longrightarrow \left(\begin{array}{c} \text{H} \\ \diagdown \\ \text{C}\!=\!\text{C} \\ \diagup \quad\quad \diagdown \\ \text{Cl} \quad\quad \text{H} \end{array}\right) \text{BCl}_2 + \text{HgCl}_2$$

Similarly, alkenyl radicals transfer to other elements with retention of steric configuration. *Cis* and *trans* addition products of mercury salts to dimethyl- and diphenylacetylene[3] may be cited as examples.

These authors proved the rule of retention of configuration in electrophilic and homolytic substitution at olefinic carbon atoms.[4,5] This rule was also confirmed by a mercury exchange with the labeled mercury in its salts.[6]

Both fully substituted mercury compounds and organomercury salts may be involved in an isotope exchange with labeled mercury halides.

$$R_2Hg + Hg^*X_2 \rightleftharpoons R_2Hg^* + HgX_2 \tag{1}$$

$$RHgX + Hg^*X_2 \rightleftharpoons RHg^*X + HgX_2 \tag{2}$$

Both reactions were carried out in solvents. Reaction (1) was studied for $RHgR'$ containing aliphatic or aromatic organic substituents. In reaction

(2), the organomercury salts contained both saturated and unsaturated aliphatic as well as aromatic substituents.

In electrophilic reaction (1) (RHgR' with ^{203}HgBr$_2$) the comparative C—Hg bond strengths were defined by ^{203}Hg distribution between RHgBr and R'HgBr. For aliphatic compounds, that strength decreases in the order: $C_3H_7 > C_2H_5 > CH_3$.[7]

The reaction of cis-2-methoxycyclohexylneophylmercury with ^{203}HgCl$_2$ proceeds with retention of configuration and, according to the authors,[8] leads to equal distribution of ^{203}Hg between RHgCl and R'HgCl. However, this work does not consider an exchange reaction identified later[9]:

$$R_2Hg + R_2'Hg^* \rightleftharpoons R_2Hg^* + R_2'Hg$$

In the compound AlkHgAr, the strength of the Alk—Hg bond is higher than that of Ar—Hg. Thus, an exchange of ArHgAlk with Hg*Cl$_2$ gives AlkHgHal and ArHg*—Hal as shown by N. Nesmeyanov and Reutov,[10] contrary to Dessy[11] and Brodersen.[7] In the reaction of ArHgC$_6$H$_5$ with Hg*Cl$_2$, the phenyl radical is eliminated more easily than p-BrC$_6$H$_4$ (which is eliminated more easily than p-CH$_3$C$_6$H$_4$).[7]

The rate and degree of isotope exchange in reaction (2) depend on the nature of R and X. Among the aliphatic mercury derivatives β-chlorovinylmercuryhalides (reaction with Hg*Cl$_2$ in acetone at room temperature) and also α-mercuriated oxocompounds (at 25–50°C in various solvents) react very easily. α-Mercurated oxocompounds may be arranged according to the decrease in reactivity with ^{203}HgHal$_2$: α-bromomercurycyclohexanone > 3-bromomercury-3-benzylcamphor > α-bromomercury phenylacetic acid ethyl ester > α-bromomercuryphenylacetic acid l-menthyl ester > 3-bromomercurycamphor.[12]

The rate of exchange of AlkHgBr with Hg*Br$_2$ decreases in the order $CH_3 > C_2H_5 > $ neo-$C_5H_{11} > $ sec-C_4H_9 (rate values differ slightly for neo-C_5H_{11} and C_2H_5) which indicates some steric influence and is basically in agreement with a cyclic model for the transition state of this reaction[13] (the mechanism is proposed to be S_Ei according to Ingold).

The rate of exchange of p-YC$_6$H$_4$CH$_2$HgBr with Hg*Br$_2$ (in quinoline at 70°C) shows the following order depending on the nature of the Y-substituent; $(CH_3)_2CH > CH_3 > H > Cl$,[14] which is probably explained by the generation of a partial negative charge on C-atom bonded with mercury by the electronegative substituent. On the other hand, electropositive substituents cause a positive charge as compared with unsubstituted benzylmercury bromide. In the first case, the attack of the organomercury compound on the electrophile complex, B:HgBr$_2$, (where B is quinoline, see below) is facile, while in the latter case the attack is fairly difficult:

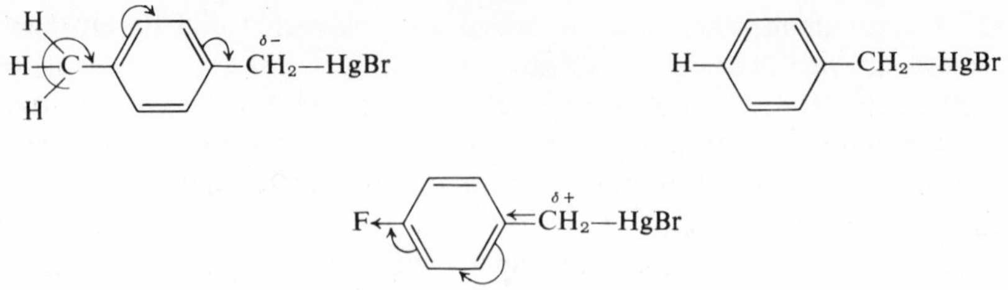

The kinetics, mechanism, and stereochemistry of reaction (2) depend on the medium. The reaction is generally of second-order kinetics (first in each component, compare also with ref. 21) in the following solvents: water,[15] alcohol,[16-20] acetic acid,[20] benzene,[20,21] toluene,[22] pyridine,[23] quinoline,[14,24] dimethylformamide[25] when R's in RHgX are CH_3[16-18], C_2H_5,[18] C_3H_7,[18,19] sec-C_4H_9,[14,24] $(CH_3)_3CHCH_2$,[16] Y-$C_6H_4CH_2$,[14,24] $C_6H_5CHCO_2C_2H_5$,[23,25,26] C_6H_5,[20-22] and p-$HO_2CC_6H_4$,[15] and X's are various anions.

The rate of reaction (2) increases with increasing solvent polarity. Thus an exchange of $C_6H_5CH(HgBr)CO_2R$ with Hg*Br_2 in dioxane containing 30% water[26-28] is more rapid than in benzene,[25,29] p-Y$C_6H_4CH_2HgBr$ reacts faster in dimethyl sulfoxide than in quinoline.[30]

Addition of acids[12,20] or bases[12,25] increases the rate of reaction of RHgHal with Hg*Hal_2. The catalytic influence of the base is probably due to the fact that its complex with mercury halide (B:HgX_2) is a stronger electrophilic reagent than the mercury halide itself which is of low polarity.[14] Although the free mercury salt probably gives no reaction in quinolinelike solvents, its complex may react.

One cannot exclude the possibility that the complex RHgX:B participates in the reaction. Acid catalytic influence for α-mercurated oxocompounds has been explained[12] by hydrogen bonding followed by a weakening of the C—Hg bond:

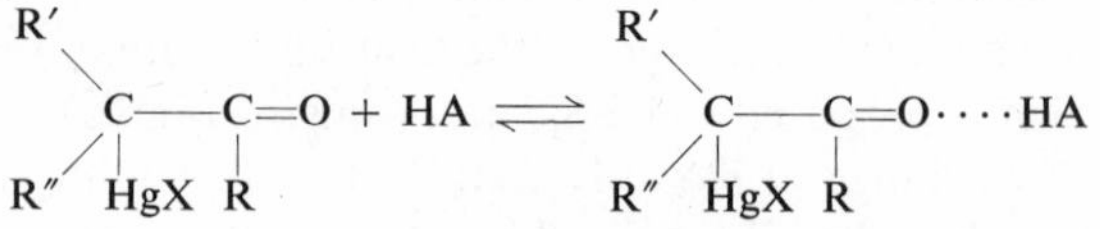

In addition, acid also catalyzes the reaction of C_6H_5HgX with Hg*X_2, which might be caused by acid anion influence on HgX_2, analogous to the base influence.

Reaction (2), where R = sec-C_4H_9 (in alcohol),[16] cis- and trans-β-ClCH=CH (in acetone),[6] cis- and trans-cyclo-2-$CH_3OC_6H_{11}$ (in dioxane)[31] proceeds through the radical mechanism with retention of configuration which excludes an S_E1 mechanism.

A sharp increase in the rate of Hg^*X_2 exchange with sec-C_4H_9HgX and especially with CH_3HgX (in alcohol) indicates the following X rate order: $Br < I \ll CH_3CO, \ll NO_3$, which, along with an increase in the ionic strength of the anion, indicates an open form of a transition state for the reactions mentioned (S_E2 mechanism, Ingold[16]). Reutov has suggested a four-centered transition state for the S_E2 reactions of isotope exchange and for analogous reactions studied by him (cf. for example Chapter 15):

$$RHgR' + Hg^*Hal_2 \rightleftharpoons \left[\begin{array}{c} R' \\ \diagup \\ Hg \\ \diagup \quad \cdots \\ R \qquad Hal \\ \cdots \quad \diagup \\ Hg \\ \diagdown \\ Hal \end{array} \right] \rightleftharpoons RHg^*R' + HgHal_2$$

where R is Alk, Ar, $R' = R$ for reaction (1) and Hal for reaction (2).

A similar four-centered transition state was suggested by Dessy for reaction (1) (cf. also Chapter 15).

If solvent polarity decreases, the cyclic transition state is more probable. In polar solvents (quinoline, for example) the reaction may proceed through an open transition state[14]:

$$RHgBr + B:Hg^*Br_2 \rightleftharpoons \left[\begin{array}{c} HgBr \\ \diagup \\ R \\ \diagdown \\ Hg^*{-}Br \\ \diagup \; | \\ B: \; Br \end{array} \right] \rightleftharpoons RHg^*Br + \underbrace{[HgBr]^+ + Br^- + :B}_{B:HgBr_2}$$

Reaction (2) of monoalkyl exchange for $R = CH_3$ and sec-C_4H_9 is catalyzed by metal halides (LiHal),[32] which may be explained by the formation of considerably less stable $HgCl_3^-$ bonding than $HgHal_2$ bonding. Two types of catalysis were established in this case: mono- and dianionic, depending on the number of Hal ions (contributed by LiHal), participating in the transition state. The retention of steric configuration of optically active sec-C_4H_9 groups in catalysis of both types indicates a cyclic transition state. For monoanionic (2) and dianionic (3) catalysis a transition state may be defined as follows (Ingold, et al.[32]):

$$\left[\begin{array}{c} Hal \\ \diagup \\ Hg \\ \diagup \quad \diagdown \\ R \qquad Hal \\ \diagdown \quad \diagup \\ Hg^* \\ | \quad \diagdown \\ Hal \quad Hal \end{array} \right] \qquad \left[\begin{array}{c} Hal \\ \diagup \\ Hg{-}Hal \\ \diagup \quad \diagdown \\ R \qquad Hal \\ \diagdown \quad \diagup \\ Hg^* \\ \diagup \quad \diagdown \\ Hal \quad Hal \end{array} \right]$$

$$\text{(2)} \qquad\qquad\qquad \text{(3)}$$

Similarly, reaction (2) is catalyzed by addition of KBr as in the case of $R = p\text{-}YC_6H_4CH_2$ in dimethylsulfoxide solution.[30] However, in the absence of KBr, the rate effect of the substituents increases as follows: $Cl < H < CH_3 < CH(CH_3)_2$ (in quinoline,[14] see above) or $F, Cl < H < CH_3, CH(CH_3)_2$ (in dimethyl sulfoxide[30]) while in the presence of KBr this effect is reversed: $(CH_3)_2CH < CH_3, H < F < Cl$. This supports a formation of organomercury complex with Br^- in the transition state, similar to that proposed by Ingold (see **2** and **3**). Electronegative substituents, attracting the electrons from the mercury atom, should enhance complex formation with the Br^- (compared to unsubstituted benzyl-mercury bromide) (see **4**):

(4)   attack is easier

(5)

(6)   attack is more difficult

On the other hand, electron-donating substituents decrease bromo anion-Hg atom covalent bonding (**6**).

Reaction (2) (in solvents of high dielectric constants) is first-order and follows an S_E1 mechanism, for example in dimethylsulfoxide when $R = p\text{-}NO_2C_6H_4CH_2$[33] (but not for some other $p\text{-}YC_6H_4CH_2$), and $C_6H_5C(H)CO_2C_2H_5$[34] as well as in dioxane when $R = C_6H_6C(H)\text{-}CO_2C_2H_5$.[26,27,34]

In these solvents when $R = C_6H_5CHCO_2C_2H_5$ product racemization results.[34]

The S_E1 mechanism is probably as follows:

$$RHgBr \underset{\longleftarrow}{\overset{slowly}{\longrightarrow}} \overset{\delta-}{R:} \overset{\delta+}{HgBr} \underset{\longleftarrow}{\overset{Hg^*Br_2}{\longrightarrow}} \overset{(-)}{R:} + \overset{(+)}{Hg^*Br} \underset{\longleftarrow}{\longrightarrow} RHg^*Br$$

According to decreasing rate, the effect of Y-substituents in $YC_6H_4\text{-}C(H)HgCOOC_2H_5$ in dioxan form the series[26]:

$$p\text{-}I, Br, Cl > p\text{-}F > H > o\text{-}CH_3 > p\text{-}CH_3 > p\text{-}tert\text{-}C_4H_9$$

which agrees with an S_E1 mechanism since the stability of the carbanion should increase with increasing negative charge delocalization:

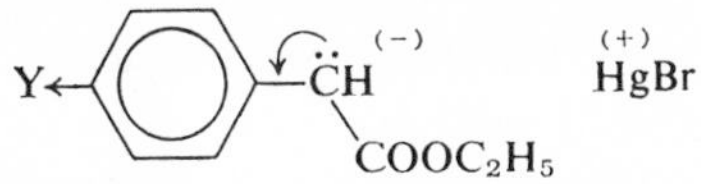

The same might be said for $p\text{-}O_2NC_6H_4CH_2HgBr$.[33]

The monomolecular reaction rate of $p\text{-}O_2NC_6H_4CH_2HgBr$ with Hg^*Br_2 is increased by KBr as well,[35] which is probably caused by formation of the $RHgBr\text{---}Br^{(-)}$ complex, giving the following charge distributions:

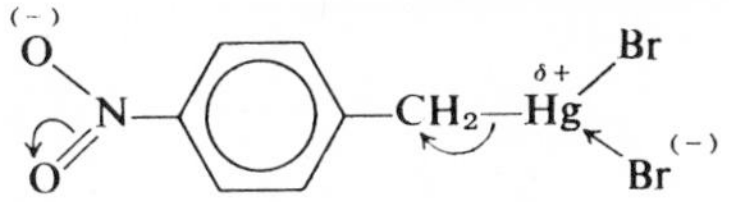

Alkylation of organothallium compounds by organomercury derivatives is due to the polarizing effect of the solvent used.[36]

The reaction of $(C_2{}^{14}H_5)_2TlBr + (s\text{-}C_4H_9)_2Hg \rightarrow (C_2{}^{14}H_5)_2Tl\text{-}s\text{-}C_4H_9 + s\text{-}C_4H_9HgBr$ (in dimethylformamide at 70°C) was found to be of general first order [first in $(\text{sec-}C_4H_9)_2$ and zero in $(C_2^*H_5)_2TlBr$] and led to complete racemization of dibutylmercury.

Under the experimental conditions described, the reaction follows an S_E1 mechanism. Ionization of $(\text{sec-}C_4H_9)_2Hg$, producing a carbanion, is the rate-determining step:

$$(s\text{-}C_4H_9)_2Hg \xrightarrow{\text{Slow}} s\text{-}C_4H_9Hg^{(+)} + s\text{-}C_4H^{(-)}$$

$$s\text{-}C_4H_9Hg^{(+)} + Br^{(-)} \xrightarrow{\text{Fast}} s\text{-}C_4H_9HgBr$$

$$s\text{-}C_4H^{(-)} + (C_2^*H_5)_2Tl^{(+)} \xrightarrow{\text{Fast}} s\text{-}C_4H_9Tl(C_2^*H_5)_2$$

III. REACTION CONDITIONS

Organomercury compounds react with halides of various elements at different rates.

ICl_3 (see Chapter 11 about reaction with ICl) reacts very rapidly with a suspension of an organomercury compound in water at room temperature.

Reactions of organomercury compounds with boron, arsenic, and thallium halides and also with thallium acylates proceed at room or at slightly elevated temperatures obtained by use of low-boiling solvents (chloroform, benzene). With PCl_3, the reaction requires more drastic conditions and is usually carried out with heating, sometimes to temperatures as high as 200–300°C, and even in sealed tubes. The aim to attain the higher alkylated or arylated organoelement compounds requires more drastic reaction conditions. Substituents bonded to mercury also split off with varying ease.

Generally, aromatic mercury compounds undergo exchange with metal halides, especially those containing electron-releasing substituents in the aromatic nuclei.

Mercury derivatives of superaromatic heterocycles (α-furyl, α-thienyl mercury derivatives) also undergo exchange. β-Chlorovinyl and other alkenyl groups containing a mercury atom at the double bond split off readily. More drastic conditions are required for exchange of saturated aliphatic organomercurials with salts of elements, and even some salts fail to exchange.

IV. APPLICATION OF THE REACTIONS

A. Reactions with Salts of Group III Elements

1. Boron Halides

The exchange reactions of saturated aliphatic mercury derivatives with boron halides have not been reported. Alkenyl and aryl boron compounds were prepared from boron halides and organomercury derivatives. In many experiments no solvents were used in the reaction. Boron halides BF_3 (with divinylmercury) the reaction product was RBF_2[37,38]), BCl_3, and BBr_3 react with alkenylmercury compounds (R_2Hg or $RHgHal$) with no or little heating. Depending on the molar ratio of the reactants, mono[2,38,39] and dialkenyl boron derivatives[2] were obtained from the reaction with BCl_3.

Since Michaelis' work[40–42] of the last century, the reactions of fully substituted aromatic mercury derivatives with boron halides have been widely used in the syntheses of aromatic boron compounds, mainly for the synthesis of $ArBHal_2$:

$$Ar_2Hg + 2BHal_3 \longrightarrow 2ArBHal_2 + HgCl_2 \qquad (3)$$

An excess of Ar_2Hg leads to Ar_2BHal:

$$ArBHal_2 + Ar_2Hg \longrightarrow Ar_2BHal + ArHgHal \qquad (4)$$

$ArHgHal$ also exchanges with $BHal_3$:

$$ArHgHal + BHal_3 \longrightarrow ArBHal_2 + HgHal_2 \qquad (5)$$

If Ar contains alkoxylike substituents, reaction (3) proceeds at room temperature, but if Ar is a hydrocarbon, only the reaction with BCl_3 is initiated at room temperature. Meanwhile Michaelis has shown that the reaction may be carried to completion by heating to 200°C in a sealed tube. High temperatures were not needed in the reaction of $(C_6H_5)_2Hg$ with

BCl_3.[42] Phenylboric acid was prepared by hydrolysis of $C_6H_5BCl_2$ at room temperature in chlorobenzene. BBr_3 is preferable to BCl_3, since it reacts more easily with organomercury compounds. The aromatic derivatives $ArBBr_2$ and Ar_2BBr,[42] in poorer yield, are obtained in boiling benzene.

In some cases, reaction (4) was carried out under more drastic conditions (in a sealed tube at 300–320°C). The author proposes the following reaction scheme for the reaction of o-phenylenemercury with BCl_3 refluxing in toluene[44]:

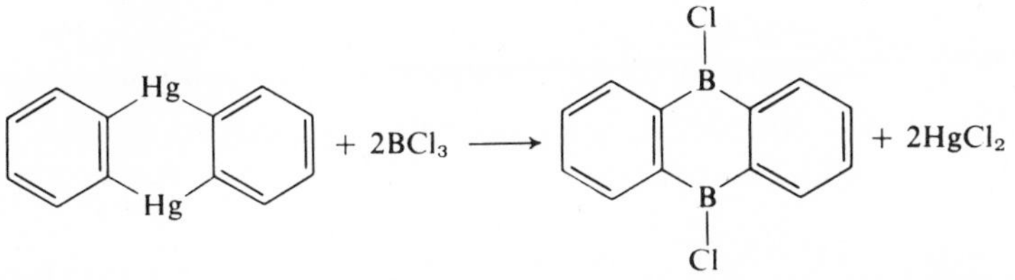

(7)   (8)

Since o-phenylene mercury (7) has a hexameric structure, the same structure or some polymeric one for the organoboron compound (8) may not be excluded.

Heating of p-chloromercuryphenol, p-chloromercurybenzoic acid, and carbomethoxyphenylmercury[41] for many hours with boron trichloride failed to give any organoboron compounds.

Again, no reaction was observed in the case of diphenylmercury and tri-n-butoxyboron under various conditions.[43] The reaction of YC_6H_4-$HgHal$ (Hal = Cl or Br, Y = H, CH_3 or Hal) with $BHal_3$ (Hal = Cl or Br) (several hours in boiling benzene) produces a good yield of the corresponding $YC_6H_4BHal_2$.[45]

2. Indium Halides

The sole reaction follows: Diphenylmercury and water-free $InCl_3$ (37 hr of refluxing in xylene) led to diphenylindium chloride.[46]

3. Thallium Salts

Among saturated aliphatic organomercury derivatives, only di-n-propylmercury reacts with thallium trichloride.[47]

$$(n\text{-}C_3H_7)_2Hg + TlCl_3 \longrightarrow n\text{-}C_3H_7TlCl + 2n\text{-}C_3H_7HgCl$$

Fully substituted organomercury derivatives containing an Hg atom at the olefinic carbon atom[1,48–56] react with both thallium trichloride and tribromide under mild conditions (ether solution) resulting in the corresponding dialkenylthallium halides. At a molar ratio of $R_2Hg:TlCl_3 = 1:2$

either di-*cis,cis*- or di-*trans,trans*-1-methyl-2-acetoxy-1-propen-1-ylmercury brought about the corresponding *cis*- or *trans* $RTlCl_2$ derivatives, also in good yields.[53] *trans* Isomers react faster than *cis* isomers. Thus, for example, when the ethereal solutions of *cis*- or *trans*-di-β-chlorovinylmercury were mixed with thallium trichloride, di-*trans*-β-chlorovinylthallium precipitated immediately while the *cis*-derivative precipitated in 4 hr.[1] *cis,trans*-Dipropenyl mercury reacts with thallium trichloride to form di-*trans*-propenylthallium chloride[50] under the reaction conditions previously described. All the reactions retain the steric configuration of the transferred radical.

Di-*o*-acetoxyphenylmercury, refluxed in alcohol solution with $TlBr_3 \cdot 4H_2O$, gives a small yield of R_2TlBr.[57] In the reaction of fully substituted unsymmetrical mercury compounds with thallium trichloride, *trans*-alkene splits off more easily from the mercury atom than the *cis*-alkene:

$$2RHgR' + TlCl_3 \longrightarrow R_2TlCl + 2R'HgCl$$

The same may be said for aryls, which indicates that aryls react more rapidly than alkyls.[58] These observations are also true with other salts or in acidic media.

Furthermore, the preparation of thallium aromatic derivatives is not easily obtained by the reaction of organomercury compounds and $TlHal_3$.[59] Far better results were obtained when thallium halide was replaced by the thallium salt of an organic acid (thallium isobutyrate). The reaction of Ar_2Hg with thallium isobutyrate is an excellent synthetic method for the preparation of $ArTlX_2$ and Ar_2TlX.[60-62] The use of equimolar amounts of Ar_2Hg and $Tl(O_2CR)_3$ leads to the monoaryl thallium salt:

$$Ar_2Hg + Tl(OCOR)_3 \longrightarrow ArTl(OCOR)_2 + ArHg(OCOR)$$

The diarylthallium salt was obtained from the reaction of 2 moles of diarylmercury with 1 mole of $Tl(O_2CCR)_3$[60,61]:

$$2Ar_2Hg + Tl(O_2CR)_3 \longrightarrow Ar_2Tl(OCOR) + 2ArHgOCOR$$

Both reactions proceed under mild conditions (short time heating in chloroform) giving high product yields and were used for the synthesis of aromatic and thienyl thallium derivatives.

The same method has been used for the preparation of thallium derivatives of radicals containing nitrogen.[62a] (See tables.)

Dicyclopentadienylmercury reacts with thallium(I) hydroxide (in water–methanol solution at $+2°C$) giving cyclopentadienylthallium, which has no analogy among other organomercury derivatives.[63]

$$(C_5H_5)_2Hg + TlOH \longrightarrow C_5H_5Tl$$

B. Reactions with Salts and Other Compounds of Group IV Elements

1. Silicon Halides and Alkyl-Silicon Halides

These reactions are not of wide preparative importance. Saturated aliphatic mercury compounds probably give no reaction with silicon halides. Divinylmercury and $SiCl_4$ produce vinyltrichlorosilane.[38] Reactions of mercury oxo-derivatives with tetrachlorosilane or chloroalkyl (aryl) or alkoxysilane (in boiling isopentane) produced compounds containing Si—O bonding, since the halogens are completely replaced by vinyloxygroups; tetravinyloxysilane, vinyloxyalkylaryl, and alkoxysilanes are the reaction products[64]:

$$n\text{-}Hg(CH_2CHO)_2 + (R_{4-n})SiCl_{4n} \longrightarrow (R_{4-n})Si(OCH{=}CH_2)_n + ClHgCH_2CHO$$

On the other hand, mercury diacetic acid esters and trialkylsilane iodides, depending on the medium, either react without transfer of the reaction center in the organomercury molecule, giving both Si—C bonding and trialkylsilane acetic acid esters, or the reaction center may be transferred to an oxygen atom (in a solvent of low ionizing power such as benzene or carbon tetrachloride):

$$Hg(CH_2COOCH_3)_2 \xrightarrow[-\,IHgCH_2COOCH_3]{+\,Et_3SiI} \begin{array}{c} \xrightarrow{C_6H_6,CCl_4} CH_2{=}C(OCH_3)(OSiEt_3) \\[2mm] \xrightarrow{CHCl_3,C_2H_4Cl_2} Et_3SiCH_2COOCH_3 \end{array}$$

resulting in the formation of α-trialkylsilane oxyvinylalkyl ethers, e.g., α-triethylsilane oxyvinylmethyl ether was prepared in about 50% yield.[65-67]

The synthesis of organosilicon compounds from the reaction of diarylmercury with tetrachlorosilane occurred for only two species (diphenyl- and di-p-tolylmercury). Heating for many hours in a sealed tube caused an exchange of only one halide atom in $SiCl_4$ with aryl radical, monoaryltrichlorosilane being the final product.[68,69]

2. Tetrachlorogermanium and Alkylgermanium Halides

Germanium tetrahalides as well as silicon tetrachlorides do not react with dialkylmercury when the alkyl group is a saturated radical. A yield of 70% of vinylgermanium trichloride was produced[38,70] when divinylmercury and germanium tetrachloride were heated at $80°C$ for 4, 5 hr:

$$(CH_2{=}CH)_2Hg + GeCl_4 \longrightarrow CH_2{=}CHGeCl_3 + CH_2{=}CHHgCl$$

Under drastic conditions (see tables) $RGeCl_3$[71] was obtained from the reaction of $GeCl_4$ with aralkyl mercury derivatives. Reacting diaryl-mercury with $GeCl_4$ led to a monoarylated germanium compound $ArGeCl_3$. The reaction requires two days heating at 140–190°C in a sealed tube.[71] Mild conditions in the reaction of germanium(II) halides and diarylmercury produce highly arylated germanium compounds giving a synthetic approach for the latter (cf. Chapter 4).

Dipropylgermanium iodide and mercury diacetic acid methyl ester refluxed in petroleum ether gave dipropylgermanium diacetic acid[72]:

$$(n\text{-}C_3H_7)_2GeI_2 + Hg(CH_2CO_2CH_3)_2 \longrightarrow (n\text{-}C_3H_7)_2Ge(CH_2CO_2CH_3)_2 + HgI_2$$

Trialkylgermanium iodide and mercury diacetic acid esters under mild conditions (refluxing in tetrahydrofuran or petroleum ether) produce trialkylgermanium acetic acid esters[65] in good yield:

$$2Alk_3GeI + Hg(CH_2CO_2R)_2 \longrightarrow Alk_3GeCH_2CO_2R + HgI_2$$

The reaction was carried out in the presence of pyridine, which precipi-tated the mercury iodide as the insoluble complex $HgI_2 : 2C_5H_5N$, which is readily isolated by filtration.

Reactions of mercury diacetic acid esters with di- and trialkylgermanes are discussed in Chapter 7.

3. Tin(IV) Chloride

Data for the synthesis of tin aliphatic compounds from tin(IV) halides and mercury aliphatic derivatives are unreliable. Divinylmercury and tin(IV) chloride gave vinyltin trichloride.[38]

If diphenylmercury is used, 12 hr of heating in ligroin led to a mixture of diphenyltin(IV) chloride and to its partial hydrolysis products.[73] Tin(II) halides are used for the synthesis of organotin compounds but not tin(IV) halides (cf. Chapter 4).

4. Lead Tetraacylate

The instability of lead tetrahalides as well as the insolubility of lead dihalides preclude their use in the syntheses of organolead derivatives in exchange-type reactions with organomercury compounds.

However, organomercury derivatives exchange with lead tetraacylates and an excellent synthetic method for the preparation of organolead derivatives is thus available.

An example is the reaction of dialkyl (aryl) mercury and lead tetra-acetate (in equimolar amounts) resulting in the formation of lead diaryl (dialkyl) diacetate.[74]

$$R_2Hg + Pb(OCOCH_3)_4 \longrightarrow R_2Pb(OCOCH_3)_2 + Hg(OCOCH_3)_2$$

The same reaction was used for the preparation of di-(*trans-β*-chloro-vinyl) lead.[75] Retention of configuration of the transferred radical was observed. Reaction of diaryl mercury and an excess of lead tetraacylate

$$Ar_2Hg + Pb(OCOR)_4 \longrightarrow ArPb(OCOR)_3 + ArHg(OCOR)$$

is a synthetic method for the preparation of monoaryl lead derivatives.[76,77]

The above reactions occur in chloroform and require mild conditions giving good yields. The last reaction also proceeds in benzene, though with poorer yield.[78]

The monoarylated lead derivative may be arylated further by diaryl-mercury to produce diaryllead diacetate.[79]

$$ArPb(OCOR)_3 + Ar_2Hg \longrightarrow Ar_2Pb(OCOR)_2 + ArHg(OCOR)$$

This reaction proceeds readily in chloroform without heating.

5. Titanium(IV) Chloride and $(C_5H_5)_2TiCl_2$

Organomercury compounds and titanium(IV) chloride usually react as follows[80]:

$$2(C_6H_5)_2Hg + 2TiCl_4 \longrightarrow 2C_6H_5HgCl + 2TiCl_3 + C_6H_5C_6H_5$$

The reaction was carried out in either CCl_4 or a sealed tube without solvent. Unstable organotitanium compounds are probably the intermediates. This phenomenon was observed for $(C_6H_5)_nTiX_{4-n}$ when $(C_6H_5)_2Hg$ and $TiCl_4$ were allowed to react in various organic solvents (chloroform and benzene) at different temperatures. In tetrahydrofuran, the intermediate could be $(C_6H_5)_4Ti$.[81]

Upon heating, the reaction of $(C_5H_5)_2TiCl_2$ with $(C_6H_5)_2Hg$ was assumed to proceed as follows since C_6H_5HgCl was formed, and after acidification the oxidation products of organotitanium compounds[82] were identified:

$$(C_5H_5)_2TiCl_2 + (C_6H_5)_2Hg \longrightarrow (C_5H_5)_2Ti(C_6H_5)_2 + 2C_6H_5HgCl$$

C. Reactions with Halides, Alkylhalides, and Other Salts in Group V Elements

1. Phosphorus Halides and Alkylhalides

One of the most valuable preparative reactions of organomercury compounds is the reaction with phosphorus trichloride or tribromide and alkylphosphine dichlorides, which is a valuable synthetic method for the preparation of alkyl and arylphosphine dichlorides and diarylphosphine chlorides.

Dialkylmercury may alkylate phosphorus trichloride to form only the monoalkyl derivative:

$$Alk_2Hg + PHal_3 \longrightarrow AlkPHal_2 + AlkHgHal$$

Drastic conditions (many hours of heating at 200–250°C in a sealed tube) were required for this reaction.[83,84] No dialkylphosphine halides were formed from dialkylmercury and $PHal_3$ or $AlkPHal_2$. Vinylphosphine dihalides were prepared from divinylmercury and PCl_3[37,85] or PBr_3[85,85a] under mild conditions (at 80°C). Other alkenyl, dialkenyl, or divinyl-phosphine halides were not prepared from organomercury compounds. However, fully substituted aromatic derivatives of mercury, Ar_2Hg, arylate phosphorus trichloride and arylphosphine dihalides as well.[86–90] Secondary phosphine halides may be prepared from arylmercury halides and monosubstituted phosphine halides if Ar is more electronegative than phenyl, for example p-tolyl:[91]

$$ArHgHal + RPHal_2 \longrightarrow RArPHal + HgHal_2$$

Diphenylmercury, reacting with phenylphosphine dichloride, gave diphenylphosphine chloride and in excess $C_6H_5PCl_2$ at elevated temperatures,[88,92] mercury was dearylated only into phenylmercury chloride (compare with $(C_6H_5)_2Hg$ and $C_6H_5AsCl_2$ in next paragraph).

Repeated arylation of $RPCl_2$ may result in mixed diarylphosphines $RR'PCl$.[91,93–95] All the reactions require high temperatures (200–250°C). No triarylphosphines were obtained from the reactions.

Organophosphorus derivatives are usually prepared without any solvent. Best results were obtained in excess PX_3 or RPX_2. A limitation of the above methods is the difficulty in separating the phosphorus from the mercury compounds which requires multiple distillations, thereby decreasing the first satisfactory yield of 50–60%.

2. Halides and Alkyl Halides of Arsenic(III)

Reactions of organomercury compounds with halides and alkyl halides of arsenic are general synthetic methods.

However, these reactions, as well as those for phosphorus, were mostly worked out for aromatic derivatives. Mercury derivatives and arsenic trichloride react under milder conditions than PCl_3. Thus, diethylmercury and $AsCl_3$ react even at room temperature, however the product is usually the monoalkyl derivative:[96,97]

$$(C_2H_5)_2Hg + AsCl_3 \longrightarrow C_2H_5AsCl_2 + C_2H_5HgCl$$

Perfluorovinylarsine dichloride[40] was prepared in a similar manner. $CF_2{=}CFAsCl_2$ and $RHgCl$[98] were prepared from $CF_2{=}CFHgR$ (R =

$CH_2=CH—$, C_2H_5, or C_6H_5) and $AsCl_3$. Meanwhile β-chlorovinyl-arsine dichloride (Lewisite 1) was obtained from the reaction of $AsCl_3$ and β-chlorovinylmercury chloride.[99]

The reaction of aromatic mercury compounds, Ar_2Hg, with arsenic trichloride led to a mixture of $ArAsCl_2$ and Ar_2AsCl.[100,101] In a large excess of $AsCl_3$ at 160–250°C, it is possible to react all the diarylmercury obtaining aryldichloroarsine and $HgCl_2$.[102] At lower temperatures this reaction does not go to completion.[102] Diarylarsine chloride was obtained from reacting diarylmercury with arylarsine dichloride.[103] If both reactants have different substituents, mixed secondary arsine chlorides, $RR'ArCl$, result.[104] In this case dearylation of diarylmercury is complete if excess arylarsine dichloride as well as high temperatures were used.[89,103] The yield of triphenylarsine in the reaction of diphenylmercury with phenyl-arsine dichloride was fairly small.[102] An excess of $RHgCl$ ($R = $ 2-furyl) or R_2Hg ($R = $ 1-thienyl) produces R_3As from $AsCl_3$ (cf. tables).

Aromatic mercury salts ($ArHgX$) also arylate $AsCl_3$[95,105] in contrast to aliphatic salts. The reaction:

$$ArHgCl + AsCl_3 \longrightarrow ArAsCl_2 + HgCl_2$$

is carried out at 100°C. However an experiment with C_6H_5HgCl has shown that not only $C_6H_5AsCl_2$ forms but also $(C_6H_5)_2AsCl$:

$$2RHgCl + AsCl_3 \longrightarrow R_2AsCl + 2HgCl_2$$

or

$$RHgCl + RAsCl_2 \longrightarrow R_2AsCl + HgCl_2$$

Since diphenylarsine chloride in the presence of mercury chloride re-arranges to phenylarsine dichloride and triphenylarsine at about 150°C, its preparation should be carried out for short reaction times.

Organomercury salts containing aryl—phenyl[105] or some group, such as α-thienyl or α-furyl are readily involved in a reaction with $AsCl_3$. On the contrary, o-nitrophenylmercury gives no reaction with $AsCl_3$ even at 200°C.[105] Trifurylarsine was obtained in 36% yield from the reaction of 3-furylmercury chloride and $AsCl_3$[106] (see tables).

Since the arsenic compound acts as a solvent, the reaction of $AsCl_3$ or $RAsCl_2$ usually requires no other solvent for the reaction with mercury compounds. However, solvents such as benzene or xylene are sometimes used.[106] In some cases exchange of the HgCl group by the $AsHal_2$ group occurs only in the reaction of $AsBr_3$ with $RHgCl$. An example is the ClHg group in the β-position in a furan derivative which has both α positions substituted. Such compounds may react only with $AsBr_3$, not with $AsCl_3$.[106]

3. Antimony Trichloride

Reactions of mercury compounds with antimony trichloride have not been widely used as a synthetic method for the preparation of organo-antimony derivatives. In contrast to the preparation of phosphorus and arsenic derivatives, their preparation is not simple and is accompanied by side reactions. In these reactions, oxidation of antimony from the trivalent to pentavalent state may occur. Only a few examples are given to illustrate these reactions. Dimethylmercury and antimony trichloride yield methyl-mercury chloride and trimethylantimony dichloride in rather small yields.[107] Metallic mercury was also isolated in this reaction.

Under mild conditions (in boiling benzene) diphenylmercury reacts with antimony trichloride yielding diphenylantimony chloride,[108] while under more drastic conditions (sealed tube, xylene, 130°C) a mixture of anti-mony(V) derivatives, $(C_6H_5)_3SbCl_2$ and $(C_6H_5)_2SbCl_3$, are produced.[109]

The reaction of phenylmercury chloride and antimony trichloride failed to give any organoantimony compounds.[110]

4. Bismuth Halides and Aryl Halides

The reactions of organomercury derivatives with bismuth trichloride and bismuth tribromide have been studied. No aliphatic mercury derivatives were found to react with bismuth halides and alkyl and arylhalides. Diphenylmercury does not react with bismuth trichloride to produce organobismuth compounds,[111,112] but bismuth tribromide unexpectedly gave a quantitative yield of triphenylbismuth under mild conditions (12 hr of heating in ether).[113] Furthermore, di-o-carboxyphenylmercury and bismuth trichloride hardly react in ether solution but when no solvent was used a bismuth derivative resulted.[114]

The exchange product is mainly obtained in the reaction of diphenyl-mercury with $(\alpha\text{-}C_{10}H_7)BiBr_2$. C_6H_5HgBr and a small amount of $(\alpha\text{-}C_{10}H_7)Bi(C_6H_5)_2$ were also formed.[113]

Mixing $(C_6H_5)_2Hg$ with $(\alpha\text{-}C_{10}H_7)_3BiBr_2$ did not result in a more arylated product. Heating the mixtures caused decomposition and led to a mixture of bismuth(III) organic derivatives.[113]

5. Vanadium and Vanadyl Halides

Diphenylmercury reacts with vanadium halides and vanadyl halides in cyclohexane solution at room temperature to give quantitatively phenyl-mercury chloride and phenylvanadium derivatives which are red in solution[115]:

$$(C_6H_5)_2Hg + VOCl_3 \rightleftharpoons C_6H_5VOCl_2 + C_6H_5HgCl$$
$$(C_6H_5)_2Hg + VCl_4 \rightleftharpoons C_6H_5VCl_3 + C_6H_5HgCl$$

The first derivative was identified by spectral data and by benzene formation upon hydrolysis with hydrogen chloride.

D. Reactions with Alkyl Halides, Halides, and Some Other Inorganic Compounds of Group VI Elements

1. Halides and Inorganic Compounds of Sulfur

Reactions of mercury derivatives with the above sulfur compounds are not typical ones and were reported only in some instances. These reactions cannot be used as synthetic methods for the preparation of organosulfur compounds.

Only mercarbides were found to react with sulfur chloride. No reaction was observed in the case of hexachloromercuryethane $(C_2(HgCl)_6)$. This compound was assigned a methanic structure[116] but not an ethanic one; in another paper,[117] this compound was alleged to have a polymeric structure. In the Hofmann derivative of the structure

$$
\begin{array}{ccc}
\text{HO—Hg} & & \text{Hg—OH} \\
\diagdown & & \diagup \\
& \text{Hg—C—C—Hg} & \\
\diagup \diagdown & & \diagup \diagdown \\
\text{O} \quad \diagdown & & \diagup \quad \text{O} \\
\text{Hg} & & \text{Hg}
\end{array}
$$

the mercury was partially split off by S_2Cl_2 in benzene producing a mercury-containing disulfide[118]:

$$
\begin{array}{c}
(ClHg)_2(C\!\!-\!\!C)(HgCl)_2 \\
\quad | \quad | \\
\quad S\!\!-\!\!S
\end{array}
$$

The reactions of Ar_2Hg with ArSCl (cf. in Chapter 10) and $SOCl_2$ with Ar_2Hg failed to form any organosulfur compound; splitting of the C—Hg bond[119,120] led to ArHgCl or to metallic mercury and ArCl. Sulfoacetic acid[121] was obtained when chloromercuryacetaldehyde was treated with sulfuric anhydride. Sulfonic derivatives of anthraquinones were prepared from its mercury derivatives by reactions with sulfuric anhydride or other sulfonating agents (cf. Chapter 9). Diferrocenylmercury and sulfocyanic acid gave a complex from which diferrocenyl sulfide[122] was obtained after treatment with sodium thiosulfate.

2. Selenium Halides, Alkyl Halides, and Dioxide

Diarylselenium was obtained from the reaction of diarylmercury and selenium tetrabromide[123]:

$$SeBr_4 + 3R_2Hg \longrightarrow R_2Se + RBr + 3RHgBr$$

Use of excess $SeBr_4$ led to complete dearylation of the diarylmercury[123]:

$$3SeBr_4 + 3R_2Hg \longrightarrow 3HgBr_2 + 2RBr + 2R_2Se$$

The reactions were carried out in carbon disulfide. A yield of 21% diferrocenylselenium was obtained in the reaction of diferrocenylmercury with selenium tetrabromide (refluxed in chloroform for 1 hr followed by sodium thiosulfate addition).[122] Diarylselenium was prepared by the reaction of dihalodiarylselenium and diarylmercury in cold solution (CS_2, acetone, CCl_4)[124,125]:

$$Ar_2SeCl_2 + Ar_2Hg \longrightarrow Ar_2Se + ArCl + ArHgCl$$

Ar_2Se, formed initially, is further arylated when heated in acetone or without any solvent, and a selenium(IV) derivative was isolated as a double salt with mercury chloride[124,125]:

$$Ar_2SeCl_2 + ArHgCl \longrightarrow Ar_3SeCl \cdot HgCl_2$$

When arylselenium monobromide was reacted with Ar_2Hg, a mixed selenium derivative ($ArSeAr'$) was prepared if Se and Hg were bound to different aryl groups[126]:

$$Ar'SeBr + Ar_2Hg \longrightarrow ArSeAr' + ArHgBr$$

Careful warming of dialkylmercury and SeO_2 results in the formation of dialkylselenium.[127] R_2SeO may also be a reaction product.

3. Tellurium Halides and Aryl Halides

Aliphatic mercury compounds do not react with tellurium halides or arylhalides, similar to the behavior of sulfur derivatives.

Diphenylmercury and $TeCl_2$ (many hours of heating at 200°C in a sealed tube) gave no tellurium derivative but did react in the following manner[128]:

$$(C_6H_5)_2Hg + TeCl_2 \longrightarrow 2C_6H_5Cl + TeHg$$

If RHgCl, having the structure

is refluxed with $TeCl_4$ in $CHCl_3$ (where X is CH_3) or in acetonitrile (when X is COOH) the corresponding aryltellurium trichlorides were prepared[129]:

$$RHgCl + TeCl_4 \longrightarrow RTeCl_3 + HgCl_2$$

Both symmetrical (Ar = Ar') and mixed diaryltellurium derivatives were obtained by refluxing in dioxane in the absence of moisture[130]:

$$ArTeCl_3 + ClHgAr' \longrightarrow HgCl_2 + ArAr'TeCl_2 \longrightarrow ArTeAr'$$

The tellurium is reduced in this case.

4. Uranium Tetrafluoride

Heating of diethylmercury and uranium tetrafluoride at 150–170°C failed to provide any organic derivatives of uranium.[131]

E. Reactions with Halides and Alkyl Halides of Group VII Elements

1. Iodine Chlorides and $RICl_2$

Both fully substituted mercury derivatives and organomercury salts arylate iodine in iodine chloride:

$$R_2Hg + 2ICl \longrightarrow 2RI + HgCl_2$$
$$RHgX + ICl \longrightarrow RI + HgXCl$$

(also cf. Chapter 11). R_2Hg as well as $RHgCl$ (R = β-ClCH=CH— or Ar) react with iodine trichloride to form diaryl-[132] of di-β-chlorovinyl-[133] iodonium chloride, which is a convenient method for the preparation of iodonium salts:

$$2RHgCl + ICl_3 \longrightarrow R_2ICl + 2HgCl_2$$
$$R_2Hg + ICl_3 \longrightarrow R_2ICl + HgCl_2$$

The iodonium salt was isolated as a double salt with mercury chloride which may be separated by hydrogen sulfide addition, precipitating mercury sulfide from the water solution.

In excess iodine trichloride, the slow addition of diarylmercury led to aryl iodine chloride:

$$Ar_2Hg + ICl_3 \longrightarrow ArICl_2 + ArHgCl$$

β-Chlorovinyliodine chloride was a product of the reaction of equimolar amounts of $RHgCl$ and ICl_3.[134]

The reaction of diarylmercury or an organomercury salt with aryl-iodine chloride is a well-known procedure for preparing aryliodonium salts (Willgerodt reaction)[135,136]:

$$ArICl_2 + Ar_2Hg \longrightarrow Ar_2ICl + ArHgCl$$
$$ArICl_2 + ArHgCl \longrightarrow Ar_2ICl + HgCl_2$$

In the above reaction, diarylmercury usually reacts under mild conditions (water suspension or in some organic solvent at room temperature).

2. RAtCl$_2$

Analogous to the Willgerodt reaction, organoastatine derivatives corresponding to iodonium salts were prepared.[137] The reaction proceeds in a short time upon heating in chloroform.

F. Reactions with Halides of Group VIII Elements

Included in these reactions is an interesting one of chloromercury acetaldehyde with anhydrous FeCl$_3$ in dry acetone. The reaction requires transfer of the reaction center in the organomercury molecule and leads to an oxygen-bound metallic iron derivative, iron hydroxychloride vinylate[138]:

$$ClHg-CH_2-C(H){=}O \xrightarrow[\text{acetone}]{FeCl_3} CH_2{=}CHOFe(OH)Cl$$

An interesting example of alkylation with organomercury compounds is the following reaction[139]:

$$\pi\text{-}C_5H_5Fe(CO)_2I + Ar_2Hg \longrightarrow \pi\text{-}C_5H_5Fe(CO)_2Ar\text{-}\sigma$$
$$Ar = C_6H_5, p\text{-}C_2H_5O_2CC_6H_4$$

The reactions are carried out in benzene solution at room temperature and many hours of UV irradiation in an argon atmosphere. They probably proceed by homolytic cleavage of the arylmercury.

V. PREPARATIVE SYNTHESES

A. Preparation of Organoboron Derivatives

1. Preparation of Di-β-chlorovinylboron Chloride from Boron Trichloride and trans,trans-Di-β-chlorovinylmercury[2]

A solution of 17 g (0.052 mole) of *trans, trans*-di-β-chlorovinylmercury and 6.16 g (0.052 mole) of boron trichloride in 8 ml of benzene reacts in a sealed glass tube with heat evolution. Boron trichloride was introduced by distillation into a tube cooled with liquid nitrogen. After the reaction was completed, the mixture was heated at 50°C for 1 hr. The solid was filtered under a dry nitrogen atmosphere, washed with benzene, and dried. Fourteen grams (98%) of mercuric chloride was obtained. The solvent from the filtrate was distilled off and the liquid residue was twice distilled *in vacuo*, which gave 4.5 g of di-β-chlorovinylboron chloride, b.p. 80–82°C (25 mm), n_D^{20} 1.5452, d_4^{20} 1.2759, and 1.6 g of di-β-chlorovinylboron chloride mixed with di-β-chlorovinylboric acid. β-Chlorovinylboric acid (0.2 g) was isolated from the distilled solvent treated with several drops of water.

2. *Preparation of* ArBBr$_2$ *Type Compounds*[42]

An equal (by weight) mixture of boron tribromide and benzene was placed in a flask fitted with a reflux condenser closed by a calcium chloride drying tube.

One-third of the required diphenylmercury (calculated from the reaction $(C_6H_5)_2Hg + 2BBr_3 \rightarrow 2C_6H_5BBr_2$) was added into the flask, which caused moderate heat evolution. The mixture was heated on a water bath in order to initiate refluxing and the remaining diphenylmercury was added slowly, which caused vigorous boiling. The reaction was heated for an additional hour and then cooled.

Prior to vacuum distillation, the dark liquid obtained was separated from the solid mercury bromide, which was washed twice with benzene.

3. *Preparation of Phenylboron Dibromide* $C_6H_5BBr_2$[42]

The above method was used for the reaction of 42 g of BBr$_3$ with 30 g of diphenylmercury. A liquid product (80–150°C/20 mm) from which pure phenylboron dibromide (b.p. 99–101°C/20 mm, m.p. 32–34°C) was isolated by fractional distillation. The product has a strong odor, fumes in the air, and is vigorously hydrolyzed by water giving phenylboron acid.

4. *Preparation of Diphenylboron Bromide,* $(C_6H_5)_2BBr$[42]

A fairly small yield of the compound was obtained. Thirty-five grams of boron tribromide in 70 g of benzene were reacted with 50 g of diphenylmercury as described above. The resulting diphenylboron bromide was a colorless liquid, b.p. 150–160°C/8 mm which solidified at 24–25°C. The substance is hydrolyzed by water and forms diphenylboron acid.

5. *Preparation of Diphenylboron Chloride*[41]

Eighteen grams of diphenylmercury and 8 g of phenylboron dichloride were heated at 300–320°C for 24 hr in a sealed tube. Some pressure was observed in the tube. The contents were filtered and washed twice with petroleum ether (b.p. below 100°C). The solid was phenylmercury chloride. The filtrate was evaporated and the residue distilled at atmospheric pressure. The first fraction was unreacted $C_6H_5BCl_2$ and the next fraction (220–340°C), after repeated distillation, gave 3.5 g of diphenylboron chloride, b.p. 270–271°C and the second fraction (b.p. 271–272°C) diphenylboron chloride contaminated with $(C_6H_5)_2BOH$.

B. Preparation of Organothallium Derivatives

1. *Preparation of Phenylthallium Diisobutyrate*[60]

Thallium triisobutyrate (4.5 g) and 3.6 g of diphenylmercury (equimolar amounts, each reactant dissolved in 7 ml of warm, dry chloroform) were mixed and allowed to react for 30 min at room temperature. A solid precipitate was filtered off and washed with petroleum ether. Four grams (87.9%) of phenylthallium diisobutyrate was obtained which crystallized from dichloroethane as white crystals, m.p. 221–222°C. In contrast to thallium triisobutyrate, the product was stable in air.

2. Preparation of Diphenylthallium Isobutyrate[60]

Thallium triisobutyrate (4.65 g), dissolved in 7 ml of warm, dried chloroform, was added to 7.1 g of diphenylmercury in 12 ml of the same warm, dry solvent and allowed to react for 30 min at room temperature. A solid precipitate was filtered off and washed with petroleum ether. Diphenylthallium isobutyrate (3.6 g, 80.9%) was obtained, which crystallized from dichloroethane in white crystals, m.p. 241°C. The product was stable in air.

C. Preparation of Organosilicon Compounds

1. Reaction of Triethylsilane Iodide with bis(carbomethoxymethyl)-mercury[66]

a. *In Benzene: O-Derivative*, CH_2=$C(OCH_3)OSiEt_3$, (Ia). Triethylsilane iodide (43.5 g, 0.18 mole) in 30 ml of benzene was added dropwise to a stirred solution of 69.3 g (0.2 mole) of bis(carbomethoxymethyl)-mercury in 300 ml of benzene at room temperature. The mixture was treated with 500 ml of petroleum ether and cooled to −30°C. An insoluble material was filtered off and 72 g (90%) of iodo-(carbomethoxymethyl)mercury was obtained, m.p. 107–108°C. The organic layer, after removal of the solvent, was distilled. The yield of (Ia) was 17.7 g (53%), b.p. 65–66.5°C, 37 mm, n_D^{20} 1.4355, d_4^{20} 0.8870.

b. *In Dichloroethane: C-Derivative*, $Et_3SiCH_2COOCH_3$ (IIa). Triethylsilane iodide (14.6 g, 0.06 mole) in 10 ml of dichloroethane was added at room temperature to a stirred solution of 22.8 g (0.066 mole) of bis-(carbomethoxymethyl) mercury in 50 ml of dichloroethane. A similar treatment as in preparation of the *O*-derivative gave 6.8 g (60%) of product (IIa), b.p. 75–77°C, 7 mm, n_D^{20} 1.4399, d_4^{20} 0.9019.

D. Preparation of Organolead Compounds

1. Preparation of Di-β-chlorovinyllead Diacetate[75]

Fifteen grams (0.046 mole) of di-(*trans*-β-chlorovinyl) mercury in 15 ml of dry chloroform was added to a solution of 10.3 g (0.02 mole) of lead tetraacetate containing a small amount of acetic acid in dried chloroform. A yellow-orange color appeared. After 30–40 sec a crystalline solid began to precipitate with heat evolution and the color of the solution began to disappear. In 2.5–3 hr the completely colorless mixture was ice-cooled and the solid was filtered, washed with 10 ml of dry, cool chloroform, and 11.8 g (78%) of β-chlorovinylmercury acetate was obtained, which at 110°C turned grey and at 118–120°C blackened without any melting.

2. Preparation of Di-o-anisyllead Diacetate, $(o\text{-}CH_3OC_6H_4)Pb(OOCCH_3)_2$[74]

Di-o-anisylmercury (4.19 g, 0.01 mole) in 15 ml of dry chloroform was added to a solution of 2.21 g (0.005 mole) of lead tetraacetate in 20 ml of the same solvent. The orange solution obtained was allowed to stand at room temperature for 4 days. The solvent was evaporated on a water bath. A yellow crystalline material (a mixture of di-o-anisyllead diacetate and o-anisylmercury acetate) was extracted with hot petroleum ether. The remaining solid was di-o-anisyllead diacetate, which was recrystallized twice from alcohol to give 1.7 g of product, m.p. 191–193°C.

3. Preparation of α-Naphthyllead Triisobutyrate[79]

One gram of di-α-naphthyl mercury was added in small portions to a solution of 1.26 g (1 mole + 15% excess) of lead tetraisobutyrate in 15 ml of chloroform, which was acidified by a drop of isobutyric acid. No part of the mercury derivative was added until the previous one had dissolved completely. The addition lasted for 40–45 min. The organomercury derivative was then separated as before and the solution was left in air until the solvent evaporated. A solid crystallized, and 1.2 g of the product was obtained (m.p. 95–97°C). It was recrystallized from hexane-isopentane, and the melting point increased to 95.5–101°C. Repeated recrystallizations gave no melting point increase. The final yield was 0.6 g (46%).

4. Preparation of Di-α-naphthyllead Diacetate[79]

Di-α-naphthylmercury (0.35 g) was added to 0.4 g of α-naphthyl lead triacetate in 15 ml of dry chloroform. The mixture was heated until all the mercury compound dissolved, the solution was allowed to stand for 3 days, and was then filtered. The solvent was evaporated on a water bath and the residue was dried in air. The crystalline solid was washed twice with hot acetone and once with absolute alcohol acidified with a drop of acetic acid, m.p. 235–236°C.

E. Preparation of Organophosphorus Compounds

1. Preparation of n-Butylphosphine Dichloride[84]

Fifty grams of di-n-butylmercury was placed in a large Pyrex bomb tube, and the tube was flushed with nitrogen. One hundred grams of phosphorus trichloride was added slowly. A silky white solid precipitated almost immediately. The tube was sealed and heated in an electric oven at 200°C for 9 hr. A considerable amount of carbonaceous material and free mercury was produced during the heating. The tube was cooled and the contents of the tube were diluted with phosphorus trichloride to aid in transferring the material to a distilling flask. Fractional distillation gave 17 g (61.4%) of a product boiling at 157–160°C at 750 mm. The product fumed in air. Analysis indicated that the product was contaminated with some nonphosphorus derivative, probably *n*-butylmercury chloride.

2. Preparation of Vinylphosphine Dichloride [85]

Freshly distilled phosphorus trichloride (420 g, 3.06 mole) and 60 ml of dry, degassed mineral oil were placed under nitrogen in a flat-bottomed flask, suitable for magnetic stirring, and equipped with a thermometer well, dropping funnel, and reflux condenser. A 145-g (0.57 mole) sample of divinylmercury was placed in a pressure-equalized addition funnel, and the flask was warmed by a heating mantle until gentle refluxing occurred. Divinylmercury was then added slowly, and stirring begun. The temperature in the reaction flask was maintained from 65 to 85°C during addition of divinylmercury, and refluxing was continued for 1 hr after all the mercury compound had been added. The color of the solution in the flask became dark brown and was cooled. The reaction flask was attached to a distillation apparatus and about 250 ml of unreacted phosphorus trichloride was removed with care at 200 mm. As shown from previous experience, complete distillation in the presence of the brown solid results in only a small amount of desired material. Therefore, the remaining liquid was transferred *in vacuo* at room temperature into a 100-ml distillation flask, containing 15–25 ml of degassed mineral oil cooled to −78°C. Toward the end of the transfer, the reaction flask was heated to 100°C to ensure complete transfer of liquid product. The contents of the 100-ml distillation flask were carefully fractionated yielding 36 g of vinyl phosphine dichloride (50%), b.p. 64.4 ± 0.2°C/200 ± 1 mm.

3. Preparation of m-Chlorophenylphosphine Dichloride [89]

Di-(*m*-chlorophenyl)mercury (0.5 g, 0.022 mole) and 25 g (0.18 mole) of PCl$_3$ were heated in a sealed ampoule at 230–250°C for 24 hr. After opening the ampoule, the liquid was quickly filtered on a Buchner filter. The ampoule and solid were washed with absolute petroleum ether and the solvent was evaporated. Vacuum distillation of the liquid residue gave 6.8 g of product, b.p. 101–103°C (5 mm) which was again filtered through a sintered-glass filter No. 4 (to remove distilled-over impurity) and repeatedly distilled *in vacuo*. The product has an unpleasant odor.

4. Preparation of Diphenylphosphine Chloride[90]

$C_6H_5PCl_2$ and excess $(C_6H_5)_2Hg$ react at 230–240°C to form diphenylphosphine chloride.[92] It was separated from by-products by fractional distillation. The main product was obtained in 55% yield, b.p.179–180°C (16 mm).

5. Preparation of p-Tolylphenylphosphine Chloride[94]

p-Tolylphosphine dichloride (32.8 g) and 38 g of diphenylmercury were heated as quickly as possible to the boiling point. After an hour the mixture was heated to 270°C. The thick, brown oil obtained was extracted five times with boiling benzene with thorough stirring.

The largest part of the solvent was evaporated at dry atmospheric pressure avoiding contact with moisture. The residue was distilled *in vacuo*. Eight grams of p-tolylphenylphosphine chloride was obtained, b.p. 230–240°C (100 mm).

F. Preparation of Organoarsenic Compounds

1. Preparation of Ethylarsine Dichloride[97]

Diethylmercury (150 g) was added to 150 g of ice-cooled $AsCl_3$. After an exothermic reaction followed by crystal precipitation, the mixture was heated on a water bath for 30 min, treated with absolute ether, and filtered to remove ethylmercury chloride. The solvent was evaporated and 90 g of ethylarsine dichloride (b.p. 145–150°C) was separated from excess $AsCl_3$ by fractional distillation.

2. Preparation of CF_2=$CFAsCl_2$[98]

CF_2=$CFHgCH$=CH_2 (0.9 g, 0.0029 mole) and 0.53 g (0.0029 mole) of $AsCl_3$ were placed in a small ampoule fitted with a reflux condenser. After a vigorous reaction followed by white crystal precipitation, the mixture was heated for 2 hr on a water bath. The liquid contents were distilled off. Repeated distillation gave 0.53 g (85%) of CF_2=$CFAsCl_2$ and from the residue 0.6 g (82%) of CH_2=$CHHgCl$ was isolated. In a similar manner, CF_2=$CFAsCl_2$ and C_2H_5HgCl were also prepared from CF_2= $CFHgC_2H_5$ and $AsCl_3$.

3. Preparation of α-Naphthylarsine Dichloride[97]

Thirty grams of di-α-naphthylmercury and 200 g of $AsCl_3$ were heated for one hour at 160°C on an oil bath, cooled, and the precipitate filtered. Excess arsenic trichloride was removed *in vacuo* at 70–80°C, the residue was extracted with ether, the solvent evaporated, and the oil remaining crystallized slowly. Thirty-four grams (94%) of the final product was obtained, m.p. 68°C.

4. Preparation of Phenyl-p-tolylarsine Chloride[104]

Thirty grams of di-p-tolymercury was added to 30 g of boiling phenylarsine dichloride. The mixture was refluxed for 5 hr, cooled, and diluted with petroleum ether, p-tolylmercury chloride was filtered out, and the filtrate evaporated. The residue distilled *in vacuo* gave 41% of product, b.p. 215–237°C (29 mm). Diphenylarsine chloride and di-p-tolylarsine chloride were prepared in a similar manner.

5. Preparation of Diphenylarsine Chloride[103]

Diphenylmercury and no less than three times excess phenylarsine dichloride were refluxed for many hours at 270°C. Significant amounts of phenylmercury chloride and a small quantity of $HgCl_2$ were obtained along with diphenylarsine chloride and some yield of triphenylarsine which increased above 254°C—its boiling point. Diphenylarsine chloride is a yellow oil of weak odor, d_4^{14} 1.42231, b.p. 333°C (without decomposition).

G. Preparation of Organoselenium Compounds

1. Preparation of Diethylselenium[127]

Fifty grams of diethylmercury and 21 g of selenium dioxide were refluxed at 45–50°C on a water bath for 2 hr with vigorous stirring. The mixture became viscous during the reaction and selenium dioxide was gradually replaced by another solid. It is noteworthy that small overheating caused a rapid reaction with formation of metallic mercury and decomposition of the reaction products. After the reaction was finished, the mixture was extracted with several portions of ether. The solvent was evaporated and the liquid product was fractionated into two fractions: 105–130 and 130–160°C. The last fraction was mainly unreacted diethylmercury and the first one was repeatedly distilled yielding diethylselenide, b.p. 107–112°C, d 1.3, and n^{20} 1.4790.

H. Preparation of Organic Derivatives of Group VII Elements

1. Preparation of Diphenyliodonium Chloride[132]

C_6H_5HgCl (5.3 g, 0.017 mole) was added to 2 g (0.008 mole) of iodine trichloride in 10 ml of dilute hydrochloric acid. The mixture was heated on a water bath for one hour, cooled, and the solid was filtered off. The oily layer of iodobenzene was isolated from the filtrate. The solid was suspended in a large volume of water and hydrogen sulfide was carefully bubbled through. Mercury sulfide was filtered off and the filtrate evaporated. Recrystallization of the solid residue from water gave 1.12 g (42%) of product, m.p. 230°C (decomposition).

TABLE I

Reactions with Halides and Other Salts as well as with Alkyl- and Aryl-halides of Some Elements

R in R_2Hg or $RHgR'$ and $RHgX$	Reagent	Reaction conditions	Products	Yield, %	Ref.
		Boron Halides			
$CH_2{=}CH$	BF_3	$-196°$ to $20°C$	$CH_2{=}CHBF_2$	—	38
$CH_2{=}CH$	BF_3	—	$CH_2{=}CHBF_2$	—	37
$CH_2{=}CH$	BBr_3	$20°C$	$CH_2{=}CHBBr_2$	82	85[a]
$CF_2{=}CF$	BCl_3	—	$CF_2{=}CFBCl_2$	45	39
trans-$ClCH{=}CH$	BCl_3	Sealed tube, C_6H_6, $50°C$, 1 hr	(*trans*-$ClCH{=}CH)_2BCl$ $HgCl_2$	51 98	2
trans-$ClCH{=}CHHgCl$, $1M$	BCl_3, $1M$	Sealed tube, kerosene, $50°C$, 2 hr	*trans*-$ClCH{=}CHBCl_2$, $HgCl_2$	27	2
trans-$ClCH{=}CHHgCl$, $2M$	BCl_3, $1M$	Sealed tube, C_6H_6, $50°C$, 2 hr	*trans*-$(ClCH{=}CH)_2BCl$ *trans*-$(ClCH{=}CH)_2BOH$[a], $HgCl_2$	—	2
C_6H_5	BCl_3	Sealed tube, $180–200°C$	$C_6H_5BCl_2$	—	40
C_6H_5	BCl_3	Sealed tube, $180–200°C$	$C_6H_5BCl_2$	—	140
C_6H_5	BCl_3	Sealed tube, $180–200°C$	$C_6H_5BCl_2$	good	41
C_6H_5	$C_6H_5BCl_2$	C_6H_5Cl, 0.5–3 hr	$C_6H_5B(OH)_2$[a]	52–70	43
C_6H_5	BBr_3, $1M$	Sealed tube, $300–320°C$, 24 hr	$(C_6H_5)_2BCl$	34	41
C_6H_5, $1M$	BBr_3, $1M$	Benzene, reflux	$C_6H_5BBr_2$	—	42
C_6H_5, $2M$	BBr_3, $1M$	Benzene, reflux	$(C_6H_5)_2BBr$	—	42
o-$CH_3C_6H_4$	BCl_3	Benzene, $150–180°C$, 12 hr	o-$CH_3C_6H_4BCl_2$	—	41
p-$CH_3C_6H_4$	BBr_3	Benzene, reflux	p-$CH_3C_6H_4BBr_2$	—	42
o-$(CH_3)_2C_6H_3$	BCl_3	Sealed tube, $150–180°C$, 12 hr	o-$(CH_3)_2C_6H_3BCl_2$	—	42
m-$(CH_3)_2C_6H_3$	BCl_3	Sealed tube, $200°C$, 24 hr	m-$(CH_3)_2C_6H_3BCl_2$	—	42
m-$(CH_3)_2C_6H_3$	BBr_3	Benzene, reflux	m-$(CH_3)_2C_6H_3BBr_2$	—	42
p-$(CH_3)_2C_6H_3$	BCl_3	Sealed tube, $180°C$	p-$(CH_3)_2C_6H_3BCl_2$	—	42
$1,2,4$-$(CH_3)_3C_6H_2$	BBr_3	Benzene, reflux	$1,2,4$-$(CH_3)_3C_6H_2BBr_2 \cdot 5$	—	42
o-$CH_3OC_6H_4$	BCl_3	Benzene, reflux, $20°C$, 6 hr	o-$CH_3OC_6H_4BCl_2$	—	41

TABLE I (Continued)

R in R₂Hg or RHgR' and RHgX	Reagent	Reaction conditions	Products	Yield, %	Ref.
$p\text{-}CH_3OC_6H_4$	BCl_3	Benzene, reflux, 20°C, 6 hr	$p\text{-}CH_3OC_6H_4BCl_2$	—	41
$o\text{-}C_2H_5OC_6H_4$	BCl_3	Benzene, reflux, 20°C, 6 hr	$o\text{-}C_2H_5OC_6H_4B(OH)_2$[a]	—	41
$p\text{-}C_2H_5OC_6H_4$	BCl_3	Benzene, reflux, 20°C, 6 hr	$p\text{-}C_2H_5OC_6H_4BCl_2$	—	41
$\alpha\text{-}C_{10}H_7$	BCl_3	Benzene, reflux, 120–180°C, 11 hr	$\alpha\text{-}C_{10}H_7BCl_2$	—	41
$\beta\text{-}C_{10}H_7$	BCl_3	Benzene, reflux, 150°C, 11 hr	$\beta\text{-}C_{10}H_7BCl_2$	—	41
o-Diphenylene (hexamer)	BCl_3	Toluene, reflux, several hrs	$(C_6H_4BCl)_2$	80	44
C_6H_5HgCl	BCl_3	C_6H_6, reflux, 4 hr	$C_6H_5BCl_2$	76	45
C_6H_5HgCl	BBr_3	C_6H_6, reflux, 4 hr	C_6H_5BClBr	54	45
			$HgBr_2$	97	45
C_6H_5HgBr	BBr_3	C_6H_6, reflux, 4 hr	$C_6H_5BBr_2$	43	45
$C_6H_5HgNO_3$	BCl_3	C_6H_6, reflux, 12 hr	$C_6H_5BCl_2$	15	45
			$C_6H_5NO_2$	7	45
			$HgCl_2$	quant.	
$o\text{-}CH_3C_6H_4HgCl$	BCl_3	C_6H_6, reflux, 12 hr	$o\text{-}CH_3C_6H_4BCl_2$	67	45
$m\text{-}CH_3C_6H_4HgCl$	BCl_3	C_6H_6, reflux, 12 hr	$m\text{-}CH_3C_6H_4BCl_2$	57	45
$p\text{-}CH_3C_6H_4HgCl$	BCl_3	C_6H_6, reflux, 12 hr	$p\text{-}CH_3C_6H_4BCl_2$	74	45
$p\text{-}CH_3C_6H_4HgBr$	BBr_3	C_6H_6, reflux, 12 hr	$p\text{-}CH_3C_6H_4BBr_2$	26	45
$p\text{-}ClC_6H_4HgCl$	BCl_3	C_6H_6, reflux, 12 hr	$p\text{-}ClC_6H_4BCl_2$	67	45
$p\text{-}BrC_6H_4HgCl$	BCl_3	C_6H_6, reflux, 4 hr	$p\text{-}BrC_6H_4BCl_2$	59	45
Indium Halides					
C_6H_5	$InCl_3$	Xylene, reflux, 37 hr	$(C_6H_5)_2InCl$	4	46
Thallium Salts					
C_3H_7	$TlCl_3$	$(C_2H_5)_2O$, 20°C, overnight	$(C_3H_7)_2TlCl$, TlCl	37 / 19	47
$i\text{-}C_5H_{11}$	$TlCl_3$	$(C_2H_5)_2O$, reflux, 15 min	TlCl / $i\text{-}C_5H_{11}HgCl$	57 / 55	47

Reactant	Reagent	Conditions	Products	Yield (%)	Ref.
CH₂=CH—	TlCl₃	—	(CH₂=CH)₂TlCl CH₂=CHHgCl	— —	54
cis,trans-CH₃—CH=CH—CH=CH	TlCl₃	(C₂H₅)₂O, 20°C, 0.5 hr	(trans-CH₃CH=CH—CH₂)₂TlCl cis-CH₃CH=CHHgCl	93 93	50
CH₂=C(CH₃)	TlBr₃	(C₂H₅)₂O, 20°C, 1 hr	[CH₂=C(CH₃)]₂TlBr CH₂=C(CH₃)HgBr	89 70	52
di-cis-CH₃—CH=CH—	TlBr₃	(C₂H₅)₂O, 20°C, 1 hr	CH₃CH=CHHgBr(cis)	62	49
di-trans-CH₃CH=CH	TlBr₃	(C₂H₅)₂O, 20°C, fast	trans-(CH₃CH=CH—CH₂)TlBr trans-CH₃CH=CHHgBr	55 94	49
CH₂C=(OCOCH₃)CH(CH₃) 1M	TlCl₃, 1M	(C₂H₅)₂O, 20°C, fast	[CH₂=C(OCOCH₃)CH(CH₃)]₂TlCl CH₂=C(OCOCH₃)CH(CH₃)HgCl	84 93	53
cis-CH₃C(OCOCH₃)=C(CH₃), 1M	TlCl₃, 2M	(C₂H₅)₂O, 20°C, 40 min	cis-CH₃C(OCOCH₃)=C(CH₃)TlCl₂ HgCl₂	98 80	53
trans-CH₃C(OCOCH₃)=C(CH₃), 1M	TlCl₃, 1M	(C₂H₅)₂O, 20°C, fast	[trans-CH₃—C(OCOCH₃)=C(CH₃)]₂TlCl [trans-CH₃—C(OCOCH₃)=C(CH₃)]HgCl	94 96	53
trans-CH₃C(OCOCH₃)=C(CH₃), 1M	TlCl₃, 2M	(C₂H₅)₂O, 20°C, 10 min	trans-CH₃-C(OCOCH₃)=C(CH₃)TlCl₂ HgCl₂	95 93	53
cis-CHCl=CH	TlCl₃	(C₂H₅)₂O, 20°C, 10 hr	(cis-ClCH=CH)₂TlCl CHCl=CHHgCl	90 22	1
(cis-trans-ClCH=CH)₂Hg	TlCl₃	(C₂H₅)₂O, 20°C, 1–25 hr	trans-(ClCH=CH₂)₂TlCl cis-(ClCH=CH)₂TlCl cis-CHCl=CHHgCl	49 13 46	1
trans-ClCH=CH	TlCl₃	(C₂H₅)₂O, 20°C, 3 hr	trans-(ClCH=CH₂)₂TlCl ClCH=CHHgCl	71 91	48
C₆H₅CH=CH	TlCl₃	(C₂H₅)₂O, stir	(C₆H₅CH=CH)₂TlCl C₆H₅CH=CHCl	80 	56
C₆H₅CH=CH	TlCl₃	(C₂H₅)₂O, stand 1 hr	(α-C₆H₅CH=CH)₂TlCl	70	53
C₆H₅CH=CH	TlBr₃	(C₂H₅)₂O, stand 1 hr	(α-C₆H₅CH=CH)₂TlBr α-C₆H₅CH=CHHgBr	67 78	55
(cis-C₆H₅CH=C(C₆H₅))	TlCl₃	Dioxane, 70–80°C, 1 hr	TlCl cis-C₆H₅CH=C(C₆H₅)HgCl	98 91	141

(continued)

157

TABLE I (*Continued*)

R in R_2Hg or $RHgR'$ and $RHgX$	Reagent	Reaction conditions	Products	Yield, %	Ref.
trans-$C_6H_5CH=C(C_6H_5)$	$TlCl_3$	Dioxane, 70–80°C, 1 hr	$C_6H_5CH=C(C_6H_5)HgCl$	91	141
			$C_6H_5CH=C(C_6H_5)Cl$	87	
			$TlCl$	97	
$C_6H_5CH_2$	$TlCl_3$	$(C_2H_5)_2O$, reflux, stand, several days	$TlCl$	quant.	47
			$C_6H_5CH_2HgCl$	quant.	
			$(C_6H_5CH_2)_2$	traces	
$C_4H_9HgCH_2C_6H_5$	$TlCl_3$	$(C_2H_5)_2O$, reflux, stand, several days	$TlCl$	50	58
			C_4H_9HgCl	—	
			$C_6H_5CH_2HgCl$	—	
			$C_6H_5CH_2HgCl$		
$C_6H_5CH_2HgCl$	$TlCl_3 + C_2H_5MgBr$	$(C_2H_5)_2O$, reflux, stand, several days	$TlCl$	33	47
			C_2H_5HgCl	2	
			(C_6H_5CHO)		
C_6H_5	$TlOH$	CH_3OH, 0–2°C, 30 min	C_5H_5Tl	31	63
C_6H_5	$TlCl_3$	$(C_2H_5)_2O$, 20°C	$(C_6H_5)_2TlCl$	—	59
C_6H_5, $1M$	$Tl(i\text{-}C_3H_7OCO)_3$, $1M$	$CHCl_3$, stir, 30 min	$(C_6H_5)Tl(OCOC_3H_7\text{-}i)_2$	87	61
C_6H_5 (2 mol)	$Tl(i\text{-}C_3H_7OCO)_3$, $1M$	$CHCl_3$, stir, 30 min	$(C_6H_5)_2Tl(OCOC_3H_7\text{-}i)$	80	61
$C_2H_5HgC_6H_5$	$TlCl_3$	$(C_2H_5)_2O$	$(C_6H_5)_2TlCl$	75	58
			C_2H_5HgCl	81	
$p\text{-}CH_3C_6H_4$, $1M$	$Tl(i\text{-}C_3H_7OCO)_3$, $1M$	$CHCl_3$, overnight	$(p\text{-}CH_3C_6H_4)Tl(OCOC_3H_7\text{-}i)_2$	46	62
$p\text{-}CH_3C_6H_4$, $2M$	$Tl(i\text{-}C_3H_7OCO)_3$, $1M$	$CHCl_3$, overnight	$(p\text{-}CH_3C_6H_4)_2Tl(OCOC_3H_7\text{-}i)$	76	62
$p\text{-}ClC_6H_4$	$Tl(i\text{-}C_3H_7OCO)_3$	$CHCl_3$, overnight	$p\text{-}ClC_6H_4Tl(OCOC_3H_7\text{-}i)_2$	73	62
$p\text{-}BrC_6H_4$	$Tl(i\text{-}C_3H_7OCO)_3$	$CHCl_3$, overnight	$p\text{-}BrC_6H_4Tl(OCOC_3H_7\text{-}i)_2$	50	62
$p\text{-}CH_3OC_6H_4$, $2M$	$Tl(OCOC_3H_7\text{-}i)_3$	$CHCl_3$	$(p\text{-}CH_3OC_6H_4)_2Tl(OCOC_3H_7\text{-}i)$	66	62
$p\text{-}CH_3OC_6H_4$, $1M$	$Tl(OCOC_3H_7\text{-}i)_3$	$CHCl_3$	$p\text{-}CH_3OC_6H_4Tl(OCOC_3H_7\text{-}i)_2$	65	60, 62
$o\text{-}CH_3OC_6H_4HgC_4H_9$ (2 mol)	$TlCl_3$ (1 mol)	$(C_2H_5)_2O$, 30 min	$(o\text{-}CH_3OC_6H_4)_2TlCl$	83	58
			$(C_4H_9)HgCl$	85	
$o\text{-}CH_3COOC_6H_4$	$TlBr_3$	C_2H_5OH, 3 hr reflux	$o\text{-}(CH_3COOC_6H_4)_2TlBr$	21	57
			$CH_3COOC_6H_4HgBr$	—	
$\alpha\text{-}C_{10}H_7$	$Tl(OCOC_3H_7\text{-}i)$	$CHCl_3$, overnight	$\alpha\text{-}C_{10}H_7Tl(OCOC_3H_7\text{-}i)_2$	47	62

(continued)

Compound	Reagent	Conditions	Product	Yield (%)	Ref.
β-C₁₀H₇ C₂H₅HgC₁₀H₇-α	Tl(OCOC₃H₇-i) TlCl₃	CHCl₃, overnight (C₂H₅)₂O	β-C₁₀H₇Tl(OCOC₃H₇-i)₂ (α-C₁₀H₇)₂TlCl C₂H₅HgCl	91 97 88	62 58
C₆H₅HgC₁₀H₇-α, 1M	TlCl₃, 1M	(C₂H₅)₂O, 30 min	C₆H₅TlCl₂ C₆H₅HgCl α-C₁₀H₇HgCl	62 64 79	58
C₆H₅HgC₁₀H₇-α, 2M	TlCl₃, 1M	Dioxane, overnight	(C₆H₅)₂ TlCl α-C₁₀H₇HgCl	76 75	58
(CH₃)₃C₆H₂HgC₁₀H₇-α, 2M	TlCl₃, 1M	(C₂H₅)₂O + toluene, 30 min	(α-C₁₀H₇)₂TlCl (CH₃)₃C₆H₂HgCl	89 80	58
o-CH₃OC₆H₄HgC₁₀H₇-α 2M	TlCl₃ 1M	Dioxane, 40°C, overnight	(o-CH₃OC₆H₄)₂TlCl α-C₁₀H₇HgCl	86 67	58
α-Thienyl	Tl(O₂CC₃H₇-i)₃	CHCl₃	α-C₄H₃STl(O₂CC₃H₇-i)₂	94	62
α-Thienylmercuriiodide	TlCl₃	C₂H₅OH, reflux, 15 min	TlCl α-Thienylmercurichloride	65	47
Radical of 8-acetyloxymercuri-2,2,4,6-tetramethyl-1,2,3,4-tetrahydroquinolinenitrogenoxide	Tl(O₂CC₃H₇-i)₃	Dioxane	Radical of 8-tri(i-butyloxy)thallium-2,2,4,6-tetramethyl-1,2,3,4-tetrahydroquinoline-nitrogenoxide	—	62[a]

Silicon Halides and Alkylsilicon Halides

Compound	Reagent	Conditions	Product	Yield (%)	Ref.
CH₂CHO	SiCl₄	isopentane, reflux, 2 hr	Si(OCH=CH₂)₄	72	64
CH₂CHO	(CH₃)₃SiCl	isopentane, reflux, 2 hr	(CH₃)₃Si(OCH=CH₂)	72	64
CH₂CHO	(C₂H₅)₂SiCl₂	(C₂H₅)₂O, reflux, 2 hr	(C₂H₅)₂Si(OCH=CH₂)₂	69	64
CH₂CHO	C₆H₅SiCl₃	(C₂H₅)₂O, reflux, 2 hr	(C₆H₅)Si(OCH=CH₂)₃	70	64
CH₃COCH₂HgCl	SiCl₄	C₅H₅N, C₆H₆, reflux, 1.5 hr	Si[OC(CH₃)=CH₂]₄	49	64
CH₂COOCH₃	(C₂H₅)₃SiI	C₆H₆, stir, 20°C	CH₂=C(OCH₃)OSi(C₂H₅)₃	53	67
CH₂COOCH₃	(C₂H₅)₃SiI	C₂H₄Cl₂, 20°C	(C₂H₅)₃SiCH₂COOCH₃	60	67
CH₂COOCH₃	(C₂H₅)₃SiI	C₆H₆, 20°C	CH₂=C(OCH₃)OSi(C₂H₅)₃ IHgCH₂COOCH₃	53 90	66
CH₂COOCH₃	(C₂H₅)₃SiI	C₂H₄Cl₂, 20°C	(C₂H₅)₃SiCH₂COOCH₃ (C₂H₅)₃SiCH₂COOCH₃	60 52	66 65
CH₂=CH	SiCl₄	Sealed tube, 300°C, several hours	(CH₂=CH)SiCl₃	—	38
C₆H₅	SiCl₄		(C₆H₅)SiCl₃	28	68

TABLE I (*Continued*)

R in R₂Hg or RHgR′ and RHgX	Reagent	Reaction conditions	Products	Yield, %	Ref.
C_6H_5	$(C_6H_5)_3SiLi$	Tetrahydrofuran, 30 min	$(C_6H_5)_4Si$	69	142
C_6H_5HgBr (1 mol)	$(C_6H_5)_3SiLi$ (1 mol)	Tetrahydrofuran, 30 min	$(C_6H_5)_4Si$	73	
			Hg	78	142
			$(C_6H_5)_6Si_2$	5	
			$(C_6H_5)_3SiOH$[a]	2	
C_6H_5HgBr (1 mol)	$(C_6H_5)_3SiLi$ (2 mol)	Tetrahydrofuran, $-70°C$	$(C_6H_5)_4Si$	15	142
			$(C_6H_5)_6Si_2$	2	
			$(C_6H_5)_2SiOH$[a]	55	
			Hg	50	
$p\text{-}CH_3C_6H_4 1M$	$(C_6H_5)_3SiLi$, $1M$	Tetrahydrofuran	$(C_6H_5)_3SiOH$[a]	6	142
			$(C_6H_5)_3Si(p\text{-}CH_3C_6H_4)$	88	
			$CH_3C_6H_4COOH$[a,b]	71	
$p\text{-}CH_3C_6H_4$, $2M$	$(C_6H_5)_3SiLi$, $1M$	Tetrahydrofuran, 0°C	$(C_6H_5)_3Si(p\text{-}CH_3C_6H_4)$[a,b]	46	142
			$CH_3C_6H_5COOH$[a,b]	51	
			$(C_6H_5)_3SiOH$[a]	46	
$p\text{-}CH_3C_6H_4$, $2M$	$SiCl_4$	Sealed tube, 320°C	$(C_6H_5)_6Si_2$	4	
			$CH_3C_6H_4SiCl_3$	—	69
Germanium tetrachloride and Alkylgermanium Halides					
CH_2COOCH_3	$(n\text{-}C_3H_7)_2Ge(H)I$	C_6H_6 + petroleum ether, 20°C, 1.5 hr	$(n\text{-}C_3H_7)_2Ge(H)CH_2COOCH_3$	74	143
CH_2COOCH_3	$(n\text{-}C_3H_7)_2GeI_2$	Petroleum ether + C_5H_5N, reflux, 45 min	$(n\text{-}C_3H_7)_2Ge(CH_2COOCH_3)_2$	73	72
CH_2COOCH_3	$(n\text{-}C_3H_7)_3GeI$	Petroleum ether + C_5H_5N, reflux, 30 min	$(n\text{-}C_3H_7)_3GeCH_2COOCH_3$	71	144
$CH_2COOC_2H_5$	$(n\text{-}C_3H_7)_3GeI$	Petroleum ether + C_5H_5N, reflux, 30 min	$(n\text{-}C_3H_7)_3GeCH_2COOC_2H_5$	64	144
CH_2COOCH_3	$(n\text{-}C_4H_9)_2Ge(H)I$	—	$(n\text{-}C_4H_9)_2Ge(H)\text{—}CH_2COOCH_3$	71	143
CH_2COOCH_3	$(n\text{-}C_4H_9)_3GeI$	Petroleum ether + C_5H_5N, reflux, 30 min	$(n\text{-}C_4H_9)_3GeCH_2COOCH_3$	85	144

160

R	Reagent	Product	Conditions	Yield (%)	Ref.
CH₂COOC₂H₅	$(n\text{-}C_4H_9)_3GeI$	$(n\text{-}C_4H_9)_3GeCH_2COOC_2H_5$	Petroleum ether + C_5H_5N, reflux, 30 min	60	144
CH₂COOC₃H₇	$(n\text{-}C_4H_9)_3GeI$	$(n\text{-}C_4H_9)_3GeCH_2COOC_3H_7$	Petroleum ether + C_5H_5N, reflux, 30 min	90	65,144
CH₂=CH	$GeCl_4$	$CH_2{=}CHGeCl_3$	Reflux, 4.5 hr	67	38, 70
C₆H₅CH₂	$GeCl_4$	$(C_6H_5CH_2GeO)_2O$[a]	Sealed tube, xylene, 115–120°C, 2 days	—	71
C₆H₅	$GeCl_4$	$(C_6H_5GeO)_2O$[a]	Sealed tube, xylene, 140°C, 2 days	—	71
m-CH₃C₆H₄	$(m\text{-}CH_3C_6H_4)_3GeI$	$m\text{-}CH_3C_6H_4HgI$ $[(m\text{-}CH_3C_6H_4)_3Ge]_2O$[a]	Xylene, reflux, 12 hr	73 51	145
p-CH₃C₆H₄	$GeCl_4$	$(p\text{-}CH_3C_6H_4GeO)_2O$[a]	Sealed tube, xylene, 160–190°C, 2 days	—	71

Tin (IV) Chlorides

R	Reagent	Product	Conditions	Yield (%)	Ref.
CH₃	$SnCl_4$	CH_3HgCl	—	—	146
C₂H₅	$SnCl_4$	C_2H_5HgCl	—	—	107
C₂H₅	$(C_2H_5)_2SnCl_2$	$(C_2H_5)_3SnCl$ C_2H_5HgCl	—	—	107
CH₂=CH	$SnCl_4$	$(CH_2{=}CH)SnCl_3$	Ligroin, reflux, 12 hr	—	38
C₆H₅	$SnCl_4$	$(C_6H_5)_2SnCl_2$		33–34	73

Lead Tetraacylates

R	Reagent	Product	Conditions	Yield (%)	Ref.
C₂H₅	$Pb(OCOCH_3)_4$	$(C_2H_5)_2Pb(OCOCH_3)_2$	$CHCl_3$	60	114
C₂H₅	$Pb(OCOCH_3)_4$	$(C_2H_5)_2Pb(OCOCH_3)_2$ $C_2H_5HgOCOCH_3$	$CHCl_3$, 20°C, 3 months	92 87	74
n-C₄H₉	$Pb(OCOCH_3)_4$	$Pb(OCOCH_3)_2$ $n\text{-}C_4H_9HgOCOCH_3$ $n\text{-}C_4H_9OCOCH_3$	C_6H_6, 20°C, 4 days	98 98 69	78
(CH₃)₃CCH₂	$Pb(OCOCH_3)_4$	$C_5H_{11}OCOCH_3$ $Pb(OCOCH_3)_2$ $C_5H_{11}HgOCOCH_3$	C_6H_6, reflux, 2 hr	26 quant. 97	78

161

(continued)

TABLE I (Continued)

R in R₂Hg or RHgR' and RHgX	Reagent	Reaction conditions	Products	Yield, %	Ref.
$(CH_3)_2C(C_6H_5)CH_2$	$Pb(OCOCH_3)_4$	C_6H_6, shake, 3 months	$Pb(OCOCH_3)_3)$	quant.	
			$(CH_3)_2C{-}CH_2{-}HgOCOCH_3$ with C_6H_5	56	78
			$C_6H_5CH{=}C(CH_3)_2$ $OCOCH_3$	12	78
			$(CH_3)_2CCH_2{-}C_6H_5$	26	
$ClCH{=}CH$	$Pb(OCOCH_3)_4$	$CHCl_3$, shake, 2.5–3 hr	$(ClCH{=}CH)_2Pb(COOCH_3)_2$	72	75
			$ClCH{=}CHHgCOOCH_3$	78	
$C_6H_5CH_2$	$Pb(OCOCH_3)_4$	C_6H_6, 20°C, 2 weeks	$Pb(OCOCH_3)_2$	97	78
			$C_6H_5CH_2HgOCOCH_3$	75	
			$C_6H_5CH_2OCOCH_3$	75	
$C_6H_5CH_2CH_2$	$Pb(OCOCH_3)_4$	C_6H_6, shake 24 hr, stand 4 weeks	$Pb(OCOCH_3)_2$	95	78
			$C_6H_5CH_2CH_2HgOCOCH_3$	60	
			$C_6H_5CH_2CH_2OCOCH_3$	72	
$C_6H_5HgCH_2C_6H_5$	$Pb(OCOCH_3)_4$	C_6H_6, shake, several hours	$C_6H_5CH_2HgOCOCH_3$	73	78
			$C_6H_5Pb(OCOCH_3)_3$	87	
C_6H_5	$Pb(OCOCH_3)_4$	$CHCl_3$	$(C_6H_5)_2Pb(OCOCH_3)_2$	60	114
C_6H_5	$Pb(OCOCH_3)_4$	C_6H_6, 20°C, shake 3 days	$C_6H_5HgOCOCH_3$	75	78
			$C_6H_5Pb(OCOCH_3)_3$	81	
C_6H_5	$C_6H_5Pb(OCOCH_3)_3$	C_6H_6, 1 week	$(C_6H_5)_2Pb(OCOCH_3)_2$	64	78
			$C_6H_5HgOCOCH_3$	90	
C_6H_5	$Pb(OCOCH_3)_4$	$CHCl_3$, 20°C, 24 hr	$(C_6H_5)_2Pb(OCOCH_3)_2$	31	74
			$C_6H_5HgOCOCH_3$	96	
C_6H_5	$Pb(OCOCH_3)_4$	$CHCl_3$, 30 min	$C_6H_5Pb(OCOCH_3)_3$	98	76
			$C_6H_5HgCl^d$	84	
C_6H_5	$Pb(OCOC_3H_7{-}i)_4$	$CHCl_3$, 30 min	$(C_6H_5)Pb(OCOC_3H_7{-}i)_3$	50	76
			$C_6H_5Hg(OCOC_3H_7{-}i)$		
$p{-}CH_3C_6H_4$	$Pb(OCOCH_3)_4$	$CHCl_3$, 24 hr	$p{-}CH_3C_6H_4Pb(OCOCH_3)_3$	66	76
			$p{-}CH_3C_6H_4HgOCOCH_3$		

162

Titanium Tetrachloride and $(C_5H_5)_2TiCl_2$ section and preceding lead tetraacetate reactions:

Ar	Reagent	Conditions	Product	Yield (%)	Ref.
$o\text{-}CH_3OC_6H_4$	$Pb(OCOCH_3)_4$	$CHCl_3$, 24 hr	$(o\text{-}CH_3OC_6H_4)_2Pb(OCOCH_3)_2$	60	114
$o\text{-}CH_3OC_6H_4$	$Pb(OCOCH_3)_4$	$CHCl_3$, 4 days	$(o\text{-}CH_3OC_6H_4)_2Pb(OCOCH_3)_2$	63	74
			$o\text{-}CH_3OC_6H_4HgOCOCH_3$	87	78
$p\text{-}CH_3OC_6H_4$	$Pb(OCOCH_3)_4$	C_6H_6, 2 days	$p\text{-}CH_3OC_6H_4Pb(OCOCH_3)_3$	90	74
			$p\text{-}CH_3OC_6H_4HgOCOCH_3$	86	
$p\text{-}C_2H_5OCOC_6H_4$	$Pb(OCOCH_3)_4$	$CHCl_3$, 5 days	$(p\text{-}C_2H_5OCOC_6H_5)_2Pb(OCOCH_3)_2$	51	74
			$p\text{-}C_2H_5COOC_6H_4HgOCOCH_3$	—	
$C_2H_5OCOC_6H_4$	$Pb(OCOCH_3)_4$	$CHCl_3$, 5 days	$(C_2H_5OOCOC_6H_4)_2Pb(OCOCH_3)_2$	60	114
$\alpha\text{-}C_{10}H_7$	$Pb(OCOCH_3)_4$	$CHCl_3$, 1.5 hr	$\alpha\text{-}C_{10}H_7Pb(OCOCH_3)_3$	55	77
$\alpha\text{-}C_{10}H_7$	$Pb(OCOC_3H_7\text{-}i)_4$	$CHCl_3$, 1.5 hr	$\alpha\text{-}C_{10}H_7Pb(OCOC_3H_7\text{-}i)_3$	46	77
$\beta\text{-}C_{10}H_7$	$Pb(OCOCH_3)_4$	$CHCl_3$, 1.5 hr	$(\beta\text{-}C_{10}H_7)_2Pb(OCOCH_3)_2$	60	114
$\beta\text{-}C_{10}H_7$	$Pb(OCOCH_3)_4$	$CHCl_3$, 3 days	$(\beta\text{-}C_{10}H_7)_2Pb(OCOCH_3)_2$	31	74
$\beta\text{-}C_{10}H_7$	$Pb(OCOCH_3)_4$	$CHCl_3$, 20°C, 1 hr	$(\beta\text{-}C_{10}H_7)_2Pb(OCOC_2H_5)_3$	38	79
$\beta\text{-}C_{10}H_7$	$Pb(OCOC_2H_5)_4$	$CHCl_3$, 20°C, 2 hr	$\beta\text{-}C_{10}H_7Pb(OCOC_2H_5)_3$	61	79
			$\beta\text{-}C_{10}H_7HgOCOC_2H_5$		
$\beta\text{-}C_{10}H_7$	$\beta\text{-}C_{10}H_7Pb(OCOCH_3)_3$	$CHCl_3$, heat, then 20°C, 1 day	$(\beta\text{-}C_{10}H_7)_2Pb(OCOCH_3)_2$		79
			$\beta\text{-}C_{10}H_7HgOCOCH_3$		
$p\text{-}IC_6H_4$	$Pb(OCOCH_3)_4$	$CHCl_3$	$p\text{-}IC_6H_4Pb(O_2CR)_3$ $R = CH_3$	35	77
$p\text{-}IC_6H_4$	$Pb(OCOC_2H_5)_4$	$CHCl_3$	$p\text{-}IC_6H_4Pb(O_2CR)_3$ $R = C_2H_5$	29	77
$p\text{-}IC_6H_4$	$Pb(OCOC_3H_7\text{-}i)_4$ $Pb(OCO\text{-}i\text{-}C_3H_7)_4$	$CHCl_3$	$p\text{-}IC_6H_4Pb(O_2CR)_3$ $R = i\text{-}C_3H_7$	45	77
$p\text{-}IC_6H_4$	$Pb(OCOC_6H_5)_4$	$CHCl_3$	$p\text{-}IC_6H_4Pb(O_2CR)_3$ $R = C_6H_5$	26, 3	77
	Titanium Tetrachloride and $(C_5H_5)_2TiCl_2$				
C_6H_5	$TiCl_4$	CCl_4, 100°	C_6H_5HgCl C_6H_5—C_2H_5	—	80
C_6H_5	$TiCl_4$	Sealed tube, 200°C, 5 hr	C_6H_5HgCl C_6H_5—C_6H_5	—	80
C_6H_5	$TiCl_4$	C_6H_6 or CCl_4 or $CHCl_3$ or THF, 90°C, 10–15 hr	C_6H_5HgCl C_6H_5—C_6H_5	Low quant.	81

(continued)

TABLE I (Continued)

R in R_2Hg or $RHgR'$ and $RHgX$	Reagent	Reaction conditions	Products	Yield, %	Ref.
C_6H_5	$(C_5H_5)_2TiCl_2$	CH_2Cl_2, 35°C, 20 hr	C_6H_5HgCl	85	82
C_6H_5	$(C_5H_5)_2TiCl_2$	C_6H_6, 70°C, 24 hr	C_6H_5HgCl	22	82
Phosphorus Halides and Arylphosphorus Halides					
CH_3	PCl_3	—	CH_3HgCl	—	146
C_2H_5	PCl_3	Sealed tube, 230°C, 4 hr	$C_2H_5PCl_2$	—	147
C_2H_5	PCl_3	Sealed tube, 250°C, 6 hr	$C_2H_5PCl_2$	—	83
$n\text{-}C_3H_7$	PCl_3	Sealed tube, 250°C, 6 hr	$(n\text{-}C_3H_7)PCl_2$	—	83
$n\text{-}C_3H_7$	PCl_3		$(n\text{-}C_3H_7)_3P$	—	148
$i\text{-}C_3H_7$	PCl_3	Sealed tube, 250°C, 6 hr	$(i\text{-}C_3H_7)PCl_2$	—	83
$n\text{-}C_4H_9$	PCl_3	Sealed tube, 200°C, 9 hr	$n\text{-}C_4H_9PCl_2$	61	84
$i\text{-}C_4H_9$	PCl_3	Sealed tube, 250°C, 6 hr	$i\text{-}C_4H_9PCl_2$	—	83
$i\text{-}C_5H_{11}$	PCl_3	Sealed tube, 250°C, 6 hr	$i\text{-}C_5H_{11}PCl_2$	—	83
$(CH_3)_3SiCH_2$	PCl_3	Hexane, reflux, 24 hr	$(CH_3)_3SiCH_2PCl_2$	96	149
$CH_2{=}CHO$, $3M$	PCl_3, $1M$	isopentane + $(C_2H_5)_3N$, 20°C, stir 4 hr	$(CH_2{=}CHO)_3P$	46	150
$CH_2{=}CHO$	$Cl\,P(OR)_2$	as above, 15 hr	$CH_2{=}CHOP(OR)_2$	46–80	150
$CH_2{=}CHO$, $2M$	$Cl_2\,POR$, $1M$	as above, 5 hr	$(CH_2{=}CHO)_2POR$	46–80	150
$CH_2{=}CH$	PCl_3	Nujol, reflux, 1 hr	$CH_2{=}CHPCl_2$	50	85
			$CH_2{=}CHHgCl$		
$CH_2{=}CH$	PCl_3	80°C	$(CH_2{=}CH)PCl_2$	—	37
$CH_2{=}CH$	PBr_3	Nujol, 80°C, 12 hr	$(CH_2{=}CH)PBr_2$	78	85
			$HgBr_2$		
$CH_2{=}CH$	PBr_3	Nujol, 80°C, 12 hr	$(CH_2{=}CH)PBr_2$	78	85[a]
			$HgBr_2$		
$C_6H_5CH_2$	PCl_3	Sealed tube, 220°C, 6–8 hr	$C_6H_5CH_2PCl_2$	31	151
			$HgCl_2$		
C_6H_5	PCl_3	Sealed tube, 180°C, several hours	$(C_6H_5)_3P$	High	152
			C_6H_5HgCl		
C_6H_5	PCl_3	Sealed tube, 180°C, several hours	$C_6H_5PCl_2$	16	153

C_6H_5	PCl_3	Sealed tube, 180C°	$C_6H_5PCl_2$	33	86
C_6H_5	PBr_3	Heat	$C_6H_5PBr_2$	—	154
C_6H_5	$C_6H_5PCl_2$	220–230°C, 1 hr	$(C_6H_5)_2PCl$	64	155
C_6H_5	$C_6H_5PCl_2$	Reflux, 1 hr	$(C_6H_5)_2PCl$	—	88
C_6H_5	$C_6H_5PCl_2$	Heat	$(C_6H_5)_2PCl$	72	156
C_6H_5	$C_6H_5PCl_2$	230–240°C, 1.5 hr	$(C_6H_5)_2PCl$	55	92
C_6H_5	$C_6H_5PCl_2$	230–240°C, 1.5 hr	$(C_6H_5)_2PCl$	—	90
C_6H_5	$C_6H_5PCl_2$	270°C, 1–2 hr	$(C_6H_5)_2PCl$	—	93
C_6H_5, 1M	p-$CH_3C_6H_4PCl_2$, 1M	Reflux, 1 hr, 270°C	(p-$CH_3C_6H_4$)(C_6H_5)PCl	21	94
C_6H_5, 1M	p-$CH_3C_6H_4PCl_2$, 1.5M	Reflux, 1 hr, 270°C	(C_6H_5)(p-$CH_3C_6H_4$)PCl	33	94
C_6H_5, 1M	p-$CH_3C_6H_4PCl_2$, 1.5M		(C_6H_5)(p-$CH_3C_6H_4$)PCl	—	157
C_6H_5	(p-BrC_6H_4)PCl_2	210°C, 1.5 hr	C_6H_5(p-BrC_6H_4)PCl	49–56	158
C_6H_5	(p-$CH_3OC_6H_4$)PCl_2	170–180°C, 1.5 hr	C_6H_5(p-$CH_3OC_6H_4$)PCl	51	158
p-$CH_3C_6H_4$	PCl_3	Sealed tube, 220–230°C, 48 hr	p-$CH_3C_6H_4PCl_2$	66	159
o-$CH_3C_6H_4$	PCl_3	Sealed tube, 180–190°C, 48 hr	o-$CH_3C_6H_4PCl_2$	78	87
m-$CH_3C_6H_4$	PCl_3	Sealed tube, 200°C, 12 hr	m-$CH_3C_6H_4PCl_2$	—	87
p-$CH_3C_6H_4HgBr$	$C_6H_5PCl_2$	Sealed tube, 270°C, 2–3 hr	(p-$CH_3C_6H_4$)(C_6H_5)PCl	63	91
2,4-$(CH_3)_2C_6H_3$—	PCl_3	Sealed tube, 230–240°C	[2,4-$(CH_3)_2C_6H_3$]PCl_2	50	160
$(CH_3)_3C_6H_2$	PCl_3	Sealed tube, 230–240°C	$(CH_3)_3C_6H_2PCl_2$	—	87
m-ClC_6H_4	PCl_3	Sealed tube, 230–250°C, 24 hr	m-$ClC_6H_4PCl_2$	96	89
p-$(CH_3)_2NC_6H_4$	PCl_3	Sealed tube, 120°C, several hours	p-$(CH_3)_2NC_6H_4PCl_2$	—	161
p-$(CH_3)_2NC_6H_4$	PCl_3	Sealed tube, 130°C, 3 hr	p-$(CH_3)_2NC_6H_4PCl_2$	—	162
α-$C_{10}H_7$	PCl_3	Sealed tube, 200°C	α-$C_{10}H_7PCl_2$	—	163
α-$C_{10}H_7$	PCl_3	Sealed tube, 180–200°C, 3–5 days	α-$C_{10}H_7PO(OH)H$[a]	—	164
β-$C_{10}H_7$	PCl_3	Sealed tube, 200°C, 48 hr	(β-$C_{10}H_7$)PCl_2	—	165

Halides and Alkylhalides of Arsenic(III)

C_2H_5	$AsCl_3$	Heat	$C_2H_5AsCl_2$	—	96
C_2H_5	$AsCl_3$	100°C, 0.5 hr	$C_2H_5AsCl_2$	88	97
C_2H_5	$(C_6H_5)_2As_2$	Sealed tube, 150°C	$C_6H_5As(C_2H_5)_2$	—	166
n-C_3H_7	$AsCl_3$	—	n-$(C_3H_7)_3As$	—	148

TABLE I (*Continued*)

R in R_2Hg or $RHgR'$ and $RHgX$	Reagent	Reaction conditions	Products	Yield, %	Ref.
CH_2COCH_3	C_6H_5AsS	Xylene, reflux	$C_6H_5As(CH_2COCH_3)_2$	50–60	179
$CH_2COC_2H_5$	C_6H_5AsS	Xylene, reflux	$C_6H_5As(CH_2COC_2H_5)_2$	50–60	179
$CH_2CO_2CH_3$	C_6H_5AsS	Xylene, reflux	$C_6H_5As(CH_2CO_2CH_3)_2$	50–60	179
$CG_2CO_2C_2H_5$	C_6H_5AsS	Xylene, reflux	$C_6H_5As(CH_2CO_2C_2H_5)_2$	50–60	179
$CF_2{=}CFHgC_2H_5$	$AsCl_3$	100°C, 2 hr	$CF_2{=}CFAsCl_2$ C_2H_5HgCl	82 81	98
$CF_2{=}CFHgCH{=}CH_2$	$AsCl_3$	100°C, 2 hr	$CF_2{=}CFAsCl_2$ $CH_2{=}CHHgCl$	85 82	98
$CF_2{=}CF$	$AsCl_3$	—	$CF_2{=}CFAsCl_2$ $CF_2{=}CFHgCl$	70 51	39
$ClCH{=}CHHgCl$	$AsCl_3$	CH_3COCH_3, 25–70°C	$ClCH{=}CHAsCl_2$ $(ClCH{=}CH)_3As$	80–85 54	99
C_6H_5	$AsCl_3$	Sealed tube, 170°	$C_6H_5AsCl_2$	—	100
C_6H_5	$AsCl_3$	Heat, 1 hr	$C_6H_5AsCl_2$	—	101
C_6H_5	$AsCl_3$	270°C, several hours	$(C_6H_5)_2AsCl$	—	103
C_6H_5	$AsCl_3$	170–250°C, 4 hr	$C_6H_5AsCl_2$ $HgCl_2$	60–100	102
C_6H_5	$AsCl_3$	Petroleum ether, 200–210°C, 4 hr	$\left.\begin{array}{l}(C_6H_5)_2AsCl\\ C_6H_5AsCl_2\\ C_6H_5HgCl\end{array}\right\}$	Low High	102
C_6H_5	$AsCl_3$	210°C, several hours	$(C_6H_5)_2AsCl$	Low	102
C_6H_5	$AsCl_3$	130–140°C, 3 hr	$C_6H_5AsO^a$	60	167
C_6H_5	$C_6H_5AsCl_2$	270°C, 4 hr	$(C_6H_5)_2AsCl$	64	102
C_6H_5	$C_6H_5AsCl_2$	Sealed tube, 300°C, 4 hr	$(C_6H_5)_2AsCl$ $As(C_6H_5)_3$	High Low	102
C_6H_5	$C_6H_5AsCl_2$	Reflux, 1 hr	$(C_6H_5)_2AsCl$	—	88
C_6H_5	$C_6H_5AsCl_2$	Reflux, 1 hr	$(C_6H_5)_2AsCl$	Quant.	104
C_6H_5HgCl	$AsCl_3$	100°C, 4–5 hr	$C_6H_5AsCl_2$	Low	105
C_6H_5HgCl	$AsCl_3$	—	$(C_6H_5)_2AsCl$	75	95

166

Starting material	Reagent	Conditions	Product	Yield (%)	Ref.
$o\text{-}CH_3C_6H_4$	$AsCl_3$	Sealed tube, 250–260°C	$(o\text{-}CH_3C_6H_4)AsCl_2$	—	102
$o\text{-}CH_3C_6H_4$	$AsCl_3$	Heat	$o\text{-}CH_3C_6H_4AsCl_2$	—	168
$p\text{-}CH_3C_6H_4$	$AsCl_3$	Heat	$(p\text{-}CH_3C_6H_4)AsCl_2$	—	168
$p\text{-}CH_3C_6H_4$	$AsCl_3$	Sealed tube, 250–260°C	$p\text{-}CH_3C_6H_4AsCl_2$	—	102
$p\text{-}CH_3C_6H_4$	$C_6H_5AsCl_2$	Reflux, 5 hr	$C_6H_5(p\text{-}CH_3C_6H_4)AsCl$	41	104
$m\text{-}(CH_3)_2C_6H_3$	$AsCl_3$	Stand, 20 hr	$m\text{-}(CH_3)_2C_6H_3AsCl_2$	—	169
$p\text{-}ClC_6H_4HgCl$	$AsCl_3$	—	$p\text{-}ClC_6H_4AsCl_2$	—	95
$p\text{-}CH_3OC_6H_4HgCl$	$AsCl_3$	100°C, heat, 5 hr	$p\text{-}CH_3OC_6H_4AsCl_2$	—	105
$\alpha\text{-}C_{10}H_7$	$AsCl_3$	Reflux	$\alpha\text{-}C_{10}H_7AsCl_2$	—	164
$\alpha\text{-}C_{10}H_7$	$AsCl_3$	Sealed tube, C_6H_6, 145°C, 6 hr	$\alpha(C_{10}H_7)_2AsCl$ } $\alpha(C_{10}H_7)_3As$	Mixture	169
$\alpha\text{-}C_{10}H_7$	$AsCl_3$	130–140°C, 3 hr	$\alpha\text{-}C_{10}H_7AsO$[a]	60	167
$\alpha\text{-}C_{10}H_7HgCl$	$AsCl_3$	Sealed tube, 160°C, 1 hr	$\alpha\text{-}C_{10}H_7AsCl_2$	High	170
$C_6H_5OC_6H_4HgCl$	$AsCl_3$	100°C, 5 hr + H_2O_2	$C_6H_5OC_6H_4AsO_3$[c]	—	171
2-furyl–HgCl (structure)	$AsCl_3$	C_6H_6, reflux, 4 hr	$(C_4H_3O)AsCl_2$ / $(C_4H_3O)_2AsCl$ / $(C_4H_3O)_3As$ / $HgCl_2$	Quant. 50–60 (Pure)	173
2-furyl–HgCl (structure)	$AsCl_3$	C_6H_6, reflux, 4 hr	$(C_4H_3O)_2AsO_2H$[a]	8	174
2-furyl–HgCl (structure)	$AsCl_3$	C_6H_6, reflux, 3 hr	$(C_4H_3O)_3As$	36	106
Methyl 4-chloromercuri-5-bromo-2-furoate	$AsBr_3$	Xylene, Reflux, 3 hr	Br_2As– furan –$COOCH_3$ (with Br)	50	106
Ethyl 4-chloromercuri-5-bromo-2-furoate	$AsBr_3$	Xylene, reflux, 3 hr	Br_2As– furan –$COOC_2H_5$ (with Br)	39	106

167

(continued)

TABLE I (*Continued*)

R in R₂Hg or RHgR' and RHgX	Reagent	Reaction conditions	Products	Yield, %	Ref.
 (thiophene)	$AsCl_3$	Stir, 5 hr, stand, 24 hr	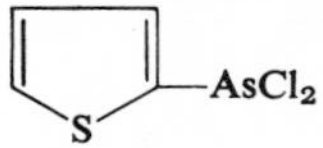 2-thienyl–$AsCl_2$	40	
			(2-thienyl)$_2$–AsCl	8	172
			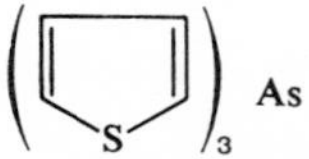 (2-thienyl)$_3$–As	36	
2-Chloromercuri-5-benzothienone	$AsCl_3$	50–70°C, 2–3 min	5-Benzothienone-2 dichloroarsine	Quant.	175
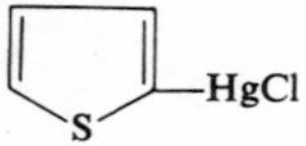	$AsCl_3$	—	$(C_4H_3S)AsCl_2$	—	176
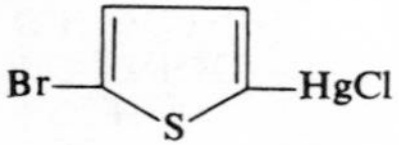 (2-Br-5-chloromercurithiophene)	$AsCl_3$	—	$BrC_4H_2SAsCl_2$	—	177

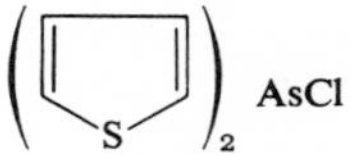

Antimony Trichloride

R in R₂Hg or RHgR' and RHgX	Reagent	Reaction conditions	Products	Yield, %	Ref.
CH_3	$SbCl_3$	—	$CH_3HgCl \cdot (CH_3)_2SbCl_2$	—	107
C_2H_5	$SbCl_3$	—	C_2H_5HgCl	—	107
			$(C_2H_5)_3Sb$	—	
C_6H_5	$SbCl_3$	Sealed tube, xylene, 130°C, several hours	$(C_6H_5)_2SbCl_3$	—	109
C_6H_5	$SbCl_3$	C_6H_6, heat	$(C_6H_5)_2SbO^a$	—	108
C_6H_5HgBr	$SbCl_3$		No reaction		110

168

C_6H_5	$BiCl_3$	$CHCl_3$, reflux, 4 hr	C_6H_5HgCl	71	111
			$HgCl_2$	22	
C_6H_5	$BiCl_3$	$CHCl_3$, reflux, 5 hr	C_6H_6	—	112
C_6H_5	$BiBr_3$	$(C_2H_5)_2O$, reflux, 12 hr	C_6H_6	—	113
			C_6H_5HgBr		
			$(C_6H_5)_3Bi$	Quant.	
C_6H_5	$\alpha\text{-}C_{10}H_7BiBr_2$	$(C_2H_5)_2O$, stir, overnight	C_6H_5HgBr	—	113
			$(\alpha\text{-}C_{10}H_7)_2Hg$		
			$(C_6H_5)_2(\alpha\text{-}C_{10}H_7)Bi$	Low	
C_6H_5	$(\alpha\text{-}C_{10}H_7)_3BiBr_2$	C_6H_6, reflux, 2.5 hr	C_6H_5HgBr	Low	113
			C_6H_5Br		
			$(C_6H_5)_2(\alpha\text{-}C_{10}H_7)Bi$		
$o\text{-}C_2H_5OCOC_6H_4$	$BiCl_3$	Heat	$(o\text{-}C_2H_5OCOC_6H_4)_3Bi$	Low	114

Vanadium Halides and Oxohalides

C_6H_5	$VOCl_3$	C_6H_{12}, 30°, 1 hr	C_6H_5HgCl	95	115
			$C_6H_5C_6H_5$	97–100	
			$VOCl_2$		
C_6H_5	VCl_4	C_6H_{12}, 20°C, 20 min.	C_6H_5HgCl	95	115
			$C_6H_5C_6H_5$		
			VCl_3		

Halides, Arylhalides, and Some Other Inorganic Compounds of Sulfur

$(ClHg)_3C{-}C(HgCl_3)_3$	S_2Cl_2	C_6H_6, 20°C, several days	No reaction		118
$[OHg_2C(HgOH)]_2$	S_2Cl_2		$[(ClHg)_2SCl]$		118
$ClHgCH_2CHO$	SO_3	$C_2H_4Cl_2$, 100°C, 12 hr	HSO_3CH_2COOH	41	121
C_6H_5	$SOCl_2$	Heat	C_6H_5HgCl	Quant.	119
$C_6H_4N(CH_3)_2$	$SOCl_2$	C_6H_6, 20°C	$[C_6H_4N(CH_3)_2]_3SCl$[d]	—	120
$C_6H_5HgC_6H_4CH_3\text{-}p$	C_6H_5SCl	CCl_4, 200°C, 48 hr	$p\text{-}CH_3C_6H_4S{-}C_6H_5$	Quant.	178
R[e]C_6H_4	$R'C_6H_4SCl$	CCl_4, reflux, 1 hr	$R'C_6H_4SC_6H_5HR$	Quant.	178
$o\text{-}CH_3C_6H_4$	$o\text{-}NO_2C_6H_4SCl$	CCl_4, reflux, 1 hr	$o\text{-}NO_2C_6H_4SC_6H_4CH_3\text{-}o$	25	178
$m\text{-}CH_3C_6H_4$	$o\text{-}NO_2C_6H_4SCl$	CCl_4, reflux, 1 hr	$o\text{-}NO_2C_6H_4SC_6H_4CH_3\text{-}m$	40	178
$p\text{-}CH_3C_6H_4$	$o\text{-}NO_2C_6H_4SCl$	CCl_4, reflux, 1 hr	$p\text{-}CH_3C_6H_4SC_6H_4NO_2\text{-}o$	15	178
$p\text{-}ClC_6H_4$	$p\text{-}NO_2C_6H_4SCl$	CCl_4, reflux, 1 hr	$p\text{-}NO_2C_6H_4SC_6H_4Cl\text{-}p$	50	178
$C_5H_5FeC_5H_4$	$(CNS)_2$	C_2H_5OH, heat, 10–15 min	$(C_5H_5FeC_5H_4{-}S)_2$[g]	15	122

(continued)

TABLE I (*Continued*)

Halide, Alkyl(aryl)halides, and Dioxide of Selenium

R in R_2Hg or $RHgR'$ and $RHgX$	Reagent	Reaction conditions	Products	Yield, %	Ref.
C_2H_5	SeO_2	Reflux, 100°C, 3 hr	$(C_2H_5)_2Se$	—	127
C_2H_5	$SeBr_4$	CS_2	$(C_6H_5)_2Se$	95	123
C_6H_5	$(p\text{-}CH_3C_6H_4)_2SeCl_2$	Fuse	C_6H_5HgBr	—	125
			mixt. $(C_6H_5)(p\text{-}CH_3C_6H_4)_2SeCl$ and $(C_6H_5)_3SeCl$		
$o\text{-}CH_3C_6H_4$	$(C_6H_5)_2SeCl_2$	Fuse	C_6H_5HgCl	—	125
			$(o\text{-}CH_3C_6H_4)_2(C_6H_5)SeCl$		
$m\text{-}CH_3C_6H_4$, $1M$	$(C_6H_5)_2SeCl_2$, $1M$	Fuse	C_6H_5HgCl	—	125
			$(m\text{-}CH_3C_6H_4)_2(C_6H_5)SeCl$		
$m\text{-}CH_3C_6H_4$, $1M$	$(m\text{-}CH_3C_6H_4)_2SeCl_2$, $2M$	Fuse, 75–80°C	mixture $(m\text{-}CH_3C_6H_4)_3SeCl \cdot HgCl_2$ and $(m\text{-}CH_3C_6H_4)_3SeCl$	—	125
$p\text{-}CH_3C_6H_4$	$SeBr_4$	CS_2	$p\text{-}CH_3C_6H_4HgBr$	82	123
			$(p\text{-}CH_3C_6H_4)_2Se$		
			$CH_3C_6H_4Br$		
$p\text{-}CH_3C_6H_4$	$(C_6H_5)_2SeCl_2$	Fuse	$(C_6H_5)_2(p\text{-}CH_3C_6H_4)SeCl$	—	125
$p\text{-}CH_3C_6H_4$, $1M$	$(p\text{-}CH_3C_6H_4)_2SeCl_2$, $1M$	Fuse, 140–150°C	$p\text{-}CH_3C_6H_4HgCl$	—	124
			$(p\text{-}CH_3C_6H_4)_3SeCl \cdot HgCl_2$		
$p\text{-}CH_3C_6H_4$, $1M$	$(p\text{-}CH_3C_6H_4)_2SeCl_2$, $2M$	Fuse, 140–150°C	$(p\text{-}CH_3C_6H_4)_3SeCl \cdot HgCl_2$	98	124
$p\text{-}CH_3C_6H_4$	$(p\text{-}CH_3C_6H_4)_2SeCl_2$	CCl_4, 1 hr	$(p\text{-}CH_3C_6H_4)_3SeCl$	—	125
			$(p\text{-}CH_3C_6H_4)_2Se$	Quant.	126
$\alpha\text{-}C_{10}H_7$	$\alpha\text{-}C_{10}H_7SeBr$	CCl_4, stand 1 hr	$(\alpha\text{-}C_{10}H_7)_2Se$	Quant.	126
R_2Hg[h]	$R'SeBr$[i]	CCl_4, stand 1 hr	$RSeR'$	Quant.	126
$\beta\text{-}C_{10}H_7$	$SeBr_4$	100°C, stir	$\beta\text{-}C_{10}H_7HgBr$	87	123
			$\beta\text{-}C_{10}H_7)_2Se$		
			$\beta\text{-}C_{10}H_7Br$		
$C_6H_5C_6H_4$	$SeBr_4$	100°C, stir	$(C_6H_5C_6H_4)HgBr$	100	123
			$(C_6H_5C_6H_4)_2Se$		
			$C_6H_5C_6H_4Br$		
C_6H_5HgBr	$SeBr_4$	Shake 6 days	$HgBr_2$	—	123
			$(C_6H_5)_2Se$		
			C_6H_5Br		

				Yield (%)	Ref.
p-CH₃C₆H₄HgCl	(C₆H₅)₂SeCl₂	Fuse, 150–160°C, 1 min	(C₆H₅)₂(p-CH₃C₆H₄)SeCl·HgCl₂	73	124
p-CH₃C₆H₄HgCl, 1M	(p-CH₃C₆H₄)₂SeCl₂, 1M	Fuse, 150–160°C, 1 min	(p-CH₃C₆H₄)₃SeCl·HgCl₂	21	124
C₅H₅FeC₅H₄	SeBr₄	CHCl₃, reflux, 1 hr	(C₅H₅FeC₅H₄)₂Se		122

Halides and Arylhalides of Tellurium

C₆H₅	TeCl₂	Sealed tube, 200°C, several hours	C₆H₅Cl		128
CH₃C₆H₄OC₆H₄HgCl	TeCl₄	CHCl₃, reflux, 1.5 hr	CH₃C₆H₄OC₆H₄TeCl₃	80	129
C₆H₅HgCl	p-CH₃OC₆H₄TeCl₃	Dioxane, reflux, 1.5 hr	C₆H₅Te(CH₃OC₆H₄)ᵏ	87	130
C₆H₅HgCl	p-C₆H₅O·C₆H₄TeCl₃	Dioxane, reflux, 1.5 hr	C₆H₅TeC₆H₄OC₆H₅-pᵏ	70	130
4-HOOCC₆H₄OC₆H₄HgCl-2'	TeCl₄	CH₃CN or dioxane, reflux, 4–5 hr	4-HOOCC₆H₄OC₆H₄TeCl₃-2'	—	129
α-C₁₀H₇HgCl	C₆H₅TeCl₃	Dioxane, reflux, 1.5 hr	C₆H₅(α-C₁₀H₇)Teᵏ	80	130
α-C₁₀H₇HgCl	4-CH₃OC₆H₄·TeCl₃	Dioxane, reflux, 1.5 hr	4-CH₃OC₆H₄Te·C₁₀H₇-αᵏ	95	130
α-C₁₀H₇HgCl	4-C₆H₅O·C₆H₄TeCl₃	Dioxane, reflux, 1.5 hr	4-C₆H₅OC₆H₄Te·C₁₀H₇-αᵏ	70	130
α-C₁₀H₇HgCl	β-C₁₀H₇TeCl₃	Dioxane, reflux, 1.5 hr	α-C₁₀H₇TeC₁₀H₇-βᵏ	91	130
β-C₁₀H₇HgCl	TeCl₄	Dioxane, reflux 2 hr	β-C₁₀H₇TeCl₃	85	130
β-C₁₀H₇HgCl	C₆H₅TeCl₃	Dioxane, reflux, 1.5 hr	C₆H₅Te(β-C₁₀H₇)ᵏ	75	130
β-C₁₀H₇HgCl	4-C₆H₅OC₆H₄TeCl₃	Dioxane, reflux, 1.5 hr	p-C₆H₅OC₆H₄TeC₁₀H₇-βᵏ	87	130

Tetrafluoride of Uranium

C₂H₅	UF₄	150–170°C	Hg C₂H₅HgF		131

Iodine Chlorides and ArICl₂

C₂H₅	C₆H₅ICl₂	H₂O, 20°C, 6 days	C₂H₅HgCl C₂H₅Cl C₆H₅I	— — —	136
C₂H₅HgCl	C₆H₅ICl₂	H₂O, heat	C₂H₅Cl C₆H₅I	—	136
trans-ClCH=CHHgCl, 1M	ICl₃, 1M	15% HCl, 20°C, 30 min	ClCH=CHICl₂	64	134
trans-ClCH=CHHgCl, 2M	ICl₃, 1M	3% HCl, 20°C, 20–30 min	(ClCH=CH)₂ICl·HgCl₂	6	133
C₆H₅	ICl₃	Dilute HCl, 20°C, 15–20 min	(C₆H₅)₂ICl·HgCl₂	51	132

(continued)

TABLE I (Continued)

R in R_2Hg or $RHgR'$ and $RHgX$	Reagent	Reaction conditions	Products	Yield, %	Ref.
C_6H_5	$C_6H_5ICl_2$	H_2O, 20°C, 12 hr	$(C_6H_5)_2ICl$	—	136
C_6H_5	$C_6H_5ICl_2$	H_2O, 20°C, 15 min then reflux	$(C_6H_5)_2ICl$	—	135
C_6H_5	$C_6H_5ICl_2$	H_2O, 20°C, 4 hr	C_6H_5HgCl $C_9H_5IO_2$ C_6H_5HgCl $(C_6H_5)_2ICl$	—	135
C_6H_5	$o\text{-}CH_3C_6H_4ICl_2$	H_2O, 20°C, 12 hr	$(o\text{-}CH_3C_6H_4)(C_6H_5)ICl$	—	136
C_6H_5	$p\text{-}CH_3C_6H_4ICl_2$	H_2O, shake 12 hr	$(p\text{-}CH_3C_6H_4)(C_6H_5)ICl$	—	136
C_6H_5	$\beta\text{-}C_{10}H_7ICl_2$	H_2O, shake 12 hr	$(\beta\text{-}C_{10}H_7)(C_6H_5)ICl$	—	136
C_6H_5HgCl	$o\text{-}CH_3C_6H_4ICl_2$	H_2O, shake 12 hr	$(o\text{-}CH_3C_6H_4)(C_6H_5)ICl$	—	136
C_6H_5HgCl	ICl_3	100°C, 1 hr	$(C_6H_5)_2ICl$	42	132
		ArAtCl₂			
C_6H_5	$C_6H_5I(At)Cl_2$	$CHCl_3$, reflux, few minutes	$(C_6H_5)_2I(At)Cl$	—	137
		Halides of Group VIII Elements			
$ClHgCH_2CHO$	$FeCl_3$	CH_3COCH_3, 20°C	$CH_2{=}CHOFe(OH)Cl$	83	138
C_6H_5	$\pi\text{-}C_5H_5Fe(CO)_2I$	C_6H_6, 20°C	$\pi\text{-}C_5H_5Fe(CO)_2C_6H_5\text{-}\sigma$	—	139
$p\text{-}C_2H_5O_2CC_6H_4$	$\pi\text{-}C_5H_5Fe(CO)_2I$	C_6H_6, 20°C	$\pi\text{-}C_5H_5Fe(CO)_2C_6H_4CO_2C_2H_5\text{-}\sigma$	—	139

[a] Hydrolysis.
[b] Treatment CO_2.
[c] Treatment H_2O_2.
[d] Treatment HCl.
[e] R = H, o-, m-, p-CH_3; o-, m-, p-Cl; p-CH_3O—.
[f] R' = H; o-, p-CH_3, o-, m-, p-Cl; p-NO_2.
[g] Treatment $Na_2S_2O_3$.
[h] R = which R is unknown.
[i] R' = which R' is unknown.
[j] R or R' = C_4H_9—, $C_6H_5CH_2$—, 4-$CH_3C_6H_4$—, 4-BrC_6H_4—, 4-ClC_6H_4—, 4-$CH_3OC_6H_4$—, $\alpha\text{-}C_{10}H_7$— by R' or R = C_6H_5; and 4-$ClC_6H_4SeC_6H_4CH_3$-4; 2-$C_6H_5C_6H_4SeC_6H_4CH_3$-4.

172

2. Preparation of Diphenyliodonium Chloride[135,136]

Five grams of diphenylmercury, 5 g of phenyliodonium chloride, and water were ground in a mortar, followed by the addition of more water. The mixture was shaken for 12 hr and the solid was filtered off. Evaporation of the filtrate at not too high a temperature gave crystalline diphenyliodonium chloride, m.p. 230°C. The double salt, $(C_6H_5)_2I \cdot HgCl_2$, was present in addition to C_6H_5HgCl in the solid. The former was gradually dissolved in boiling water and HgS precipitated from the filtrate by hydrogen sulfide addition. Evaporation of the filtrate gave an additional quantity of diphenyliodonium chloride.

REFERENCES

1. A. N. Nesmeyanov, A. E. Borisov, and R. I. Shepeleva, *Izv. Akad. Nauk SSSR, Otd. Khim. Nauk*, **1949**, 582.
2. A. E. Borisov, *Izv. Akad. Nauk SSSR, Otd. Khim. Nauk*, **1951**, 402
3. A. N. Nesmeyanov, A. E. Borisov, and V. D. Vil'chevskaya, *Dokl. Akad. Nauk SSSR*, **90**, 383 (1953).
4. A. E. Nesmeyanov and A. E. Borisov, *Tetrahedron*, **1**, 158 (1957).
5. A. E. Nesmeyanov and A. E. Borisov, *Dokl. Akad. Nauk SSSR*, **60**, 67 (1948).
6. A. N. Nesmeyanov, O. A. Reutov, and P. G. Knoll, *Dokl. Akad. Nauk SSSR*, **118**, 99 (1958).
7. K. Brodersen and U. Schlenker, *Chem. Ber.*, **94**, 3304 (1961).
8. S. Winstein, T. G. Traylor, and C. S. Garner, *J. Am. Chem. Soc.*, **77**, 3741 (1955).
9. O. A. Reutov, T. A. Smolina, and K.-V. Khu, *Izv. Akad. Nauk SSSR, Otd. Khim. Nauk*, **1959**, 559.
10. N. A. Nesmeyanov and O. A. Reutov, *Dokl. Akad. Nauk SSSR*, **144**, 126 (1962); *Tetrahedron*, **20**, 2803 (1964).
11. R. E. Dessy, I. K. Lee, and J.-Y. Kim, *J. Am. Chem. Soc.*, **83**, 1163 (1961).
12. O. A. Reutov, T. A. Smolina, Y.-T. U, and Ju. I. Bubnov, *Nauchn. dokl. Vyschei shkoly, Khim. i khim. technol.*, **1958**, N2, 324.
13. E. D. Hughes and H. C. Volger, *J. Chem. Soc.*, **1961**, 2359.
14. O. A. Reutov, T. A. Smolina, and V. A. Kalyavin, *Dokl. Akad. Nauk SSSR*, **139**, 389 (1961).
15. H. Cerfontain and G. M. F. Van Aken, *J. Am. Chem. Soc.*, **78**, 2094 (1956).
16. E. D. Hughes, C. K. Ingold, F. G. Thorpe, and H. C. Volger, *J. Chem. Soc.*, **1961**, 1133.
17. V. D. Nefedov, E. N. Sinotova, and N. Ja. Frolov, *Zh. Fiz. Khim.*, **30**, 2356 (1956).
18. V. D. Nefedov and E. N. Sinotova, *Sb. Rabot po Radiokhim. Izdat. Leningradsk. Gos. Univer.*, 1955, p. 113.
19. E. N. Sinotova, *Zh. Neorg. Khim.*, **2**, 1205 (1957).
20. J. M. Cross and J. J. Pinajian, *J. Am. Pharm. Assoc.*, **40**, 95 (1951); *Chem. Abstr.*, **45**, 3694 (1951).
21. A. Ansaloni, G. Belmondi, and U. Croatto, *Ric. Sci.*, **26**, 3315 (1956); *Chem. Abstr.*, **51**, 5517 (1957).

22. T. A. Smolina, T. Me-Tchy, and O. A. Reutov, *Izv. Akad. Nauk SSSR, Ser. Khim.*, **1966**, 413.
23. O. A. Reutov, V. I. Sokolov, and I. P. Beletskaya, *Izv. Akad. Nauk SSSR, Otd. Khim. Nauk*, **1961**, 1213.
24. E. D. Hughes, C. K. Ingold, and R. M. Roberts, *J. Chem. Soc.*, **1964**, 3900.
25. O. A. Reutov, V. I. Sokolov, and I. P. Beletskaya, *Izv. Akad. Nauk SSSR, Otd. Khim. Nauk*, **1961**, 1561.
26. O. A. Reutov, V. I. Sokolov, and I. P. Beletskaya, *Izv. Akad. Nauk SSSR, Otd. Khim. Nauk*, **1961**, 1217.
27. O. A. Reutov, V. I. Sokolov, and I. P. Beletskaya, *Dokl. Akad. Nauk SSSR*, **136**, 631 (1961).
28. O. A. Reutov, V. I. Sokolov, and I. P. Beletskaya, *Izv. Akad. Nauk SSSR, Otd. Khim. Nauk*, **1961**, 1427.
29. O. A. Reutov and I. P. Beletskaya, *Dokl. Akad. Nauk SSSR*, **131**, 853 (1960).
30. V. A. Kalyavin, T. A. Smolina, and O. A. Reutov, *Dokl. Akad. Nauk SSSR*, **155**, 596 (1964).
31. O. A. Reutov and P. G. Knoll, J.T.U, *Dokl. Akad. Nauk SSSR*, **120**, 1052 (1958).
32. H. B. Charman, E. D. Hughes, C. Ingold, and H. C. Volger, *J. Chem. Soc.*, **1961**, 1142.
33. V. A. Kaljavin, T. A. Smolina, and O. A. Reutov, *Dokl. Akad. Nauk SSSR*, **156**, 95 (1964).
34. O. A. Reutov, B. Praisnar, I. P. Beletskaya, and V. I. Sokolov, *Izv. Akad. Nauk SSSR, Otd. Khim. Nauk*, **1963**, 970.
35. V. A. Kaljavin, T. A. Smolina, and O. A. Reutov, *Dokl. Akad. Nauk SSSR*, **157**, 919 (1964).
36. C. R. Hart and C. K. Ingold, *J. Chem. Soc.*, **1964**, 4372.
37. B. Bartocha, F. E. Brinckman, H. D. Kaesz, and F. G. A. Stone, *Proc. Chem. Soc.*, **1958**, 116.
38. USP 3100217; *Chem. Abstr.*, **60**, 551 (1964).
39. R. N. Sterlin, V. H. Li, and I. L. Knunyants, *Zh. Vses. Khim. Obsh.*, **6**, 108, 1961.
40. A. Michaelis and P. Becker, *Chem. Ber.*, **13**, 58 (1880).
41. A. Michaelis, *Chem. Ber.*, **27**, 244 (1894).
42. A. Michaelis, *Ann. Chem.*, **315**, 19 (1901).
43. H. Gilman and L. Moore, *J. Am. Chem. Soc.*, **80**, 3609 (1958).
44. R. Clement, *Compt. Rend.*, **261**, 4436 (1965).
45. W. Gerard, M. Howard, E. F. Moonev, and D. E. Pratt, *J. Chem. Soc.*, **1963**, 1582.
46. A. E. Goddard and D. Goddard, *Organometallic Compounds*, Newton Friends Text Book of Inorganic Chemistry, Vol. XI. London 1928, part I, p. 235.
47. A. E. Goddard, *J. Chem. Soc.*, **123**, 1161 (1923).
48. A. N. Nesmeyanov, R. Ch. Freidlina, and A. K. Kochetkov, *Izv. Akad. Nauk SSSR, Otd. Khim. Nauk*, **1948**, 445.
49. A. N. Nesmeyanov, A. E. Borisov, and N. V. Novikova, *Izv. Akad. Nauk SSSR, Otd. Khim. Nauk*, **1959**, 1216.
50. A. E. Borisov, *Izv. Akad. Nauk SSSR, Otd. Khim. Nauk*, **1961**, 1036.
51. A. N. Nesmeyanov, A. E. Borisov, and N. V. Novikova, *Dokl. Akad. Nauk SSSR*, **94**, 289 (1954).

52. A. N. Nesmeyanov, A. E. Borisov, and N. V. Novikova, *Izv. Akad. Nauk SSSR, Otd. Khim. Nauk*, **1959**, 259.
53. A. E. Borisov, V. D. Vilchevskaya, and A. N. Nesmeyanov, *Izv. Akad. Nauk SSSR, Otd. Khim. Nauk*, **1954**, 1008.
54. A. N. Nesmeyanov, A. E. Borisov, I. S. Saveljeva, and E. I. Golubeva, *Izv. Akad. Nauk SSSR, Otd. Khim. Nauk*, **1958**, 1490.
55. A. E. Borisov and N. V. Novikova, *Izv. Akad. Nauk SSSR, Otd. Khim. Nauk*, **1957**, 1258.
56. A. N. Nesmeyanov and T. A. Kudryavtseva, *Uchen. Zap. Moskov. Gos. Univer.*, **151**, 57 (1951).
57. N. N. Melnikov and M. S. Rokitskaya, *Zh. Obshch. Khim.*, **7**, 1472 (1937).
58. A. E. Borisov and M. A. Osipova, *Izv. Akad. Nauk SSSR, Otd. Khim. Nauk*, **1961**, 1039.
59. A. E. Goddard, *J. Chem. Soc.*, **121**, 36 (1922).
60. V. P. Gluschkova and K. A. Kocheshkov, *Dokl. Akad. Nauk SSSR*, **103**, 615 (1955).
61. V. P. Gluschkova and K. A. Kocheshkov, *Dokl. Akad. Nauk SSSR*, **116**, 233 (1957).
62. V. P. Gluschkova and K. A. Kocheshkov, *Izv. Akad. Nauk SSSR, Otd. Khim. Nauk*, **1957**, 1193.
62a. A. P. Schapiro and E. G. Rozantsev, *Izv. Akad. Nauk SSSR, Ser. Khim.*, **1966**, 1650.
63. A. N. Nesmeyanov, G. G. Dvorjantseva, N. S. Kochetkova, R. B. Materikova, and Yu. N. Scheinker, *Dokl. Akad. Nauk SSSR*, **159**, 847 (1964).
64. A. N. Nesmeyanov, I. F. Lutsenko, and V. A. Bratzev, *Dokl. Akad. Nauk SSSR*, **128**, 551 (1959).
65. Yu. I. Baukov and I. F. Lutsenko, *Zh. Obshch. Khim.*, **32**, 2746 (1962).
66. Yu. I. Baukov, G. S. Burlatchenko, and I. F. Lutsenko, *Dokl. Akad. Nauk SSSR*, **157**, 119 (1964).
67. I. F. Lutsenko, Yu. I. Baukov, G. S. Burlatchenko, and B. N. Khasapov, *J. Organometallic Chem.*, **5**, 20 (1966).
68. A. Ladenburg, *Ann. Chem.*, **173**, 143 (1873); *Ber.*, **6**, 379 (1873).
69. A. Ladenburg, *Chem. Ber.*, **7**, 387 (1874).
70. F. E. Brinckman and F. G. A. Stone, *J. Inorg. Nucl. Chem.*, **11**, 24 (1959).
71. W. K. Orndorff, D. L. Tabern, and L. M. Dennis, *J. Am. Chem. Soc.*, **49**, 2512 (1927).
72. Yu. I. Baukov and I. F. Lutsenko, *Zh. Obshch. Khim.*, **34**, 3453 (1964).
73. B. Aronheim, *Ann. Chem.*, **194**, 145 (1878).
74. M. M. Nadj and K. A. Kocheshkov, *Zh. Obshch. Khim.*, **12**, 409 (1942).
75. A. N. Nesmeyanov, R. Ch. Freidlina, and A. K. Kochetkov, *Izv. Akad. Nauk SSSR, Otd. Khim. Nauk*, **1948**, 127.
76. E. M. Panov, E. I. Lodochnikova, and K. A. Kocheshkov, *Dokl. Akad. Nauk SSSR*, **111**, 1042 (1956).
77. E. I. Lodochnikova, E. M. Panov, and K. A. Kocheshkov, *Izv. Akad. Nauk SSSR, Otd. Khim. Nauk*, **1957**, 1484.
78. R. Criegee, P. Dimroth, and R. Schempf, *Chem. Ber.*, **90**, 1337 (1957).
79. E. I. Lodochnikova, E. M. Panov, and K. A. Kocheshkov, *Zh. Obshch. Khim.*, **29**, 2253 (1959).
80. G. A. Razuvaev and I. F. Bogdanov, *Zh. Obshch. Khim.*, **3**, 367 (1933).

81. G. A. Razuvaev, V. N. Latjaeva, A. V. Malyscheva, and G. A. Kiljakova, *Dokl. Akad. Nauk SSSR*, **150**, 566 (1963).
82. G. A. Razuvaev, V. I. Latjaeva, and L. I. Vyschinskaya, *Zh. Obshch. Khim.*, **31**, 2667 (1961).
83. F. Guichard, *Chem. Ber.*, **32**, 1572 (1899).
84. L. R. Drake and C. S. Marvel, *J. Org. Chem.*, **2**, 387 (1937).
85. H. D. Kaesz and F. G. A. Stone, *J. Org. Chem.*, **24**, 635 (1959).
85a. B. Bartocha, C. M. Douglas, and M. Y. Gray, *Z. Naturforsch.*, **14b**, 809 (1959).
86. A. Michaelis, *Ann. Chem.*, **181**, 288 (1876).
87. A. Michaelis, *Ann. Chem.*, **293**, 261 (1896).
88. A. Michaelis and A. Link, *Ann. Chem.*, **207**, 193 (1881).
89. L. M. Yagupolsky and P. A. Yufa, *Zh. Obshch. Khim.*, **28**, 2853 (1958).
90. H. Hartman, C. Beerman, and H. Czempik, *Z. Anorg. Allg. Chem.*, **287**, 261 (1956).
91. W. Pope and C. S. Gibson, *J. Chem. Soc.*, **101**, 735 (1912).
92. J. Meisenheimer, J. Casper, M. Höring, W. Lauter, L. Lichtenstadt, and W. Samuel, *Ann. Chem.*, **449**, 213 (1926).
93. A. Michaelis, *Ann. Chem.*, **315**, 43 (1901).
94. E. Wedekind, *Ber.*, **45**, 2935 (1912).
95. C. D. Nenitzescu, D. A. Isacescu, and C. Gruescu, *Bull. Soc. Chim. România*, **20**, 127 (1938); through Zbl., **1940**, I, 532.
96. W. La Coste, *Ann. Chem.*, **208**, 3 (1881).
97. W. Steinkopf and W. Mieg, *Ber.*, **53**, 1013 (1920).
98. R. N. Sterlin, V. H. Li, and I. L. Knunyants, *Dokl. Akad. Nauk SSSR*, **140**, 137 (1961).
99. W. E. Jones, R. J. Rosser, and F. N. Woodward, *J. Soc. Chem. Ind.*, **68**, 258 (1949).
100. A. Michaelis, *Ber.*, **8**, 1316 (1875).
101. A. Michaelis, *Ber.*, **9**, 1566 (1876).
102. W. La Coste and A. Michaelis, *Ann. Chem.*, **201**, 184 (1880).
103. W. La Coste and A. Michaelis, *Ber.*, **11**, 1883 (1878).
104. A. Michaelis, *Ann. Chem.*, **321**, 141 (1902).
105. G. Roeder and N. Blasi, *Ber.*, **47**, 2748 (1914).
106. W. W. Beck and C. S. Hamilton, *J. Am. Chem. Soc.*, **60**, 620 (1938).
107. G. B. Buckton, *J. Chem. Soc.*, **16**, 22 (1863).
108. A. Michaelis and A. Güntner, *Ber.*, **44**, 2316 (1911).
109. I. Hasenbäumer, *Ber.*, **31**, 2911 (1898).
110. L. W. Woods, *Iowa State Coll. J. Sci.*, **19**, 611 (1944); *Chem. Abstr.*, **39**, 693 (1945).
111. Z. M. Manulkin, A. N. Tatarenko, and F. Yusupov, *Sb. Stat. po Obshchei Khim.*, **2**, 1308 (1953).
112. Z. M. Manulkin, *Uzbeksk. Khim. Zh.*, **2**, 41 (1958).
113. F. Challenger and C. F. Allpress, *J. Chem. Soc.*, **1921**, 923.
114. K. A. Kocheshkov, T. K. Kozminskaya, and M. M. Nadj, *Bull. Vses. Khim. Obshch.*, **N9**, 21 (1940).
115. W. L. Carrick, W. T. Reichle, F. Pennella, and J. J. Smith, *J. Am. Chem. Soc.*, **82**, 3887 (1960).
116. S. V. Bleschinsky and M. Usubakunov, *Sb. Fis.-Mat. Konferenz., Posvyashch. 100-let. akad. Kurnakova, Frunse*, 1963.
117. A. Weiss and A. Weiss, *Z. Anorg. Allgem. Chem.*, **282**, 324 (1955).

118. K. A. Hofmann and H. Feigel, *Ber.*, **38**, 3655 (1905).
119. K. Heumann and P. Köchlin, *Ber,*, **16**, 1625 (1883).
120. A. Michaelis and E. Godchaux, *Ber.*, **24**, 757 (1891).
121. A. P. Terentjev, A. N. Kost, A. M. Yurkevitch, E. E. Khasskina, and L. I. Obreimova, *Vestn. Mosk. Gos. Univer. Ser. Fis. mat. i estesstv. nauk* **N4**, 121 (1953).
122. A. N. Nesmeyanov, E. G. Perevalova, and O. A. Nesmeyanova, *Dokl. Akad. Nauk SSSR*, **119**, 288 (1958).
123. H. M. Leicester, *J. Am. Chem. Soc.*, **60**, 619 (1938).
124. H. M. Leicester, *J. Am. Chem. Soc.*, **57**, 1901 (1935).
125. H. M. Leicester and J. W. Bergstrom, *J. Am. Chem. Soc.*, **53**, 4428 (1931).
126. T. W. Campbell and J. D. McCullough, *J. Am. Chem. Soc.*, **67**, 1965 (1945).
127. N. N. Melnikov and M. S. Rokitskaya, *Zh. Obshch. Khim.*, **8**, 834 (1938).
128. F. Krafft and R. E. Lyons, *Ber.*, **27**, 1768 (1894).
129. I. G. M. Campbell and E. E. Turner, *J. Chem. Soc.*, **1938**, 37.
130. H. Rheinboldt and G. Vicentini, *Chem. Ber.*, **89**, 624 (1956).
131. A. E. Comyns, Atomic Energy Research Estab. (Gt. Brit). C/m-258 7p. (1957); *Chem. Abstr.*, **51**, 17557 (1957).
132. R. Ch. Freidlina and A. N. Nesmeyanov, *Dokl. Akad. Nauk SSSR*, **29**, 566 (1940).
133. R. Ch. Freidlina, E. M. Brainina, and A. N. Nesmeyanov, *Izv. Akad. Nauk SSSR, Otd. Khim. Nauk*, **1945**, 647.
134. A. N. Nesmeyanov and R. Ch. Freidlina, *Izv. Akad. Nauk SSSR, Otd. Khim. Nauk*, **1943**, 152; *Dokl. Akad. Nauk SSSR*, **31**, 892 (1941).
135. C. Willgerodt, *Chem. Ber.*, **30**, 56 (1897).
136. C. Willgerodt, *Chem. Ber.*, **31**, 915 (1898).
137. V. D. Nefedov, Yu. V. Norseev, Kh. Savlevitch, E. N. Sinotova, M. A. Toropova, and V. A. Khalkin, *Dokl. Akad. Nauk SSSR*, **144**, 806 (1962).
138. A. N. Nesmeyanov, I. F. Lutsenko, and R. M. Khomutov, *Dokl. Akad. Nauk SSSR*, **120**, 1049 (1958).
139. A. N. Nesmeyanov, I. V. Polovyanyuk, B. V. Lokshin, Yu. A. Chapovski, and L. G. Makarova, *Zh. Obshch. Khim.*, **37**, 2015 (1967).
140. A. Michaelis and P. Becker, *Ber.*, **15**, 180 (1882).
141. A. N. Nesmeyanov, A. E. Borisov, and N. A. Vol'kenau, *Izv. Akad. Nauk SSSR, Otd. Khim. Nauk*, **1956**, 162.
142. M. V. George, G. D. Lichtenwalter, and H. Gilman, *J. Am. Chem. Soc.*, **81**, 978 (1959).
143. Yu. I. Baukov, I. Yu. Belavin, and I. F. Lutsenko, *Zh. Obshch. Khim.*, **35**, 1092 (1965).
144. I. F. Lutsenko, Yu. I. Baukov, and B. N. Khasapov, *Zh. Obshch. Khim.*, **33**, 2724 (1963).
145. L. I. Emel'anova, V. N. Vinogradova, L. G. Makarova, and A. N. Nesmeyanov, *Izv. Akad. Nauk SSSR, Otd. Khim. Nauk*, **1962**, 53.
146. G. B. Buckton, *Ann. Chem.*, **108**, 103 (1858).
147. A. Michaelis, *Ber.*, **13**, 2174 (1880).
148. A. Cahours, *Compt. Rend.*, **76**, 1383 (1873); *Jahresber.*, **1873**, 520.
149. D. Seyferth and W. Freyer, *J. Org. Chem.*, **26**, 2604 (1961).
150. A. N. Nesmeyanov, I. F. Lutsenko, Z. S. Kraïts, and A. P. Bokovoï, *Dokl. Akad. Nauk SSSR*, **124**, 1251 (1959).
151. D. Jerchel, *Ber.*, **76**, 600 (1943).

152. E. Dreher and R. Otto, *Ann. Chem.*, **154**, 93 (1870).
153. A. Michaelis and F. Graeff, *Ber.*, **8**, 922 (1875).
154. A. Michaelis and H. Rohler, *Ber.*, **9**, 519 (1876).
155. A. Michaelis, *Ber.*, **10**, 1109 (1877).
156. A. Michaelis and W. La Coste, *Ber.*, **18**, 2109 (1885).
157. L. G. Radcliffe and W. H. Brindley, *Chem. Ind.*, **42**, 64 (1923).
158. W. C. Davies and F. G. Mann, *J. Chem. Soc.*, **1944**, 276.
159. A. Michaelis and Cl. Paneck, *Ann. Chem.*, **212**, 203 (1882).
160. J. Weller, *Ber.*, **20**, 1718 (1887).
161. A. Schenk and A. Michaelis, *Ber.*, **21**, 1497 (1888).
162. A. Michaelis and A. Schenk, *Ann. Chem.*, **260**, 1 (1890).
163. W. Kelbe, *Ber.*, **9**, 1051 (1876).
164. W. Kelbe, *Ber.*, **11**, 1499 (1878).
165. J. Linder and M. Strecker, *Monatsh. Chem.*, **53/4**, 274 (1929).
166. A. Michaelis and C. Schulte, *Ber.*, **15**, 1953 (1882).
167. F. F. Blicke and F. D. Smith, *J. Am. Chem. Soc.*, **51**, 3479 (1929).
168. W. La Coste and A. Michaelis, *Ber.*, **11**, 1888 (1878).
169. A. Michaelis, *Ann. Chem.*, **320**, 271 (1901).
170. K. Matsumiya, Memoirs of the College of Science Kyoto Imp. Univ. Serie A. 8. 6 Seiten. 1925; through *Chem. Zentr.*, **1926**, I, 652.
171. W. C. Davies and C. W. Othen, *J. Chem. Soc.*, **1936**, 1236.
172. W. Steinkopf, *Ann. Chem.*, **413**, 310 (1917).
173. W. G. Lowe and C. S. Hamilton, *J. Am. Chem. Soc.*, **57**, 1081 (1935).
174. W. G. Lowe and C. S. Hamilton, *J. Am. Chem. Soc.*, **57**, 2314 (1935).
175. A. W. Weitkamp and C. S. Hamilton, *J. Am. Chem. Soc.*, **59**, 2699 (1937).
176. C. Finzi, *Gazz. Chim. Ital.*, **45**, II, 280 (1915).
177. C. Finzi, *Gazz. Chim. Ital.*, **55**, 824 (1925).
178. D. Spinelli and A. Salvemini, *Ann. Chim.* (*Rome*), **51**, 389 (1961).
179. V. V. Kudinova, V. L. Foss, and I. F. Lutzenko, *Zh. Obshch. Khim.*, **36**, 1863 (1966).

Chapter 2

Methylene Insertion and Addition Reactions by Organomercury Compounds

I. INTRODUCTION

Organomercury compounds of the type $C_6H_5HgCR'R''R'''$, have been shown by Seyferth to decompose according to the following scheme under mild conditions:

$$C_6H_5HgCR'R''R''' \longrightarrow C_6H_5HgR' + CR''R'''$$
$$(R' = R'' = Hal, R''' = H \text{ or } R' = R'' = R''' = Hal)$$

This reaction generates either a mono or dihalocarbene. The generated carbene can be inserted into the following bonds: C—H in hydrocarbons, Si—H in alkylsilanes, Ge—H in alkylgermanes, O—H of carboxylic acids, and alcohols, H—Cl in hydrochloric acid, and to replace the oxygen in aldehydes and ketones by the $CR'R'''$ group. $:CHal_2$ formed from $C_6H_5HgCHal_3$ adds to a C=C bond.

Both $(HalCH_2)_2Hg$ and $HalCH_2HgHal$ [in the presence of $(C_6H_5)_2Hg$] generate $:CH_2$ which also adds to a C=C bond. See Chapter 13 for reactions of $C_6H_5HgCCl_2Br$ with carbodiimides.

II. REACTION MECHANISMS

Since the reaction of cyclohexene with the $CHCl_2Br$-tert-C_4H_9OH system produced a mixture of 7,7-dichloro- and 7-chloro-7-bromo-bicyclo(4,1,0)heptanes,[1] while with $C_6H_5HgCCl_2Br$ produced only the 7,7-dichloro derivative,[2] it may be assumed that the latter reaction proceeds by a different mechanism than the former, i.e., not through preliminary formation of CCl_3Br, then generation of $CHal_2$ as is typical for methylenation by means of the first system. See, for example, ref. 2.

However the similarity of the relative reactivities of $C_6H_5HgCCl_2Br$ and CCl_3COONa with respect to different olefins suggests a uniform reaction mechanism of dihalomethylation by both reagents.

Seyferth[1] believes that these reactions are effected by free dihalocarbenes. Both reagents cause carbene insertion into a C—H bond, thus suggesting similar reaction mechanisms with respect to the last sources of $:CHal_2$. The authors[3] believe that the reactions involving acids proceed according to a mechanism involving free carbenes, which in this case behave as nucleophiles:

$$:CHal_2 + HX \longrightarrow HHal_2C^+X^- \longrightarrow CHHal_2X$$

179

The reactions of $C_6H_5HgCHal_3$ with olefins carried out in the presence of NaI proceed with participation of $CHal_3^-$, which is shown by the formation of $CHHal_3$ when the reaction is conducted in acetone, evidently due to the extraction of hydrogen from the acetone by the $CHal_3^-$ ion.[4]

III. REACTION CONDITIONS

The previously mentioned reactions of methylene insertion by $C_6H_5HgCR'R''R'''$ are usually carried out under mild conditions, for example, in boiling benzene. If the reactant is a gas, the reaction is carried out in an autoclave. The reaction may also occur if the gas is passed into a heated solution of the organomercury compound.

IV. APPLICATION OF THE REACTION

A. Introduction of Methylene (Halo- and Dihalomethylene) in Compounds Containing a Mobile Atom of H

1. Methylene Insertion into a C—H Bond

Dihalocarbene (dichloro- or dibromocarbene) generated during the decomposition of phenyltrihalomethylmercury:

$$C_6H_5HgCHal_3 \longrightarrow C_6H_5HgHal + :CHal_2$$

reacts with a CH bond in hydrocarbons with the formation of the dihalo-derivatives of the hydrocarbons[5]:

$$RCHR_2' + :CHal_2 \longrightarrow RR_2'CHal_2H$$
$$RCH_2R' + :CHal_2 \longrightarrow RR'CHCHal_2H$$

The reaction occurs by heating the carbene–forming species in an excess of hydrocarbons.

2. Methylene Insertion into a Si—H Bond

Dihalocarbene produced during decomposition of phenyltrimercuric halide reacts with the Si—H bond in the triaryl and diaryl derivatives of silane producing a 77–80% yield of dihaloderivatives of organosilicon compounds, $R_3Si(CHal_2)H$ and $Ar_2Si(CHal_2H)H$ or $Ar_2Si(CHal_2H)_2$ depending on the relative percentages of $C_6H_5HgCHal_3$ and Ar_2SiH_2.[5] The reaction is conducted by heating the reactants to 80°C for 2–4 hr in benzene. $(C_6H_5)_3SiCH_2Br$ is produced in 72% yield according to the following reaction[6]:

$$C_6H_5HgCHBr_2 + (C_6H_5)_3SiH \longrightarrow (C_6H_5)_3SiCH_2Br + C_6H_5HgBr$$

3. *Methylene Insertion into a* Ge—H *Bond*

:CCl_2 produced by heating $C_6H_5HgCCl_2Br$ in benzene at 80°C for 2–4 hr is also inserted into a Ge—H bond producing $(C_6H_5)_3GeCCl_2H$ in 88% yield.[5]

4. *Reaction with Trialkyltin(IV) Hydride*

Instead of the expected methylene insertion reaction into the Sn—H bond of tributyltin(IV) hydride by the reaction of excess tin compound with $C_6H_5HgCCl_2Br$, an exchange reaction of H by Br occurs[7]:

$$(n\text{-}C_4H_9)_3SnH + C_6H_5HgCCl_2Br \longrightarrow$$
$$C_6H_5HgCCl_2H + (n\text{-}C_4H_9)_3SnBr$$

The reaction does not occur with equimolar amounts of reagents.

5. *Methylene Insertion into the* O—H *Bonds of Carboxylic Acids and Alcohols and the* HCl *Bond*

$C_6H_5HgCHal_3$ esterifies carboxylic acids[3]:

$$C_6H_5HgCCl_2Br + RCOOH \longrightarrow RCOOCHCl_2 + C_6H_5HgBr$$

The reactions proceed very quickly at 60–80°C in benzene with approximately quantitative yields.

$ROCHCl_2$ is produced by the reaction of $C_6H_5HgCCl_2Br$ with alcohols[3] carried out at 80–85°C in ethylbenzene in addition to other typical reaction and side-reaction products formed during the preparation of chloromethyl ethers.

In addition to :CCl_2 insertion into the HCl bond, protodemercuration and C_6H_6 formation occurs in the reaction of $C_6H_5HgCHal_3$ with HCl:

$$C_6H_5HgCCl_2Br + HCl \longrightarrow C_6H_5HgBr + CHCl_3$$

Chloroform is produced in 69% yield (in toluene at 80°C). With $C_6H_5HgCBr_3$ (in C_6H_5Cl, at 80–85°C), along with the expected $CHBr_2Cl$ (5.3%), $CHCl_2Br$ and $CHCl_3$ were found as well.

B. Mono- and Dihalomethylene Addition to a C=C Bond. Synthesis of Mono- and Gem-Dihaloderivatives of Cyclopropane

1. *Action of* $C_6H_5HgCHal_3$ *and* $C_6H_5HgCHHal_2$

a. *Olefins.* The reaction of phenyl(trihalomethyl)mercury compounds ($C_6H_5HgCCl_3$, $C_6H_5HgCCl_2Br$, $C_6H_5HgCClBr_2$, $C_6H_5HgCBr_3$) with olefins is suggested as a method for the synthesis of gem-dihalocyclo-propanes[1,8-11]:

$$C_6H_5HgCHal_3 + \quad \diagdown C{=}C \diagup \quad \longrightarrow \quad \diagdown C{-}C \diagup \quad + \quad C_6H_5HgHal$$

This method has an advantage over the reaction of olefins with dihalo-carbenes from other sources since yields are higher and the sensitivity of the olefin to alkalis is not a factor.

If the $C_6H_5HgCHal_3$ used is $C_6H_5HgCCl_2Br$ or $C_6H_5HgCClBr_2$, :CCl_2 is produced in the first case but not :$CClBr$, and $CClBr$ but not :CBr_2 is produced in the second case. In both cases C_6H_5HgBr is produced.

Addition to olefins is strongly facilitated by the presence of a bromine group, thus $C_6H_5HgCCl_2Br$ reacts more readily with olefins than $C_6H_5HgCCl_3$. Various olefins such as ethylene,[8] tetrachloroethylene,[10] trimethylvinylsilane,[10] and others may be used as reactants. The reaction with ethylene is conducted in an autoclave, with tetrachloroethylene, heating to 80°C and with *trans*-stilbene, boiling benzene. As a rule, yields of gem-dihalocyclopropanes are excellent. Thermal decomposition of $C_6H_5HgCCl_3$ and formation of $CCl_2{=}CCl_2$ and hexachlorocyclopropane occurred in tetrachloroethylene.

$C_6H_5HgCHBr_2$ and $C_6H_5HgCClHBr$ react with cyclohexene (at 130°C, for approximately 10 hr, in a sealed flask with N_2) and produce mono-bromosubstituted derivatives of cyclopropane (a mixture of *cis* and *trans*)[6,7]:

$$C_6H_5HgCHBr_2 + \quad \longrightarrow \quad + \quad C_6H_5HgBr$$

In contrast to other methods producing silylsubstituted cyclopropanes with very low yields, $C_6H_5HgCHal_3$ reacts with vinylsilanes to produce 1,1-dihalo-2-trimethylsilylcyclopropane with approximately an 80% yield.[10] The reaction with dimethylvinylsilane chloride produces 1,1-dichloro-2-(chlorodimethylsilyl)-cyclopropane,[10] i.e., proceeds without side reactions of the Si—Cl bond which has a high reactivity.

The reaction was successfully carried out with a number of olefins containing groups sensitive to bases such as acrylonitrile, allyl bromide, allyl isocyanate, and mesityl oxide and produced good yields of the respective substituted gem-dihalocyclopropanes.[11]

b. *Unsaturated Acids.* The reactions of $C_6H_5HgCCl_2Br$ with unsaturated acids, acrylic, *trans*-crotonic, 3-butenic acids (1:1), results in dichloromethylene insertion into the OH bond[11]:

$$RCH{=}CHCOOH + C_6H_5HgCCl_2Br \longrightarrow RCH{=}CHCO_2CHCl_2 + C_6H_5HgBr$$

The carbene addition to a $C{=}C$ bond in these acids is observed only after formation of the dichloromethyl ester in an organomercury compound-to-acid ratio of 2:1.[11]

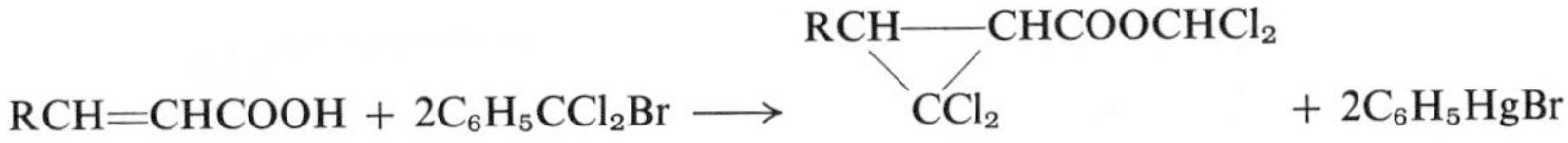

$$RCH{=}CHCOOH + 2C_6H_5CCl_2Br \longrightarrow \quad + 2C_6H_5HgBr$$

c. *Dienes.* The reaction of $C_6H_5HgCCl_2Br$ with dienes, for example, 1,3-butadiene and allene, advantageously results in the formation of the monoadduct.[11] The diadduct is produced in a substantially smaller amount, for example:

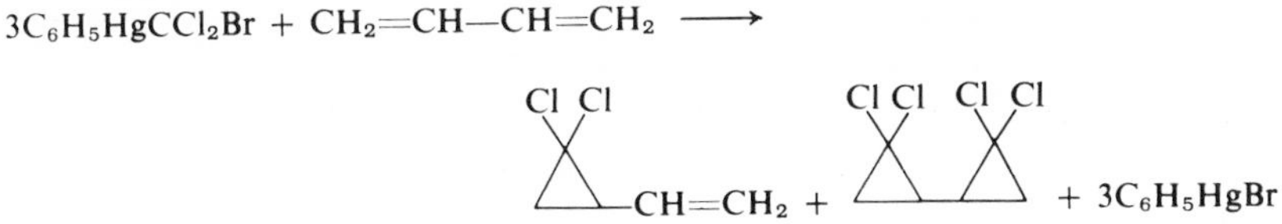

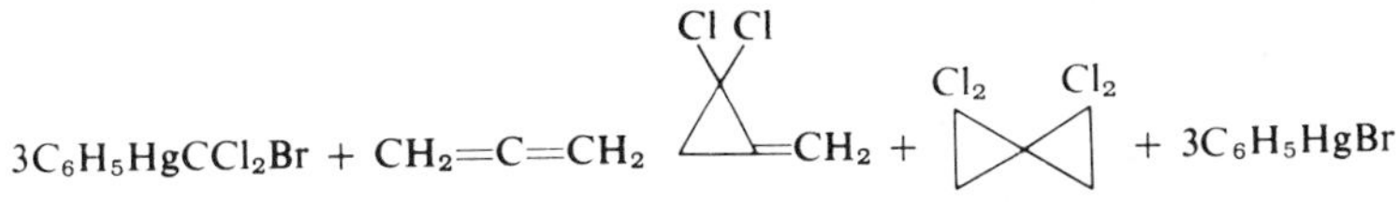

$$3C_6H_5HgCCl_2Br + CH_2{=}CH{-}CH{=}CH_2 \longrightarrow$$

$$3C_6H_5HgCCl_2Br + CH_2{=}C{=}CH_2$$

However, the reactions of the monoadducts with $C_6H_5HgCCl_2Br$ produce good yields of the diadducts. The reactions of $C_6H_5HgCCl_2Br$ with allenes constitute a very valuable way for the synthesis of alkenylcyclopropanes and substituted spiropentanes.[11]

Formation of **1** in 82.5% yield as compared to **2** in a 9.4% yield during the following reaction proves higher reactivity of the ring double bond

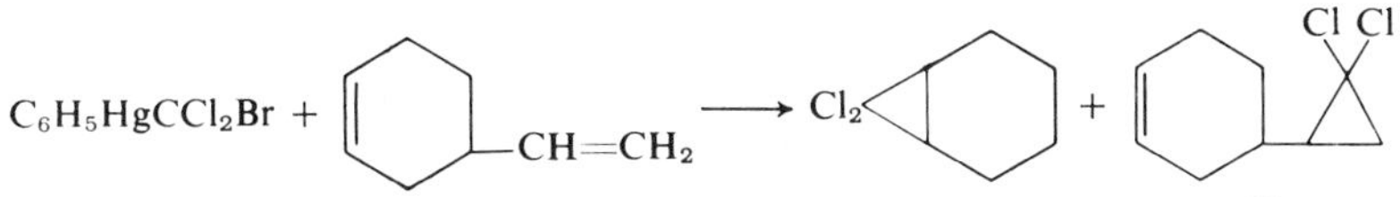

with respect to $C_6H_5HgCCl_2Br$. In this case, steric factors also play an additional part in lowering the reactivity of the vinyl group. Cyclohexene is more reactive with $C_6H_5HgCCl_2Br$ than 1-heptene.[11]

d. *Unsaturated Ethers.* In the reaction with 2,5-dihydrofuran (in benzene, 80°C)

(3)   (4)

3 is produced in 44% yield and **4** in 52% yield in the relationship similar to that obtained during dichloromethylene insertion by the action of CCl_3CO_2Na, which again suggests a similar mechanism in both reactions. A 67% yield of the compound similar to **4**, 2-dichloromethyltetrahydrofuran, is produced during the reaction of $C_6H_5HgCCl_2Br$ with tetrahydrofuran.[11] In allyl ethyl ether the double bond is substantially higher in reactivity than CH bond with respect to $:CCl_2$ produced from $C_6H_5HgCCl_2Br$.[11]

$$C_6H_5HgCCl_2Br + C_2H_5OCH_2CH{=}CH_2 \longrightarrow$$

$$C_2H_5OCH_2{-} \triangle + C_2H_5OCH(CHCl_2)CH{=}CH_2 + C_6H_5HgBr$$
      82.3%   13.7%

Addition of dihalocarbenes from $C_6H_5CHal_3$ to the double bond occurs stereospecifically; refer, for example, to the addition of *cis-* and *trans-*2-butene which discounts a free-radical mechanism.[11]

2. Reactions of $C_6H_5HgCHal_3 + NaI$

Addition of $:CHal_2$ from $C_6H_5HgCHal_3$ to a double bond in the presence of NaI is another method for carrying out the reaction.[4]

The reaction proceeds at room temperature and is useful in cases where thermally unstable olefins are involved. Thus, cyclohexene reacts with $C_6H_5HgCCl_3$ and NaI (30°C, 48 hr) to produce a 72% yield of 7,7-dichlorobicyclo-(4,1,0)-heptane and 14% $CHCl_3$, reactants which do not form under the same conditions in the absence of NaI. At 80°C for four hours, a 72% yield of dichloronorcarane and a 4% yield of $CHCl_3$ are produced. In the absence of NaI a 16% yield of dichloronorcarane is produced. In a similar manner, the reaction of $C_6H_5HgCCl_2Br + NaI$ with cyclohexene (4 hours at room temperature) produces a 75% yield of 7,7-dichloro- and a 4% yield of 7-bromo-7-chlorobicyclo-(4,1,0)-heptane.[4] Under the same conditions in the absence of NaI only a 1.5% yield of dichloronorcarane is produced.

3. Reactions of Halomethylmercury Compounds

The methylene group generated during decomposition of $(HalCH_2)_2Hg$ or $HalCH_2HgHal$ adds to the double bond of cyclohexene producing a norcarane.[12]

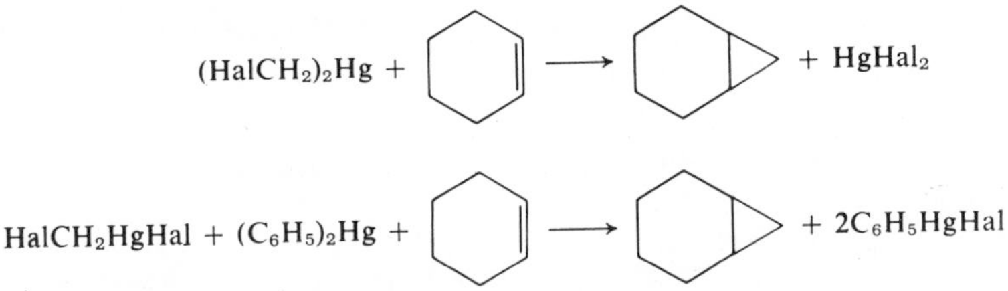

$$(HalCH_2)_2Hg + \quad \longrightarrow \quad + \ HgHal_2$$

$$HalCH_2HgHal + (C_6H_5)_2Hg + \quad \longrightarrow \quad + \ 2C_6H_5HgHal$$

C. Preparation of 1-Haloolefins from Aldehydes and Ketones

The reaction of $C_6H_5HgCRHalBr$ with aldehydes or ketones carried out in the presence of $(C_6H_5)_3P$ is a common method for the synthesis of 1-haloolefins[6]:

$$C_6H_5HgCRHalBr + (C_6H_5)_3P + R'R''C{=}O \longrightarrow R'R''C{=}CRHal$$
$$+ \ C_6H_5HgBr + (C_6H_5)_3PO$$
$$(R = H \ or \ Hal; \ R' = H \ or \ Alk; \ R'' = Alk)$$

The reaction is usually carried out in boiling benzene under nitrogen or in boiling o-xylene with a cyclohexanone. A 50–60% yield of the haloolefin is obtained.

The authors[6] propose the following reaction scheme:

$$C_6H_5HgCX_2Br \ \longrightarrow \ CX_2 \ \xrightarrow{(C_6H_5)_3P} \ (C_6H_5)_3P{=}CX_2 \ \xrightarrow{R'R''C=O} \ R'R''C{=}CX_2$$

They do not discount, however, direct reaction without the intermediate formation of a phosphonium ylide.

D. Reaction of $C_6H_5HgCRHalBr$ with $P(OAlk)_3$ (R = H or Hal)

During the reaction of $C_6H_5HgCRHalBr$ with trimethyl or triethyl phosphite, the following reactions occur[13]:

$$C_6H_5HgCCl_2Br + P(OAlk)_3 \longrightarrow C_6H_5HgCCl_2P(O)(OAlk)_2 + AlkBr$$
$$C_6H_5HgCHClBr + P(OAlk)_3 \longrightarrow C_6H_5HgCHClP(O)(OAlk)_2 + AlkBr$$

No α-elimination was observed.

E. Miscellaneous Reactions

CCl_2 which is obtained from $C_6H_5HgCCl_3$ is inserted into the β—CH bond of tetraalkylsilicon and -tin compounds.[14] CCl_2 is also inserted into the Ge—H and Si—H bond.[15,41]

CYX (Y = H or Cl) (X = halogen) is inserted into the Hg—Cl bond of ArHgCl (Ar = aromatic group) or $HgCl_2$,[16] into the Sn—X bond of R_3SnX[17] (R = alkyl group); into the Sn—Sn bond of R_3Sn—SnR_3,[18] into Si—Hg and GeHg bonds of $(Alk_3M)_2Hg$ (M = Si, Ge) giving $R_3MHg(CX_2)_2MR_3$ which undergoes β-elimination to form R_3MCX= CX_2.[19] Insertion of CCl_2 into the N—H bond of secondary amines produces trichlorovinylamines and 1,1-dichloro-2,2-bisamines.[20] CCl_2 reacts with tertiary amines to form dialkyltrichlorovinylamines.[21] Addition of CX_2 to the triple bond of acetylene leads to formation of cyclopropenes and their derivatives.[22–24]

New data have become available on the reaction of $C_6H_5HgCCl_2Br$ with carboxylic acids,[25] alcohols,[26] HCl,[27] aldehydes, and ketones[28] and the addition of CX_2 to an olefin in the presence of NaI.[29] A kinetic study of the reaction of $ArHgCCl_2Br$ with some olefins has been carried out.[30,36] The reaction of $C_6H_5HgCCl_2Br$ with alkenylcarboranes produces *gem*-dichlorocyclopropylcarboranes.[32] The reaction of organoboron compounds with $C_6H_5HgCCl_2Br$ is used as a method to convert C_n terminal olefins to C_{2n+1} internal olefins, for example[33]:

$$RCH{=}CH_2 \xrightarrow{\ B_2H_6\ } (RCH_2CH_2)_3B \xrightarrow{\ C_6H_5HgCCl_2Br\ }$$
$$RCH_2CH_2CH{=}CHCH_2R\cdot \longleftarrow\rceil$$

Substituted cyclopropanes are not formed by thermal decomposition of $RR'CHHgCCl_2CHRR'$ (R = alkyl, R' = H or alkyl) in the presence of cyclohexene or 2,3-dimethyl-2-butene.[34]

CX_2 from $C_6H_5HgCX_2Br$ has been shown to add to polyene macromolecules.[35]

The thermal decomposition of $C_6H_5HgO_2CCF_2X$ (X = Cl or F) does not insert CXF into the double bond of cycloolefins.[31] $(Me_3SiCCl_2)_2Hg$, in the presence of $(C_6H_5)_2Hg$, yields Me_3SiCCl_3 which in turn produces CCl_2. The CCl_2 then reacts with cycloalkenes and the Si—H bond.[37]

$C_6H_5HgCCl_3$ and cyclohexene react to form dichloronorcarane.[38] N_2 in some phosphorus-substituted diazoalkanes is substituted by CX_2.[39]

CCl_2 from $C_6H_5HgCBrCl_2$ adds to the double bond of vinylene carbonates.[40]

A Hammett study of the methylenation reaction of the Si—H bond confirms the electrophilic attack by CCl_2.[41]

V. PREPARATIVE SYNTHESES

A. Preparation of *gem*-Dihalocyclopropane[10]

$C_6H_5HgCBr_3$ (0.073 mole) in 100 ml of benzene reacts with ethylene (50 atm) in an autoclave, first at 100°C for a short time, then at 80°C for

24 hr, to form 1,1-dibromocyclopropane, b.p. 48°C at 5 mm in 53% yield. A similar reaction using $C_6H_5HgCCl_2Br$ and chlorobenzene as the solvent resulted in the formation of 1,1-dichlorocyclopropane in 65% yield, b.p. 75–76°C.

B. Preparation of Hexachlorocyclopropane[11]

A solution of 0.10 mole of phenyl(bromodichloromethyl)mercury in 1.0 mole of redistilled tetrachloroethylene in a 300-ml, three-necked flask, equipped with reflux condenser, was slowly heated to 90°C with stirring; this temperature was maintained for 1 hr. After 3 min light gray flakes began to precipitate. The reaction mixture was cooled and filtered. The residue was washed with 25 ml of tetrachloroethylene and dried *in vacuo*, giving 33.4 g (93%) of phenylmercury bromide as gray flakes, m.p. 285–287°C. A second crop of 0.6 g was obtained by removal of some of the excess olefin from the filtrate at 170–180 mm. The filtrate was concentrated further at 170–180 mm, cooled to 0°C, and filtered. The residue was washed with 11 ml of tetrachloroethylene leaving 20.6 g (83%) of white, crystalline solid, m.p. 100–104°C.

C. Preparation of 1,1-Dichloro-2-vinylcyclopropane[11]

This reaction is described to illustrate the procedure which may be used when the olefin is a gas but is not applicable if the gas is ethylene. A solution of 0.02 mole of phenyl(bromodichloromethyl)mercury in 60 ml of benzene was heated to about 70°C in a three-necked flask equipped with a thermometer, gas inlet tube, magnetic stirrer, and water-cooled reflex condenser topped with a Dewar-type condenser filled with Dry Ice–acetone. The system was flushed with nitrogen and gaseous 1,3-butadiene was then allowed to bubble slowly into the stirred solution for 30 min. Phenylmercury bromide precipitated and the solution became deep yellow in color. The reaction mixture was cooled and 5.7 g (81%) of phenylmercury bromide was filtered. GC analysis of the filtrate showed that 1,1-dichloro-2-vinylcyclopropane had been formed in 58% yield and 2,2,2′,2′-tetrachlorobicyclopropyl in 10% yield. To a quantity of the above reaction mixture calculated to contain 5 moles of 1,1-dichloro-2-vinyl-cyclopropane was added 10 moles of phenyl(bromodichloromethyl)-mercury. This mixture was heated at 80°C for 8 hr and then filtered from C_6H_5HgBr (87.5%). Gas chromatography showed that 2,2,2′,2′-tetrachlorobicyclopropyl (*d,l* and meso forms in 1:1 ratio) were produced in 91% yield. The higher melting form had a shorter retention time on an SE-30 silicone gum column than the lower melting form. A similar procedure was used in the reaction of allene with $C_6H_5HgCCl_2Br$ and in the reactions of *cis*- and *trans*-2-butene with $C_6H_5HgCBr_3$.

D. Preparation of 7,7-Dichloronorcarane[11]

The procedure used to prepare this compound is typical of the procedure used when both olefin and product were liquids. In a 100-ml flask equipped with reflux condenser and magnetic stirrer, 0.03 mole of phenyl(bromodichloromethyl)mercury, 0.09 mole of cyclohexene, and 35 ml of benzene were placed. All of the organomercury compound dissolved by the time reflux temperature was reached, and precipitation of a white, flaky solid began within 15 min. After the reaction mixture had been stirred at reflux for 2 hr, TLC analysis indicated that all of the starting mercury compound had decomposed and that only phenylmercury bromide was present. The reaction mixture was filtered, leaving 10.5 g (98%) of phenylmercury bromide, m.p. 284–286°C. The solvent and excess cyclohexene were removed by trap-to-trap distillation at 25° (5 mm). The residue was distilled rapidly *in vacuo* into a receiver at −78°C and the distillate was fractionally distilled to give 7,7-dichloronorcarane in 87% yield, b.p. 73.5–75°C (10 mm).

E. Reactions with *trans*-Crotonic Acid[11]

Anhydrous, crystalline, sublimed *trans*-crotonic acid (0.01 mole) and phenyl(bromodichloromethyl)mercury (0.01 mole) were heated at 80°C in 25 ml of benzene for 50 min. The mixture was filtered leaving 3.48 (97%) of phenylmercury bromide. The filtrate was trap-to-trap distilled *in vacuo*. GC analysis of the distillate (20% SE-30 on Chromosorb P, jacket at 112°C, 11.2 psi of helium) showed that dichloromethyl *trans*-crotonate had been formed in 87% yield.

REFERENCES

1. D. Seyferth and J. M. Burlitch, *J. Am. Chem. Soc.*, **86**, 2730 (1964).
2. W. E. Parham and R. R. Twelves, *J. Org. Chem.*, **22**, 730 (1957).
3. D. Seyferth, J. Y.-P. Mui, and L. J. Todd, *J. Am. Chem. Soc.*, **86**, 2961 (1964).
4. D. Seyferth, J. Y.-P. Mui, M. E. Gordon, and J. M. Burlitch, *J. Am. Chem. Soc.*, **87**, 681 (1965).
5. D. Seyferth and J. M. Burlitch, *J. Am. Chem. Soc.*, **85**, 2667 (1963).
6. D. Seyferth, H. D. Simmons, Jr., and G. Singh, *J. Organometal. Chem. (Amsterdam)*, **3**, 337 (1965).
7. D. Seyferth and L. J. Todd, *J. Organometal. Chem. (Amsterdam)*, **2**, 282 (1964).
8. D. Seyferth, J. M. Burlitch, and J. K. Heeren, *J. Org. Chem.*, **27**, 1491 (1962).
9. D. Seyferth and J. M. Burlitch, *J. Am. Chem. Soc.*, **84**, 1757 (1962).
10. D. Seyferth, K. J. Minasz, A. J.-H. Treiber, J. M. Burlitch, and S. R. Dowd, *J. Org. Chem.*, **28**, 1163 (1963).
11. D. Seyferth, J. M. Burlitch, R. J. Minasz, J. Y.-P. Mui, H. D. Simmons, Jr., A. J.-H. Treiber, and S. R. Dowd, *J. Am. Chem. Soc.*, **87**, 4259 (1965).
12. D. Seyferth, M. A. Eisert, and L. J. Todd, *J. Am. Chem. Soc.*, **86**, 121 (1964).

13. D. Seyferth, J. Y.-P. Mui, and G. Singh, *J. Organometal. Chem. (Amsterdam)*, **5**, 185 (1966).
14. D. Seyferth and S. S. Washburne, *J. Organometal. Chem.*, **5**, 389 (1966).
15. D. Seyferth, J. M. Burlitch, H. Dertouzos, and H. D. Simmons, Jr., *J. Organometal. Chem.*, **7**, 405 (1967).
16. M. E. Gordon, K. V. Darragh, and D. Seyferth, *J. Am. Chem. Soc.*, **88**, 1831 (1966).
17. D. Seyferth, F. M. Armbrecht, Jr., B. Prokai, and R. J. Cross, *J. Organometal. Chem.*, **6**, 573 (1966).
18. D. Seyferth and F. M. Armbrecht, Jr., *J. Am. Chem. Soc.*, **89**, 2790 (1967).
19. D. Seyferth, R. J. Cross, and B. Prokai, *J. Organometal. Chem.*, **7**, P20 (1967).
20. W. E. Parham and J. R. Potoski, *Tetrahedron Lett.*, **21**, 2311 (1966).
21. D. Seyferth, M. E. Gordon, and R. Damrauer, *J. Org. Chem.*, **32**, 469 (1967).
22. E. V. Dehmlow, *J. Organometal. Chem.*, **6**, 296 (1966).
23. D. Seyferth and R. Damrauer, *J. Org. Chem.*, **31**, 1660 (1966).
24. H. M. Cohen and A. H. Keough, *J. Org. Chem.*, **31**, 3428 (1966).
25. D. Seyferth and J. Y.-P. Mui, *J. Am. Chem. Soc.*, **88**, 4672 (1966).
26. D. Seyferth, V. A. Mai, J. Y.-P. Mui, and K. V. Darragh, *J. Org. Chem.*, **31**, 4079 (1966).
27. D. Seyferth, J. Y.-P. Mui, L. J. Todd, and K. V. Darragh, *J. Organometal. Chem.*, **8**, 29 (1967).
28. D. Seyferth, J. K. Heeren, G. Singh, S. O. Grim, and W. B. Hughes, *J. Organometal. Chem.*, **5**, 267 (1966).
29. D. Seyferth, M. E. Gordon, J. Y.-P. Mui, and J. M. Burlitch, *J. Am. Chem. Soc.*, **89**, 959 (1967).
30. D. Seyferth, J. Y.-P. Mui, and J. M. Burlitch, *J. Am. Chem. Soc.*, **89**, 4953 (1967).
31. D. Seyferth, B. Prokai, and R. L. Cross, *J. Organometal. Chem.*, **13**, 169 (1968).
32. D. Seyferth and B. Prokai, *J. Organometal. Chem.*, **8**, 366 (1967).
33. D. Seyferth and B. Prokai, *J. Am. Chem. Soc.*, **88**, 1834 (1966).
34. J. A. Landgrebe and R. D. Mathis, *J. Am. Chem. Soc.*, **88**, 3552 (1966).
35. C. Pinazzi and G. Levesque, *Compt. Rend.*, **264**, 288 (1967).
36. D. Seyferth, J. Y.-P. Mui, and R. Damrauer, *J. Am. Chem. Soc.*, **90**, 6182 (1968).
37. D. Seyferth and E. M. Hanson, *J. Am. Chem. Soc.*, **90**, 2438 (1968).
38. G. A. Razuvaev, V. I. Tshcherbakov, and S. T. Zhilzov, *Izv. Akad. Nauk SSSR, Ser. Khim.*, **1968**, 2803.
39. D. Seyferth, J. D. H. Paetsch, and R. S. Marmor, *J. Organometal. Chem.*, **16**, 185 (1969).
40. F. W. Breitbeil, D. T. Dennerlein, A. E. Fiebig, and R. E. Kuznicki, *J. Org. Chem.*, **33**, 3389 (1968).
41. D. Seyferth, R. Damrauer, J. Y.-P. Mui, and T. F. Jula, *J. Am. Chem. Soc.*, **90**, 2944 (1968).

Chapter 3

Exchange Reactions with Metals and Metal Alloys

I. INTRODUCTION

As far as preparative purposes are concerned, one of the most important reactions of organomercury compounds is their ability to alkylate metals (alkyl group exchange).

For many metals this reaction is a rather important method for the synthesis of their organometallic compounds. The reaction is usually carried out with completely substituted mercury compounds and can be generally expressed by the following equation:

$$\frac{n}{2}\,R_2Hg + Me^{(n)} \longrightarrow R_nMe^{(n)} + \frac{n}{2}\,Hg$$

The advantage of using organomercury compounds in these syntheses is that the organometallic compounds are produced in a pure state, free of halides.

Using this reaction (R = Alk or Ar) completely substituted organometallic compounds of the following metals were produced: Li, Na, K, Be, Mg, Zn, Cd, Al, Ga, In, Sn, Sb, and Bi. With R = cyclo-C_5H_5 and Me = Fe, ferrocene was produced.

In some cases, for example, in the synthesis of organotin compounds, the metals were alkylated by using organomercury salts.

In addition to the use of metals, metal alloys may also be used. Na—Sn and Na—Bi alloys react with R_2Hg and RHgCl to produce the respective organotin and bismuth compounds. Transition metals from Sc to Cu (except for Fe; refer to synthesis of ferrocene), Y to Ag, La to Au, and probably Ca, Sr, and Ba as well as Pb do not form organometallic compounds upon reaction with organomercury compounds. Heating of R_2Hg with some of the aforementioned elements catalyzes decomposition of the R_2Hg to form R_2 and Hg (comprehensively described in Chapters 7 and 8).

II. REACTION MECHANISM AND STEREOCHEMISTRY

For the first time regularities relating to stereochemistry and mechanism of the exchange reactions of organomercury compounds with metals were reported by Nesmeyanov and Borisov[1] by investigating the reactions of β-chlorovinyl and other alkenyl compounds of mercury with various metals and metal alloys. The reactions of *cis*- or *trans*-β-chlorovinyl-mercury chloride with a tin–sodium alloy resulted in the formation of β-chlorovinyltin compounds which had the same stereochemical configuration as the original organomercury compounds:

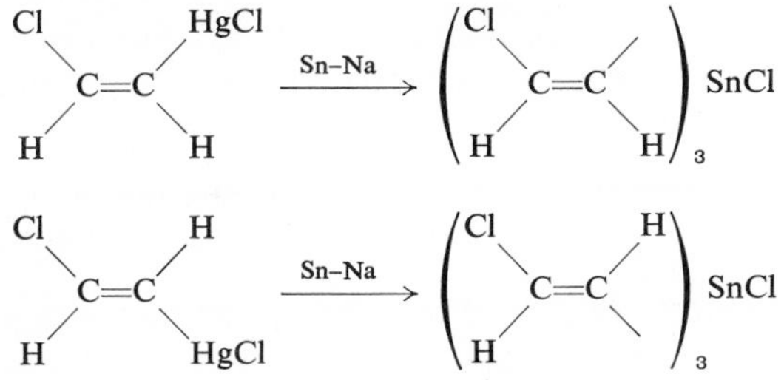

The reverse reaction from β-chlorovinyltin compounds to organomercury occurred as well with retention of configuration of the transferred radical.

On the basis of these and other reactions, Nesmeyanov and Borisov formulated a retention-of-configuration rule which is followed during electrophilic and homolytic exchange reactions. (See Chapter 1 for references.) The latter are exemplified by the reaction of β-chlorovinyl-mercury compounds with a tin–sodium alloy.

The simplest models of alkylation (arylation) reactions of organomercury compounds with other metals described in the present chapter are the reactions of organomercury compounds with radioactive mercury. Both completely substituted compounds of mercury or some organomercury salts enter into exchange reactions with labeled mercury (^{203}Hg):

$$R_2Hg + Hg^* \rightleftharpoons R_2Hg^* + Hg$$

$$RHgX + Hg^* \rightleftharpoons RHg^*X + Hg$$

Di-*trans*- and di-*cis*-β-chlorovinylmercury readily react with finely divided labeled mercury (^{203}Hg) in acetone under cold conditions. Isotope equilibrium is obtained after several hours[2] and stereochemical configuration of the chlorovinyl group is preserved. Among the aliphatic organomercury salts, the compounds having a carbonyl in an α-position readily exchange with ^{203}Hg.[3,4] For some of them, provided the reaction proceeds for a sufficient length of time, 100% exchange is obtained. The exchange reactions of the 1-methyl ether of α-bromomercuryphenylacetic acid and *cis*- and *trans*-β-chlorovinylmercury chloride occur stereospecifically. These two reactions were the experimental foundations on which Nesmeyanov and Borisov based the rule for retention of configuration during homolytic reactions involving saturated carbon atoms.[2]

Under mild conditions isotope exchange with radioactive mercury occurs:

$$(\bar{p}\text{-RC}_6\text{H}_4)_2\text{Hg} + *\text{Hg} \rightleftharpoons (\bar{p}\text{-RC}_6\text{H}_4)_2*\text{Hg} + \text{Hg}$$

The rate of the reaction depends on the nature of the substituent R in the aromatic moiety. The following series shows this effect: NO_2, Cl < H < CH_3 < OCH_3. (The rates were determined in pyridine at 60°C.[5]) The reaction

$$\text{Ar}_2\text{Hg} + \text{Hg}^{203} \longrightarrow \text{Ar}_2\text{Hg}^{203} + \text{Hg}$$
$$(\text{Ar} = \text{C}_6\text{H}_5,\ p\text{-CH}_3\text{C}_6\text{H}_4)$$

has a first-order rate constant. The reaction rate decreases as the Ar_2Hg concentration increases, evidently due to reaction inhibition caused by the products formed. In addition, the rate increases as the concentration of metallic mercury increases.[6]

The exchange reaction of ArHgX with Hg occurs faster than that of Ar_2Hg.[3,7] The effect on the rate of exchange by para substituents is identical to that observed with Ar_2Hg: O_2N < C_2H_5OCO < Cl < H < CH_3,[7] i.e., in this case, also, the electropositive substituents increase, while the electronegative substituents decrease, the rate of the isotope exchange reaction.

Exchange of RHgX with ^{203}Hg usually occurs under cold conditions in such solvents as acetone, benzene, $CHCl_3$ and, for R = Ar, in pyridine, xylene, and dioxane.

α-Thienylmercury chloride reacts with metallic mercury faster than all other arylmercury salts. Ferrocenylmercury chloride reacts at approximately the same rate as tolylmercury chloride.

The rate of exchange of various RHgBr compounds such as phenylmercury bromide and 3-camphormercury bromide with ^{203}Hg were determined.[3] In the cases studied, retention of configuration was observed. Although the main factor in determining the rate of exchange was electronic, steric factors were also found to be important.

Isotope exchange reactions of R_2Hg and RHgHal with metallic mercury does not proceed by the formation of the free radicals R· and ·Hg, since in many cases these reactions occur under such mild conditions that free radicals are not produced from organomercury compounds. Also, during the isotope exchange of unsymmetrical, completely substituted mercury compounds and after isotope equilibrium has been obtained, only the original unsymmetrical mercury compound is left in the reaction mixture. If the reaction occurred through the formation of the free radicals ·Ar and ·Ar', there should be found with ArHgAr' in the reaction mixture, Ar_2Hg and Ar'_2Hg. Finally, retention of stereochemical configuration contradicts the idea of a free-radical mechanism of these exchange reactions. Thus, the isotope exchange reactions with metallic mercury go by a homolytic mechanism, but not through free radicals. A transition state such as **1** was proposed by Reutov[7,8]:

$$\text{RHgR}' + {}^{203}\text{Hg} \rightleftharpoons R \underset{{}^{203}\text{Hg}}{\overset{\text{Hg}}{\diagup\diagdown}} R' \rightleftharpoons R^{203}\text{HgR}' + \text{Hg}$$

(1)

where R = Alk, Ar or ClCH = CH—; R' = R or Cl.
Thus exchange proceeds by a concerted mechanism.

III. REACTION CONDITIONS

Organomercury compounds alkylate free metals over a wide temperature range which depends on the metal and, to a lesser extent, on the mercury substituents.

Organomercury compounds transfer organic groups to alkali metals at room or slightly higher temperatures. In many cases with other metals, heating is required, sometimes up to rather high temperatures.

Many exchange reactions of organomercury compounds with metals, particularly those described in the previous publications, were conducted in sealed tubes without solvents. It is often convenient to carry out these reactions with solvents. Naturally, the solvent should be unreactive with respect to the reactant metal and products of the reaction. Therefore, aliphatic and, in many cases, aromatic hydrocarbons may be used. Other solvents such as ether, dioxane, and tetrahydrofuran may also be used. Aromatic compounds of mercury usually react more readily than aliphatic compounds and thus the latter were used to a lesser extent. Upon heating, the reactions of organomercury compounds with metals are often complicated by decomposition:

$$\text{R}_2\text{Hg} \xrightarrow{\text{Me}} \text{R}_2 + \text{Hg}$$

This decomposition occurs particularly with organic groups such as benzyl (refer to Chapters 7 and 8).

To produce organometallic compounds sensitive to oxygen and moisture the reaction should be conducted in a dry medium with use of special equipment under inert gases such as nitrogen and argon also in the cases of zinc and indium these gases are preferred. Use of the inert gas is advisable in other cases when organomercury compounds interact with the metal for a long period of time.

Metals taking part in a reaction in which mercury separates can produce amalgams; therefore, they should be used in excess amounts.

IV. APPLICATION OF THE REACTION

A. Reactions of Organomercury Compounds with Alkali Metals

The reactions of organomercury compounds with alkali metals should be carried out in nitrogen or argon absolutely free of oxygen. Hydrocarbons such as pentane, hexane, and gasoline are usually used as solvents. Other solvents are used for specific synthesis; i.e., aromatic compounds—benzene, and ethers, diethyl ether or dioxane. However, it should be taken into account that aromatic hydrocarbons and ethers are reactive to alkali metals (see reactions with sodium).

The reactions were conducted in sealed flasks (refer to the example in ref. 9), but they may be carried out in standard equipment used for Grignard synthesis (three-necked flask with stirrer, mercury seal, and reflux condenser).

The alkali metal used in the reaction should have a clean surface (stir with glass, etc.).

1. Metallic Lithium

The reactions of organomercury compounds with metallic lithium produced aliphatic compounds of lithium with Li at the saturated[9-11], unsaturated[12-14] carbon and aromatic compounds of lithium.[9,15-17]

These reactions take place with heating under an inert gas in the following solvents: pentane, ligroin, and benzene. The vapor of the boiling solvent may be used as the inert gas. Various narrow-boiling, purified fractions of kerosene were also used as the boiling solvent.[10]

Continuous shaking of the hexamer o-phenylenemercury,[18] tetramer o-diphenylenemercury,[19] and the dimer o-terphenylenemercury[20] with metallic lithium in ether produced, respectively, o-dilithiumbenzene, o-lithiumdiphenyl, and o,o'-dilithiumdiphenyl.

Synthesis of the organolithium compounds by reacting R_2Hg with lithium is particularly useful since the products are not contaminated with alkyl halides.

However, in many cases the organolithium compounds produced are used in the next reaction without isolating the organolithium compound from solution (for detailed information, refer to the tables).

Refer to Chapter 12 for the preparation of RLi by the following reaction:

$$R_2Hg + 2R'Li \longrightarrow 2RLi + R'_2Hg$$

Lithium vinylate is produced[21] by the reaction of mercurybisacetalde-hyde with lithium solution in liquid ammonia:

$$Hg(CH_2CHO)_2 + nLi \xrightarrow{NH_3} Li_nHg + CH_2{=}CHOLi$$

2. Sodium

Metallic sodium reacts more slowly with completely substituted mercury compounds than metallic lithium when shaken for several days at room temperature in a Schlenk flask or special Grignard equipment.

Products and solvents best suited for the preparative reactions are as follows: pentane, ligroin (aliphatic compounds)[9,22-24]; benzene (phenyl- and benzylsodium)[9]; pentane (vinylsodium).[12] Benzylsodium can also be produced by heating tolylsodium resulting in the transfer of the sodium to the side chain.

Although benzene is used as a solvent during the synthesis of organo-sodium compounds from organomercury compounds and sodium, it should be noted that sodium alkyls react with aromatic hydrocarbons[25,26] and with heterocyclics like furan and thiophene[27] according to the following scheme:

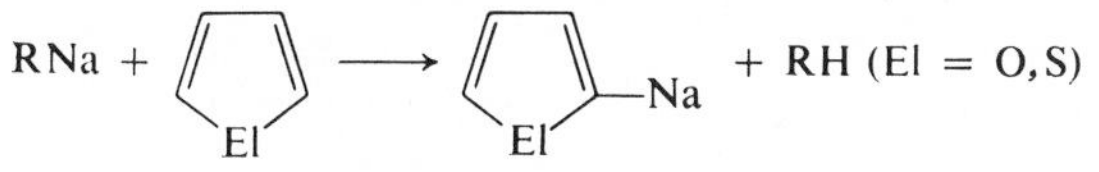

$$RNa + \text{(ring with El)} \longrightarrow \text{(ring with El)}{-}Na + RH \quad (El = O,S)$$

$$RNa + C_6H_6 \longrightarrow C_6H_5Na + RH$$
$$RNa + C_6H_5CH_3 \longrightarrow C_6H_5CH_2Na + RH$$

These reactions provide a method for the synthesis of aryl, aralkyl, and heterocyclic derivatives of sodium.

It is not necessary to isolate the organosodium compound from the reaction mixture. The second component may be introduced directly into the mixture of dialkylmercury and sodium[28]; for example:

$$Alk_2Hg + Na + CO_2 \xrightarrow{H_2O} AlkCOOH$$

$$Alk_2Hg + Na + RCHO \xrightarrow{H_2O} AlkRCHOH$$

$$Alk_2Hg + Na + RCOR' \xrightarrow{H_2O} RAlkR'COH$$

$$Alk_2Hg + Na + RCOOR' \xrightarrow{H_2O} RAlk_2COH$$

These reactions are usually carried out in ether.

When carbon dioxide is used as the inert gas for the reaction of sodium with dialkylmercury (alkyl, ethyl, and amyl) in pentane or ligroin, in addition to the expected monocarboxylic acid, a dialkylmalonic acid is also produced, probably as a result of carboxylation of the disodium derivative produced according to the proposed equation:

$$(C_5H_{11})_2Hg + 3Na \longrightarrow C_5H_{10}Na_2 + C_5H_{12} + NaHg$$

The mechanism of formation of the disodium derivative, however, is not clear. Increasing the amount of solvent increases the formation of the dicarboxylic acid.

If a mixture of dialkylmercury and sodium is placed in an aromatic hydrocarbon under carbon dioxide, solvent participation results in the formation of arylcarboxylic[23] acids at room temperature, at lower temperatures, the dicarboxylic acid[29] (in benzene, a mixture of iso- and terephthalic acids) is formed.

When using benzene homologs, the sodium is incorporated into the side chain: therefore, toluene treated by CO_2 produces phenylacetic acid[25] and, at lower temperatures, phenylmalonic acid[29]; xylenes give their respective tolylacetic acids[25,26]; mesitylene gives 3,5-dimethylacetic acid[26]; and ethylbenzene gives hydroatropic acid.[25] Reduction of the reaction temperature increases the yield. If the reaction is carried out in thiophene, furan, α-methylfuran, or α-methylthiophene, increased carboxylation occurs as compared to carboxylation in aromatic hydrocarbons and results in the formation of α-thiophenecarboxylic,[26] α-furancarboxylic,[27] or 2-methylfuran-5-carboxylic[27] and 2-methylthiophene-5-carboxylic acids.[27]

It is evident that in all these reactions, the AlkNa produced initially reacts with the solvent yielding the sodium derivatives which then carboxylate.

The reactions of dialkylmercury with sodium may be conducted in ether as a solvent, but it should be noted that ethylsodium decomposes the ethers producing sodium alcoholates[30]:

$$C_2H_5Na + (C_2H_5)_2O \longrightarrow C_2H_5ONa + C_2H_6 + C_2H_4$$

As lithium does, sodium dissolved in liquid ammonia reacts with mercurybisacetaldehyde producing sodium vinylate.[21]

Production of complex compounds $(R_2Li)Na$ also occurs; refer to Chapter 12.

3. Potassium

Reactions of organomercury compounds with metallic potassium have been studied to a lesser extent than those with lithium or sodium. Vinylpotassium was produced by the reaction of divinylmercury with potassium in pentane.[12] As in the case with benzylsodium benzylpotassium can be produced from the reaction of any isomer of ditolylmercury with metallic potassium.

B. Reactions with Metals of Group II

1. Beryllium

Synthesis of the organoberyllium compounds by reactions of organo-mercury compounds with metallic beryllium should be carried out under an inert atmosphere of nitrogen or argon.

The reaction of Alk_2Hg with metallic beryllium was carried out in sealed tubes over a temperature range of 100–130°C; however, there is a description of the synthesis of dimethylberyllium in an open flask.[31] In these methods the mercury was isolated by entrapment on gold foil[32] or powder[31] (see tables for other syntheses of Alk_2Be).

Diarylmercury reacts with metallic beryllium at high temperatures in sealed tubes and also without solvents in the presence of catalysts: $HgCl_2$ (225°C for 6 hr)[33]; in the presence of $HgCl_2$ and $BeBr_2$ (170°C, 72 hr)[34]; heating in a sealed tube with xylene (150°C for 72 hr),[16,35] and heating to 210–220°C *in vacuo* for 1–2 hr.[36] Ar_2Be yields were good.

2. Magnesium

Synthesis of organomagnesium compounds should be conducted in an inert atmosphere of nitrogen or argon.

Catalysts such as mercury(II) chloride or ethyl acetate,[39] sealed tubes,[38,39] and high temperatures (120–200°C) are usually needed for the magnesium reactions; however, the reaction of $(CH_3)_2Hg$ and Mg proceeds upon heating at 100°C for 24 hr.[37]

Reactions of Alk_2Hg with Mg produced good yields of Alk_2Mg.[37,40] Continuous shaking of R_2Hg in ether or benzene solutions with magnesium at room temperature produced R_2Mg. In particular, $(C_2H_5)_2Mg$ and $(C_6H_5)_2Mg$ were produced in this manner.[41] When the reaction is conducted in ether, R_2Mg is isolated as the etherate, $R_2Mg \cdot (C_2H_5)_2O$. The reaction is catalyzed by the addition of trace amounts of $HgCl_2$. Variable results are obtained if the reaction takes place in benzene; the mercury may amalgamate the magnesium and separation is simple; at other times a dispersion of mercury forms and does not settle for many days.

3. Zinc

Reactions of organomercury compounds with metallic zinc resulting in organozinc compounds are usually conducted under carbon dioxide. In most cases, particularly in the synthesis of aliphatic zinc compounds, heating at 120–130°C in sealed tubes with no solvent was sufficient for the reaction to occur.[42] A convenient and simple method for the synthesis of arylzinc compounds (including those which contain substituents in the benzene ring) consists of boiling the diarylmercury with zinc in xylene.[43]

The reaction is complete in 2–3 hr for diphenyl, di-*p*-fluorophenyl-, di-*p*-chlorophenyl-, and di-*p*-dimethylaminophenylzinc. Di-*o*-tolylzinc was produced after boiling for 5–6 hr. Di-*p*-bromophenyl and di-*p*-iodophenyl-mercury do not react with zinc upon boiling in xylene solution. If the reaction is conducted in xylene at temperatures of 220–230°C in sealed tubes, the organomercury compounds are decomposed without formation of organozinc compounds. Dibenzyl but not the dibenzylzinc are produced by heating dibenzylmercury with zinc to 110°C in xylene. Di-*p*-carbethoxy-phenylmercury[43] does not react with zinc even when heated to 210°C in a sealed tube without a solvent.

4. Cadmium

Reactions of metallic cadmium with alkyl and aromatic mercury compounds cannot be used as a synthetic procedure for organocadmium compounds since these reactions result mainly in the formation of iso-morphous mixtures of organocadmium and the initial organomercury compound. These reactions with Alk_2Hg were conducted in sealed tubes.[44] The highest yield of diphenylcadmium from the reaction of di-phenylmercury with cadmium was produced in the following manner: excess cadmium was continuously added (portionwise) to the boiling reactants.[45]

C. Reactions with Metals of Group III

The organometallic compounds of group III were obtained for the first time from organomercury compounds by the reaction:

$$3R_2Hg + 2Me \longrightarrow 2R_3Me + 3Hg$$

This reaction is an important method for producing completely substi-tuted aliphatic and aromatic compounds of gallium and indium and aromatic compounds of aluminum.

1. Aluminum

Alkyl and arylaluminum compounds were produced by reacting the respective organomercury compound with aluminum. The simplest ali-phatic aluminum compounds were produced in this manner more than 100 years ago. The reactions take place in sealed tubes with continuous heating to 100–120°C (sometimes higher) of the dialkylmercury with an excess amount of aluminum shavings.[46,47] The reactions are usually carried out without pressure buildup in the absence of a solvent or by contact with no heating, as is exemplified by the preparation of trimethyl-aluminum.[48] Hydrocarbons may also be used as solvents.

It should be noted that alkyl isomerizations take place upon heating alkyl-aluminum compounds.[49] Thus the reaction of diisopropyl- or disecondarybutylmercury with aluminum produces a mixture of the respective secondary and tertiary trialkylaluminum compounds. However, no isomerization of optically active substituents occurred upon heating $(+)$-$[C_3H_7CH(CH_3)CH_2]_2Hg$ or racemic R_2Hg with aluminum[50] for three days.

Fluoroalkylaluminum compounds were not prepared from the usual reactions of R_2Hg ($R = CF_3$, C_2F_5, i-C_3F_7, CF_3CHF) with aluminum (see Chapter 7),

The reaction of diarylmercury with aluminum produced triarylaluminum compounds. The reactions were carried out in sealed equipment without solvents at temperatures of 130–140°C,[51] or in sealed equipment in ether.[52] A simple and convenient method for the preparation of triarylaluminum compounds is the reaction of diarylmercury with aluminum in boiling xylene or other inert solvents such as toluene, octane, or nonane.[53] By this method, Ar_3Al (Ar = phenyl, o-, and m-tolyl, pseudocumene, o-anisyl, p-fluoro and p-chlorophenyl) were produced in good yield. Some Ar_2Hg compounds did not react with aluminum under these conditions.

2. Gallium

The reaction of R_2Hg with metallic gallium produced R_3Ga (R = alkyl, alkenyl, or aryl). An inert atmosphere of nitrogen is needed in these reactions. If R = Alk, heating for several hours in sealed flasks at temperatures of 130–150°C without pressure[55] is needed. The reaction is catalyzed by traces of metallic mercury[54] or mercury chloride.[54] Heating of diethylmercury with gallium for 200 hr produces diethylgallium.[56]

Only prolonged contact and not heating is necessary if R is alkenyl (vinyl,[54a,57] iso-, cis-, and trans-propenyl[58]).

Only one arylgallium compound, triphenylgallium, has been prepared and heating at 130°C for 50 hr without solvent was necessary.[59,60]

$$3(C_6H_5)_2Hg + 2Ga \longrightarrow 2(C_6H_5)_3Ga + 3Hg$$

3. Indium

Few reactions of organomercury compounds with metallic indium are known. Trimethylindium was produced by the reaction of dimethylmercury with indium and a small amount of mercury(II) chloride[61] or, better, mercury(II) chloride and traces of iodine and magnesium[62] at a temperature of 100°C in an atmosphere of carbon dioxide.

Heating of diphenylmercury and metallic indium (130°C, 48 hr) in a sealed flask under CO_2,[63] or better, under nitrogen[59,64] produced an 80% yield of triphenylindium. If precautions for exclusion of air are not taken, oxidized indium compounds are produced: $C_6H_5InO \cdot In_2O_3$.[65]

D. Reactions with Metals and Metal Alloys of Group IV

1. Tin and Tin Alloys

Saturated aliphatic tin compounds were not formed by the reactions of aliphatic mercury compounds with metallic tin or with tin alloys. Organotin compounds were produced by the reaction of tin with β-chlorovinylmercury compounds. Retention of substituent configuration was observed. Acetylene was also produced in this reaction. The cis-isomer, as usual, reacts more slowly than the trans-isomer. The reaction of the trans-isomer with metallic tin results in the quantitative isolation of metallic mercury and in formation of the product mixture, R_3SnCl, R_2SnCl_2, and $RSnCl_3$ (R = trans-ClCH=CH). Prolonged heating facilitates formation of tri-trans-β-chlorovinyl tin(IV) chloride. The reaction of di-cis-β-chlorovinylmercury with metallic tin requires longer heating as compared to the trans-isomer. The organotin compound produced was not isolated, however, its cis-configuration has been proved.[66]

Tetraphenyltin was quantitatively produced by heating diphenylmercury with tin at 200°C in sealed tubes without a solvent.[67] If xylene is used as a solvent only a 30% yield of tetraphenyltin was obtained.[68] $(C_6F_5)_4Sn$ was obtained in 60% yield upon heating of $(C_6F_5)_2Hg$ with tin in a sealed tube (260°C, 9 days).[69]

Triaryltin and tribenzyltin chlorides were produced by boiling the necessary reactants in xylene[68]:

$$6ArHgCl + 3Sn \longrightarrow 3Ar_2SnCl_2 + 6Hg$$

$$3Ar_2SnCl_2 \longrightarrow 2Ar_3SnCl + SnCl_4$$

$$SnCl_4 + Sn \longrightarrow 2SnCl_2$$

The reaction of β-chlorovinylmercury with a tin–sodium alloy (15% Na) in dry benzene at a temperature of 50°C produced tris-β-chlorovinyltin chloride with retention of configuration. Acetylene was also obtained as a product.

$$4ClCH=CHHgCl + NaSn \longrightarrow (ClCH=CH)_3SnCl + CH\equiv CH$$
$$+ 4NaCl + 4Hg$$

The reaction of cis- or trans-β-stilbenemercury chloride (3 hours boiling in xylene) with a tin–sodium alloy failed to produce organotin compounds.[70]

Boiling a mixture of diphenylmercury and tin–sodium alloy (15% sodium) in xylene produced a good yield of tetraphenyltin.[68] The arylmercury chloride and hexaarylditin under similar conditions produce a 40–50% yield of tetraaryltin as shown by the following reactions[68]:

$$6ArHgCl + 3Na_2Sn \longrightarrow 3Ar_2Sn + 6NaCl + 6Hg$$

$$3Ar_2Sn \longrightarrow Ar_3Sn{-}SnAr_3 + Sn$$

$$Ar_3Sn{-}SnAr_3 \longrightarrow Ar_4Sn + Ar_2Sn$$

No reaction occurs between diphenylmercury and lead.[67]

E. Reactions with Metals and Metal Alloys of Group V

The reactions of organomercury compounds with metallic antimony, bismuth, and a bismuth–sodium alloy are usually carried out under severe conditions and are not considered to be general syntheses of the organic derivatives.

Antimony reacts with dimethylmercury (sealed tube, 150°C, 160 hr and a trace of mercury(II) chloride) to produce trimethylantimony.[71] Triethylantimony was also produced using the same method.[72]

Triphenylbismuth (47% yield) and biphenyl were produced by the reaction of diphenylmercury with bismuth (250°C, H_2 flow). The reaction does not go to completion.[45] A very low yield of triphenylbismuth was obtained by the reaction of phenylmercury chloride with Bi–Na alloy for 18 hr in boiling xylene.[73]

F. Reactions with Elements of Group VIII

Only iron was found to react with a mixture of dicyclopentadienylmercury and cyclopentadienylmercury chloride to form the organoiron compound, ferrocene.[75] The reaction occurs instantaneously in a cold solution of tetrahydrofuran. Decomposition of the organomercury compound results from the attempted reactions of other group VIII metals (see Chapters 7 and 8).

G. Reactions with Actinides

Heating of diethylmercury and uranium to 150–160°C produced no organouranium compounds.[74]

V. PREPARATIVE SYNTHESIS

A. Preparation of Soluble Lithium Alkyls[10]

The appropriate dialkylmercury is placed in the reaction flask (Fig. 1) and covered with a small amount of solvent (various fractions of petroleum which boil over a 20°C range). Finely sliced lithium and more solvent is added until the flask is 80% full. The flask is connected to a suction pump

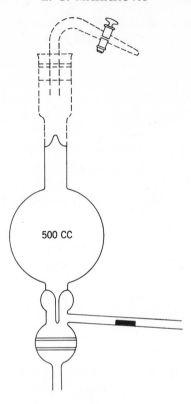

Fig. 1

and heated near the top. Fifty milliliters of solvent is distilled to completely remove air from the flask. While the solvent is boiling, the flask is
sealed. The flask is then put aside until the reaction is complete. For the
preparation of ethyllithium the reaction flask should be kept at 70°C for
3 days, for n-butyl, isoamyl or n-heptyllithium, the flask is kept at 50°C for
three days or 25°C for six days.

The reaction flask is then sealed to the calibrated measuring flask
(Fig. 2). The sealed glass capsule of mercury, which is later used to open
the reaction flask, is put in place and the air is expelled from the measuring
flask through the valve sidearm by displacement with solvent vapor or by
evacuating to 10^{-4} mm or less by means of a mercury-vapor pump. The
sidearm is then sealed, the tip of the reaction flask is broken by dropping
the mercury slug and the alkyllithium solution is filtered into the measuring
flask. This filtration may be hastened by gently warming the reaction flask.
After filtration is complete, the measuring flask is cooled until the internal

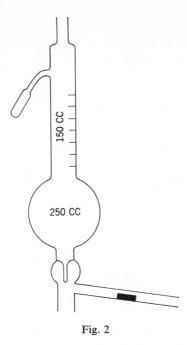

150 CC

250 CC

Fig. 2

pressure is below atmospheric pressure and is then sealed just below the filter. After thoroughly mixing and measuring the contents of the flask, the sampling bulb is filled to the mark with the solution and sealed off. This sample bulb is broken under water and the lithium hydroxide thus formed is titrated. This gives an accurate measure of the concentration of alkyllithium.

B. Preparation of Vinylalkali-metal Compounds[12]

Vinylalkali-metal compounds are prepared by reacting an alkali metal with divinylmercury: Lithium metal (6.9 g, 1 mole), as a dispersion in mineral oil, is placed into a reaction flask fitted with stirrer, dropping funnel, and reflux condenser. Pentane (150 ml) is added and the divinylmercury (125 g, 0.49 mole) introduced dropwise. The mixture is then refluxed for one hour, allowed to settle, and the mercury filtered off. The resulting solution is used for vinylation reactions. Lithium wire or small pieces of metal may also be used. A light yellow, solid material which could not be obtained in a crystalline form remained after the removal of the solvent. Yields based on the formation of ethylene upon hydrolysis were between 70 and 80%. Vinylsodium and vinylpotassium are also prepared in the same manner.

C. Preparation of p-Tolyllithium Solution[17]

Seven grams of (p-CH$_3$C$_6$H$_4$)$_2$Hg as a powder were placed in 50 ml of absolute ether and shaken with 0.3 g of Li shavings and some pieces of glass in a Shlenck tube for 20 hr. In order to start the reaction, dip the flask, with shaking, into a water bath at 50°C, for 10–15 min. After the suspended lithium amalgam settled, the solution was transferred to a measuring burette. The yield (determined by titration) is about 100%.

D. Preparation of Ethylsodium[77]

The organoalkali metal compound is obtained by the addition of 0.04 g atom of sodium cut into small pieces to the solution containing 2.59 g (0.01 mole) diethylmercury in 25 ml of petroleum ether (b.p. 68–77°C) and stirring for 1.5 day. Gas evolution occurs. The ethylsodium suspension obtained may be used for other reactions.

E. Preparation of Benzylsodium[23]

Sodium (10 g, 0.43 mole) and 7.7 g (0.02 mole) of di-p-tolylmercury are added to 160 ml petroleum ether (b.p. 85–155°C). The di-p-tolylmercury does not completely dissolve until the end of the reaction. Nineteen hours refluxing of the mixture yields a yellow-brown solution of benzylsodium which may be used as a reactant for further preparations. Carboxylation with carbon dioxide produced phenylacetic acid (57% yield).

F. Preparation of Diphenylberyllium[36]

Diphenylmercury (20 g, m.p. 123–124°C) and 1 g of beryllium (as foil) are heated to 210–220°C under nitrogen *in vacuo* for 1–2 hours. Benzene (200 ml) is added and the solution is filtered through a sintered glass filter under nitrogen at reduced pressure. The solution is reduced to approximately 50 ml from which diphenylberyllium precipitates. The mother liquor is decanted and the product is recrystallized from benzene. The yield is 80% of hexahedral prisms with m.p. 244–248°C (decomposition in a sealed capillary). The reaction of beryllium with di-p-tolylmercury produces di-p-tolylberyllium.

G. Preparation of Diethylmagnesium[38]

A reaction occurs immediately when fresh diethylmercury and a small amount of mercury(II) chloride are heated with magnesium powder at 130°C. The reaction product is a light gray-brown powder, spontaneously inflammable in the air. Considerable pressure developed in the sealed tube.

H. Preparation of the Etherate of Diphenylmagnesium [41]

Diphenylmercury (8 g) in 100 ml of absolute ether was shaken with 8 g of magnesium shavings in a Schlenk tube under nitrogen. Metallic mercury was observed after one day. The reaction goes to completion after 9 days shaking and 5 days of additional standing. After cooling in a Dry Ice–acetone bath, the concentrated solution of the etherate of diphenylmagnesium (decanted from sediment) was salted out and the reaction products melted at room temperature. The viscous mass was heated to a temperature of 100°C for 1 hr and then cooled. A brown mass was obtained which did not melt at room temperature.

I. Preparation of Dimethylzinc [42]

Dimethylmercury and 20-mesh granular zinc were maintained at the reflux temperature of dimethylmercury by periodic distillation of the accumulated product, b.p. 45–46°C, through a 25 cm Vigreux column. By this procedure, 35 g of dimethylmercury was converted into dimethylzinc in 48 hr.

J. Preparation of Di-p-chlorophenylzinc [43]

Di-p-chlorophenylmercury (2 g) and 1.4 g of zinc wool were placed in a three-necked flask fitted with a condenser, dry CO_2 inlet, and sampling opening. Xylene (10 ml) was added and the mixture was heated for 3 hr. The products were transferred to a thick-walled flask, flushed with carbon dioxide, and the flask was sealed. Upon cooling, small needles crystallized out and were filtered, washed with petroleum ether, and dried in a stream of carbon dioxide, 1.05 g (75%), m.p. 212–214°C. Mercury was obtained in approximately 98% yield.

K. Preparation of Trimethylaluminum [42]

Trimethylaluminum was prepared by refluxing dimethylmercury with three times the stoichiometric quantity of aluminum under a nitrogen atmosphere. About one-third of the aluminum was foil and the remainder was thin turnings. The product was removed from the reaction flask by vacuum distillation to a mercury-sealed storage container.

L. Preparation of Triphenylaluminum [53]

Diphenylmercury (3.01 g), 1 g of freshly cut aluminum shavings and 19 ml of dry xylene were boiled for 5 hr. The solution was perfectly clear throughout the experiment. Cooling caused precipitation of a white material which had a m.p. of 237°C after repeated washings with dry petroleum ether (fraction to 35°C).

M. Preparation of Triethylgallium[56]

Three small distilling bulbs, each of about 25-cc capacity, were fused together in series, and the last bulb was attached to a fourth by a normal slip-joint and a glass stopcock, which permitted this fourth bulb to be closed and detached. Ten grams of metallic gallium were placed in the first bulb, and 2 g in each of the next two. Fifty-two grams of diethylmercury were placed in the first bulb, the apparatus was filled with nitrogen, and the necks of the three flasks were sealed. The first bulb, surrounded by a small air bath, was very gradually raised to a temperature of 165°C over 200 hr. The bulb was then cooled, the nitrogen in the apparatus was pumped out, and the liquid in the first bulb was distilled under reduced pressure into the second bulb on the 2 g of gallium that it contained. The first bulb was then sealed off from the rest of the bulbs. Examination of the residual metal in the first bulb showed that it was nearly pure mercury, and its amount indicated that the reaction had proceeded practically to completion. To make sure that all of the diethylmercury had reacted with the gallium, the contents of the second bulb were heated for 24 hr at 120–125°C and then 18 hr longer at about 150°C. No change in the appearance of the colorless distillate or the bright metallic gallium resulted. The procedure was repeated by distillation into the third bulb, the second bulb was sealed off, and the third bulb was heated for 24 hr longer. Again the absence of change in the appearance of the liquid or metallic gallium furnished evidence that the distillate contained no diethylmercury. The liquid was then distilled into the fourth bulb, the stopcock of this bulb was closed, and it was connected to a vacuum fractionation apparatus by the slip-joint. Its contents were held at 0°C and were separated into three fractions by distillation. The middle fraction was $(C_2H_5)_3Ga$, m.p. 82.3°C.

N. Preparation of Triphenylgallium[60]

A mixture of 1 g (0.0143 moles) of gallium metal and 5.45 g (0.0153 mole) of diphenylmercury was heated in a Schlenk tube which had been throughly flushed out with nitrogen. Heating was effected at 130°C in an oven, and after 50 hr the molten diphenylmercury had been replaced by a white crystalline mass. The tube was opened under nitrogen and 10 cc of dry chloroform were added. After warming the chloroform to dissolve all of the solid, the solution was filtered and then subjected to solvent evaporation (under nitrogen) until crystals began to appear in the hot solution. The mass of shiny, needlelike crystals which separated on cooling were filtered on a sintered glass plate, and washed with petroleum ether (b.p. 65–68°C.) The product was redissolved in 3 cc of hot chloroform, and the crystals which separated on cooling were again filtered, washed with petroleum ether, and dried under nitrogen. The triphenylgallium (1.85 g) melted at 166°C; the clear colorless melt quickly solidified on cooling and remelted at the same temperature. Better yields of triphenylgallium were obtained when the theoretically required quantities of gallium and diphenylmercury were used. Thus, from 1.92 g (0.0275 g atom) of gallium and 14.5 g (0.0405 mole) of diphenylmercury, heated at 130°C for 75 hr, 6.68 g or an 82% yield of triphenylgallium was obtained which melted at 165–166°C without recrystallization.

O. Reaction of *trans*-β-Chlorovinylmercury Chloride with a Tin and Sodium Alloy[66]

Into a cylindrical, three-necked flask fitted with a gas outlet, mercury-sealed stirrer, and a condenser guarded by a calcium chloride tube, 19g of *trans*-β-chlorovinylmercury chloride and 30 ml of dry benzene were introduced. A sodium–tin alloy (25 g, 15% sodium content) was added to the solution. The reactants were heated in a stream of hydrogen on a water bath at 45–48°C for 3 hr with vigorous mixing.

The reaction started rapidly, as discernible by the formation of a voluminous gray precipitate and slight heat evolution. Upon termination of the reaction, the still-warm reaction mass was filtered from excess alloy, the metallic Hg released was washed, and the slurry was also washed on the filter with dry benzene. This solution was combined with the filtrate, the solvent evaporated under reduced pressure, and the dry residue twice recrystallized from petroleum ether (b.p. 50–75°C). The dry crystals melted at 120–121°C. Additional recrystallizations did not change the melting point. The yield of tris-*trans*-β-chlorovinyl) tin chloride was 2.9 g (40%).

P. Preparation of Tetra-*p*-tolyltin[68]

p-Tolymercury chloride (2.5 g) and 7.5 g of tin–sodium alloy (15% Na) were heated to boiling in 20 ml of xylene for 18 hr. Crystals (0.12 g) salted out from the xylene solution. The alloy residue was extracted with benzene which produced 0.25 g more of the substance. Both products melt at 236–237°C. The yield was 41%.

Q. Preparation of Triphenyltin Chloride[68]

Phenylmercury chloride (3,1 g) and 8.5 g of crushed tin were heated in 20 ml of xylene on a bath for 18 hr until the mixture began boiling. After one hour, mercury deposition on the flask walls was observed. Separation and concentration of the xylene solution resulted in crystals which melted over the range 104–106°C. More product was isolated by benzene extraction of the residue.

TABLE II

Exchange Reactions with Metals and Metal Alloys

R in R_2Hg, or RHgX	Reaction conditions	Products	Yield, %	Ref.
		Metallic Lithium		
CH_3	$(C_2H_5)_2O$	CH_3Li	—	17
CH_3	Petroleum ether, several days	CH_3Li	—	9,76
C_2H_5	Petroleum ether, 4 days	CH_3Li	—	77
		$COOH$[a,b]		
C_2H_5	Sealed tube, C_6H_6, shake, 8 weeks	C_2H_5COOH[b]	quant.	77
C_2H_5	C_6H_6, 65°C, 3 days	C_2H_5Li	—	9,76
C_2H_5	Ligroin, 70°C, 3 days	C_2H_5Li	—	10
C_2H_5	Sealed tube, C_6H_6, 65°C, 24 hr	C_2H_5Li	93	78
C_2H_5	C_6H_6, heat, several hours	$(C_6H_5)_2CHCH_2COC_6H_5$[c]	13	79
		$C_6H_5CH{=}CH{-}C(C_6H_5)_2$[c] $-OH$	69	
$n\text{-}C_3H_7$	C_6H_6 or ligroin, 65°C, 3 days	$n\text{-}C_3H_7Li$	—	9
$n\text{-}C_4H_9$	Petroleum ether, 50 or 70°C, 3 days; 25°C, 6 days	$n\text{-}C_4H_9Li$	quant.	10
$i\text{-}C_5H_{11}$	Ligroin, 50°C, 3 days; 25°C, 6 days	$i\text{-}C_5H_{11}Li$	—	10
$n\text{-}C_7H_{15}$	Petroleum ether	$n\text{-}C_7H_{15}Li$	quant.	10
CH_2CHO	Liquid NH_3	$CH_2{=}CHOLi$	78	21
$CH_2{=}CH$	$(C_2H_5)_2O$	$BrCH_2CHBrCOOH$[b,d]	—	13

208

CH₂=CH	Pentane, reflux, 1 hr	C_2H_4[e]	70–80	12
C₄H₉C≡C	Dioxane, reflux, 30 min–2 hr	$C_4H_9C{\equiv}CLi$	74–93	14
C₆H₅CH₂	(C₂H₅)₂O, shake, 36 hr	$C_6H_5CH_2Li$	—	16
C₆H₅	(C₂H₅)₂O, reflux, 10 min; stand overnight; shake 1 hr	C_6H_5Li	—	15
C₆H₅	(C₂H₅)₂O, shake 2 days	C_6H_5Li	—	16
C₆H₅	(C₂H₅)₂O, shake 20 hr	C_6H_5Li	—	80
C₆H₅	C₆H₆, 65°C, 5–6 days	C_6H_5Li	—	9
C₆H₅	C₆H₆	No reaction		45
o-CH₃C₆H₄	(C₂H₅)₂O, 20 hr	$[(o\text{-}CH_3C_6H_4)_4B]Li$[f]	86	19
m-CH₃C₆H₄	(C₂H₅)₂O, 20 hr	$[(m\text{-}CH_3C_6H_4)_4B]Li$[f]	72	19
p-CH₃C₆H₄	(C₂H₅)₂O, 20 hr	$[(p\text{-}CH_3C_6H_4)_4B]Li$[f]	74	19
p-CH₃C₆H₄	(C₂H₅)₂O, shake 20 hr, then heat 50°C, 10–15 min	$p\text{-}CH_3C_6H_4Li$	100	17
o-C₆H₅C₆H₄	(C₂H₅)₂O, shake 16 hr	$o\text{-}C_6H_5C_6H_4Li$	98	19
Hexamer (o-C₆H₄)₂	(C₂H₅)₂O, shake 16 hr	$[(C_{12}H_8)_2B]Li \cdot LiF$[g]	44	19
Hexamer (o-C₆H₄)₂	(C₂H₅)₂O, shake 3 days	$o\text{-}C_6H_4Li_2$	70–80	18
Dimer (o-C₆H₄)₃	(C₂H₅)₂O, shake 6 days	$o\text{-}C_6H_5C_6H_4C_6H_5$[e]	63	20
(p-CH₃)₂NC₆H₄	(C₂H₅)₂O, shake 13 hr	$[(p\text{-}CH_3)_2NC_6H_4]Li$	94	17

Metallic Sodium

CH₃	(C₂H₅)₂O	CH_3COOH[b]	16	28
CH₃	Sealed tube, ligroin, 65°C	CH_3Na	—	9
C₂H₅	20° or heat	$C_2H_4 + C_2H_6$[e]	—	81
C₂H₅	(C₂H₅)₂O, 20°, 2 days	$(C_6H_5)_2C(OH)C_2H_5$[h]	40	28
C₂H₅	(C₂H₅)₂O, 20°, 2 days	$C_6H_5C(OH)(C_2H_5)_2$[i]	24	28
C₂H₅	(C₂H₅)₂O, 20°C, 2 days	$C_6H_5CH(OH)C_2H_5$[j]	15	28

(continued)

209

TABLE II (*Continued*)

R in R$_2$Hg, or RHgX	Reaction conditions	Products	Yield, %	Ref.
C$_2$H$_5$	(C$_2$H$_5$)$_2$O. 20°C, reflux	C$_2$H$_5$COOH[b]	20	28
C$_2$H$_5$	C$_6$H$_6$, heat, several hours	C$_6$H$_5$COOH[b]	12	25
C$_2$H$_5$	Toluene, heat	C$_6$H$_5$CH$_2$COOH[b]	7	25
C$_2$H$_5$	Xylene, heat	CH$_3$C$_6$H$_4$CH$_2$COOH[b]	6	25
C$_2$H$_5$	C$_2$H$_5$C$_6$H$_5$, heat	C$_6$H$_5$CH(CH$_3$)COOH[b]	1	25
C$_2$H$_5$	C$_6$H$_6$, 70–80°C	C$_2$H$_6$[e]	quant.	26
		C$_6$H$_5$COOH[b]	Trace	26
C$_2$H$_5$	Thiophene, 70–80°C	C$_2$H$_6$[e]	96	26
		COOH[b]	35–40	26
C$_2$H$_5$	o-Xylene, 80°C	o-CH$_3$C$_6$H$_4$CH$_2$COOH[b]	8	26
C$_2$H$_5$	Mesitylene	3,5-(CH$_3$)$_2$C$_6$H$_3$CH$_2$COOH[b]	14	26
C$_2$H$_5$	p-Cymene	p-(CH$_3$)$_2$CHC$_6$H$_4$CH$_2$COOH[b]	6	26
C$_2$H$_5$	(C$_6$H$_5$)$_2$CH$_2$, heat	(C$_6$H$_5$)$_2$CHCOOH[b]	0.49	26
C$_2$H$_5$	Ligroin, 65°C, 1–2 hr	C$_2$H$_5$Na	—	9
C$_2$H$_5$	100°C, 2.5–3 hr	C$_2$H$_4$ + C$_2$H$_6$[e]	—	30
C$_2$H$_5$	(C$_2$H$_5$)$_2$O, heat, 1 hr	C$_2$H$_5$OH[e]	—	30
C$_2$H$_5$		C$_2$H$_4$ + C$_2$H$_6$[e]	80	30
C$_2$H$_5$	Hexane, anisole or phenetole, 80–140°C, 2 hr	C$_6$H$_5$OH[e]	54–65	30
C$_2$H$_5$	Sealed tube, 20°C, 24–48 hr, heat to 142°C	C$_2$H$_4$ + C$_2$H$_6$[e]	94	82
C$_2$H$_5$	C$_6$H$_6$, heat several hours	(C$_6$H$_5$)$_2$CHCH$_2$COC$_6$H$_5$[c]	3	79
		C$_6$H$_5$CH=CHC(OH)(C$_6$H$_5$)$_2$[c]	39	79

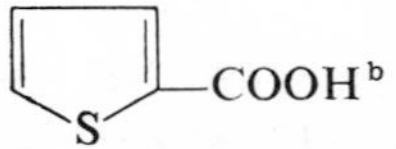

210

	Conditions	Product		
C_2H_5	Furan, 60°C, 3 days	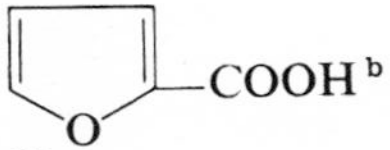 furan–$COOH^b$	15–20	27
C_2H_5	Pentane, 48°C, 24 hr	Hg; $CH_3CH(COOH)_2{}^b$	80	29
C_2H_5	C_6H_6, 20°C, 46 hr	$C_6H_5COOH^b$	—	83
C_2H_5	Petroleum ether, 20°C, 1.5 days	o- and p- $C_6H_4(COOH)_2{}^b$	low	77
		dibenzofuran–$COOH^{a,b}$	Mixture	
		dibenzofuran ($COOH$)($COOH$)^{a,b}		
C_2H_5	Cumene, 20°C, 12 hr	$C_6H_5C(CH_3)_2COOH^b$	41	23
C_2H_5	Pentane	C_2H_5Na	80	22
C_2H_5	2-Methylfuran, 60°C, 10 hr	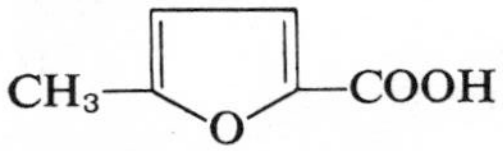 CH_3–furan–$COOH$	low	27
n-C_3H_7	Ligroin, 65°C	n-C_3H_7Na	—	9
n-C_3H_7	Octane, 20°C	n-C_3H_7Na	88	22
n-C_4H_9	Petroleum ether, 20°C, 26 hr	β-$C_{10}H_7CH_2COOH^k$	31	23
n-C_4H_9	$(C_2H_5)_2O$, 20°C	C_6H_5–C=$C_6H_5{}^h$ / HC–C_3H_7	57	17

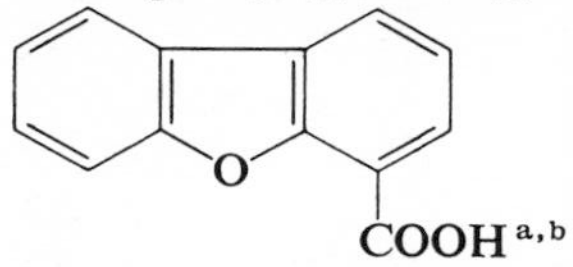

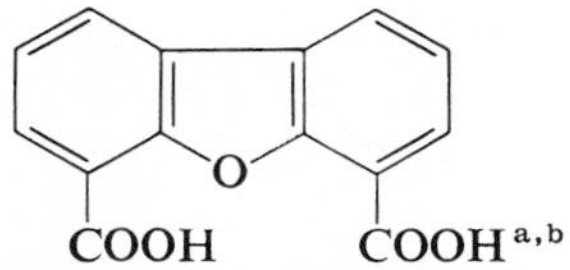

(continued)

211

TABLE II (Continued)

R in R_2Hg, or RHgX	Reaction conditions	Products	Yield, %	Ref.
$n\text{-}C_4H_9$	Sealed tube, petroleum ether, shake 2 days	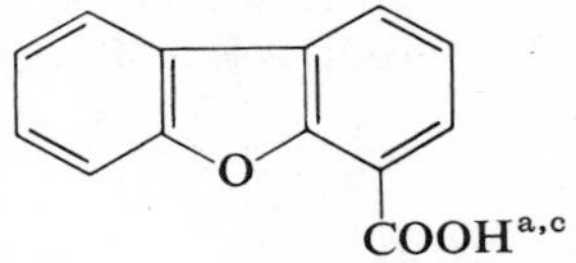 [dibenzofuran] COOH[a,c]	36–58	84
$s\text{-}C_4H_9$	n-Dodecane, 20–25°C, 4 hr	Hg		
		CH_3CH_2—CHCOOH[b] (CH_3)	95	24
		C_2H_5, CH_3—C—COOH, COOH[b]	<5	24
		C_2H_5—C—COOH	low	
		C_4H_8[e]	low	24
$s\text{-}C_5H_{11}$	n-Dodecane, 20°C, 8 days	$C_4H_9CH(COOH)_2$[b]	40–50	29
$n\text{-}C_5H_{11}$	Pentane, 35°C, 2 hr	$C_5H_{11}COOH$[b]	19	
$n\text{-}C_5H_{11}$	C_6H_6, 20°C, 1 hr	Hg	93	29
		$C_6H_4(COOH)_2$[b]	24	
		C_6H_5COOH[b]	75	
$n\text{-}C_5H_{11}$	Toluene, 20° C, 1.5 hr	$C_6H_5CH(COOH)_2$[b]	45	29
		$C_6H_5CH_2COOH$[b]	13	29
$i\text{-}C_5H_{11}$	$(C_2H_5)_2O$, 20°C, 2 days	$i\text{-}C_5H_{11}COOH$[b]	Trace	28
$i\text{-}C_5H_{11}$	$(C_2H_5)_2O$, reflux, 1 hr	C_2H_5OH[e]	34	30
C_8H_{17}	Benzine,	$C_8H_{17}Na$	—	9

212

Substituent	Conditions	Product	Yield (%)	Ref.
CH_2CHO	Liquid NH_3	$CH_2=CHONa$	80	21
$CH_2=CH$	Pentane, reflux, 1 hr	C_2H_4[e]	70–80	12
$C_6H_5CH_2$	C_6H_6, 20°C, 2 days	$C_6H_5CH_2Na$	—	9
$C_6H_5CH_2$	Furan, 60°C, 3 days	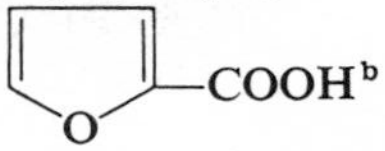 [furan]—$COOH$[b]	58	27
		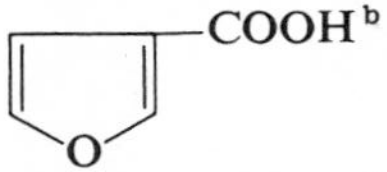 [furan]—$COOH$[b]	Trace	
		$C_6H_5CH_2COOH$[b]	Very low	
$C_6H_5CH_2$	α-Methylthiophene + heptane, 60°C, 144 hr	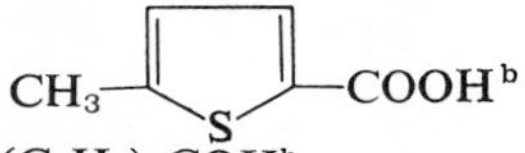 CH_3—[thiophene]—$COOH$[b]	21	27
C_6H_5	C_6H_6 or ligroin, reflux	$(C_6H_5)_3COH$[h]	quant.	85
C_6H_5	C_6H_6 or gasoline	C_6H_5Na	—	9
C_6H_5	Furan, 60°C, 3 days	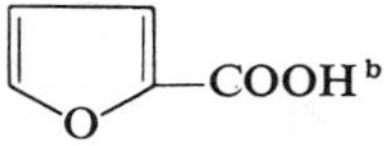 [furan]—$COOH$[b]	5–15	27
C_6H_5	2-Methylfuran, 60°C, 10 hr	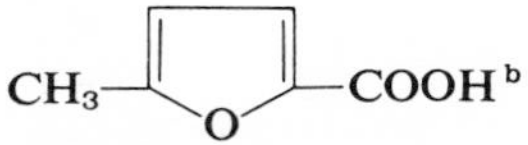 CH_3—[furan]—$COOH$[b]	7.5	27
C_6H_5	C_6H_6, reflux or 20°C, 24 hr	C_6H_5COOH[b]	62–86	23
C_6H_5	$(C_2H_5)_2O$, C_6H_5Li, 20°C, 2 hr	$[(C_6H_5)_2Li]Na$	—	80
C_6H_5	C_6H_6	No reaction	—	45
C_6H_5	C_6H_6, reflux	Na/Hg	—	86
C_6H_5HgI	Heat	Unknown comp.	—	86
		$(C_6H_5)_2Hg$	—	
		Hg, NaI	—	

(continued)

213

TABLE II (Continued)

R in R_2Hg, or RHgX	Reaction conditions	Products	Yield, %	Ref.
p-CH$_3$C$_6$H$_4$	Petroleum ether or toluene, reflux, 19 hr	C$_6$H$_5$CH$_2$COOH[b]	57–66	23
p-CH$_3$C$_6$H$_4$	C$_6$H$_6$, 20°C, 1 month	p-CH$_3$C$_6$H$_4$COOH[b]	55	87
		C$_6$H$_5$CH$_2$COOH[b]	6	
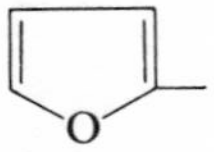 (furan)	Furan or hexane, or cyclohexane, or ether, 60°C, 52 hr	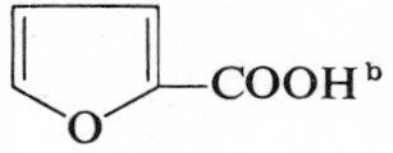 (furyl)—COOH[b]	20–26	27
(thiophene)	Thiophene, 60°C, 12 hr	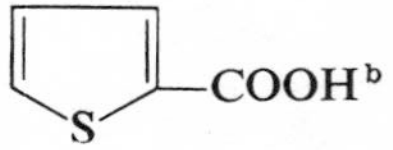 (thienyl)—COOH[b]	16	27

Metallic Potassium

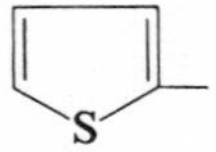

R in R_2Hg, or RHgX	Reaction conditions	Products	Yield, %	Ref.
C$_2$H$_5$	Petroleum ether, 20°C, 2 days	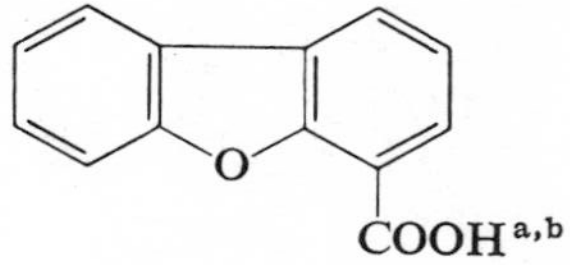 (dibenzofuran)—COOH[a,b]	—	77
		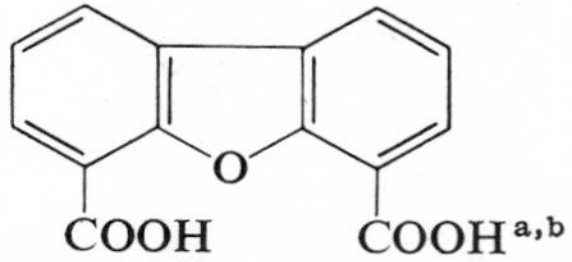 (dibenzofuran)—COOH[a,b]	—	
C$_2$H$_5$	C$_6$H$_6$ (30–75 ml), 20°C, 20–48 hr	C$_6$H$_5$COOH[b]	33–45	83
		p-C$_6$H$_4$(COOH)$_2$[b]	11–17	
C$_2$H$_5$	C$_6$H$_6$, heat, several hours	C$_6$H$_5$CH=CHC(OH)(C$_6$H$_5$)$_2$[c]	52	79
n-C$_4$H$_9$	Toluene, reflux, 6 hr	C$_6$H$_5$CH$_2$COOH[b]	44	23

(continued)

$CH_2{=}CH$	C_2H_4[e]	Pentane, reflux, 1 hr	70–80	12
C_6H_5	No reaction	C_6H_6		45
$p\text{-}CH_3C_6H_4$	$C_6H_5CH_2COOH$[b]	Petroleum ether, reflux, 18.5 hr	16	23
Metallic Beryllium				
CH_3	$(CH_3)_2Be$	—	quant.	32
CH_3	$(CH_3)_2Be$	Sealed tube, 120–130°C, 24 hr	—	88,89
CH_3	$(CH_3)_2\,Be$	Sealed tube, 125°C, 72 hr	comp.	33
C_2H_5	$(C_2H_5)_2Be$	Sealed tube, 130–135°C	75	31
C_2H_5	$(C_2H_5)_2Be$	Sealed tube, 130°C	—	90
$n\text{-}C_3H_7$	$(n\text{-}C_3H_7)_2Be$	Sealed tube, 130°C	—	91
$n\text{-}C_3H_7$	$(n\text{-}C_3H_7)_2Be$	Sealed tube, 130–135°C	—	91
C_6H_5	$(C_6H_5)_2Be$	Sealed tube, 225°C, 6 hr, $HgCl_2$	—	90
C_6H_5	$(C_6H_5)_2Be$	Sealed tube, xylene, 150°C, 72 hr	86	33
C_6H_5	$(C_6H_5)_2Be$	Sealed tube, 170°C, 72 hr, $HgCl_2$, $BeBr_2$	—	16,35
C_6H_5	$(C_6H_5)_2Be$	Sealed tube, 210–220°C, 1–2 hr	80	34
$p\text{-}CH_3C_6H_4$	$(p\text{-}CH_3C_6H_4)_2Be$	Sealed tube, 225°C, 6 hr, $HgCl_2$	—	36
				33
Metallic Magnesium				
CH_3	$(CH_3)_2Mg$	Sealed tube, 95–105°C, 24 hr	—	37
CH_3	$(CH_3)_2Mg$	Sealed tube, 130°C, 36 hr	16	44
	Hg		88	38
C_2H_5	No reaction	Sealed tube, 130°C, 3 days	—	38
C_2H_5	$(C_2H_5)_2Mg$	Sealed tube, 115–120°C, 24 hr	81	40
C_2H_5	$(C_2H_5)_2C(OH)CH_3$[l,e]	Sealed tube, 130°C, $HgCl_2$	—	38

TABLE II (*Continued*)

R in R$_2$Hg, or RHgX	Reaction conditions	Products	Yield, %	Ref.
C$_2$H$_5$	C$_6$H$_5$ or (C$_2$H$_5$)$_2$O, HgCl$_2$, 20°C, 14 days	(C$_2$H$_5$)$_2$Mg	—	41
C$_2$H$_5$	Sealed tube, 130°C, 36 hr	(C$_2$H$_5$)$_2$Mg	—	44
n-C$_3$H$_7$	Sealed tube, 115–120°C, 24 hr	(n-C$_3$H$_7$)$_2$Mg	89	40
i-C$_3$H$_7$	Sealed tube, (C$_2$H$_5$)$_2$O, 130°C, 24 hr	(i-C$_3$H$_7$)$_2$Mg	42	40
n-C$_4$H$_9$	Sealed tube, 120–130°C, 10 hr	(n-C$_4$H$_9$)$_2$Mg	—	92
i-C$_5$H$_{11}$	Sealed tube, 130°C, 10 hr	(i-C$_5$H$_{11}$)$_2$Mg	71	93
C$_6$H$_5$	Sealed tube, 200°C, 20–40 hr	No reaction		39
C$_6$H$_5$	Sealed tube, 180–185°C, several hours, HgCl$_2$, CH$_3$COOEt	(C$_6$H$_5$)$_2$Mg	—	39
C$_6$H$_5$	C$_6$H$_6$ + (C$_2$H$_5$)$_2$O, 20°C, 14 days, HgCl$_2$	(C$_6$H$_5$)$_2$Mg	—	41
C$_6$H$_5$	Sealed tube, 200°C	(C$_6$H$_5$)$_2$Mg	—	45, 94
C$_6$H$_5$	Sealed tube, 200–210°C, 5–6 hr	(C$_6$H$_5$)$_2$Mg	90	95
Metallic Zinc				
CH$_3$	Reflux, 48 hr	(CH$_3$)$_2$Zn	—	42
CH$_3$	Sealed tube, 120°C, 24 hr	(CH$_3$)$_2$Zn	—	72
C$_2$H$_5$	Sealed tube, 100°C, 36 hr	(C$_2$H$_5$)$_2$Zn	—	72
n-C$_3$H$_7$	Sealed tube, heat	(n-C$_3$H$_7$)$_2$Zn	—	90
n-C$_3$H$_7$	Sealed tube, 100–120°C	(n-C$_3$H$_7$)$_2$Zn	—	96
n-C$_3$H$_7$	Sealed tube, 120–130°C	(n-C$_3$H$_7$)$_2$Zn	—	97
n-C$_3$H$_7$	Sealed tube, 110–120°C, 84 hr	(n-C$_3$H$_7$)$_2$Zn	—	98
n-C$_4$H$_9$	Sealed tube, 120–130°C, several hours	(n-C$_4$H$_9$)$_2$Zn	—	99

216

R	Organometallic product	Conditions	Yield (%)	Reference
$n\text{-}C_4H_9$	$(n\text{-}C_4H_9)_2Zn$	Sealed tube, 130–150°C, 36 hr	—	100
$i\text{-}C_4H_9$	$(i\text{-}C_4H_9)_2Zn$	100°C, 60 hr	—	101,102
$i\text{-}C_4H_9$	$(i\text{-}C_4H_9)_2Zn$	120–130°C	—	90
$i\text{-}C_5H_{11}$	$(i\text{-}C_5H_{11})_2Zn$	Sealed tube, 130–150°C, 36 hr	—	72,100
$i\text{-}C_5H_{11}$	$(i\text{-}C_5H_{11})_2Zn$	—	30	102
$C_6H_5CH_2$	$(C_6H_5CH_2)_2$	Xylene, reflux 1 hr; no solvent, heat 15 min	—	43
C_6H_5	$(C_6H_5)_2Zn$	Xylene, reflux 2 hr	70	43
C_6H_5	$(C_6H_5)_2Zn$	Fuse, 2 min	quant.	45
C_6H_5	$(C_6H_5)_2Zn$	Sealed tube, 180°C	—	86
C_6H_5	$(C_6H_5)_2CHCH_2COC_6H_5$[c]	Xylene, reflux, 2 hr	91	76
$o\text{-}CH_3C_6H_4$	$(o\text{-}CH_3C_6H_4)_2Zn$	Xylene, reflux, 6 hr	39	43
$p\text{-}FC_6H_4$	$(p\text{-}FC_6H_4)_2Zn$	Xylene, reflux, 6 hr	38	43
$p\text{-}ClC_6H_4$	$(p\text{-}ClC_6H_4)_2Zn$	Xylene, reflux, 3 hr	75	43
$p\text{-}BrC_6H_4$	No organozinc compounds	Sealed tube, xylene, 220–230°C, 18 hr		43
$p\text{-}IC_6H_4$	No organozinc compounds	Sealed tube, xylene, 200–250°C, 20 hr		43
$p\text{-}C_2H_5OCOC_6H_4$	No organozinc compounds	Sealed tube, xylene, 200–210°C, 30 hr		43
$p\text{-}(CH_3)_2NC_6H_4$	$[p\text{-}(CH_3)_2NC_6H_4]_2Zn$	Xylene, reflux, 1.5–3 hr	30	43
$\alpha\text{-}C_{10}H_7$	No reaction	Xylene, reflux, 1.5–3 hr		43
$\beta\text{-}C_{10}H_7$	$(\beta\text{-}C_{10}H_7)_2Zn$	Xylene, reflux, 3 hr	67	43

Metallic Cadmium

R	Organometallic product	Conditions	Yield (%)	Reference
CH_3	Hg	Sealed tube, 200–210°C	—	44
	C_2H_6		—	
C_2H_5	$(C_2H_5)Cd$	Sealed tube, 100–130°C	Trace	72
C_2H_5	No reaction	—		103

(continued)

217

TABLE II (*Continued*)

R in R₂Hg, or RHgX	Reaction conditions	Products	Yield, %	Ref.
C₆H₅	Fuse, 10 sec	(C₆H₅)₂Cd	9–51	
		(C₆H₅)₂Hg	91–49	45
		(C₆H₅)₂	Trace	
	Aluminum			
CH₃	Reflux	(CH₃)₃Al	—	42
CH₃	Sealed tube, 100°C, 24 hr	(CH₃)₃Al	quant.	46
CH₃	Sealed tube, 100°C	(CH₃)₃Al	—	47,104, 105
CH₃	20°C, 24 hr	(CH₃)₃Al	—	48
C₂H₅	Sealed tube, 100°C, 24 hr	(C₂H₅)₃Al	quant.	46
C₂H₅	Sealed tube	(C₂H₅)₃Al	—	76,104, 105
C₂H₅	Sealed tube, 110°C, 30 hr	(C₂H₅)₃Al	high	106
C₂H₅	Sealed tube, 100–110°C, 48 hr	(C₂H₅)₃Al	—	107
C₃H₇	Sealed tube, 130°C	(C₃H₇)₃Al	—	90,96
				97
C₃H₇	Sealed tube, 110°C, 30 hr	(C₃H₇)₃Al	high	106
i-C₃H₇	Sealed tube, 60–70°C, 24 hr	(*i*-C₃H₇)₃Al	—	106
i-C₃H₇	—	(*i*-C₃H₇)₃Al ⎫ mixture	—	49
		(C₃H₇)₃Al ⎭		
n-C₄H₉	120–130°C	[*n*-C₄H₉)₃Al]₂	—	99
sec-C₄H₉	—	(sec-C₄H₉)₃Al	—	49
[(+)-*i*-C₅H₁₁]	100–110°C	[(+)-*i*-C₅H₁₁)₃Al	56	108

218

i-C_5H_{11}	$(i$-$C_5H_{11})_3Al$	Sealed tube, 180°C, 16 hr	—	109
$(+)$-$CH_3CH_2CH_2CHCH_2$—CH_3	$[(+)$-$CH_3CH_2CH_2CHCH_2]_3Al$ CH_3	2,2,4-Trimethylpentane, reflux, 3 days	70	50
$(\pm)$-$CH_3CH_2CH_2CHCH_2$—CH_3	$[(\pm)$-$CH_3CH_2CH_2CHCH_2]_3Al$ CH_3	2,2,4-Trimethylpentane, reflux, 3 days	75	50
CH_2=CH	$(CH_2$=$CH)_3Al$	Pentane, reflux, 4 hr	76	110
C_6H_5	$(C_6H_5)_3Al$	No solvent or xylene, heat	—	51
C_6H_5	$(C_6H_5)_3Al$	Xylene, reflux, 5 hr	—	53
C_6H_5	$(C_6H_5)_3Al$	Xylene, reflux, 130°C, 5 hr	—	59
C_6H_5	$(C_6H_5)_3Al$	140°C, 10–15 sec	—	65,111
C_6H_5	$(C_6H_5)_3Al$	No solvent, or with C_6H_6, or xylene, 125–130°C	—	112
C_6H_5	$(C_6H_5)_3Al$	Xylene, reflux	—	113
$^{14}C_6H_5$	$(^{14}C_6H_5)_3Al$		—	114
C_6H_5	$(C_6H_5)_2CHCH_2COC_6H_5$c	Xylene, heat	94	79
o-$CH_3C_6H_4$	$(o$-$CH_3C_6H_4)_3Al$	Xylene, reflux, 6.5 hr	—	53
m-$CH_3C_6H_4$	$(m$-$CH_3C_6H_4)_3Al$	Nonane–octane, reflux, 5.5 hr	—	53
p-$CH_3C_6H_4$	$(p$-$CH_3C_6H_4)_3Al$	No solvent or with xylene, heat	—	51
p-$CH_3C_6H_4$	$(p$-$CH_3C_6H_4)_3Al\cdot(C_2H_5)_2O$	Sealed tube, $(C_2H_5)_2O$, heat	22	52
$(CH_3)_3C_6H_2$	$[(CH_3)_3C_6H_2]_3Al$	Nonane–octane, reflux, 4–5 hr	—	53
C_6H_5—C_6H_4	$[(C_6H_5C_6H_4)_3Al$	Xylene, reflux	—	53
p-FC_6H_4	$(p$-$FC_6H_4)_3Al$	Nonane, reflux, 5 hr	—	53
o-ClC_6H_4	No reaction	Xylene, reflux, 2–3 hr	—	53
p-ClC_6H_4	$(p$-$ClC_6H_4)_3Al$	Xylene, reflux, 4 hr	—	53
o-$CH_3OC_6H_4$	$(o$-$CH_3OC_6H_4)_3Al$	Xylene	—	53
p-$CH_3OC_6H_4$	No reaction	Xylene, 147–150°C, 6 hr	—	53

(continued)

219

TABLE II (Continued)

R in R_2Hg, or $RHgX$	Reaction conditions	Products	Yield, %	Ref.
o-$C_2H_5OC_6H_4$	Octane–nonane, reflux, 5 hr	No organoaluminum compounds		53
p-$C_2H_5OC_6H_4$	Xylene, reflux, 6 hr	No reaction		53
o-$CH_3COOC_6H_4$	Xylene, reflux	No reaction		53
p-$C_2H_5COOC_6H_4$	Decalin or nonane, reflux, 4 hr	No reaction		53
$(CH_3)_2NC_6H_4$	Xylene, reflux, 6 hr	No reaction		53
α-$C_{10}H_7$	Decalin, reflux, 8 hr	$(\alpha$-$C_{10}H_7)_3Al$		53
β-$C_{10}H_7$	Sealed tube, 150°C, or xylene, or decalin, or octane–nonane, reflux	Hg No organoaluminum compounds	quant.	53
		Metallic Gallium		
CH_3	Sealed tube, 130°C, Hg	$(CH_3)_3Ga$	quant.	54
CH_3	Sealed tube, heat, 3 days, $HgCl_2$	$(CH_3)_3Ga$	—	55
C_2H_5	165°C, 200 hr	$(C_2H_5)_3Ga$	—	56
C_2H_5	—	$(C_2H_5)_3Ga$	—	104
n-C_3H_7	Sealed tube, 100°C, 3 days	$(n$-$C_3H_7)_3Ga$	80	54[a]
i-C_3H_7	Sealed tube, 150°C, 8 hr	$(i$-$C_3H_7)_3Ga$	75	54[a]
$CH_2{=}CH$	Sealed tube, 20°C, 16 hr	$(CH_2{=}CH)_3Ga$	87	54[a]
$CH_2{=}CH$	Sealed tube, 24 hr	$(CH_2{=}CH)_3Ga$	40	57
cis-$CH{=}CH{-}CH_3$	CCl_4, 20°C, 10 hr	cis-$(CH{=}CH{-}CH_3)_3Ga$		58
$trans$-$CH{=}CH{-}CH_3$	CCl_4, 20°C, 10 hr	$trans$-$(CH{=}CH{-}CH_3)_3Ga$		58
$H_2C{=}C(CH_3)$	CCl_4, 20°C, 10 hr	$[(H_2C{=}C(CH_3)]_3Ga$		58
C_6H_5	Sealed tube, 130°C, 50 hr	$(C_6H_5)_3Ga$		59
C_6H_5	Sealed tube, 130°C, 75 hr	$(C_6H_5)_3Ga$	82	60

220

Metallic Indium

CH$_3$	Sealed tube, 100°C, 8 days, HgCl$_2$	(CH$_3$)$_3$In	—	61
CH$_3$	Sealed tube, 10 hr, HgCl$_2$, I$_2$, Mg	(CH$_3$)$_3$In	30	62
C$_6$H$_5$	Sealed tube, 130°C, 50 hr	(C$_6$H$_5$)$_3$In	—	59
C$_6$H$_5$	Sealed tube, 130°C, overnight	(C$_6$H$_5$)$_3$In	45	63
C$_6$H$_5$	Sealed tube, 130°C, 48 hr	(C$_6$H$_5$)$_3$In	65–81	64
C$_6$H$_5$	Sealed tube, 270°C, 30 min	C$_6$H$_5$InO·In$_2$O$_3$	—	65

Metallic Tin and Tin Alloys

cis-ClCH=CH	Sn	C$_2$H$_5$OH, 50°C, 8.5 hr	Hg	78	
			(ClCH=CH)$_2$SnCl$_2$ ⎱ mixture		1,66
			(ClCH=CH)$_3$SnCl ⎰		
trans-ClCH=CH	Sn	C$_2$H$_5$OH, 50°C, 1.5–4.5 hr	Hg	quant.	
			trans-(ClCH=CH)$_3$SnCl	11–29	1,66
			trans-(ClCH=CH)$_2$SnCl$_2$	23–31	
			ClCH=CHSnCl$_3$	1–28	
cis-ClCH=CHHgCl	Na/Sn	C$_2$H$_5$OH, 50°C, 4.5–5 hr	cis-(ClCH=CH)$_3$SnCl	36–58	1,66
trans-ClCH=CHHgCl	Na/Sn	C$_6$H$_6$, 45–50°C, 1.5–4 hr	trans-(ClCH=CH)$_3$SnCl	20–40	1,66
cis-C$_6$H$_5$CH=C(C$_6$H$_5$)HgCl	Na/Sn	C$_6$H$_6$, reflux, 3 hr	cis[C$_6$H$_5$CH=C(C$_6$H$_5$)]$_2$Hg	97	70
trans-C$_6$H$_5$CH=C(C$_6$H$_5$)HgCl	Na/Sn	C$_6$H$_6$ or xylene, reflux, 3–8 hr	trans-[C$_6$H$_5$CH=C(C$_6$H$_5$)]$_2$Hg	70	70
C$_6$H$_5$CH$_2$HgCl	Na/Sn	Xylene, reflux, 18 hr	(C$_6$H$_5$CH$_2$)$_3$SnCl	71	68
			(C$_6$H$_5$CH$_2$)$_2$	low	
C$_6$H$_5$	Sn	Sealed tube, xylene or no solvent, 150–325°C, 24 hr	Hg	0–40	67
C$_6$H$_5$	Sn(Na)	Xylene, reflux, 18 hr	(C$_6$H$_5$)$_4$Sn	0–90	67
C$_6$H$_5$	Sn	Xylene, reflux, 18 hr	(C$_6$H$_5$)$_4$Sn	Low	68
C$_6$H$_5$	Sn	Xylene, reflux, 18 hr	(C$_6$H$_5$)$_4$Sn	28	68

(continued)

TABLE II (Continued)

R in R₂Hg or RHgX	Reagent	Reaction conditions	Products	Yield, %	Ref.
C_6H_5HgCl	Sn(Na)	Xylene, reflux, 18 hr	Hg	—	68
			$(C_6H_5)_4Sn$	50	
			$[(C_6H_5)_3Sn]_2$	20	
C_6H_5HgCl	Sn	Xylene, reflux, 18 hr	$(C_6H_5)_3SnCl$	47	68
C_6F_5	Sn	Sealed tube, 260°C, 9 days	$(C_6F_5)_4Sn$	60	69
$p\text{-}CH_3C_6H_4HgCl$	Sn(Na)	Xylene, reflux, 18 hr	$(p\text{-}CH_3C_6H_4)_4Sn$	41	68
			$[(p\text{-}CH_3C_6H_4)_3Sn]_2$	Low	
$p\text{-}CH_3C_6H_4HgCl$	Sn	Xylene, reflux, 18 hr	$(p\text{-}CH_3C_6H_4)_3SnCl$	44	68
$p\text{-}ClC_6H_4HgCl$	Sn(Na)	Xylene, reflux, 18 hr	$[(p\text{-}ClC_6H_4)_3Sn]_2$	—	68
			$(p\text{-}ClC_6H_4)_4Sn$	High	
$C_2H_5OCOC_6H_4HgCl$	Sn	Xylene, reflux, 18 hr	$(C_2H_5OCOC_6H_4)_3SnCl$	10	68

Metals and Metal-Base Alloys of Group V

R in R₂Hg or RHgX	Reagent	Reaction conditions	Products	Yield, %	Ref.
CH_3	As	Sealed tube, 150°C, 160 hr, $HgCl_2$	$(CH_3)_3As$	—	71
CH_3	Sb	Sealed tube, 150°C, 160 hr, $HgCl_2$	$(CH_3)_3Sb$	—	71
C_6H_5	Sb	Sealed tube, 150–200°C, 24 hr	Hg	0–74	67
			$(C_6H_5)_3Sb$	38–97	
CH_3	Bi	Sealed tube, 150°C, 160 hr, $HgCl_2$	$(CH_3)_3Bi$	—	71
C_2H_5	Bi	Sealed tube, 120–140°C	$(C_2H_5)_3Bi$	Low	72
C_6H_5	Bi	Fuse, 250°C, 10 min	$(C_6H_5)_3Bi$	41	45
			$(C_6H_5)_2Hg$	57	
			$C_6H_5C_6H_5$	2	

222

C6H5HgCl	Bi(Na)	Xylene, 145–150°C, 18 hr	(C6H5)3Bi (C6H5)2Hg	Low High	73
		Element of Group VIII			
C5H5	Fe	THF, 100°C, 20 min	C5H5FeC5H5 FeCl3, Hg	24–30 —	75
C6H5	Fe	Sealed tube, 200°C, 6–8 hr	C6H5C6H5	High	86
		Element of Actinides			
C2H5	U	150°C, 2 hr	Hg Resin	—	74
		Additional Data			
trans-CH3CH=CHHgBr	Li	(C2H5)2O, then (CH3)3SiCl	(CH3)3SiCH=CHCH3	64 (87% trans, 12% cis)	115
cis-CH3CH=CHHgBr	Li	(C2H5)2O, then (CH3)3SiCl	(CH3)3SiCH=CHCH3	67 (60% trans, 40% cis)	115
C2H5HgSi(C2H5)3	Li	Benzene, 20°, 30 days	(C2H5)SiLi + C2H5Li	—	116
C2H5HgSi(C2H5)3	Li	THF	(C2H5)3SiLi + C2H6 + C2H4	—	116
(CH3)2Hg	K/Na	Pentane	CH3K	—	117

[a] Treat with dibenzofuran. [b] Treat with CO2. [c] Treat with benzalacetophenone. [d] Treat with Br2. [e] Hydrolyze. [f] Treat with (CH3C6H4)3B. [g] Treat with (C2H5)2O·BF3. [h] Treat with Benzophenone. [i] Treat with C6H5COOCH3. [j] Treat with C6H5CHO. [k] Treat with β-CH3C10H7. [l] Treat with CH3COCl.

REFERENCES

1. A. N. Nesmeyanov, A. E. Borisov, and A. N. Abramova, *Izv. Akad. Nauk SSSR, Otd. Khim. Nauk*, **1949**, 570.
2. A. N. Nesmeyanov, O. A. Reutov, and P. G. Knoll, *Dokl. Akad. Nauk SSSR*, **118**, 99 (1958).
3. O. A. Reutov and Y.T.U, *Dokl. Akad. Nauk SSSR*, **117**, 1003 (1957).
4. O. A. Reutov, P. G. Knoll, and Y.T.U, *Dokl. Akad. Nauk SSSR*, **120**, 1052 (1958).
5. O. A. Reutov and G. M. Ostapchuk, *Dokl. Akad. Nauk SSSR*, **117**, 826 (1957).
6. D. R. Pollard and J. V. Westwood, *J. Am. Chem. Soc.*, **87**, 2809 (1965).
7. O. A. Reutov and G. M. Ostapchuk, *Zh. Obshch. Khim.*, **29**, 1614 (1959).
8. O. A. Reutov, XVIIth *Intern. Congr. Pure Appl. Chem.*, Vol. I, Inorganic Chemistry, Butterworths. London, Verlag Chemie GMBH. Weinheim/Bergstr. 1960, p. 11.
9. W. Schlenk and J. Holtz, *Ber.*, **50**, 262 (1917).
10. F. D. Hager and C. S. Marvel, *J. Am. Chem. Soc.*, **48**, 2689 (1926).
11. C. S. Marvel, F. D. Hager, and D. D. Coffman, *J. Am. Chem. Soc.*, **49**, 2323 (1927).
12. B. Bartocha, C. M. Douglas, and M. I. Gray, *Z. Naturforsch.*, **14B**, 809 (1959).
13. A. N. Nesmeyanov, A. E. Borisov, I. S. Savel'eva, and E. I. Golubeva, *Izv. Akad. Nauk SSSR, Otd. Khim. Nauk*, **1958**, 1490.
14. W. J. Gensler and A. P. Mahadevan, *J. Org. Chem.*, **21**, 180 (1956).
15. T. V. Talalaeva and K. A. Kocheshkov, *Zh. Obshch. Khim.*, **8**, 1831 (1938).
16. G. Wittig, F. J. Meyer, and G. Lange, *Ann. Chem.*, **571**, 167 (1951).
17. G. Wittig and F. Bickelhaupt, *Chem. Ber.*, **91**, 865 (1958).
18. G. Wittig and F. Bickelhaupt, *Chem. Ber.*, **91**, 883 (1958).
19. G. Wittig and W. Herwig, *Ber.*, **88**, 962 (1955).
20. G. Wittig, E. Hahn, and W. Tochtermann, *Chem. Ber.*, **95**, 431 (1962).
21. A. N. Nesmeyanov, I. F. Lutsenko, and R. M. Khomutov, *Dokl. Akad. Nauk SSSR*, **120**, 1049 (1958).
22. F. C. Whitmore and H. D. Zook, *J. Am. Chem. Soc.*, **64**, 1783 (1942).
23. H. Gilman, H. A. Pacevitz, and O. Baine, *J. Am. Chem. Soc.*, **62**, 1514 (1940).
24. J. Lane and S. E. Ulrich, *J. Am. Chem. Soc.*, **73**, 5470 (1951).
25. P. P. Schorygin, *Ber.*, **41**, 2723 (1908).
26. P. P. Schorygin, *Ber.*, **43**, 1938 (1910).
27. H. Gilman and F. Breuer, *J. Am. Chem. Soc.*, **56**, 1123 (1934).
28. P. P. Schorygin, *Ber.*, **41**, 2717 (1908).
29. A. A. Morton and I. Hechenbleikner, *J. Am. Chem. Soc.*, **58**, 1024 (1936).
30. P. P. Schorygin, *Ber.*, **43**, 1931 (1910).
31. R. Muxart, R. Mellet, and R. Jaworski, *Bull. Soc. Chim. France*, **1956**, 445.
32. Sh. W. Rabideau, M. Alei, Jr., and Ch. E. Holley, Jr., U.S. Atomic Energy Comm. LA-1687, IIpp. (1954); through *Chem. Abstr.*, **50**, 3992 (1956).
33. H. Gilman and F. Schulze, *J. Chem. Soc.*, **1927**, 2663.
34. R. E. Dessy, *J. Am. Chem. Soc.*, **82**, 1580 (1960).
35. G. Wittig and H. Hornberger, *Ann. Chem.*, **577**, 11 (1952).
36. G. Wittig and D. Wittenberg, *Ann. Chem.*, **606**, 1 (1957).
37. R. E. Dessy, F. Kaplan, G. R. Coe, and R. M. Salinger, *J. Am. Chem. Soc.*, **85**, 1191 (1963).
38. H. Gilman and F. Schulze, *J. Am. Chem. Soc.*, **49**, 2328 (1927).

39. F. Waga, *Ann. Chem.*, **282**, 320 (1894).
40. D. O. Cowan and H. S. Mosher, *J. Org. Chem.*, **27**, 1 (1962).
41. W. Schlenk, Jr., *Ber.*, **64**, 736 (1931).
42. C. R. McCoy and A. L. Allread, *J. Am. Chem. Soc.*, **84**, 912 (1962).
43. K. A. Kocheshkov, A. N. Nesmeyanov, and V. I. Potrosov, *Ber.*, **67**, 1138 (1934); *Uchen. Zap. Moskov. Gos. Univer.*, **3**, 305 (1934).
44. P. Löhr, *Ann. Chem.*, **261**, 48 (1891).
45. S. Hilpert and G. Grüttner, *Ber.*, **46**, 1675 (1923).
46. C. H. Bamford, D. L. Levi, and D. M. Newitt, *J. Chem. Soc.*, **1946**, 468.
47. E. Roux and E. Louise, *Bull. Soc. Chim. France* (2), **50**, 497 (1888).
48. T. Wartick and H. J. Schlesinger, *J. Am. Chem. Soc.*, **75**, 835 (1953).
49. L. I. Zakharkin and O. Yu. Okhlobystin, *Izv. Akad. Nauk SSSR, Otd. Khim. Nauk*, **1958**, 1278.
50. G. I. Fray and R. Robinson, *Tetrahedron*, **18**, 261 (1962).
51. H. Gilman and K. Marple, *Rec. Trav. Chim.*, **55**, 133 (1936).
52. E. Krause and P. Dittmar, *Ber.*, **63**, 2401 (1930).
53. A. N. Nesmeyanov and N. N. Novikova, *Izv. Akad. Nauk SSSR, Otd. Khim. Nauk*, **1942**, 372
54. E. Wiberg, T. Johannsen, and O. Stecher, *Z. Anorg. Allgem. Chem.*, **251**, 114 (1943).
54a. J. P. Oliver and L. G. Stevens, *J. Inorg. Nuclear Chem.*, **24**, 953 (1962).
55. G. E. Coates, *J. Chem. Soc.*, **1951**, 2003.
56. L. M. Dennis and W. Patnode, *J. Am. Chem. Soc.*, **54**, 182 (1932).
57. J. P. Oliver and L. G. Stevens, *J. Inorg. Nuclear Chem.*, **19**, 378 (1961).
58. D. Moy, J. P. Oliver, and M. T. Emerson, *J. Am. Chem. Soc.*, **86**, 371 (1964).
59. W. Strohmeier and K. Humpfner, *Chem. Ber.*, **90**, 2339 (1957).
60. H. Gilman and R. G. Jones, *J. Am. Chem. Soc.*, **62**, 980 (1940).
61. L. M. Dennis, R. W. Work, E. G. Rochow, and E. M. Chamot, *J. Am. Chem. Soc.*, **54**, 1047 (1934).
62. G. P. Van der Kelen, *Bull. Soc. Chim. Belge*, **65**, 343 (1956).
63. W. C. Schumb and H. I. Crane, *J. Am. Chem. Soc.*, **60**, 306 (1938).
64. H. Gilman and R. G. Jones, *J. Am. Chem. Soc.*, **62**, 2353 (1940).
65. A. E. Goddard and D. Goddard, Organometallic Compounds, Newton Friend, Textbook of Inorganic Chemistry, vol. XI, London, 1928, Part I, p. 235.
66. A. N. Nesmeyanov, A. E. Borisov, and A. N. Abramova, *Uchen. zap. Moskov. Gos. Univer.*, **132**, 73 (1950).
67. A. E. Shurov and G. A. Razuvaev, *Ber.*, **65**, 1507 (1932).
68. M. M. Nadj and K. A. Kocheshkov, *Zh. Obshch. Khim.*, **8**, 42 (1938).
69. J. Burdon, P. L. Coe, and M. Fulton, *J. Chem. Soc.*, **1965**, 2094.
70. A. N. Nesmeyanov, A. E. Borisov, and N. A. Vol'kenau, *Izv. Akad. Nauk SSSR, Otd. Khim. Nauk*, **1956**, 162.
71. L. H. Long and J. F. Sackman, Research Correspondence, Suppl. to Research (London) 8, N5, 23–4 (1955); *Chem. Abstr.*, **50**, 9081 (1956).
72. E. Frankland and B. F. Duppa, *Ann. Chem.*, **130**, 117 (1864); J. Chem. Soc., **17**, 29 (1864).
73. L. A. Zhitkova, N. I. Sheverdina, and K. A. Kocheshkov, *Zh. Obshch. Khim.*, **8**, 1839 (1938).
74. A. E. Comyns, Atomic Energy Research Estab. (Gr. Brit). c/m-258, p. 7 (1957); *Chem. Abstr.*, **51**, 17557 (1957).
75. K. Issleib and A. Brack, *Z. Naturforsch.*, **11**B, 420 (1956).

76. F. Hein, E. Petzchner, K. Wagler, and F. A. Segitz, *Z. Anorg. Allgem. Chem.*, **141**, 161 (1924).
77. H. Gilman and R. Young, *J. Org. Chem.*, **26**, 2604 (1961).
78. H. I. Schlesinger and H. C. Brown, *J. Am. Chem. Soc.*, **62**, 3429 (1940).
79. H. Gilman and R. H. Kirby, *J. Am. Chem. Soc.*, **63**, 2046 (1941).
80. G. Wittig and E. Benz, *Ber.*, **91**, 873 (1958).
81. G. B. Buckton, *Ann. Chem.*, **112**, 222 (1859).
82. W. H. Carothers and D. D. Coffmann, *J. Am. Chem. Soc.*, **51**, 588 (1929).
83. H. Gilman and R. H. Kirby, *J. Am. Chem. Soc.*, **58**, 2074 (1936).
84. H. Gilman and R. V. Young, *J. Am. Chem. Soc.*, **56**, 1415 (1934).
85. S. F. Acree, *Am. Chem. J.*, **29**, 588 (1903).
86. E. Dreher and R. Otto, *Ann. Chem.*, **154**, 93 (1870).
87. W. E. Bachmann and H. T. Clarke, *J. Am. Chem. Soc.*, **49**, 2089 (1927).
88. V. Lavrov, *Zh. Russk.-Fiz. Khim. Obshch.*, **16**, 93 (1884).
89. A. B. Burg and H. I. Schlesinger, *J. Am. Chem. Soc.*, **62**, 3425 (1940).
90. A. Cahours, *Compt. Rend.*, **76**, 1383 (1873); *Jahresber.*, **1873**, 520.
91. E. Frankland and B. F. Duppa, *J. Chem. Soc.*, **13**, 181, 194 (1861).
92. H. Gilman and R. E. Brown, *Rec. Trav. Chim.*, **49**, 704 (1930).
93. H. S. Mosher and P. K. Loeffler, *J. Am. Chem. Soc.*, **62**, 3425 (1940).
94. H. P. Fleck, *Ann. Chem.*, **276**, 129 (1893).
95. H. Gilman and R. E. Brown, *Rec. Trav. Chim.*, **49**, 202 (1930).
96. A. Cahours, *Compt. Rend.*, **76**, 133 (1873); *Chem. Zentr.*, **1873**, 200.
97. A. Cahours, *Compt. Rend.*, **76**, 748 (1873); *Chem. Zentr.*, **1873**, 275.
98. A. Shcherbakov, *Zh. Russk. Fiz.-Khim. Obshch.*, **13**, 349 (1881).
99. A. Cahours, *J. Prakt. Chem.*, (2), **8**, 395 (1873).
100. A. Marquardt, *Ber.*, **21**, 2035 (1888).
101. G. Ponzio, *Chem. Zentr.*, **1900**, II, 624; *Gazz. Chim. Ital.*, **30**, II, 23 (1900).
102. E. Sokolov, *Zh. Russ. Fiz.-Khim. Obshch.*, **19**, 197 (1887).
103. G. B. Buckton, *J. Chem. Soc.*, **16**, 17 (1863).
104. A. W. Laubengayer and W. F. Gilliam, *J. Am. Chem. Soc.*, **63**, 477 (1941).
105. G. B. Buckton and W. Odling, *Ann. Chem. (Suppl.)*, **4**, 109 (1865).
106. K. S. Pitzer and H. S. Gutowsky, *J. Am. Chem. Soc.*, **68**, 2204 (1946).
107. A. Müller and A. Sauerwald, *Monatsh. Chem.*, **48**, 737 (1927).
108. S. Murahashi, S. Nozakura, and S. Takeuchi, *Bull. Chem. Soc., Japan*, **33**, 658 (1960).
109. F. A. Cermeno, *Anales Real Soc. Espan. Fis. Quim. (Madrid)*, *Ser. B.*, **60** (II), 753 (1964); *Chem. Abstr.*, **63**, 16376 (1965).
110. B. Bartocha, A. J. Bilbo, D. E. Bublitz, and M. I. Gray, *Z. Naturforsch.*, **16B**, 357 (1961).
111. S. Hilpert and G. Grüttner, *Ber.*, **45**, 2828 (1912).
112. C. Friedel and H. Kraffts, *Ann. Chim.* (6), **14**, 457 (1888).
113. W. Menzel, *Z. Anorg. Allgem. Chem.*, **269**, 52 (1952).
114. G. A. Razuvaev, E. V. Mitrofanova, and G. G. Petukhov, *Zh. Obshch. Khim.*, **30**, 1996 (1960).
115. D. Seyferth, L. S. Vaughan, *J. Organometal. Chem.*, **5**, 580 (1966).
116. N. S. Vyazankin, G. A. Razuvaev, E. N. Gladyshev, S. P. Korneva, *J. Organometal. Chem.*, **7**, 353 (1967).
117. E. Weiss, G. Sauermann, *Angew. Chem.*, **86**, 123 (1968).

Chapter 4

Reactions with Salts-Reducing Agents, and Low Valence Organometallic Compounds

I. INTRODUCTION

The reactions of organomercury compounds with low valence compounds of the metals resulting in the formation of metallic mercury are of great practical importance for synthesis of some organometallic compounds. Metallic mercury is produced by reduction using the halides of divalent tin, organic compounds of low valence tin, and germanium(II) iodide. These reactions constitute important methods for the synthesis of organometallic compounds of tin and germanium.

It is also possible that halides of other metals in their lowest valency are capable of reducing the organomercury compounds with formation of the organometallic compounds of these metals.

II. APPLICATION OF THE REACTION

A. Synthesis of Organogermanium Compounds

1. Reactions with Germanium(II) Halides

Germanium(II) iodide, the most common germanium(II) halide, is limitedly used as a reactant with organomercury compounds such as diethyl and di-n-butylmercury.[1] The reaction with diethylmercury was carried out in benzene, while ether or acetone was used as the solvent in the case of dibutylmercury. Formation of Hg_2I_2 and AlkHgI occurs, in the former reaction diethylgermanium was observed in the benzene solution, and in the latter an organogermanium compound with the following stoichiometry was isolated: $(C_4H_9)_2IGeGeI(C_4H_9)_2$.[1]

The reaction of diarylmercury compounds with germanium(II) iodide was studied more comprehensively since it is a method for the preparation of aromatic compounds of germanium.[2,3] The reaction usually occurs upon boiling the reactants in toluene for a short time. The main reaction products are diarylated compounds of germanium produced in 40–70% yield as exemplified by the following equation:

$$Ar_2Hg + GeI_2 \longrightarrow Ar_2GeI_2 + Hg$$

227

Further reactions which produce mono and triarylated germanium compounds also occur:

$$Ar_2Hg + Ar_2GeI_2 \longrightarrow Ar_3GeI + ArHgI$$
$$ArHgI + GeI_2 \longrightarrow ArGeI_3 + Hg$$

No Ar_4Ge was produced during the reaction of Ar_2Hg with GeI_2. HgI_2 and Hg_2I_2 were also produced in addition to metallic mercury. Various Ar_2Hg compounds were used as reactants with GeI_2 (Ar = phenyl, o-, m-, p-tolyl, o-, m-, p-chlorophenyl, o-, p-bromophenyl, o-anisyl, o- and p-ethylphenyl, and β-naphthyl). In a similar manner C_6H_5HgI reacts with GeI_2 producing $(C_6H_5)_2GeI_2$ and $C_6H_5GeI_2$ 50 and 30% yields, respectively. The reaction evidently proceeds according to the following equations:

$$C_6H_5HgI + GeI_2 \longrightarrow C_6H_5GeI_3 + Hg$$
$$C_6H_5HgI + C_6H_5GeI_3 \longrightarrow (C_6H_5)_2GeI_2 + HgI_2$$
$$2HgI_2 + GeI_2 \longrightarrow GeI_4 + Hg_2I_2$$

B. Synthesis of Organotin Compounds

1. Reactions with Tin(II) Salts

The most convenient and all-purpose method of synthesis of organotin compounds by use of organomercury compounds is by the reaction of completely substituted compounds of mercury with tin(II) chloride or bromide.

The reaction is represented by the general equation:

$$R_2Hg + SnX_2 \longrightarrow R_2SnX_2 + Hg$$

Nesmeyanov and Kocheshkov[4,5] first proposed this synthetic method for the preparation of diaryltin(IV) halides. The reaction of Ar_2Hg with anhydrous tin halide is conducted in boiling, dry acetone for 0.5–1 hour. The metallic mercury is poured out, the acetone is driven off, and the diaryltin halide is extracted by washing the residue with petroleum ether. This method may be used for the preparation of various Ar_2SnHal_2 compounds in good yields. If the reaction is conducted in alcohol, diphenylmercury and diarylmercury compounds such as diphenylmercury which contain radicals more positive than phenyl or dibenzylmercury, react according to the previous equation.

In alcohol, the reaction of tin(II) halide with diarylmercury compounds containing aryl groups more electronegative than phenyl does not occur. The alcohol reacts instead and hydrogen is substituted for mercury:

$$R_2Hg + SnX_2 + 2C_2H_5OH \longrightarrow 2RH + Hg + (C_2H_5O)_2SnX_2$$

In acetone also, a similar reaction occurs for aryls more electronegative, for example, p-aminophenyl. (See Chapter 7.)

Diethylmercury reacts slowly and incompletely with tin(II) chloride in alcohol and in acetone. The reaction goes in a complicated manner resulting in the formation of not only diethyltin(IV) chloride (up to 60%) but also with the formation of small amounts of triethyltin chloride, ethylmercury chloride and ethane. Triethyltin chloride is evidently produced by the following reaction:

$$(C_2H_5)_2SnCl_2 + (C_2H_5)_2Hg \longrightarrow (C_2H_5)_3SnCl + C_2H_5HgCl$$

The products of the reaction depend on the nature of the halide in SnX_2. In the reaction of $(C_2H_5)_2Hg$ or $(\alpha\text{-}C_{10}H_7)_2Hg$ with $SnBr_2$, the mercury is isolated as Hg_2Br_2. In other respects, the reaction of $(C_2H_5)_2Hg$ with $SnBr_2$ produces the same products that $SnCl_2$ does; $(\alpha\text{-}C_{10}H_7)_2Hg$ with $SnBr_2$ in alcohol and acetone produces naphthalene, whereas with $SnCl_2$ in acetone it reacts to produce R_2SnCl_2. SnI_2 does not react with R_2Hg. Although SnF_2 and $SnSO_4$ react with R_2Hg, they do so extremely slowly and are therefore unsuited as reactants for preparative methods. Both stereoisomers of di-(β-chlorovinyl)mercury react smoothly with tin(II) chloride in alcohol solution under cold conditions that result in formation of a good yield of R_2SnCl_2 with retention of configuration of the transferred radical,[6] for example:

$$trans\text{-}(\beta\text{-}ClCH{=}CH)_2Hg + SnCl_2 \longrightarrow trans\text{-}(\beta\text{-}ClCH{=}CH)_2SnCl_2 + Hg$$

However, among stereoisomeric mercury derivatives of stilbene only bis(cis-α-stilbenyl)mercury and not the $trans$ compound reacts with $SnCl_2$ in acetone producing $RSnCl_3$ as the main reaction along with R_2SnCl_2 and RH with retention of substituent configuration. Bis($trans$-α-stilbene)mercury does not react with $SnCl_2$ in alcohol or acetone as also in the solid state at temperatures between 240 and 260°C. The reactions of organomercury halides with tin(II) halides produce organotin compounds[4]. In these reactions, the reaction depends on the organic substituents, solvents, and mercury and tin halides. Phenyl- and benzylmercury chlorides react with $SnCl_2$:

$$2RHgCl + 2SnCl_2 \longrightarrow R_2SnCl_2 + 2Hg + SnCl_4 \qquad (1)$$

This reaction proceeds in either alcohol or acetone.

The reaction which could be expected here occurs to a slight extent:

$$RHgX + SnX_2 \longrightarrow RSnX_3 + Hg$$

p-$CH_3C_6H_4HgCl$ reacts with $SnCl_2$ in acetone producing R_2SnCl_2 according to Eq. (1) and in alcohol it is reduced to the hydrocarbon.

In the case with arylmercury bromide, the reaction proceeds in an unusual manner; the tin is oxidized to the tetravalent state owing to the aryl cleavage:

$$2ArHgBr + SnBr_2 \longrightarrow Ar_2SnBr_2 + Hg_2Br_2$$

Also produced in a small amount is $ArSnBr_3$. If the aryl group is similar to α-$C_{10}H_7$ and the reaction is carried out in acetone, a small amount of ArH is produced.

CH_3HgCl reacts with $SnCl_2$ extremely slowly and produces a very small amount of an organotin compound which was not identified. C_2H_5HgCl did not react with $SnCl_2$ upon boiling for 12 hr in alcohol.

The reactions of $[CH_2{=}C(CH_3)]_2Hg$ with $SnBr_2$ in acetone produced organotin compounds such as R_2SnBr_2 and R_4Sn. The mercury is isolated, not only as metallic mercury, but also as RHgBr. Bis(*trans*-α-chlorostilbenyl)mercury does not react with $SnCl_2$ to produce organotin compounds, but is isomerized.

If acetone or alcohol are replaced by inert solvents such as petroleum ether or ligroin, a great number of organomercury compounds may be reacted with tin(II)chloride, for example, alkyl and alkenyl compounds.[7] The reaction of diethylmercury with tin(II) chloride, which does not go smoothly in alcohol or acetone, proceeds readily in ligroin (90°C, 12 hr) according to the equation:

$$(C_2H_5)_2Hg + SnCl_2 \longrightarrow (C_2H_5)_2SnCl_2 + Hg$$

Trans and *cis*-isomers of dipropenylmercury in petroleum ether react with $SnCl_2$ to produce R_2SnBr_2 with retention of configuration. Diisopropenylmercury reacts with $SnBr_2$ in petroleum ether and produces the reaction products in acetone or without a solvent: $[CH_2{=}C(CH_3)]_2$-$SnBr_2$, R_4Sn, RHgBr, and Hg, however, dealkylation of R_2Hg proceeds more completely.

When the reaction of fully substituted aromatic mercury compounds with tin halides is conducted in ligroin, no replacement of mercury by hydrogen or formation of a hydrocarbon for compound such as $(\alpha$-$C_{10}H_7)_2Hg$ occurs. The latter, which reacts with $SnCl_2$, and di-o-tolylmercury, which reacts with $SnCl_2$ or $SnBr_2$, produce good yields of R_2SnHal_2. The reaction product of diphenylmercury with $SnCl_2$ and $SnBr_2$ in ligroin is triphenyltin halide.

2. Reactions with Organotin Compounds

Organic compounds of R_3Sn—SnR_3 or R_2Sn, as do tin(II) halides, reduce organomercury compounds[8]:

$$R_3'SnSnR_3' + R_2Hg \longrightarrow 2R_3'SnR + Hg$$

$$R_3'SnSnR_3' + RHgCl \longrightarrow R_3'SnR + R_3'SnCl + Hg$$

$$R_2'Sn + R_2Hg \longrightarrow R_2'SnR_2 + Hg$$

Tetrasubstituted tin compounds are produced in these cases. The reaction of R_3SnSnR_3 with R_2Hg proceeds with more difficulty (at a higher temperature and prolonged heating) than with $RHgX$. The reactions are conducted without solvents at temperatures of 170°C (for R_2Hg) and 130°C (for $RHgX$). Hexaethylditin and diethyltin were used as reactants. Introduced into reaction with the first compound was an alkylmercury compound, the methyl ester of mercurybisacetic acid.[9] The reaction was carried out at 160–180°C and proceeds without transfer of the reaction center with metallation on the carbon:

$$[(C_2H_5)_3Sn]_2 + Hg(CH_2CO_2CH_3)_2 \longrightarrow 2(C_2H_5)_3SnCH_2CO_2CH_3 + Hg$$

III. PREPARATIVE SYNTHESES

A. Preparation of Organogermanium Compounds

1. Reaction of Di-p-tolylmercury with Germanium(II) Iodide[3]

Di-p-tolylmercury (3.288 g, 0.008 mole) and 3.264 g of GeI_2 (0.01 m) were boiled in 30 ml of dry toluene for 15 min. The yellow-greyish residue was washed twice with toluene. Identified were Hg, Hg_2I_2, HgI_2, and unreacted GeI_2. The solvent was driven off from the toluene filtrate under vacuum. Added to the residue were 20 ml of C_2H_5OH and 15 ml of 20% NaOH solution. The mixture was boiled on a water bath for about 1 hr with subsequent loss of alcohol. The residue was diluted with water. The yellow crystals produced were thoroughly washed with water and dissolved in benzene. Obtained from the benzene solution was 1.72 g (79%) of colorless, needlelike crystals with m.p. 220–223°C. Recrystallization from petroleum ether–benzene gave a product with m.p. 221.5–222.5°C which was $[(p\text{-}CH_3C_6H_4)_2GeO]_2$. During acidification 0.21 g of (p-$CH_3C_6H_4GeO)_2O$ was isolated as light yellow powder. Purified by repeated dissolution in alkali and then neutralized by acetic acid, a 3.3% yield based on p-tolyl compound was obtained.

B. Preparation of Di-*cis*-β-chlorovinyltin(IV) Chloride[6]

Di-(*cis*-β-chlorovinyl)mercury (1.8 g, 0.00556 m) in 5 ml of absolute alcohol and 3.2 g (0.0168) of anhydrous tin(II) chloride in 15 ml of absolute alcohol slightly acidified by hydrogen chloride were mixed. Precipitation of metallic mercury began in 10 min. The mixture was heated to 65–70°C for 5 hours. Tin(II) chloride, 0.5 g in 3 ml of absolute alcohol, was added to the clear mixture. Formation of metallic mercury was again observed and heating was continued for 2 hr. The clear solution was decanted from the metallic mercury (isolated 1.12 g; theoretical yield: 1.14 g); and alcohol was then driven off as completely as possible. Fifteen milliliters of petroleum ether were added and the remaining alcohol was driven off. The residue was extracted with petroleum ether, the solution dried with melted $CaCl_2$, the solvent was driven off, and the remaining portion of the product was distilled *in vacuo*. Clear liquid (0.8 g) was obtained at 3 mm and 100–102°C, (50% yield).

C. Preparation of Diethyltin(IV) Chloride[7]

Diethylmercury (5 g, 0.019 mole) and 7 g (0.036 mole) of tin(II) chloride in 25 ml of ligroin (b.p. 100–110°C) were vigorously stirred at 85–90°C for 12 hr. The cooled solution was decanted and the precipitate was extracted by three portions of heated petroleum ether and then by methyl alcohol. Upon evaporation of the petroleum ether 2.5 g (52%) of diethyltin(IV) chloride were isolated which, after recrystallization from petroleum ether, melted at 83–83.5°C. One gram of ethylmercury(II) chloride with m.p. 191–192°C was obtained from the methanol solution. Metallic mercury (2.59 g, 69%) was isolated.

D. Preparation of Di-*p*-chlorophenyltin(IV) Chloride[5]

Di-*p*-chlorophenylmercury (8.46 g, 0.02 mole) and 5.7 g of anhydrous tin(II) chloride (0.03 mole) were heated on a water bath with 40 ml of absolute alcohol for 1 hour. Metallic mercury (3.9 g vs. 4 g theoretical yield) was formed. The alcohol was driven off as completely as possible and petroleum ether then added and, in turn, driven off. This operation was repeated twice with the objective of removing traces of alcohol. This residue was treated with a fresh portion of petroleum ether. After filtration from undissolved tin(II) chloride, the petroleum ether was evaporated. Aggregates of white crystals were obtained (6.5 g, 79% of theoretical yield). The product was moderately soluble in petroleum ether and ligroin and very soluble in all ordinary organic solvents; it undergoes hydrolysis in water; m.p. 86.5°C.

TABLE III

Reaction with Salts—Reducing Agents and Incomplete Valence Organic Compounds of Metals

R in R_2Hg or $RHgX$	Reactant	Reaction conditions	Products	Yield, %	Ref.
C_2H_5	GeI_2	C_6H_6	C_2H_6	—	1
			C_6H_5HgI	—	1
$n\text{-}C_4H_9$	GeI_2	CH_3COCH_3	$(C_4H_9)_2IGe\text{—}GeI(C_4H_9)_2$	—	
C_6H_5	GeI_2	Toluene, reflux, 15 min	$(C_6H_5)_2GeI_2$	39	3
			$(C_6H_5)_3GeI$	37	
				0.5	
C_6H_5HgI	GeI_2	Toluene, reflux, 15 min	$(C_6H_5)_2GeI_2$	51	3
			$C_6H_5GeI_3$	32	
$o\text{-}CH_3C_6H_4$	GeI_2	Toluene, reflux, 15 min	$(o\text{-}CH_3C_6H_4)_2GeI_2$	59	3
$m\text{-}CH_3C_6H_4$	GeI_2	Toluene, reflux, 15 min	$(m\text{-}CH_3C_6H_4)_2GeI_2$	27	3
			$(m\text{-}CH_3C_6H_4)_3GeI$	34	3
			$m\text{-}CH_3C_6H_4GeI_3$	1.56	
$p\text{-}CH_3C_6H_4$	GeI_2	Toluene, reflux, 15 min	$[(p\text{-}CH_3C_6H_4)_2GeO]_2$[a]	79	3
			$(p\text{-}CH_3C_6H_4GeO)_2O$[a]	3	
$o\text{-}ClC_6H_4$	GeI_2	Toluene, reflux, 15 min	$(o\text{-}ClC_6H_4)_2GeI_2$	58	3
$m\text{-}ClC_6H_4$	GeI_2	Toluene, reflux, 20 min	$(m\text{-}ClC_6H_4)_2GeI_2$	25	3
$p\text{-}ClC_6H_4$	GeI_2	Toluene, reflux, 15 min	$(p\text{-}ClC_6H_4)_3GeI$	3	3
			$(p\text{-}ClC_6H_4)_2GeI_2$	40	
			$p\text{-}ClC_6H_4GeI_3$	1	
$o\text{-}BrC_6H_4$, $1M$	GeI_2	Toluene, reflux, 15 min	$(o\text{-}BrC_6H_4)_2GeI_2$	67	3
$p\text{-}BrC_6H_4$, $1M$	GeI_2, $1M$	Toluene, reflux, 15 min	$(p\text{-}BrC_6H_4)_3GeI$	28	3
			$(p\text{-}BrC_6H_4)_2GeO$[a]	28	3
			$(p\text{-}BrC_6H_4GeO)_2O$[a]	3	
$p\text{-}BrC_6H_4$, $1M$	GeI_2, $1.5M$	Toluene, reflux, 15 min	$(p\text{-}BrC_6H_4)_3GeI$	42	3
			$(p\text{-}BrC_6H_4)_2GeI_2$	27	
			$p\text{-}BrC_6H_4GeI_3$	1	

(continued)

233

TABLE III (*Continued*)

R in R$_2$Hg of RHgX	Reactant	Reaction conditions	Products	Yield, %	Ref.
p-BrC$_6$H$_4$, 1.5M	GeI$_2$, 1M	Toluene, reflux, 15 min	(p-BrC$_6$H$_4$)$_3$GeI	19	3
			(p-BrC$_6$H$_4$)$_2$GeO^a	32	
p-BrC$_6$H$_4$, 1M	GeI$_2$, 1M	Toluene, reflux, 15 min	(p-BrC$_6$H$_4$)$_3$GeI	5^b	3
			(p-BrC$_6$H$_4$)$_2$GeI$_2$	43	
p-CH$_3$OC$_6$H$_4$	GeI$_2$	Toluene, reflux, 15 min	(p-CH$_3$OC$_6$H$_4$)$_3$GeI	62	3
			(p-CH$_3$OC$_6$H$_4$)$_2$GeO^a	9	
o-C$_2$H$_5$OC$_6$H$_4$	GeI$_2$	Toluene, reflux, 15 min	(o-C$_2$H$_5$OC$_6$H$_4$)$_2$GeI$_2$	63	3
p-C$_2$H$_5$OC$_6$H$_4$	GeI$_2$	Toluene, reflux, 15 min	(p-C$_2$H$_5$OC$_6$H$_4$)$_2$GeI$_2$	31	3
			(p-C$_2$H$_5$OC$_6$H$_4$)$_3$GeI	48	
β-C$_{10}$H$_7$	GeI$_2$, 1M	Toluene, reflux, 15 min	(β-C$_{10}$H$_7$)$_3$GeI	23	3
			(β-C$_{10}$H$_7$)$_2$GeO^a	5	
CH$_3$HgCl	SnCl$_2$	C$_2$H$_5$OH, reflux, 15 hr	Hg	Quant.	4
C$_2$H$_5$	SnCl$_2$	C$_2$H$_5$OH, reflux, 40 hr	C$_2$H$_5$HgCl	28	4
			(C$_2$H$_5$)$_2$SnCl$_2$	63	
C$_2$H$_5$	SnCl$_2$	Ligroin, 85–90°C, 12 hr	(C$_2$H$_5$)$_2$SnCl$_2$	52	7
			Hg	69	
C$_2$H$_5$	SnBr$_2$	C$_2$H$_5$OH, reflux, 15 hr	C$_2$H$_5$HgBr	—	4
			(C$_2$H$_5$)$_3$SnBr $\Big\}$ mixture (C$_2$H$_5$)$_2$SnBr$_2$	—	
C$_2$H$_5$HgCl	SnCl$_2$	C$_2$H$_5$OH, reflux, 15 hr	No reaction		4
CH$_2$COOCH$_3$	(C$_2$H$_5$)$_6$Sn$_2$	160°C, 1 hr	Hg	68	9
			(C$_2$H$_5$)$_3$SnCH$_2$COOCH$_3$	41	
cis-CH$_3$CH=CH	SnBr$_2$	Petroleum ether, 20°C, 2 hr	cis-(CH$_3$CH=CH)$_2$SnBr$_2$	31	10
trans-CH$_3$CH=CH	SnBr$_2$	Petroleum ether, 20°C, 2 hr	trans-(CH$_3$CH=CH)$_2$SnBr$_2$	77	10
CH$_2$=C(CH$_3$)	SnBr$_2$	Petroleum ether, 20°C, 5 hr	Hg	44	7,11
			[CH$_2$=C(CH$_3$)]$_2$SnBr$_2$	22	
			[CH$_2$=C(CH$_3$)]$_4$Sn	40	
CH$_2$=C(CH$_3$)	SnBr$_2$	Petroleum ether, 65°C, 5 hr	[CH$_2$=C(CH$_3$)]$_2$SnBr$_2$	21	7,11
			Hg	71	
			[CH$_2$=C(CH$_3$)]$_4$Sn	27	

234

Organomercury compound	Reagent	Conditions	Products	Yield (%)	Reference
$(CH_2{=}C(CH_3))$	$SnBr_2$	CH_3COCH_3, 20°C, 20 hr	$[CH_2{=}C(CH_3)]HgBr$	—	7,11
			$[CH_2{=}C(CH_3)]_2SnBr_2$	17	
			$[CH_2{=}C(CH_3)]_4Sn$	60	
			Hg	46	
$CH_2{=}C(CH_3)$	$SnBr_2$	20°C, 20 hr	$[CH_2{=}C(CH_3)]HgBr$	—	11
			$[CH_2{=}C(CH_3)]_2SnBr_2$	9	
			$[CH_2{=}C(CH_3)]_4Sn$	70	
			Hg	50	
cis-$ClCH{=}CH$	$SnCl_2$	C_2H_5OH, 65–70°C, 7 hr	*cis*-$(ClCH{=}CH)_2SnCl_2$	Quant.	6
			Hg	50	
trans-$ClCH{=}CH$	$SnCl_2$	C_2H_5OH, 45–50°C, 10 min, then 20°C, 1 hr	*trans*-$(ClCH{=}CH)_2SnCl_2$	97	6
			Hg	58	
$C_6H_5CH{=}CH$	$SnCl_2$	CH_3COCH_3, 30–35°C, then 20°C for 1 hr	$(C_6H_5CH{=}CH)_2SnCl_2$	24	12
cis-$C_6H_5CH{=}C(C_6H_5)HgCl$	$SnCl_2$	Sealed tube, C_2H_5OH, CH_3COCH_3, 20–80°C, 2–12 hr	*cis*-$C_6H_5CH{=}CHC_6H_5$	27.5–77	13
			cis-$[C_6H_5CH{=}C(C_6H_5)]SnCl_3$	15	
			cis-$C_6H_5CH{=}C(C_6H_5)SnOOH$[c]	24.5–47	
			cis-$[C_6H_5CH{=}C(C_6H_5)]_2SnO$[c]	20	
			Hg		
cis-$C_6H_5CH{=}C(C_6H_5)HgCl$	$SnCl_2$	Sealed tube, C_2H_5OH, CH_3COCH_3, 20–80°C, 2–12 hr	*cis*-$C_6H_5CH{=}CHC_6H_5$	31–70	13
			cis-$[C_6H_5CH{=}C(C_6H_5)]SnCl_3$	11.5	
			cis-$C_6H_5CH{=}C(C_6H_5)SnOOH$[c]	27	
			cis-$[C_6H_5CH{=}C(C_6H_5)]_2SnO$[c]	18	
			Hg		
trans-$C_6H_5CH{=}C(C_6H_5)$	$SnCl_2$	Sealed tube, C_2H_5OH or CH_3COCH_3 or dioxane, 100–180°C, 8–20 hr, or no solvent, 240–260°C, 10–15 min	No organo-Sn-compounds		13
trans-$C_6H_5CH{=}C(C_6H_5)HgCl$	$SnCl_2$	Sealed tube, C_2H_5OH or CH_3COCH_3, 70–100°C, 3–5 min	No organo-Sn-compounds		13
$C_6H_5CH_2$[d]	$SnCl_2$	C_2H_5OH, reflux, 30 min	Hg	Quant.	4
			$(C_6H_5CH_2)_2SnCl_2$	Quant.	
$C_6H_5CH_2$[e]	$SnCl_2$	C_2H_5OH, reflux, 3.5 hr	Hg	64	4
			$(C_6H_5CH_2)_2SnCl_2$	—	
			$C_6H_5CH_2HgCl$	—	

(continued)

235

TABLE III (*Continued*)

R in R_2Hg or $RHgX$	Reactant	Reaction conditions	Products	Yield, %	Ref.
$C_6H_5CH_2$[d]	$SnBr_2$	C_2H_5OH, reflux, 2 hr	Hg $(C_6H_5CH_2)_2SnBr_2$	Quant. 88	4
CH_3CH_2[e]	$SnBr_2$	C_2H_5OH, reflux, 4 hr	$C_6H_5CH_2HgBr$	33.5	4
$C_6H_5CH_2HgCl$	$SnCl_2$	C_2H_5OH, reflux, 3 hr	Hg Mixture organo-Sn-compounds	Quant. —	4
C_6H_5	$SnCl_2$	250–260°C, 5 min	Hg $(C_6H_5)_2SnCl_2$	92 71	14
C_6H_5	$SnCl_2$	Ligroin, 60–65°C, 7 hr	Hg $(C_6H_5)_3SnCl$	98.2 75	7
C_6H_5	$SnCl_2$	C_2H_5OH of CH_3COCH_3, reflux, 30 min	Hg $(C_6H_5)_2SnCl_2$	96 80	4
C_6H_5	$SnCl_2 \cdot 2H_2O$	C_2H_5OH, reflux	Hg $(C_6H_5)_2SnCl_2$	Quant. 30	4
C_6H_5	$SnBr_2$	C_2H_5OH or CH_3COCH_3, reflux, 30 min	Hg $(C_6H_5)_2SnBr_2$	Quant. 92	4
C_6H_5	$SnBr_2$	Ligroin, 60–65°C, 12 hr	Hg $(C_6H_5)_3SnBr$	89 23	7
C_6H_5	$(C_2H_5)_2Sn$	150°C, 30 min	Hg $(C_6H_5)_2Sn(C_2H_5)_2$	75 60	8
C_6H_5	$[(C_2H_5)_3Sn]_2$	160°C, 7 hr	Hg $(C_2H_5)_3SnC_6H_5$	60 40	8
C_6H_5HgCl	$[(C_2H_5)_3Sn]_2$	150–160°C, 3 hr	Hg $(C_6H_5)Sn(C_2H_5)_3$ $(C_2H_5)_3SnCl$	— 30 —	8
C_6H_5HgCl	$SnCl_2$	C_2H_5OH, reflux, 30 min	Hg $(C_6H_5)_2SnCl_2$	Quant. 75	4
C_6H_5HgBr	$SnBr_2$	CH_3COCH_3, reflux, 2 hr	$(C_6H_5)_2SnO$[a] C_6H_5SnOOH[a] HgBr	— Trace Quant.	11

Starting material	Reagent	Conditions	Products	Yield (%)	Ref.
$o\text{-}CH_3C_6H_4$	$SnCl_2$	C_2H_5OH, reflux, 30 min	Hg	Quant.	4
			$CH_3C_6H_5$	85	
			$(o\text{-}CH_3C_6H_4)_2SnCl_2$	14	
$o\text{-}CH_3C_6H_4$	$SnCl_2$	Ligroin, 70°C, 9 hr	$(o\text{-}CH_3C_6H_4)_2SnCl_2$	66	7
$p\text{-}CH_3C_6H_4$	$SnCl_2$	Ligroin, 90–95°C, 13 hr	Hg	100	7
			$(p\text{-}CH_3C_6H_4)_2SnCl_2$	72	
$p\text{-}CH_3C_6H_4$	$SnCl_2$	CH_3COCH_3, reflux, 30 min	Hg	96	4
			$(p\text{-}CH_3C_6H_4)_2SnO$[a]	Quant.	
$p\text{-}CH_3C_6H_4$	$SnCl_2$	C_2H_5OH, reflux, 30 min	Hg	—	4
			$CH_3C_6H_5$	Quant.	
$p\text{-}CH_3C_6H_4$	$SnBr_2$	CH_3COCH_3, reflux, 3 hr	Hg	85	4
			$(p\text{-}CH_3C_6H_4)_2SnBr_2$	Quant.	
$p\text{-}CH_3C_6H_4$	$SnCl_2$	CH_3COCH_3, reflux, 30 min	Hg	87	4
			$(p\text{-}CH_3C_6H_4)_2SnO$[a]	Quant.	
$p\text{-}CH_3C_6H_4HgCl$	$SnBr_2$	C_2H_5OH, reflux, 30 min	Hg	—	4
			$CH_3C_6H_5$	Quant.	
$p\text{-}CH_3C_6H_4HgCl$	$SnCl_2$	CH_3COCH_3, reflux, 1 hr	$(p\text{-}CH_3C_6H_4)_2SnO$[a]	85	4
$p\text{-}CH_3C_6H_4HgBr$	$SnBr_2$		$p\text{-}CH_3C_6H_4SnOOH$[a]	Low	4
$p\text{-}ClC_6H_4$	$SnCl_2$	C_2H_5OH, reflux, 1 hr	$(p\text{-}ClC_6H_4)_2SnCl_2$	—	5
			Hg	Quant.	
$p\text{-}ClC_6H_4$	$SnBr_2$	C_2H_5OH, reflux, 1 hr	$(p\text{-}ClC_6H_4)_2SnBr_2$	79	5
			Hg	Quant.	
$p\text{-}BrC_6H_4$	$SnCl_2$	C_2H_5OH, reflux, 1 hr	$(p\text{-}BrC_6H_4)_2SnCl_2$	92	5
			Hg	Quant.	
$p\text{-}BrC_6H_4$	$SnBr_2$	C_2H_5OH, reflux, 30 min	$(p\text{-}BrC_6H_4)_2SnBr_2$	85	5
			Hg	Quant.	
$p\text{-}IC_6H_4$	$SnCl_2$	C_2H_5OH, reflux, 45 min	$(p\text{-}IC_6H_4)_2SnCl_2$	Quant.	5
			Hg	75	
$p\text{-}IC_6H_4$	$SnBr_2$	CH_3COCH_3, reflux, 35 min	$(p\text{-}IC_6H_4)_2SnBr_2$	Quant.	5
			Hg	70–75	
$o\text{-}HOC_6H_4$	$[(C_2H_5)_3Sn]_2$	130°C, 2hr	$HOC_6H_4Sn(C_2H_5)_3$	Quant.	8
				—	

(continued)

TABLE (Continued)

R in R₂Hg of RHgX	Reactant	Reaction conditions	Products	Yield, %	Ref.
p-$C_2H_5OCOC_6H_4$	$SnCl_2$	C_2H_5OH, reflux, 40 min	Hg	86	14
			$(p$-$C_2H_5OCOC_6H_4)_2SnCl_2$	70–75	
p-$C_2H_5OCOC_6H_4$	$SnCl_2$	CH_3COCH_3, reflux, 50 min	Hg	Quant.	14
			$(p$-$C_2H_5OCOC_6H_4)_2SnCl_2$	70–75	
p-$C_2H_5OCOC_6H_4$	$SnCl_2$	Fuse, 250–260°C, 5 min	Hg	80	14
			$(p$-$C_2H_5OCOC_6H_4)_2SnCl_2$	73	
p-$C_2H_5OCOC_6H_4$	$SnBr_2$	Fuse, 225–235°C, 5 min	Hg	80	14
			$(p$-$C_2H_5OCOC_6H_4)_2SnBr_2$	57	
p-$C_2H_5OCOC_6H_4$	$SnBr_2$	CH_3COCH_3, reflux, 3 hr	p-$C_2H_5OCOC_6H_4HgBr$	—	14
p-$(CH_3)_2NC_6H_4$	$[(C_2H)_3Sn]_2$	100°C, 30 min; 150–170°C, 5 hr	Hg	Quant.	8
			$(C_2H_5)_3SnC_6H_4N(CH_3)_2$-$p$	95	
α-$C_{10}H_7$	$SnCl_2$	Ligroin, 90–95°C, 20 hr	Hg	53	7
			$(\alpha$-$C_{10}H_7)_2SnCl_2$	Quant.	
α-$C_{10}H_7$	$SnCl_2$	CH_3COCH_3, reflux, 30 min	Hg	66	4
			$(\alpha$-$C_{10}H_7)_2SnCl_2$	High	
α-$C_{10}H_7$	$SnCl_2$	C_2H_5OH, reflux	$C_{10}H_8$	Trace	4
α-$C_{10}H_7$	$SnBr_2$	CH_3COCH_3, reflux, 40 min	$C_{10}H_8$	Trace	4
			$(\alpha$-$C_{10}H_7)_2SnBr_2$	65	
			Hg	Low	
α-$C_{10}H_7HgBr$	$SnBr_2$	CH_3COCH_3, reflux, 3 hr	$C_{10}H_8$	—	4
β-$C_{10}H_7$	$SnCl_2$	CH_3COCH_3, reflux, 45 min	Hg	62–70	4
			$(\beta$-$C_{10}H_7)_2SnCl_2$	High	
β-$C_{10}H_7$	$SnCl_2$	C_2H_5OH	$C_{10}H_8$	Quant.	4
β-$C_{10}H_7$	$SnBr_2$	CH_3COCH_3, reflux, 30 min	Hg	59	4
			$(\beta$-$C_{10}H_7)_2SnBr_2$	Quant.	
β-$C_{10}H_7HgBr$	$SnBr_2$	CH_3COCH_3, reflux, 3 hr	HgBr	—	4
			Mixture organo-Sn-compounds		

a NaOH treatment. b Another order addition. c NH₄OH treatment. d Freshly prepared. e Stand, 2.5 month.

238

E. Preparation of Diphenyldiethyltin [8]

Diethyltin (8.5 g, 0.05 m) was heated for half an hour at 150°C on an oil bath with 17 g (0.05 m) of diphenylmercury. Metallic mercury (7.5 g, 75% of theoretical yield) was formed during this time. On cooling, the reactants were driven off and the residue fractionated. The main fraction, b.p. 154–156°C (4 mm), consisted of 9.3 g (60% yield) diphenyldiethyltin.

F. Preparation of Triethyltin by the Reaction of Hexaethylditin with Phenylmercury(II) Chloride [8]

Hexaethylditin (12.4 g, 0.03 mole) heated on an oil bath with 9.4 g (0.03 m) of phenylmercury chloride for 3 hr at 150–160°C. The reaction proceeds slowly. The solution was decanted from metallic mercury and fractionated *in vacuo*. The fraction boiling at 113–114°C (6 mm) was pure triethylphenyltin (30% yield). The fractions at lower temperatures were treated with potassium hydroxide and the triethyltin(II) chloride was identified as the hydroxide, m.p. 44°C.

G. Reaction of Hexaethylditin with the Methylester of Mercurybisacetic Acid [9]

Hexaethylditin (10.5 g) and 8.79 g of the methylester of mercurybisacetic acid were heated under nitrogen at 160°C for one hour. The liquid was decanted from mercury (3.4 g, 68%). Isolated by distillation were 5.6 g (41%) of the methylester of triethyltinacetic acid with b.p. 69–71°C (1.8 mm), n_D^{20} 1.4836, d_4^{20} 1.2967.

REFERENCES

1. G. Jacobs, *Compt. Rend.*, **238**, 1825 (1954).
2. L. I. Emelyanova and L. G. Makarova, *Izv. Akad. Nauk SSSR, Otd. Khim. Nauk*, **1960**, 2067.
3. L. I. Emelyanova, V. N. Vinogradova, L. G. Makarova, and A. N. Nesmeyanov, *Izv. Akad. Nauk SSSR, Otd. Khim. Nauk*, **1962**, 53.
4. A. N. Nesmeyanov and K. A. Kocheshkov, *Zh. Russk. Khim. Obshch.*, **62**, 1795 (1930); *Ber.*, **63**, 2496 (1930).
5. K. A. Kocheshkov and A. N. Nesmeyanov, *Zh. Obshch. Khim.*, **1**, 219 (1931); *Ber.* **64**, 628 (1931).
6. A. N. Nesmeyanov, A. E. Borisov, and A. N. Abramova, *Izv. Akad. Nauk SSSR, Otd. Khim. Nauk*, **1946**, 647.
7. A. N. Nesmeyanov, A. E. Borisov, N. V. Novikova, and M. A. Osipova, *Izv. Akad. Nauk SSSR, Otd. Khim. Nauk*, **1959**, 263.
8. A. N. Nesmeyanov, K. A. Kocheshkov, and V. P. Puzyreva, *Zh. Obshch. Khim.*, **7**, 118 (1937).

9. I. F. Lutsenko, Yu. I. Baukov, and B. N. Khasapov, *Zh. Obshch. Khim.*, **33**, 2724 (1963).
10. A. N. Nesmeyanov, A. E. Borisov, and N. V. Novikova, *Izv. Akad. Nauk SSSR, Otd. Khim. Nauk*, **1959**, 1216.
11. A. N. Nesmeyanov, A. E. Borisov, and N. V. Novikova, *Izv. Akad. Nauk SSSR, Otd. Khim. Nauk*, **1959**, 259.
12. A. N. Nesmeyanov and T. A. Kudryavtseva, *Uchen. zap. Moskov. Gos. Univer.*, **151**, 57 (1951).
13. A. N. Nesmeyanov, A. E. Borisov, and N. A. Vol'kenau, *Izv. Akad. Nauk SSSR, Otd. Khim. Nauk*, **1956**, 162.
14. I. T. Eskin, A. N. Nesmeyanov, and K. A. Kocheshkov, *Zh. Obshch. Khim.*, **8**, 35 (1938).

Chapter 5

Reaction with Hydrides and Alkyl and Arylhydrides of Elements and Their Metallic Derivatives

I. INTRODUCTION

The hydrides and alkyl and arylhydrides of various elements react with organomercury compounds. In these reactions, the hydride is replaced by the organic substituent of the organomercury compound while the mercury in R_2Hg or $RHgX$ is reduced by the hydride to metallic mercury.

Simple hydrides are used only in the case of aluminum in the form of their amine complexes. "Double hydrides" such as $LiAlH_4$ or alkylhydrides (and their lithium derivatives) of some elements of groups III and IV of the periodic table are also used as reactants. These reactions are not widely used since they are for particular cases only. (See also Chapter 7.)

II. APPLICATION OF THE REACTION

A. Reactions with Group III Hydrides

1. Synthesis of Organoaluminum Compounds

Aluminum compounds such as lithium aluminum hydride and aluminum hydride complexes with tertiary amines react with organomercury compounds to form compounds containing a C—Al bond (see Chapter 3 for reactions with metallic aluminum).

Divinylmercury, heated with $LiAlH_4$, produced $LiAl(CH=CH_2)_4$. Depending on the relative amounts of divinylmercury and $AlH_3 \cdot N(CH_3)_3$ taken, various products were obtained: $(CH_2=CHAlH_2 \cdot N(CH_3)_3$ or $(CH_2=CH)_3Al \cdot N(CH_3)_3$.[1] $AlH_3 \cdot N(CH_3)_3$ reacts with diperfluorvinylmercury(II) in ether, and produces $(CF_2=CF)_3Al \cdot N(CH_3)_3$.[2]

The reaction product of 4-camphylmercury(II) chloride and $LiAlH_4$ in ether was decomposed by water and produced camphane.[3] Diphenylmercury reacted with $LiAlH_4$ (ether, room temperature) and resulted in the formation of $LiAlH_2(C_6H_5)_2$ and $LiAl(C_6H_5)_4$.[4] Benzene was produced when the reaction products were decomposed by water.[5]

B. Reactions with Hydrides (and Metallic Derivatives) of Group IV Elements

1. Syntheses of Organosilicon Compounds

During the reaction of diarylmercury compounds or $ArHgBr$ with $(C_6H_5)_3SiLi$ (in tetrahydrofuran at room temperature) $(C_6H_5)_3SiAr$ and

$[(C_6H_5)_3Si]_2$ were produced. After treatment with water, hydrolysis products of $(C_6H_5)_3SiLi$ and metallic mercury were obtained.[6]

2. Syntheses of Organogermanium Compounds

Lutsenko and Baukov[7] have found that the products of the reaction of dialkylgermane with the methyl ester of mercurybisacetic acid depends on the reagent ratios. If equimolar amounts of reagents are used, a 60% yield of the methyl ester of dialkylgermaniumacetic acid was produced[7]:

$$R_2GeH_2 + Hg(CH_2COOCH_3)_2 \longrightarrow$$
$$R_2Ge(H)CH_2COOCH_3 + Hg + CH_3COOCH_3$$
$$(R = n\text{-}C_3H_7, n\text{-}C_4H_9)$$

If a 1:2 reagent ratio was used, the following reaction occurs:

$$R_2GeH_2 + 2Hg(CH_2COOCH_3)_2 \longrightarrow$$
$$R_2Ge(CH_2COOCH_3)_2 \cdot + Hg + 2CH_3COOCH_3$$

Both reactions are conducted in boiling tetrahydrofuran.[7] R_2Hg also reacts with R_2GeClH (see tables and Chapter 7).

Trialkylgermane reacts with esters of mercurybisacetic acid as well, producing esters of trialkylgermanium acetic acid under similar reaction conditions.[8]

3. Syntheses of Organotin Compounds

Trialkyltin(IV) hydride reacts with esters of mercurybisacetic acid:

$$(C_2H_5)_3SnH + Hg(CH_2COOR)_2 \longrightarrow$$
$$(C_2H_5)_3SnCH_2COOR + Hg + CH_3COOR$$

This reaction occurs at 70°C and produces a higher product yield than in the case with trialkylgermane. Under more drastic conditions (100–140°C) the reaction proceeds with other R_2Hg compounds: reaction of triethyltin hydride with diethyl- and dibenzylmercury produced RH, Hg, and $(C_2H_5)_6Sn_2$:[9,10] However, aromatic compounds of mercury arylate triethyltin hydride[10]

$$2(C_2H_5)_3SnH + Ar_2Hg \longrightarrow 2(C_2H_5)_3SnAr + Hg + H_2$$
$$Ar = C_6H_5, p\text{-}CH_3C_6H_4$$

III. PREPARATIVE SYNTHESES

A. Reactions of Dibutylgermane with the Methyl Ester of Mercurybisacetic Acid

1. 1:1 Reagent Ratio

The methyl ester of mercurybisacetic acid (17.4 g in 50 ml of tetrahydrofuran) was added dropwise to 10.6 g of dibutylgermane in 10 ml of tetrahydrofuran and boiled for 30 min. The solution was decanted from

TABLE IV

Reactions with Hydrides and Alkyl and Arylhydrides of Elements and Their Metallic Derivatives

R in R_2Hg or RHgX	Reactant	Reaction conditions	Products	Yield, %	Ref.
trans-2-(ClHg)-bicyclo[2.2.2.]-octan-3-ol	$NaBH_4$	$(C_2H_5)_2O$, 20°C, 14 hr	bicyclo[2.2.2.]-octan-2-ol	25	4
$CH_2=CH$	$AlH_3 \cdot N(CH_3)_3$	$(C_2H_5)_2O$, 25°C, 12 hr	$(CH_2=CH)_3Al \cdot N(CH_3)_3$ H_2 Hg	Quant. — —	1
$CH_2=CH$	$LiAlH_4$	$(C_2H_5)_2O$, reflux, overnight	$(CH_2=CH)_4AlLi$ H_2 Hg	Quant. — —	1
$CF_2=CF$	$AlH_3 \cdot N(CH_3)_3$	$(C_2H_5)_2O$, 20°C, 16 hr	$(CF_2=CF)_3Al \cdot N(CH_3)_3$ H_2 Hg	— — —	2
4-Camphylmercury-chloride	$LiAlH_4$	$(C_2H_5)_2O$, 20°C, 12 hr	Camphane	19	3
C_6H_5	$LiAlH_4$	$(C_2H_5)_2O$, 20°C, 2 hr. $HgCl_2$	C_6H_5HgCl Hg	60 98	4
C_6H_5	$LiAlH_4$	$(C_2H_5)_2O$, 20°C, 15 min	H_2 Hg	90 68	4
C_6H_5	$LiAlH_4$	$(C_2H_5)_2O$, reflux, 0.5 hr	Hg C_6H_6	100 —	5
C_6H_5HgCl	$LiAlH_4$	$(C_2H_5)_2O$, 20°C	C_6H_6[a]	—	4
C_6H_5	$(C_6H_5)_3SiLi$	THF, 20°C	$(C_6H_5)_4Si$ Hg	69 —	6
C_6H_5HgBr, $1M$	$(C_6H_5)_3SiLi$, $1M$	THF, 20°C	$(C_6H_5)_4Si$ Hg $[(C_6H_5)_3Si]_2$ $(C_6H_5)_3SiOH$[b]	73 78 5 2	6

(continued)

TABLE IV (*Continued*)

R in R₂Hg or RHgX	Reactant	Reaction Conditions	Products	Yield, %	Ref.
C_6H_5HgBr, $1M$	$(C_6H_5)_3SiLi$, $2M$	THF, −70°C	$(C_6H_5)_4Si$	15	6
			$[(C_6H_5)_3Si]_2$	2	
			$(C_6H_5)_3SiOH^b$	55	
			Hg	50	
$p\text{-}CH_3C_6H_4$, $1M$	$(C_6H_5)_3SiLi$, $1M$	THF, 20°C	$(C_6H_5)_3SiOH^b$	6	6
			$(C_6H_5)_3Si(C_6H_4CH_3\text{-}p)^b$	88	
			$p\text{-}CH_3C_6H_4COOH^b$	71	
$p\text{-}CH_3C_6H_4$, $1M$	$(C_6H_5)_3SiLi$, $2M$	THF, 0°C	$p\text{-}CH_3C_6H_4Si(C_6H_5)_3$	46	6
			$p\text{-}CH_3C_6H_4COOH^b$	51	
			$(C_6H_5)_3SiOH^b$	46	
CH_2COOCH_3	$(C_3H_7)_2GeH_2$	THF, reflux, 1 hr	$(C_3H_7)_2Ge(H)CH_2COOCH_3$	—	7
CH_2COOCH_3	$(C_3H_7)_2Ge(H)Cl$	THF, reflux, 2.5 hr	Hg	57	11
CH_2COOCH_3, $1M$	$(C_4H_9)_2GeH_2$, $1M$	THF, reflux, 0.5 hr	$(C_4H_9)_2Ge(Cl)CH_2COOCH_3$	45	7
			$(C_4H_9)_2Ge(H)CH_2COOCH_3$	62	
			Hg	89	
CH_2COOCH_3, $2M$	$(C_4H_9)_2GeH_2$, $1M$	THF, reflux, 1 hr	$(C_4H_9)_2Ge(CH_2COOCH_3)_2$	52	7
			Hg	98	
CH_2COOCH_3	$(C_4H_9)_2Ge(H)CH_2COOCH_3$	THF, reflux, 0.5 hr.	$(C_4H_9)_2Ge(CH_2COOCH_3)_2$	78	7
			Hg	92	
$CH_2COOC_3H_7$	$(C_3H_7)_3GeH$	THF, reflux, 1 hr	$(C_3H_7)_3GeCH_2COOC_3H_7$	41	8
$CH_2COOC_3H_7$	$(n\text{-}C_4H_9)_3GeH$	THF, reflux, 1 hr	$(n\text{-}C_4H_9)_3GeCH_2COOC_3H_7$	65	8
$CH_3HgOCOCH_3$	$(C_2H_5)_3SnH$	Sealed tube, 100°C, 2 hr	$(C_2H_5)_3SnOCOCH_3$	65	10
			CH_4	77	
			Hg	70	
C_2H_5	$(C_2H_5)_3SnH$	Sealed tube, 100°C, 3–5 hr	C_2H_6	Quant.	9, 10
			Hg	Quant.	
			$(C_2H_5)_6Sn_2$	86	
			Hg	78	
C_2H_5HgCl	$(C_2H_5)_3SnH$	C_6H_6, 20°C	C_2H_6	62	9
			$(C_2H_5)_3SnCl$	41	

Reactant	Tin reagent	Conditions	Products	Yield (%)	Ref.
C_2H_5HgCl	$(C_2H_5)_3SnH$	Toluene 100°C, 2 hr	Hg	78	10
			$(C_2H_5)_3SnCl$	41	
			C_2H_6	62	
CH_2COOCH_3	$(C_2H_5)_2Sn(H)Cl$	75°C, 1.5 hr	$(C_2H_5)_2Sn(Cl)CH_2COOCH_3$	73	11
CH_2COOCH_3	$(C_2H_5)_3SnH$	75°C, 1.5 hr	$(C_2H_5)_3SnCH_2COOCH_3$	68	8
$CH_2COOC_3H_7$	$(C_2H_5)_3SnH$	—	Hg	96	
			$(C_2H_5)_3SnCH_2COOC_3H_7$	73	8
			$CH_3COOC_3H_7$	51	
CH_2COOCH_3	$(C_3H_7)_2Sn(H)Cl$	20°C, 0.5 hr	Hg	97	
			$(C_3H_7)_2Sn(Cl)CH_2COOCH_3$	77	11
CH_2COOCH_3	$(n\text{-}C_4H_9)_2Sn(H)Cl$	—	$(n\text{-}C_4H_9)_2SnClCH_2COOCH_3$	40	11
$C_6H_5CH_2$	$(C_2H_5)_3SnH$	Sealed tube, xylene, 125°C, 17 hr	Hg	97	
			$(C_2H_5)_6Sn_2$	68	10
C_6H_5	$(C_2H_5)_3SnH$	Sealed tube, xylene, 140°C, 22 hr	H_2	52	
			Hg	93	
			$(C_2H_5)_3Sn(C_6H_5)$	46	10
			H_2	90	
$p\text{-}CH_3C_6H_4$	$(C_2H_5)_3SnH$	Sealed tube, toluene, reflux, 85 hr	Hg	80	
			$(C_2H_5)_3SnC_6H_4CH_3\text{-}p$	62	10

^a Treatment CO_2, hydrolyze.

^b Treatment dilute HCl.

the mercury produced (89%) during the reaction. The following compounds were isolated by distillation: 0.8 g of unreacted dibutylgermane, b.p. 37–40°C (2 mm), n_D^{20} 1.21420, d_4^{20} 0.9787; 8.4 g (62%) of the methyl ester of dibutylgermaneacetic acid, b.p. 73–74.5°C (1 mm), n_D^{20} 1.4552, d_4^{20} 1.0613; and 1 g (6%) of the methyl ester of dibutylgermaniumbisacetic acid, b.p. 124–128°C (2 mm), n_D^{20} 1.4640, d_4^{20} 1.1201.

2. 1:2 Reagent Ratio

In a preparation similar to that above, except that a 1:2 reagent ratio was used and the reactants were boiled for 1 hr, the following compound was isolated by distillation: 2.1 g (52%) of the methyl ester of dibutyl-germaniumbisacetic acid.

If dipropylgermane is reacted with the methyl ester of mercurybisacetic acid in a 1:1 reagent ratio, the methyl ester of dipropylgermaneacetic acid was produced (as in part 1, above).

B. Reaction of the Methyl Ester of Dibutylgermane acetic Acid with the Methyl Ester of Mercurybisacetic Acid[7]

To a warmed solution of 3.2 g of the methyl ester of mercurybisacetic acid in 3 ml of tetrahydrofuran was added dropwise 2.5 g of the methyl ester of dibutylgermaneacetic acid. The mixture was boiled for 30 min and the solution was decanted from the residue of metallic mercury (92%). During distillation, 78% yield of the methyl ester of dibutyl-germanium bisacetic acid, b.p. 123–126°C (2 mm) was isolated.

C. Reaction of Triethyltin Hydride with the Propyl Ester of Mercurybisacetic Acid[8]

The propyl ester of mercury bisacetic acid (24.3 g) and 12.5 g of tri-ethyltin(IV) hydride are heated at 90°C under nitrogen for 90 min followed by decantation of the solution from the mercury formed (11.5 g, 96%). During distillation, 3.1 g of the propyl ester of acetic acid (51%) (b.p. 99–101°C at 743 mm, n_D^{20} 1.3839, d_4^{20} 0.8856) and 13.2 g (73%) of the propyl ester of triethyltin(IV) acetic acid, b.p. 102–105°C (3.5 mm) n_D^{20} 1.4779, d_4^{20} 1.2226 were isolated.

In a similar manner, the methyl ester of triethyltin acetic acid was produced: 68% yield, b.p. 71–73°C (2 mm) n_D^{20} 1.4831, d_4^{20} 1.2962.

REFERENCES

1. B. Bartocha, A. J. Bilbo, D. E. Bublitz, and M. I. Gray, Z. Naturforsch., 16B, 357 (1961).
2. B. Bartocha and A. J. Bilbo, J. Am. Chem. Soc., 83, 2202 (1961).
3. S. Winstein and T. G. Traylor, J. Am. Chem. Soc., 78, 2597 (1956).

4. T. G. Traylor, *Chem. Ind.*, **1959**, 1223; *J. Am. Chem. Soc.*, **86**, 244 (1964).
5. D. H. R. Barton and W. J. Rosenfelder, *J. Chem. Soc.*, **1951**, 2381.
6. M. V. George, G. D. Lichtenwalter, and H. Gilman, *J. Am. Chem. Soc.*, **81**, 978 (1959).
7. Yu. I. Baukov and I. F. Lutsenko, *Zh. Obshch. Khim.*, **34**, 3453 (1964).
8. I. F. Lutsenko, Yu. I. Baukov, and B. N. Khasapov, *Zh. Obshch. Khim.*, **33**, 2724 (1963).
9. N. S. Vyazankin, G. A. Razuvaev, and S. P. Korneva, *Zh. Obshch. Khim.*, **33**, 1041 (1963).
10. N. S. Vyazankin, G. A. Razuvaev, and S. P. Korneva, *Zh. Obshch. Khim.*, **34**, 2787 (1964).
11. Yu. I. Baukov, I. Yu. Belavin, and I. F. Lutsenko, *Zh. Obshch. Khim.*, **35**, 1092 (1965).

Chapter 6

Reactions with Elements of the Sulfur Group

I. APPLICATION OF THE REACTION

Organomercury compounds are dealkylated (dearylated) by elements of the sulfur group, S, Se, and Te, producing dialkyl (diaryl)sulfides, selenides, and tellurides. In these reactions mercury is not isolated as metallic mercury but rather as mercury(II) sulfide, selenide, or telluride.

These reactions are represented by the following general equation, which constitutes a very important method of synthesis of R_2El:

$$R_2Hg + 2El \longrightarrow R_2El + MgEl \ (El = S, Se, Te)$$

These reactions are usually carried out at high temperatures (200°C and above) in sealed tubes. Dibenzylmercury heated with sulfur at 100°C does not produce R_2S but is decomposed producing dibenzyl.[1] The reaction of secondary perfluoralkyl compounds of mercury with sulfur results in the synthesis of perfluorothioketones,[2] for example:

$$[(CF_3)_2CF]_2Hg \xrightarrow{\ S\ } CF_3C(S)CF_3$$

These reactions proceed under drastic conditions, i.e., the boiling point of sulfur.

Reactions of RHgCl with sulfur proceed in suitable solvents as follows:

$$2RHgCl + 2S \longrightarrow RSSR + Hg_2Cl_2 \quad (R = Alk \text{ or } Ar)$$

with formation of disulfides with good yields.[10]

Reaction of $[o\text{-BrC}_6F_4]_2Hg$ with S or Se at high temperature leads to substitution of S or Se for Hg and Br and to the formation of octafluorothianthrene and -selenanthrene.[11] SO_2 inserts into the R—Hg bond of R_2Hg giving $RHgSO_2R$.[12,13]

II. PREPARATIVE SYNTHESES

A. Preparation of Di-o-tolylsulfide[3]

Di-o-tolylmercury (5 g) and 0.8 of finely crushed sulfur are heated in a sealed tube filled with carbon dioxide for 12 hr at a temperature of 225–230°C (for the last 4 hr at 285°C). Di-o-tolylsulfide is extracted from the reaction mixture by ether and repeatedly distilled in vacuum, wherein it is crystallized and finally salted out from alcohol, m.p. 64°C, b.p. 175°C (16 mm).

TABLE V

Reactions with Elements of the Sulfur Group

R in R_2Hg or RHgX	Element	Reaction conditions	Products	Yield, %	Ref.
C_2F_5	S	Sealed tube, 250°C, 8 hr	$(CF_3CF_2S)_2$	—	4
$(CF_3)_2CF$	S	Reflux	$(CF_3)_2C{=}S$	60	2
CF_3CFCl	S	Reflux	$CF_3C(S)F$	80	2
$C_6H_5CH_2$	S	Sealed tube, 100°C, 24 hr; 150°C, 1 hr	$(C_6H_5CH_2)_2$	quant.	1
C_6H_5	S	Sealed tube, 220–230°C, 6 hr	$(C_6H_5)_2S$	66	7, 5
C_6H_5	S	Fused	$(C_6H_5)_2S$	—	
			$(C_6H_5S)_2Hg$	—	6
			C_6H_5SH	—	
C_6F_5	S	Sealed tube, 250°C, 7 days	$(C_6F_5)_2S$	—	8
$o\text{-}CH_3C_6H_4$	S	Sealed tube, 225°C, 12 hr	$(o\text{-}CH_3C_6H_4)_2S$	—	3
C_6F_5HgCl	S	Sealed tube, 250°C, 6 days	$(C_6F_5)_2S$	—	8
C_6H_5	Se	Sealed tube, 220°C, 4 hr	$(C_6H_5)_2Se$	85	7
C_6F_5	Se	Sealed tube, 240°C, 7 days	$(C_6F_5)_2Se$	75	8
$o\text{-}CH_3C_6H_4$	Se	Sealed tube, 220°C, 12 hr	$(o\text{-}CH_3C_6H_4)_2Se$	—	3
$p\text{-}CH_3C_6H_4$	Se	Sealed tube, 225°C, 12 hr	$(p\text{-}CH_3C_6H_4)_2Se$	—	3
$\alpha\text{-}C_{10}H_7$	Se	Sealed tube, 16 mm, 190°C, 12 hr	$(\alpha\text{-}C_{10}H_7)_2Se$	54	9
C_6H_5	Te	Sealed tube, 220°C	$(C_6H_5)_2Te$	70	7
$o\text{-}CH_3C_6H_4$	Te	Sealed tube, 225–230°C, 12 hr	$(o\text{-}CH_3C_6H_4)_2Te$	—	3
$p\text{-}CH_3C_6H_4$	Te	Sealed tube, 225–230°C, 15 hr	$(p\text{-}CH_3C_6H_4)_2Te$	—	3
$\alpha\text{-}C_{10}H_7$	Te	Sealed tube, 16 mm, 190°C, 8 hr	$(\alpha\text{-}C_{10}H_7)_2Te$	53	9
			$C_{10}H_8$	Trace	

B. Preparation of Di-*o*-tolylselenide[3]

Thoroughly stirred portions of di-*o*-tolylmercury and selenium (5% excess amount) in a sealed tube filled with carbon dioxide are heated for 12 hr at 220°C and then for a short time at 235–240°C. The di-*o*-tolylselenide produced was obtained by solvent extraction (ether or benzene) and repeatedly distilled in vacuum affording a brilliant white crystalline product, m.p. 61–62°C, b.p. 186°C (16 mm).

C. Preparation of Di-*o*-tolyltelluride[3]

Di-*o*-tolylmercury and tellurium are placed in a sealed tube filled with carbon dioxide and are heated for 12 hr at 225–235°C. Extraction by ether followed by vacuum distillation yields di-*o*-tolyltelluride, m.p. 37–38°, b.p. 202.5°C (16 mm).

REFERENCES

1. G. A. Razuvaev and M. M. Koton, *Zh. Obshch. Khim.*, **5**, 361 (1935).
2. W. J. Middleton, E. J. Howard, and W. H. Sharkey, *J. Am. Chem. Soc.*, **83**, 2589 (1961).
3. H. Zeiser, *Ber.*, **28**, 1670 (1895).
4. USP 2844614; *Chem. Abstr.*, **53**, 3061 (1959).
5. E. Dreher and R. Otto, *Ber.*, **2**, 542 (1869).
6. E. Dreher and R. Otto, *Ann. Chem.*, **154**, 93 (1870).
7. F. Kraft and R. E. Lyons, *Ber.*, **27**, 1768 (1894).
8. J. Burdon, P. L. Coe, and M. Fulton, *J. Chem. Soc.*, **1965**, 2094.
9. R. E. Lyons and G. C. Bush, *J. Am. Chem. Soc.*, **30**, 831 (1908).
10. J. M. La Roy and E. C. Kooyman, *J. Organometal. Chem.*, **7**, 357 (1967).
11. S. C. Cohen and A. G. Massey, *J. Organometal. Chem.*, **12**, 341 (1968).
12. M. A. D. Carey and H. C. Clarke, *Can. J. Chem.*, **46**, 649 (1968).
13. P. J. Pollick, J. P. Bibler, and A. Wojicicki, *J. Organometal. Chem.*, **16**, 201 (1969).

Chapter 7

Reactions with Reducing Agents—Hydrogen Donors

I. INTRODUCTION

Mercury in organomercury compounds may be reduced to Hg by reducing agents; the ease of the reaction varies with the reducing agent. The reducing agents are hydrogen donors and substitute hydrogen for mercury in the organomercury compounds. This homolytic substitution may be used for determining the structure of the organomercury compound and is suitable in cases where heterolytic reactions such as acid hydrolysis cannot be used, i.e., in quasicomplex compounds of mercury.

II. APPLICATION OF THE REACTION

A. Reaction with Molecular Hydrogen

In order to study the mechanism of the decomposition of organometallic compounds, Razuvaev et al.[1] studied the reaction of organomercury compounds with molecular hydrogen. High pressures and temperatures are necessary for this reaction (50 atm hydrogen and 175–200°C). Diphenylmercury decomposes according to the following equation:

$$(C_6H_5)_2Hg + H_2 \longrightarrow 2C_6H_6 + Hg$$

Di-β-phenylethylmercury yields ethylbenzene.[2] However, under similar reaction conditions, dibenzylmercury yields only bibenzyl and not the expected toluene. This reaction difference is also observed if the reactions were carried out in a solvent such as alcohol or ligroin.

The reaction occurs in the presence of catalysts such as Group VIII metals or Group I noble metals at considerably lower temperatures although hydrogen pressure is still needed. The different products as noted above are also observed in these cases.[3]

The substitution of hydrogen for mercury in bispolyfluoroalkylmercury compounds occurs in the presence of metallic aluminum at room temperatures (long reaction times are required).[4] The reaction occurs, not due to the intermediate formation of organoaluminum compounds, but rather to the presence of water which reacts with the aluminum to form hydrogen:

$$(CF_3CHF)_2Hg \xrightarrow[25°C, 12\ hr]{Al,\ H_2O} 2CF_3CH_2F + Hg$$

This reaction (as distinct from the aforementioned reactions with molecular hydrogen, which have no preparative significance) may be considered to be a supplementary method for the synthesis of some saturated fluoro-hydrocarbons:

$$CF_2{=}CFH \xrightarrow{\ HgF_2\ } (CF_3CFH)_2Hg \xrightarrow{\ Al/H_2O\ } CF_3CFH_2$$

B. Reactions with Alcohols and Other Compounds which are Subject to Dehydration

Substances subject to dehydration reduce R_2Hg to RH at high temperatures and pressures similar to the reaction with molecular hydrogen.[1,5] The reactions are facilitated by the presence of metal catalysts, especially Pt and Pd.[6] The effect of several factors on the reaction process of hydrogen substitution for mercury by dehydrating substances has been thoroughly studied. Depending on the case of the homolytic cleavage of the organomercury compounds in the presence of catalysts in alcohol or tetralin, the substituent organic group R may be arranged in the following order[6]: $C_6H_5CH_2 > C_6H_5 > p\text{-}CH_3OC_6H_4 > p\text{-}C_2H_5OC_6H_4 > p\text{-}BrC_6H_4 > p\text{-}CH_3C_6H_4 > \alpha\text{-}C_{10}H_7$.

The effect of alcohols on the decomposition rate of $(C_6H_5)_2Hg$ (200°C, 6 hr) are arranged in the following order[7]: $CH_3OH > C_5H_{11}OH > C_6H_5CH_2OH > C_3H_7OH > C_2H_5OH > t\text{-}C_4H_9OH > n\text{-}C_4H_9OH$.

In diols, the decomposition of $(C_6H_5)_2Hg$ is considerably easier than in alcohols with the same number of carbon atoms.[5]

For alcohols which substitute hydrogen for mercury in phenylmercury halides, the ease of decomposition of mercury salts is as follows: $F > I > Br > Cl$.[8] The ease of reaction of RHgBr in ethanol varies according to the R substituent as follows[9]: $\alpha\text{-}C_{10}H_7 > n\text{-}C_3H_7 > C_2H_5 > n\text{-}C_4H_9 > C_6H_5$.

If the organic substituents are acids, the ease of decomposition of $C_6H_5HgOOCR$ in alcohols is as follows[10]: $HCOO > C_6H_4COO > C_3H_6(OH)COO > C_2H_4(OH)COO > CH_3COO > C_2H_5COO > C_3H_7COO > C_5H_{11}COO > C_{17}H_{35}COO > C_6H_5COO$. Alcohols also affect the decomposition rate of $C_6H_5HgOOCR$ as follows: $i\text{-}C_5H_{11}OH > C_2H_5OH > CH_3OH$.

Decomposition of C_6H_5HgOH in C_2H_5OD produces Hg, acetaldehyde and benzene which do not include deuterium.[11] The OH group in organomercury compounds thus is probably the hydrogen donor. During the thermal decomposition of R_2Hg containing a secondary or tertiary alcohol group in the γ-position with respect to the mercury atom, the secondary or tertiary alcohol, in addition to the respective ketone, is formed[12]:

$$(C_6H_5CH(OH)CH_2CH_2)_2Hg \longrightarrow [Hg + 2C_6H_5CHOHCH_2CH_2] \longrightarrow$$
$$Hg + C_6H_5COCH_2CH_3 + C_6H_5CHOHCH_2CH_3$$

R_2Hg ($R = C_6H_5CH_2$, $CH_2COC_2H_5$, $CH_2CO_2CH_3CH_2CHO$) reacts quite easily with phenylphosphine according to the equation[13]:

$$2C_6H_5PH_2 + 2R_2Hg \longrightarrow (C_6H_5P)_4 + 2Hg + 4RH$$

Phenylarsine substitutes hydrogen for mercury in $(C_6H_5)_2Hg$ and C_6H_5HgCl, i.e.:

$$2C_6H_5AsH_2 + 2C_6H_5HgCl \longrightarrow 2C_6H_6 + 2Hg + 2HCl + C_6H_6As = AsC_6H_5$$

The reaction with $(C_6H_5)_2Hg$ occurs in benzene at room temperature while with C_6H_5HgCl boiling in benzene for 2 hr is required.[14]

C. Reduction by Hydrides and Hydride Derivatives of Group III and IV Elements

Of preparative significance is the formation of RH by the reaction of R_2Hg or RHgCl with reducing agents such as $LiAlH_4$,[15-18] $NaBH_4$,[19-23] $LiBH_4$ or $NaBH(OCH_3)_3$[24] under mild conditions. The intermediates $LiAlR_4$, $NaBR_4$, etc., are formed which are then hydrolyzed by acid.

Oxymercuration products of cyclic olefins and their derivatives have been demercurated using $NaBH_4$ but without isolating the products.[92-94,103-106]

Regeneration of the cycloalkene rather than substitution of hydrogen for mercury takes place during the reaction of $LiAlH_4$ with products of the addition products of mercury salts to alicyclic compound double bonds.

The reaction of triethylsilane with $(C_2H_5)_2Hg$ with formation of ethyl (triethylsilyl)mercury, bis-(triethylsilyl)mercury and metallic mercury proceeds slowly under vigorous reaction conditions (130–140°C, 117 hr).[25] The primary product of the reaction is ethyl (triethylsilyl)mercury:

$$(C_2H_5)_3SiH + (C_2H_5)_2Hg \longrightarrow C_2H_6 + (C_2H_5)_3SiHgC_2H_5$$

Triethylsilane reduces RHgCl (R = furyl, thienyl, phenyl) to RH.[26] The increasing order of reactivity is as follows: $C_6H_5 < C_4H_3O < C_4H_3S$. The reaction occurs rapidly in pyridine or ethanol. Difurylmercury is not reduced by triethylsilane in pyridine.[26]

The reaction of $(C_2H_5)_3GeH$ with $(C_2H_5)_2Hg$ (100–120°C) proceeds as follows[27]:

$$2(C_2H_5)_3GeH + (C_2H_5)_2Hg \longrightarrow 2C_2H_6 + (C_2H_5)_3GeHgGe(C_2H_5)_3$$

Dialkylhalogermanes exchange hydrogen for the CH_2COOCH_3 radical (but not chlorine) with formation of the acetic acid ester by reaction with the ester of mercurybisacetic acid.[28]

$$R_2GeClH + Hg(CH_2COOCH_3)_2 \longrightarrow$$
$$R_2Ge(Cl)CH_2COOCH_3 + Hg + CH_3COOCH_3$$
$$(R = n\text{-}C_3H_7, n\text{-}C_4H_9)$$

The reaction of dipropylchlorotin hydride with the ester of mercury bisacetic acid occurs in a similar manner.[28] Analogous reactions of R_2GeH_2 were discussed in Chapter 5.

The reactions of triethyltin hydride with ethylmercury chloride or with methylmercury acetate are represented by the following equations[29]:

$$(C_2H_5)_3SnH + C_2H_5HgCl \longrightarrow Hg + (C_2H_5)_3SnCl + C_2H_6$$

$$(C_2H_5)_3SnH + CH_3HgOCOCH_3 \longrightarrow Hg + (C_2H_5)_3SnOCOCH_3 + CH_4$$

Both reactions are exothermic and are carried out in toluene by heating at $100°C$ for 2 hr.

Triethyltin hydride reacts with R_2Hg under more vigorous reaction conditions. Depending on the nature of the R group, the reaction may follow two possible directions[29,30]:

$$R_2Hg + 2(C_2H_5)_3SnH \begin{cases} \longrightarrow 2RH + Hg + (C_2H_5)_6Sn_2 & (a) \\ \longrightarrow H_2 + Hg + 2(C_2H_5)_3SnR & (b) \end{cases}$$

Reaction path (a) is followed if R is ethyl or benzyl and reaction path (b) is followed if R is phenyl or p-tolyl.

Only one hydrogen is transferred in the reaction of the esters of mercury-bisacetic acid with triethyltin hydride[31]:

$$(C_2H_5)_3SnH + Hg(CH_2COOR)_2 \longrightarrow CH_3COOR + Hg + (C_2H_5)_3SnCH_2CO_2R$$

See Chapter 5 for a discussion of reactions with alkylhydrides of Group IV elements.

D. Substitution of Hydrogen for Mercury by Action of Symmetrizators

Some organomercury salts, particularly multiple-bond addition products, under the action of reducing and symmetrizing agents such as metallic sodium, sodium amalgam, sodium stannite, and hydrazine, or during electrolysis are not symmetrized but are reduced, displacing the atom of mercury by hydrogen. β-Alkanolmercury salts produce alcohols, for example[32]:

$$HOCH_2CH_2HgCl \xrightarrow[OH^-]{Na/Hg} HOCH_2CH_3$$

Alkoxyalkylmercury salts produce ethers[33]:

$$ROCR^1R^2CR^3R^4HgX \xrightarrow{Na/Hg} ROCR^1R^2CR^3R^4H$$

Mercury addition products of olefins in the medium of secondary amines give tertiary amines, for example[34]:

$$C_5H_{10}NCH_2CH_2HgCl \xrightarrow[H_2O]{Na/Hg} C_5H_{10}NC_2H_5$$

Other organomercury salts behave in a similar manner under the action of symmetrizing-reducing agents.

Sodium amalgam (particularly in water) substitutes hydrogen for HgX in many RHgX compounds.[35-37] Ethaneoxyhexamercarbide boiled in 30% hydrazine hydrate solution reacts completely with isolation of metallic mercury and with formation of ethane (according to Hofmann[38]) and methane (according to Bleshinsky and Usubakunov[39]).

The symmetrizing effect of various reagents such as hydrazine hydrate is replaced by the reducing effect (usually if excess amounts of the reagent are used[40]). It should be kept in mind that, along with symmetrizing and reducing effects, the hydrazine can cause isomerizations.[41]

Attempts to symmetrize CF_3HgI by the action of alkali solutions of sodium stannite, $Fe(SO_4)_2$, and acetone or water solutions of KCN or NaI were unsuccessful, but forms CHF_3, for example[42]:

$$CF_3HgI + 3I^- + H_2O \longrightarrow CHF_3 + OH^- + HgI_4^{2-}$$

$(i\text{-}C_3F_7)_2Hg$, not hydrolyzed by boiling water, reacts with water at 200°C producing $[(i\text{-}C_3H_7)_2Hg]_2O$ and 2-H-heptafluoropropane. The latter is formed by the reaction of Na_2SnO_2 and Na_2S or KI with $(i\text{-}C_3F_7)_2Hg$ in boiling water.[43]

Due to the high electronegativity of bis-pentafluorophenylmercury, the nucleophilic attack of hydrazine on the latter results in substitution of hydrogen for mercury and is accompanied by displacement of an atom of fluorine by the hydrazine radical[17]:

$$2(C_6F_5)_2Hg + 3NH_2NH_2 \longrightarrow Hg + C_6F_5H$$
$$+ (NH_2NHC_6F_4)_2Hg + NH_2NHC_6F_4H$$

Substances with extremely labile C—Hg bonds such as α-halomercury-oxo and α-halomercurycarboxy compounds reacting with water or alcohol solutions of the majority of symmetrizing-complexing agents (KI, KCN, $Na_2S_2O_3$) are not symmetrized but substitute hydrogen for mercury[44]:

$$RHgX + 4I^- + H_2O \longrightarrow RH + HgI_4^{2-} + OH^- + X^-$$

Organomercury compounds such as mercurated malonic esters,[45] malonic[46] and cyanoacetic[47] acids, and mercurated phenols[48] with a free hydroxyl group react with aqueous solutions of KI or KBr and substitute hydrogen for mercury. (See also Chapter 15.)

Potassium cyanide reacts in a similar manner with mercury derivatives of monohaloacetylene,[49] trihaloethylene,[50] and with some polymercurated compounds.

Water or alcohol solutions of KI or KCN react with bis-(bromo-acetylenyl)mercury, resulting in hydrolysis or alcoholysis and isolation of the acetylene halide[51]:

$$(BrC{\equiv}C)_2Hg + 2H_2O + 4I^- \longrightarrow 2BrC{\equiv}CH + HgI_4^{2-} + 2OH^-$$

E. Other Hydrogen for Mercury Substitutions

Nesmeyanov and Kocheshkov have shown that R_2Hg reacts with $SnCl_2$ in acetone according to the following equation[52-54]:

$$R_2Hg + SnCl_2 \longrightarrow R_2SnCl_2 + Hg$$

while in alcohol, the following reaction occurs[52]:

$$R_2Hg + SnX_2 + 2R'OH \longrightarrow 2RH + Hg + (R'O)_2SnX_2$$

where $R = p$- and o-$CH_3C_6H_4$, o-IC_6H_4, and α- and β-$C_{10}H_7$. RHgCl also reacts in a similar manner[52]:

$$RHgCl + SnCl_2 + R'OH \longrightarrow RH + Hg + R'OSnCl_3$$

Di-p-aminophenylmercury,[55] bis-o-oxyphenylmercury,[55] and bis(cis-α-stilbenyl)mercury[56] react with $SnCl_2$ in acetone to form, along with RR and $RSnCl_3$, RH in the last case. Diferrocenylmercury produces small amounts of ferrocene when boiled for 15 hr with metallic sodium in benzene or with tin(II) chloride in petroleum ether.[57]

A free-radical reaction mechanism is proposed[58] for solvolysis and reduction of tetracetatemercurypyrrole to pyrrole by the action of acetic acid in the presence of ferrocene. Similar solvolyses of phenylmercury salts occur in the presence of ferrocene.[59]

The addition of alkali sulfides in water to the addition products of mercury salts and α- or β-olefinic acids do not regenerate nonsaturated compounds nor produce sulfides, but do substitute hydrogen for mercury[60]:

$$C_6H_5CH(OH)CH(HgCl)COOH \xrightarrow[H_2O]{Na_2S} C_6H_5CH(OH)CH_2COOH$$

An ether solution of $(p$-$HOC_6F_4)_2Hg$ treated with hydrogen sulfide causes substitution of hydrogen for mercury.[17] The mercury derivatives of monohaloacetylene and trihaloethylene react with hydrogen sulfide in a similar manner.[49,50]

TABLE VI

Reactions with Reducing Agents—Hydrogen Donors

Organomercury compound	Reducing agent	Reaction conditions	Products	Yield, %	Ref.
$HOCH_2CH_2HgI$	Na(Hg)	—	$HOCH_2CH_3$	—	32
$HOCH(CH_3)CH_2HgI$	Na(Hg)	—	$HOCH(CH_3)CH_3$	—	32
Ethane hexamercarbide $C_2Hg_6O_2(OH)_2$	NH_2NH_2 (30%)	H_2O, reflux, 4 hr	C_2H_6 Hg		38
Ethane hexamercarbide $C_2Hg_6O_2(OH)_2$	NH_2NH_2 (30%)	H_2O, reflux, 4 hr	CH_4		39
$CH_3CH(OCH_3)CH_2HgCl$	$NaBH(OCH_3)_3$	Aqueous NaOH, 0°C, 1 hr	$CH_3CH(OCH_3)CH_3$	69	24
$C_6H_5CH(HgCl)CH(OCH_3)CH_3$	NH_2NH_2	Aqueous NaOH, reflux, 6 hr	$C_6H_5CH_2CH(OCH_3)CH_3$	7.5	88
$C_6H_5CH(OCH_3)CH(CH_3)HgCl$	NH_2NH_2	Aqueous NaOH, reflux, 40 hr	$C_6H_5CH(OCH_3)C_2H_5$	11	88
$(CH_3)_2C(OCH_3)CH(C_6H_5)HgCl$	NH_2NH_2	Aqueous NaOH, reflux, 40 hr	$(CH_3)_2C(OCH_3)CH_2C_6H_5$ $\left[(CH_3)_2C(OCH_3)\overset{C_6H_5}{\underset{\vert}{CH}} \right]_2$ 2 diastereomers		87
$CH_3CH_2CH(OCH_3)CH(HgCl)CH_2CH_2OH$	NH_2NH_2	Aqueous NaOH	cis-Hex-3-enol 4-Methoxyhexan-1-ol	40 48	19
$C_6H_5CH=C(COOH)HgCl$	NH_2NH_2	Aqueous NaOH, steam bath, 22 hr	trans-$C_6H_5CH=CHCOOH$	16	40
$[C_6H_5CH=C(COOH)]_2Hg$	NH_2NH_2	Aqueous NaOH, steam bath, 8 hr	$C_6H_5CH_2CH_2COOH$	55	40
3β-Chloromercuri-4α-methoxycyclohexan-1β-ol	NH_2NH_2	Aqueous NaOH, reflux, 16 hr	4α-Methoxycyclohexane	40	19
1-Chloromercurimethyl-1-methoxycyclohexan	$NaBH(OCH_3)_3$	Aqueous NaOH, 0°C, 0.5 hr	1-Methoxy-1-methylcyclohexane	71	24

(continued)

257

TABLE VI (*Continued*)

Organomercury compound	Reducing agent	Reaction conditions	Products	Yield, %	Ref.
	Na(Hg)	H₂O, 0°C, 2 days		68	35
	Na(Hg)	H₂O		51	35
	Na(Hg)	H₂O, 0°C, 2 days		85	35
	Na(Hg)	H₂O, 0°C, 2 days		80	35
	NH₂NH₂	3% aqueous NaOH, reflux, 48 hr		28	36
			Carvomenthen	18	

258

Reactant	Reagent	Conditions	Product	Yield (%)	Ref.
(structure: H₃C–CH₃ HC… OH … HgCl)	Na(Hg) (3%)	H₂O, 0°C, 2 days	Carvomenthen	60	36
(structure: HgCl, OH, OH cyclohexane)	NH₂NH₂	Aqueous NaOH, reflux, 16 hr	Cyclohexan-1-4α-diol	90	19
2β-Chloromercuri-4β-cyan-1α-methoxycyclohexane (structure: CH₂OH, HgCl, OH)	NH₂NH₂	Aqueous NaOH, reflux, 38 hr	Cyclohex-3-encarboxylic acid / 4α-Methoxycyclohexan-1β-carboxylic acid	48 / 45	19
(structure: CH₂OH, HgCl, OH cyclohexane)	NH₂NH₂		(structure: CH₂OH … OH)	65	19
(structure: CH₂OH, HgCl, OCH₃ cyclohexane)	NH₂NH₂	Aqueous NaOH	Cyclohex-3-enylmethanol / 4α-Methoxy-1β-cyclohexylmethanol	59 / 32	19
4β-Benzyloxy-2β-chloromercuri-1α-methoxy-cyclohexane	NH₂NH₂	Aqueous NaOH	Cyclohexanol[a] / 4α-Methoxycyclohexan-1β-ol	58 / 32	19
4β-Acetoxy-2β-acetoxymercuri-1α-methoxy-cyclohexane	NH₂NH₂	Aqueous NaOH	4α-Methoxycyclohexan-1β-ol	50	19
3β-Chloromercuri-4α-methoxy-1β-cyclohexyl methylacetate	NH₂NH₂	Aqueous NaOH	Cyclohex-3-enylmethanol / 4α-Methoxycyclohexyl-1β-carbinol	53 / 37	19

(continued)

TABLE VI (*Continued*)

Organomercury compound	Reducing agent	Reaction conditions	Products	Yield, %	Ref.
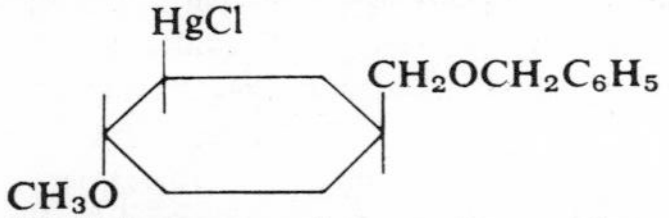 HgCl / $CH_2OCH_2C_6H_5$ / CH_3O	NH_2NH_2	Aqueous NaOH	Cyclohexylmethanol[a] 4α-Methoxy-1β-cyclohexylmethanol	60 14	19
3β-Chloromercuri-4α-methoxy-1β-cyclohexane-carboxylic acid	NH_2NH_2	Aqueous NaOH	Cyclohex-3-encarboxylic acid 4α-Methoxycyclohexanecarboxylic acid	58 30	19
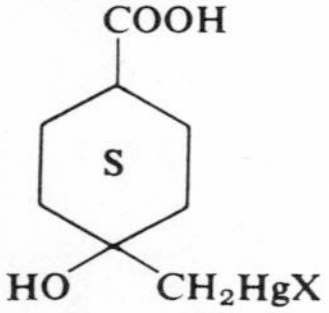 COOH / HO CH_2HgX	$NaBH_4$	Aqueous NaOH, heat 10 min	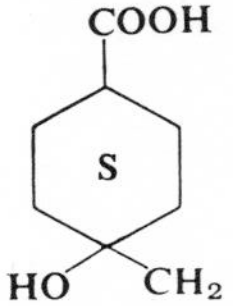 COOH / HO CH_2	90	22
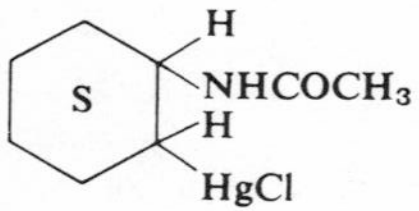 H NHCOCH₃ H HgCl	$NaBH_4$	Aqueous NaOH	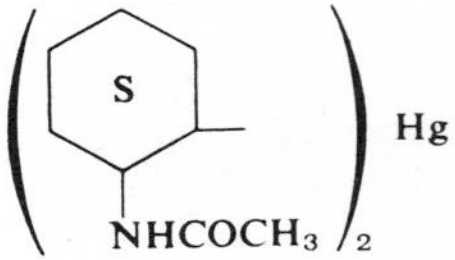 ($NHCOCH_3$)₂ Hg	99	21
CH_2HgCl NHCOR	$NaBH_4$	Aqueous CH_3OH	$NHCOCH_3$	Very low	
			CH_3 / NHCOR	79	23

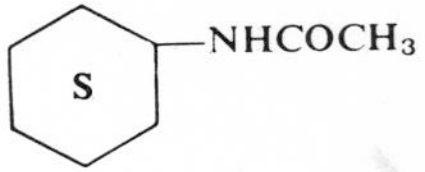

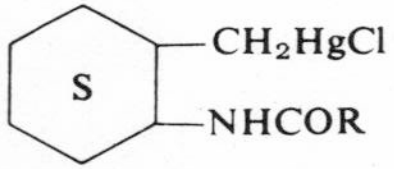

(R = CH_3 or C_6H_5)

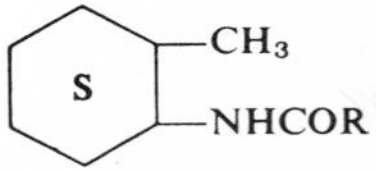

260

Organomercury compound	Reducing agent	Conditions	Products	Yield (%)	Refs.
exo-cis-3-Hydroxy-2-norbornylmercury chloride	NH$_2$NH$_2$	Aqueous CH$_3$OH NaOH, reflux, 6 hr	exo-Norborneol Cis-(exo-3-hydroxy-exo-2-norcamphanyl)-mercury	38 56	37 37
exo-cis-3-Hydroxy-2-norbornylmercury chloride	Na(Hg)	D$_2$O	(structure, labelled D and OH)		37
exo-cis-3-Hydroxy-2-norbornylmercury chloride exo-cis-3-Hydroxy-2-norbornylmercury chloride	Na(Hg) NaBH$_4$	H$_2$O —	exo-Norborneol exo-Norborneol		101 37
4β-Chloromercuri-6-oxa-tricyclo[3.2.1.1$^{3-8}$]-nonane	NH$_2$NH$_2$	Aqueous NaOH, reflux, 24 hr	6-oxa-tricyclo[3.2.1.1$^{3-8}$]-nonane	67	19
4β-Chloromercuri-6-oxa-tricyclo[3.2.1.1$^{3-8}$]-nonane	LiBH$_4$	(C$_2$H$_5$)$_2$O, reflux	6-oxa-tricyclo[3.2.1.1$^{3-8}$]-nonane	44	19
5β-Chloromercuri-6α-hydroxybicyclo[2.2.1]heptan-2-carboxylic acid lactone	NaBH$_4$	(C$_2$H$_5$)$_2$O—CH$_3$OH, 1:3, reflux, 3 hr	6-Hydroxybicyclo[2.2.1]heptane-2-carboxylic acid lactone	80	19
Mixture of two position isomers obtained by monooxymercuration of dicyclopentadiens	Na(Hg)	D$_2$O, 5 hr	(structure, labelled OH and D)		37
4-Camphylmercury chloride	LiAlH$_4$	(C$_2$H$_5$)$_2$O, 25°C, 12 hr	Camphane	19	16
cis,exo-3-Hydroxybenz-2-exo-norbornyl-mercury chloride	Na(Hg)	Aqueous NaOH, 6 hr	Benznorborneol	—	37
3-↓-Methoxy-1,4-↑↑-methylene-cyclohexyl-2-mercury chloride	Na(Hg)	H$_2$O, 12 hr	2-↓-Methoxy-1,4-methylencyclohexane	50	33
6-↓-Hydroxy-5-↓-chloromercuri-1,4-↑↑-methyleneecyclohexane-2-↓-carboxylic acid γ-lactone	NaBH(OCH$_3$)	Aqueous NaOH, −2 to 40°C	6-↓-Hydroxy-1,4-↑↑-methylenecyclo-hexane-2-↓-carboxylic acid γ-lactone	76	41
6-↓-6Hydroxy-5-↓-chloromercuri-1,4-↑↑-methylenecyclohexane-2-↓-carboxylic acid γ-lactone	NH$_2$NH$_2$	CH$_3$OH, 35–40°C, 48 hr	Second diastereomer of initial compounds 1,4-Methylencyclohexane-2-↓-carboxylic acid	17 64	41 41

(continued)

TABLE VI (*Continued*)

Organomercury compound	Reducing agent	Reaction conditions	Products	Yield, %	Ref.
6-↓-Hydroxy-5-↑-chloromercuri-1,4-↑↑-methylenecyclohexane-2-↓-carboxylic acid, γ-lactone	NaBH(OCH₃)₃	Aqueous NaOH, −2 to 40°C	6-↓-Hydroxy-1,4-↑↑-methylenecyclohexane-2-↓-carboxylic acid, γ-lactone	76	41
Hydroxy-chloromercuri-1,4-↑↑-methylene-cyclohexane-2-↑-carboxylic acid	NaBH₄	Aqueous NaOH, −5°C, 1 hr	5-Hydroxy-1,4-↑↑-methylene cyclohexane-2-↑-carboxylic acid	22	41
Hydroxy-chloromercuri-1,4-↑↑-methylene cyclohexane-2-↓-carboxylic acid (second diastereomer)	NaBH₄	Aqueous NaOH, −5°C, 1 hr	5-Hydroxy-1,4-↑↑-methylene cyclohexane-2-↑-carboxylic acid (second diastereomer)	30	41
5-↓-Hydroxy-6-chloromercuri-1,4-↑↑-methylene-cyclohexane-2-↓-3-↑-dicarboxylic acid	Na(Hg)	Aqueous NaOH, overnight	5-↓-Hydroxy-1,4-↑↑-methylenecyclohexane-2-↓-3-↑-dicarboxylic acid	72	33
5-↑-Hydroxy-6-chloromercuri-1,4-↑↑-methylenecyclohexane-2-↓-3-↑-dicarboxylic acid	Na(Hg)	Aqueous NaOH, 33°C, 7 hr	5-↑-Hydroxy-1,4-↑↑-methylenecyclohexane-2-↓-3-↑-dicarboxylic acid	90	33
cis-2-Chloromercuri-bicyclo-[2,2,2]octan-3-ol	NaBH₄	Aqueous NaOH, 20°C, 4 hr	Bicyclo[2,2,2]octan-2-ol	—	20
trans-2-Chloromercuri-bicyclo-[2,2,2]octan-3-ol	NaBH₄	Aqueous NaOH, 20°C, 14 hr	Bicyclo[2,2,2]octan-2-ol	25	20
4-Chloromercurimethyl[2,2,2]bicyclo-8-oxa-octane	NaBH₄	Aqueous NaOH, 1.5 hr	4-Methyl[2,2,2]bicyclo-8-oxaoctane	80	22
(C₆H₅)₂Hg	LiAlH₄	(C₂H₅)₂O, reflux, 0.5 hr	C₆H₆	—	15
C₆H₅HgCl	(C₂H₅)₃SiH	Pyridine or C₂H₅OH	C₆H₆	—	26
(C₆F₅)₂Hg	LiAlH₄	—	C₆F₅H	—	17
(C₆F₅)₂Hg	NH₂NH₂	C₂H₅OH, reflux, 6 hr	C₆F₅H (H₂NHNC₆F₄)₂Hg C₆HF₄NHNH₂	9 56 29	17
Di-α-Furylmercury	(C₂H₅)₃SiH	Pyridine	No reduction		26
α-Chloromercurifuran	(C₂H₅)₃SiH	Pyridine or C₂H₅OH	Furan		26
3-Chloromercuri-2-ethyltetrahydrofuran	NH₂NH₂	Aqueous NaOH, reflux, 16 hr	2-Ethyltetrahydrofuran trans-Hex-3-enol	42 42	19
2-Chloromercurithiophene	(C₂H₅)₃SiH	Pyridine or C₂H₅OH	Thiophene		26

	Reagent	Conditions		Yield	
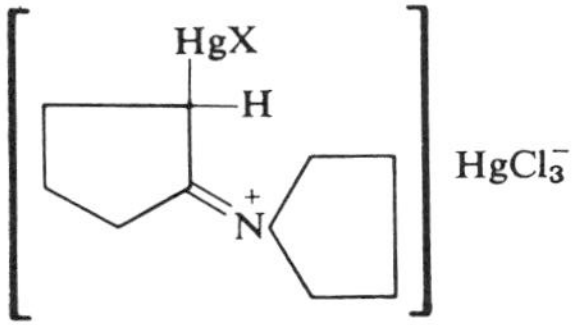 (iminium salt structure) HgX, $HgCl_3^-$	$LiAlH_4$	$(C_2H_5)_2O$, $-70°C$	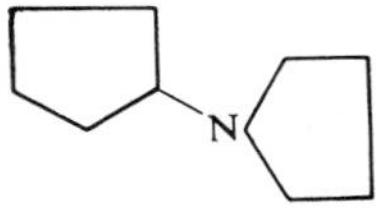 (bicyclopentylamine structure)	70	18
$CH_3CH(OCH_3)CH(HgOAc)CO_2menth$	$NaBH_4$	Aqueous NaOH,20°C,1hr	$CH_3CH(OCH_3)CH_2CO_2menth$	Small	89
$CH_3CH(OCH_3)CH(HgOAc)CO_2menth$	H_2S	Pyridine	$CH_3CH(OCH_3)CH_2CO_2menth$	High	89
$C_6H_5CH(OCH_3)CH(HgOAc)CO_2menth$	H_2S	Pyridine	$C_6H_5CH(OCH_3)CH_2CO_2menth$		89
$C_6H_5(CH_3)C(OCH_3)CH(HgOAc)CO_2menth$	H_2S	Pyridine	$C_6H_5(CH_3)C(OCH_3)CH_2CO_2menth$		89
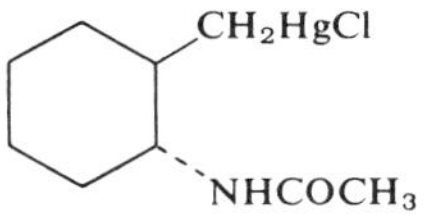 (cyclohexane) CH_2HgCl, $NHCOCH_3$	Na/Hg	Aqueous CH_3OH	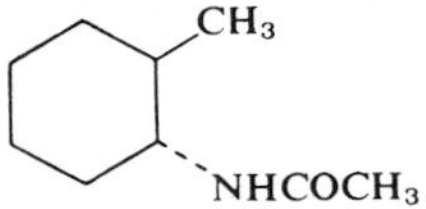 (cyclohexane) CH_3, $NHCOCH_3$	79	90
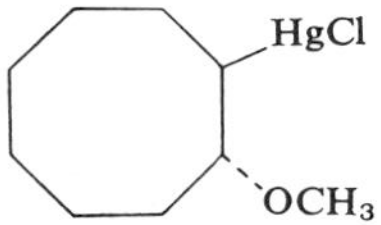 (cyclooctane) $HgCl$	$LiAlH_4$	$(C_2H_5)_2O$	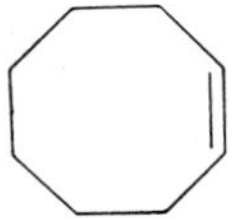 (cyclooctene)	—	91
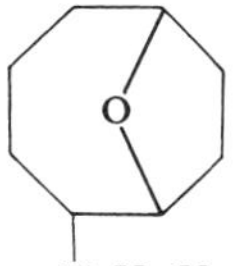 (bicyclic ether) OCH_3, HgX $(X = OCOCH_3, Cl, I)$	$NaBH_4$	pH = 7 (X = $OCOCH_3$–H_2O X = Cl–A + py X = I–A)	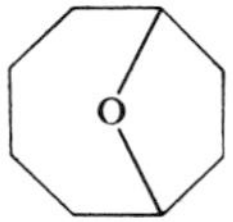 (bicyclic ether) OCH_3	—	95
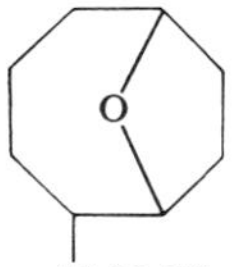 (bicyclic ether) HgX $(X = OCOCH_3, Cl, I)$	$NaBH_4$	pH = 4 (X = $OCOOH_3$–H_2O X = Cl–A + py X = I–A)	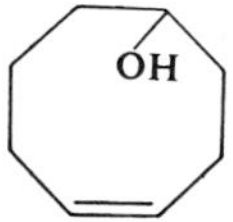 (cyclooctene) OH	—	95

(continued)

263

TABLE VI (*Continued*)

Organomercury compound	Reducing agent	Reaction conditions	Products	Yield, %	Ref.
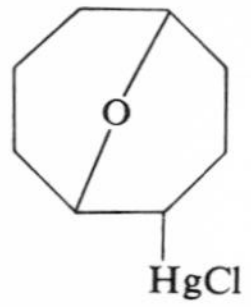 O HgCl	$NaBH_4$	Aqueous NaOH	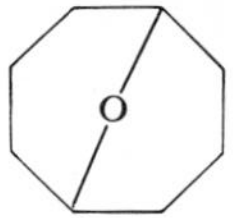 O	—	95
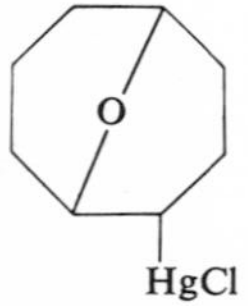 O HgCl	$NaBH_4$	$CHCl_3—C_2H_5OH$	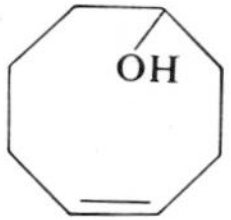 OH	—	95
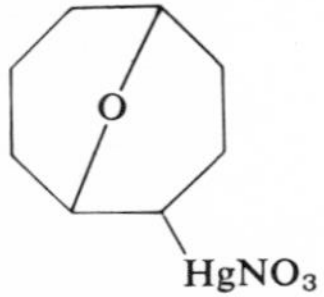 O $HgNO_3$	$NaBH_4$	$HCON(CH_3)_2$ abs. C_2H_5OH	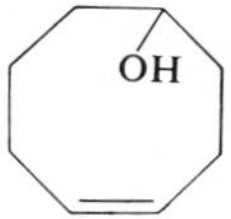 OH	—	95
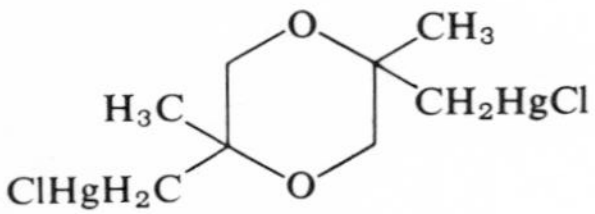 O CH_3 H_3C CH_2HgCl $ClHgH_2C$ O	$NaBH_4$	Aqueous NaOH or D_2O	2,2,5,5-Tetramethyl-1,4-dioxan	65	95
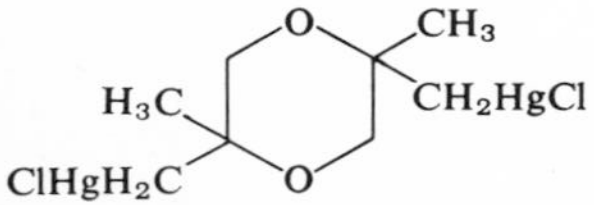 O CH_3 H_3C CH_2HgCl $ClHgH_2C$ O	$NaBD_4$	Aqueous NaOH	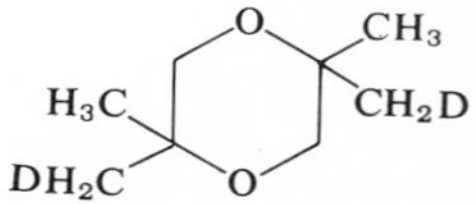 O CH_3 H_3C CH_2D DH_2C O	—	95

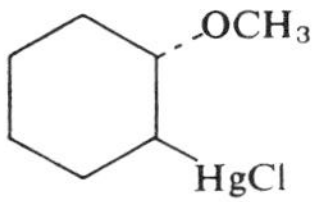

![cyclohexane with OCH3 and HgCl] (OCH3, HgCl)	NaBH4	Aqueous NaOH	C6H11OCH3	86	95
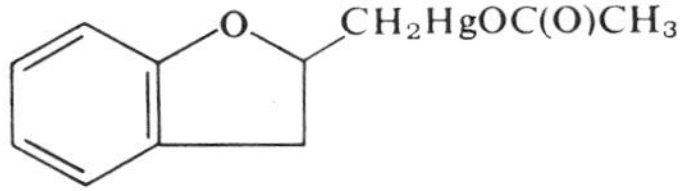 ![benzofuran with CH2HgOC(O)CH3] (CH2HgOC(O)CH3)	NaBH4	Aqueous NaOH	Bis-[2-(2,3-dihydrobenzofuranyl)methyl]-mercury + 2-methyl-2,3-dihydro-benzofuran	23 + 26	95
2,5-Bis(brommercurymethyl)tetrahydrofuran	NaBH4	Aqueous NaOH	cis-2,5-Dimethyltetrahydrofuran / trans-2,5-Dimethyltetrahydrofuran	Most traces	95 / 95
C6H5HgOCOCH3	NaBH4	Aqueous NaOH	(C6H5)2Hg	51	95
p-CH3C6H4HgCl	NaBH4	Aqueous NaOH	(p-CH3C6H4)2Hg	68	95
![norbornane with OH and HgCl] (OH, HgCl)	NaBH4	Aqueous NaOH	exo-2-Norborneole	40	95
	NaBD4	NaOH-aq	exo-3D-2-Norborneole	—	95
exo-Bicyclo-[2,2,2]-oct-2-enyl-5-HgCl	LiAlH4	$(C_2H_5)_2O$	Bicyclo-[2,2,2]-octene	26	96
Tri-cyclo-[2,2,2.0^{2,6}]-octyl-3-mercury chloride	LiAlH4	$(C_2H_5)_2O$	Bicyclo-[2,2,2]-octene	45	96
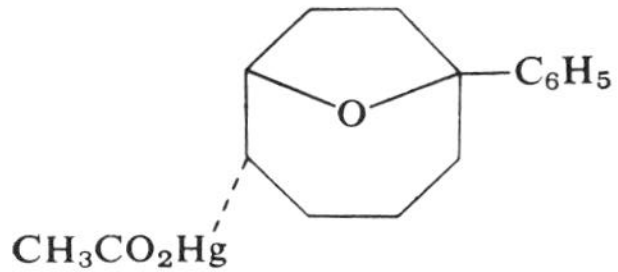 ![bicyclic structure with C6H5 and HgCl]	NaBH4	NaOH-aq	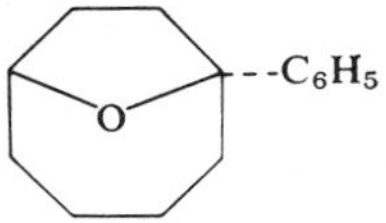 ![bicyclic structure with C6H5]	61,5	97
![structure with CH3CO2Hg]			![structure with C6H5]	> 3	
(RCB10H10C—)2Hg R = CH3, C6H5	LiAlH4	H2O	RCB10H10CH	Quant.	98

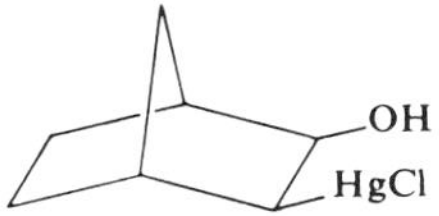

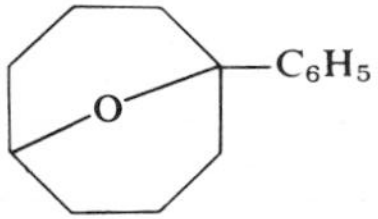

265

(continued)

TABLE VI (*Continued*)

Organomercury compound	Reducing agent	Reaction conditions	Products	Yield, %	Ref.
$C_6H_5NHCH_2CH_2HgCl$	$LiAlH_4$ or Hg/Na	—	$C_6H_5NHC_2H_5$	45 30	99
$C_6H_5N(CH_2CH_2HgOC(O)CH_3)_2$	$LiAlH_4$ or Hg/Na	—	$C_6H_5N(C_2H_5)_2$	40	99
$C_6H_5N(CH_3)CH_2CH_2HgCl$	$LiAlH_4$ or Hg/Na	—	$C_6H_5N(CH_3)C_2H_5$	40	99
$C_6H_5N(C_2H_5)CH_2CH_2HgCl$	$LiAlH_4$ or Hg/Na	—	$C_6H_5N(C_2H_5)_2$	40	99
cyclo-$C_6H_{11}N(CH_3)CH_2CH_2HgBr$	$LiAlH_4$ or Hg/Na	—	*cyclo*-$C_6H_{11}N(CH_3)C_2H_5$	40	99
cyclo-$C_5H_{11}NCH((C_6H_5)CH_2HgCl$	$LiAlH_4$	—	*cyclo*-$C_5H_{11}N(C_6H_5)CH_3$ / *cyclo*-$C_5H_{11}NCH_2CH_2C_6H_5$	75 25	99
(cyclopentane ring)–NCH(CH₂HgCl)C₆H₅	$LiAlH_4$		(cyclopentane ring)–NCH(C₆H₅)CH₃ / (cyclopentane ring)–NCH₂CH₂C₆H₅		99
Diacetoxymercuri-2-D-indole	$LiAlH_4$	$(C_2H_5)_2O$	2-D-Indole	—	100
Diacetoxymercuri-2-D-indole	$LiAID_4$	$(C_2H_5)_2O$	Indole + 3-D-indole (1:1)	—	100
2,3,4,5-Tetracetatomercuripyrrole	Ferrocene	Ice-CH_3COOH	Pyrrole	58	58
$(CH_3)_2C(OC_2H_5)C(CH_2HgCl)=CHCl$	$NaBH_4$	—	$C(CH_3)_2(OC_2H_5)C(CH_3)=CHCl$	—	102

[a] Treat with Li in $C_2H_5NH_2$.

F. Reactions of Organomercury Compounds in the Presence of Substances Susceptible to Dehydrogenation by Light Initiation and Other Initiating Agents Causing Radical Decomposition

Investigations of reactions of radicals formed during decomposition of organometallic compounds caused by the action of light have been conducted in large scale by Razuvaev and co-workers. They have shown that reactivity increases substantially when use is made of ultraviolet light.[61] Under these conditions and in alcohols,[62-72] ethers,[65,66,73] esters,[62] acetone,[62,64] morpholine,[74] other amines,[70] benzaldehyde,[70] saturated[62,65,75,76] and sometimes aromatic[77,78] hydrocarbons, the organomercury compounds are decomposed at lower temperatures (sometimes at room temperature). Metallic mercury is isolated and the formation of radicals R, which dehydrogenate the solvent forming RH, are observed; occasionally (when $R = C_6H_5CH_2$[79]) the radicals dimerize and disproportionate. Photolysis of $o\text{-}IC_6H_4HgI$ in cyclohexane produces benzene and Hg rather than iodobenzene.[80] The radical formed during the photodecomposition of the organomercury compound forms, in some cases, RH by the dehydrogenation of compounds such as $CHCl_3$,[68,71,72] trichloroethylene,[65] and monohaloalkanes.[65,81] Di-naphthyl-[82,83] and di-p-anisylmercury,[84] decomposed by ultraviolet light even in CCl_4, produce naphthalene and anisole, respectively, due to hydrogen abstraction of the $C_{10}H_7\cdot$ and $CH_3OC_6H_4\cdot$ radicals.

The decomposition of C_6H_5HgOH in ROD forms benzene free of deuterium,[61] thus suggesting that the reaction goes through a cyclic intermediate rather than through the formation of kinetically independent free radicals.

Like ultraviolet light, copper and the halides of some transition metals initiate decomposition of R_2Hg forming $RHg\cdot$ and $R\cdot$. The radicals, $R\cdot$, abstract hydrogen from many solvents (ethylcellosolve, dioxane, ethylene glycol, etc.) producing RH.[85,86]

These reactions have no preparative significance and are not listed in the following tables.

REFERENCES

1. G. A. Razuvaev and M. M. Koton, *Zh. Obshch. Khim.*, **1**, 864 (1931); *Ber.*, **65**, 613 (1932).
2. G. A. Razuvaev and A. I. Subbotina, *Uchen. Zap. Gor'kov. Gos. Univer.*, **23**, 149 (1952).
3. G. A. Razuvaev and M. M. Koton, *Ber.*, **66**, 854 (1933).
4. A. N. Stear and J. J. Lagowski, *J. Chem. Soc.*, *Suppl. I*, **1964**, 5848.
5. G. A. Razuvaev and M. M. Koton, *Zh. Obshch. Khim.*, **5**, 381 (1935).

6. G. A. Razuvaev and M. M. Koton, *Zh. Obshch. Khim.*, **4**, 647 (1934); *Ber.*, **66**, 1210 (1933).
7. M. M. Koton, *Zh. Obshch. Khim.*, **8**, 1791 (1938).
8. M. M. Koton, *Zh. Obshch. Khim.*, **11**, 179 (1941).
9. M. M. Koton and F. S. Florinsky, *Zh. Obshch. Khim.*, **9**, 2196 (1939).
10. M. M. Koton, *Zh. Obshch. Khim.*, **9**, 1622 (1939).
11. G. A. Razuvaev, G. G. Petukhov, and R. V. Kalinina, *Zh. Obshch. Khim.*, **26**, 1685 (1956).
12. R. Ja. Levina, V. N. Kostin, and V. A. Tartarkovski, *Zh. Obshch. Khim.*, **27**, 2049 (1947).
13. G. V. Postnikova and I. F. Lutsenko, *Zh. Obshch. Khim.*, **33**, 4029 (1963).
14. A. N. Nesmeyanov and R. Kh. Freidlina, *Zh. Obshch. Khim.*, **5**, 53 (1935); *Ber.*, **67**, 735 (1934).
15. D. H. R. Barton and W. J. Rosenfelder, *J. Chem. Soc.*, **1951**, 2381.
16. S. Winstein and T. G. Traylor, *J. Am. Chem. Soc.*, **78**, 2597 (1956).
17. J. Burdon, P. L. Coe, M. Fulton, and J. C. Tatlow, *J. Chem. Soc.*, **1964**, 2673.
18. K. Brodersen, G. Opitz, and D. Breitinger, *Chem. Ber.*, **97**, 2046 (1964).
19. H. H. Henbest and R. Nikolls, *J. Chem. Soc.*, **1959**, 227.
20. T. G. Traylor, *Chem. Ind.*, **1959**, 1223; *J. Am. Chem. Soc.*, **86**, 244 (1964).
21. D. Chow, J. H. Robson, and G. F Wright, *Can. J. Chem.*, **43**, 312 (1965).
22. N. L. Weinberg and G. F Wright, *Can. J. Chem.*, **43**, 24 (1965).
23. V. N. Sokolov, N. B. Rodina, and O. A. Reutov, *Zh. Obshch. Khim.*, **36**, 955 (1966).
24. J. H. Robson and G. F Wright, *Can. J. Chem.*, **38**, 21 (1960).
25. N. S. Vyazankin, G. A. Razuvaev, and E. N. Gladyshev, *Dokl. Akad. Nauk SSSR*, **155**, 830 (1964).
26. E. Lukevics and S. A. Hillers, *Latvijas PSR Zinatnu Akad. Vestis.*, **1961**, N4, 99; *Chem. Abstr.*, **56**, 10178 (1962).
27. N. S. Vyazankin, G. A. Razuvaev, and E. N. Gladyshev, *Dokl. Akad. Nauk SSSR*, **151**, 1326 (1963).
28. Ju. I. Baukov, I. Ju. Belavin, and I. F. Lutsenko, *Zh. Obshch. Khim.*, **35**, 1092 (1965).
29. N. S. Vyazankin, G. A. Razuvaev, and S. P. Korneva, *Zh. Obshch. Khim.*, **34**, 2787 (1964).
30. N. S. Vyazankin, G. A. Razuvaev, and S. P. Korneva, *Zh. Obshch. Khim.*, **33**, 1041 (1963).
31. I. F. Lutsenko, Yu. I. Baukov, and B. N. Khasapov, *Zh. Obshch. Khim.*, **33**, 2724 (1963).
32. J. Sand and F. Singer, *Ber.*, **35**, 3170 (1902).
33. M. J. Abercrombie, A. Rodgman, K. R. Bharucha, and G. F Wright, *Can. J. Chem.*, **37**, 1328 (1959).
34. R. Kh. Freidlina and N. S. Kochetkova, *Izv. Akad. Nauk SSSR, Otd. Khim. Nauk*, **1945**, 128.
35. W. R. R. Park and G. F Wright, *Can. J. Chem.*, **35**, 1088 (1957).
36. H. B. Henbest and R. S. McElhinney, *J. Chem. Soc.*, **1959**, 1834.
37. T. G. Traylor and A. W. Baker, *J. Am. Chem. Soc.*, **85**, 2746 (1963).
38. K. A. Hofmann, *Ber.*, **33**, 1328 (1900).
39. S. V. Bleschinsky and M. Usubakunov, *Sb. Phys.-mat. Konferenz, poswjashch. 100-let. akad. Kurnakova. Frunse*, 1963.
40. W. R. Park and G. F Wright, *J. Org. Chem.*, **19**, 1325 (1954).

41. M. Malaiyandi and G. F Wright, *Can. J. Chem.*, **41**, 1493 (1963).
42. H. J. Emeléus and R. N. Haszeldine, *J. Chem. Soc.*, **1949**, 2953, 2948.
43. P. E. Aldrich, E. G. Howard, W. J. Linn, W. J. Middleton, and W. H. Sharkey, *J. Org. Chem.*, **28**, 184 (1963).
44. A. N. Nesmeyanov and I. F. Lutsenko, *Dokl. Akad. Nauk SSSR*, **59**, 707 (1948).
45. E. Biilman, *Ber.*, **35**, 2581 (1902).
46. E. Biilman and J. Witt, *Ber.*, **42**, 1070 (1909).
47. L. Petterson, *J. Prakt. Chem.*, **86** (2), 464 (1912).
48. F. C. Whitmore and E. Middleton, *J. Am. Chem. Soc.*, **43**, 622 (1925).
49. K. A. Hofmann and H. Kirmreuther, *Ber.*, **42**, 4232 (1909).
50. K. A. Hofmann and H. Kirmreuther, *Ber.*, **41**, 314 (1908).
51. A. N. Nesmeyanov and N. K. Kochetkov, *Izv. Akad. Nauk SSSR, Otd. Khim. Nauk*, **1949**, 587; *Uchen. Zap. Moskov. Gos. Univer.*, **132**, 51 (1950).
52. A. N. Nesmeyanov and K. A. Kocheshkov, *Ber.*, **63**, 2496 (1930).
53. K. A. Kocheshkov and A. N. Nesmeyanov, *Zh. Obshch. Khim.*, **1**, 219 (1931); *Ber.*, **64**, 628 (1931).
54. I. T. Eskin, A. N. Nesmeyanov, and K. A. Kocheshkov, *Zh. Obshch. Khim.*, **8**, 35 (1938).
55. A. N. Nesmeyanov, A. E. Borisov, N. V. Novikova, and U. A. Osipova, *Izv. Akad. Nauk SSSR, Otd. Khim. Nauk*, **1959**, 263.
56. A. N. Nesmeyanov, A. E. Borisov, and N. V. Novikova, *Dokl. Akad. Nauk SSSR*, **119**, 504 (1958).
57. A. N. Nesmeyanov, E. G. Perevalova, and O. A. Nesmeyanova, *Izv. Akad. Nauk SSSR, Otd. Khim. Nauk*, **1962**, 47.
58. G. N. O'Connor, J. V. Crawford, and C.-H. Wang, *J. Org. Chem.*, **30**, 4090 (1965).
59. C.-H. Wang, *J. Am. Chem. Soc.*, **85**, 2339 (1963).
60. E. Biilmann, *Ber.*, **43**, 568 (1910).
61. G. A. Razuvaev, G. G. Petukhov, A. F. Rekasheva, and G. P. Miklukhin, *Dokl. Akad. Nauk SSSR*, **90**, 569 (1953).
62. G. A. Razuvaev and Ju. A. Ol'dekop, *Dokl. Akad. Nauk SSSR*, **64**, 77 (1949).
63. G. A. Razuvaev and G. G. Petukhov, *Tr. Khim. i Khim. Technol. (Gor'ky)*, **1961**, Vyp. 1, 150; *R. Zh. Khim.*, **1962**, 4Zh50.
64. G. A. Razuvaev and Ju. A. Ol'dekop, *Zh. Obshch. Khim.*, **19**, 736 (1949).
65. G. A. Razuvaev and Ju. A. Ol'dekop, *Zh. Obshch. Khim.*, **19**, 1483 (1949).
66. Ju. A. Ol'dekop and Z. B. Idelchik, *Zh. Obshch. Khim.*, **30**, 2564 (1960).
67. G. A. Razuvaev, Ju. A. Ol'dekop, and N. P. Nazarkova, *Uchen. Zap. Gor'kov. Gos. Univer.*, **23**, 117 (1952).
68. G. A. Razuvaev, Ju. A. Ol'dekop, and Z. N. Manchinova, *Zh. Obshch. Khim.*, **22**, 480 (1952).
69. G. A. Razuvaev, Ju. A. Ol'dekop, and V. N. Latjaeva, *Zh. Obshch. Khim.*, **24**, 260 (1954).
70. G. A. Razuvaev and Ju. A. Ol'dekop, *Sb. Stat. Obshch. Khim.*, **1**, 275 (1953).
71. G. A. Razuvaev and Ju. A. Ol'dekop, *Zh. Obshch. Khim.*, **20**, 181 (1950).
72. G. A. Razuvaev, Ju. A. Ol'dekop, and M. N. Koroleva, *Zh. Obshch. Khim.*, **21**, 650 (1951).
73. G. A. Razuvaev and N. F. Novatorov, *Uchen. Zap. Gor'kov. Gos. Univer.*, **17**, 197 (1951).
74. G. A. Razuvaev and N. F. Novatorov, *Uchen. Zap. Gor'kov. Gos. Univer.*, **23**, 121 (1952).

75. D. N. Derbyshire and E. W. R. Steacie, *Can. J. Chem.*, **32**, 457 (1954).
76. G. A. Razuvaev, G. G. Petukhov, Ju. A. Kaplin, and L. F. Kudryavtsev, *Dokl. Akad. Nauk SSSR*, **141**, 371 (1961).
77. B. J. McDonald, D. Bryce-Smyth, and B. W. Pengilly, *J. Chem. Soc.*, **1959**, 3174.
78. D. Bryce-Smith, *J. Chem. Soc.*, **1955**, 1712.
79. G. A. Razuvaev and Ju. A. Ol'dekop, *Zh. Obshch. Khim.*, **19**, 1487 (1949).
80. G. Wittig and H. F. Ebel, *Ann. Chem.*, **650**, 20 (1961).
81. G. A. Razuvaev and Ju. A. Ol'dekop, *Zh. Obshch. Khim.*, **21**, 1122 (1951).
82. G. A. Razuvaev and Ju. A. Ol'dekop, *Zh. Obshch. Khim.*, **20**, 506 (1950).
83. Ju. A. Ol'dekop, *Zh. Obshch. Khim.*, **22**, 2063 (1952).
84. Ju. A. Ol'dekop, *Zh. Obshch. Khim.*, **22**, 478 (1952).
85. M. S. Fedotov, *Sb. Stat. Obshch. Khim.*, **2**, 984 (1953).
86. G. A. Razuvaev and M. S. Fedotov, *Zh. Obshch. Khim.*, **22**, 484 (1952).
87. L. Berman, R. H. Hall, R. G. Pyke, and G. F Wright, *Can. J. Chem.*, **30**, 541 (1952); *Chem. Abstr.*, **48**, 1951 (1954).
88. W. R. R. Park and G. F Wright, *J. Org. Chem.*, **19**, 1435 (1964).
89. J. Oda, T. Nakagawa, and Y. Inoue, Bull. *Chem Soc. Japan* **40**, 373 (1967).
90. V. I. Sokolov, N. B. Rodira, and O. A. Reutov, *Zh. Organ. Khim.* **3**, 2089 (1967).
91. V. I. Sokolov, L. L. Troitzkaya, and O. A. Reutov, *Dokl. Akad. Nauk SSSR* **166**, 136 (1966).
92. H. C. Brown and P. Geoghegan, jr. *J. Am. Chem. Soc.* **89**, 1522 (1967).
93. H. C. Brown and W. J. Hammar, *J. Am. Chem. Soc.* **89**, 1524 (1967).
94. H. C. Brown, J. H. Kawakami, and Sh. Ikegami, *J. Am. Chem. Soc.* **89**, 1525 (1967).
95. F. C. Bordwell and M. L. Douglass, *J. Am. Chem. Soc.* **88**, 993 (1966).
96. D. S. Matteson and M. L. Talbot, *J. Am. Chem. Soc.* **89**, 1123 (1967).
97. A. C. Cope, M. A. McKervey, and N. M. Weinshenker, *J. Am. Chem. Soc.* **89**, 2932 (1967).
98. V. I. Bregadze and O. Yu. Okhlobystin, Dokl *Akad. Nauk SSSR*, **177**, 347, (1967).
99. A. Lattes and J. J. Périé, *Compt. rend.* **262**, (C) 1591 (1966).
100. G. W. Kirby and S. W. Shah, *Chem. Communicat.* **1965**, 381.
101. T. G. Traylor, A. W. Baker *Tetrahedron Lett.* **19**, 14 (1959).
102. R. B. Babb and P. D. Gardner, *Chem. Communicat.* **1968**, 1678.
103. W. C. Baird jr. and M. Buza, *J. Org. Chem.* **33**, 4105 (1968).
104. S. Moon and B. H. Waxmann, *Chem. Communicat.* **1967**, 1283.
105. M. R. Johnson and B. Rickborn, *Chem. Communicat.* **1968**, 1073.
106. J. S. Mills, *J. Chem. Soc.* (C) **1967**, 2514.

Chapter 8

Coupling Reactions

I. INTRODUCTION

A number of metals such as Pd, Pt, Ag, Au, Co, Cu, Fe, and Ni do not produce organometallic compounds with organomercury compounds but rather catalyze decomposition of the R_2Hg which results in coupling of the organic radicals[1,2]:

$$R_2Hg \longrightarrow R\text{—}R + Hg \qquad (1)$$

Pd and Pt are the best catalysts for the decomposition. Decomposition of many R_2Hg compounds in the presence of these catalysts occurs at room temperature. Decomposition of compounds such as dibenzyl-mercury according to reaction (1) occurs not only catalytically but also thermally (see Chapters 3, 7, and 16), and in the absence of metals under hydrogen pressure or in the presence of hydrogen donors (see Chapter 7) and during its photodecomposition (see Chapter 16).

II. APPLICATION OF THE REACTION

The decomposition reaction[1] is applicable to aromatic mercury compounds and to dibenzylmercury. It may be conducted by heating of the R_2Hg compound with an excess of metallic powder for several hours (up to 12 hr) in sealed tubes and at 200°C.

The higher the temperature, the greater is the percentage of decomposition which is typical for all metal catalysts. Complete decomposition of Ar_2Hg occurs at 25°C if Pt and Pd are used. Lesser amounts of Pd and Pt are necessary as compared with other metals. Dibenzylmercury decomposes at lower temperatures than those characteristic of aromatic mercury compounds. Wittig decomposed orthopolyphenylene compounds of mercury in the presence of excess silver powder. The reaction was conducted at temperatures of 260–300°C which were maintained for several minutes. The decomposition was conducted under nitrogen in unsealed vessels or in vacuum. Dependent on the nature of the organomercury compound, different hydrocarbon compounds were formed; in some cases, the products were more complicated than the expected coupled product. Heating of the hexamer orthophenylenemercury, with silver powder resulted in the formation of diphenylene and triphenylene in high vacuum[3]:

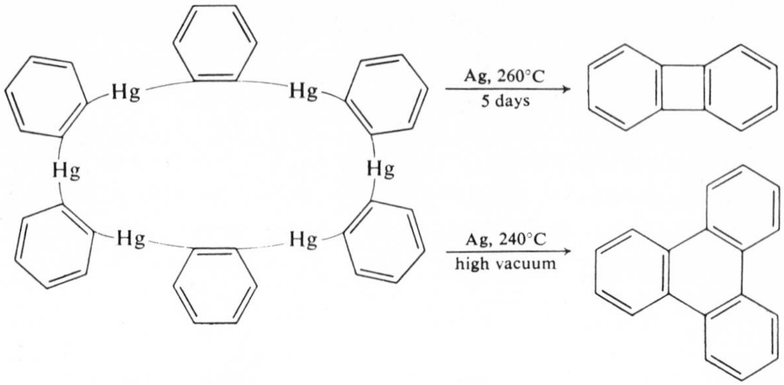

Heating of the tetramer orthodiphenylene mercury, with silver powder at 300°C resulted in the formation of diphenylene[4]:

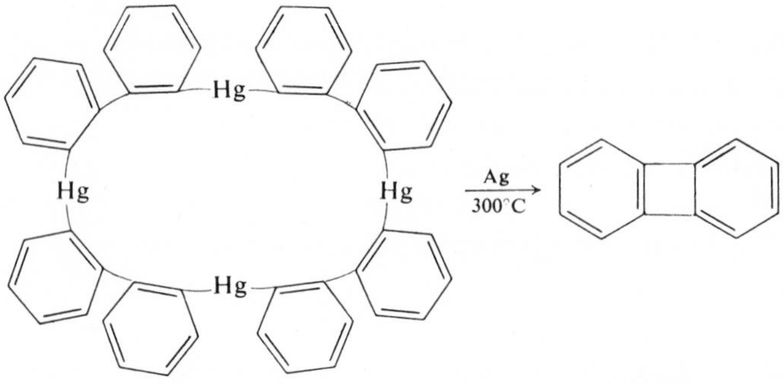

The dimer orthoterphenylenemercury heated with silver powder at 280°C for 10 min produced a 79% yield of triphenylene[5]:

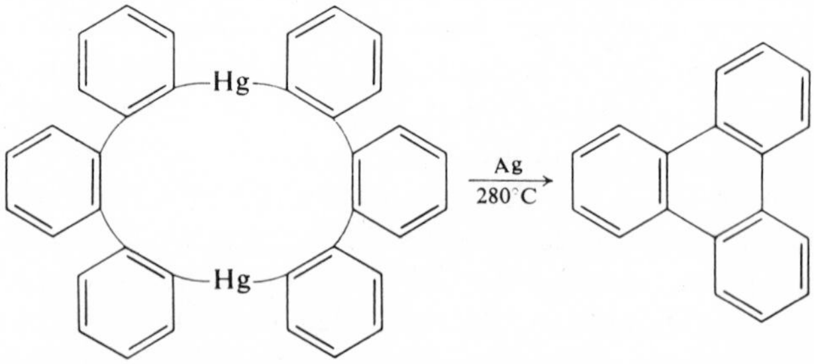

Decomposition of the polymer meta-phenylenemercury[6] (structure and molecular weight are unknown), obtained by the interaction of meta-dibromobenzene with sodium amalgam heated at 260°C for 12 hr with palladium powder in a sealed tube produced diphenyl, metaterphenyl, and metanonylphenyl (which is identical to Busch's metanonylphenyl of a linear structure). Silver powder was not as active in this case. To obtain complete decomposition of the organomercury compound during experiments involving silver, it was necessary to repeatedly heat the compounds after addition of fresh silver powder.

3,4-Benzodiphenylenemercury, heated with silver powder at 360°C, produced a 17% yield of 3,4-benzodiphenylene.[7]

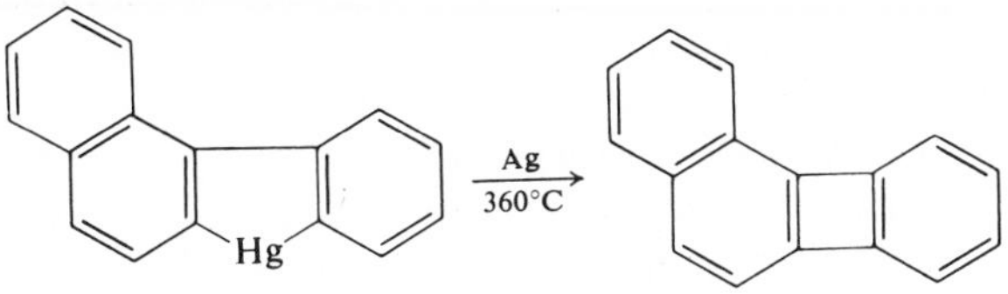

However, use of silver powder for the purpose of decomposing diferrocenylmercury according to Eq. (1) produced better results than use of palladium black, so successfully employed for the purpose of decomposing monomeric aromatic mercury compounds, Ar_2Hg.

Thus, for example, heating of diferrocenylmercury with an eightfold molar excess amount of silver at 265°C for 17 hr produced a 54% yield of diferrocenyl[8]:

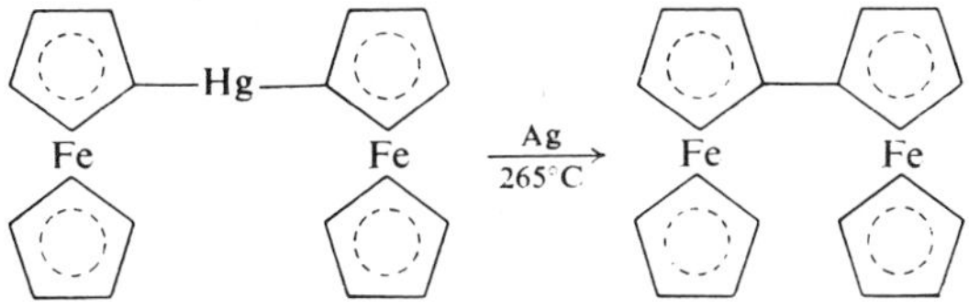

In addition, ferrocene was produced—a common product of the thermal decomposition of diferrocenylmercury. Decomposition of diferrocenylmercury, when heated with palladium powder at a temperature of 170–220°C produced, apart from ferrocene (16–49% yield), only a 1.5–6% yield of diferrocenyl.[9]

Decomposition of the mixture of R_2Hg and R'_2Hg in the presence of metals produced a mixture of all possible combinations of products, RR, R'R, and R'R' in various proportions. Decomposition of a mixture of diferrocenylmercury and Ar_2Hg in the presence of silver, in a molar ratio of 1:1:15, upon heating to temperatures ranging from 235 to 300°C for 17–22 hours have produced, in addition to RR and R'R', RR', arylferrocene.[8] Arylferrocene was obtained in a 45% yield for Ar = phenyl, 6% yield for 2-biphenyl, 2% yield for 3-biphenyl, and 20% yield for 4-biphenyl.[8]

Polymercuriferrocenylene,

obtained by a symmetrization reaction using NaI, sodium thiosulfate, and 1,1-dichloromercuryferrocene and consisting probably as a mixture of linear and cyclic polymers having a low value of X, produced mainly polyferrocenylene $(C_{10}H_8Fe)_r$ when heated with silver powder to 300°C for 16 hr.[10] Small amounts of ferrocene and biferrocenyl were also formed.

Diphenyl was formed by the reaction of a nickel–aluminum alloy and phenylmercury acetate in an aqueous alkali solution.[11]

REFERENCES

1. G. A. Razuvaev and M. M. Koton, *Zh. Obshch. Khim.*, **3**, 792 (1933).
2. G. A. Razuvaev and M. M. Koton, *Zh. Obshch. Khim.*, **4**, 647 (1934); *Ber.*, **66**, 1210 (1933).
3. G. Wittig and F. Bickelhaupt, *Chem. Ber.*, **91**, 883 (1958).
4. G. Wittig and W. Herwig, *Ber.*, **87**, 1511 (1954).
5. G. Wittig, E. Hahn, and W. Tochterman, *Chem. Ber.*, **95**, 431 (1962).
6. V. N. Vinogradova and L. G. Makarova, *Izv. Akad. Nauk SSSR. Ser. Khim.*, **1966**, in press.
7. M. P. Cava and J. F. Stucker, *J. Am. Chem. Soc.*, **77**, 6022 (1955).
8. M. D. Rausch, *Inorg. Chem.*, **1**, 1414 (1962).
9. O. A. Nesmeyanov and E. G. Perevalova, *Dokl. Akad. Nauk SSSR*, **126**, 1007 (1959).
10. M. D. Rausch, *J. Org. Chem.*, **28**, 3337 (1963).
11. E. Schwenke, D. Papa, B. Whitman, and H. Ginsberg, *J. Org. Chem.*, **9**, 1 (1944).

Chapter 9

Mercury-to-Acid Exchange Reactions

I. INTRODUCTION

A. Substitution Reactions of Hydrogen for Mercury (Protodemercuration Reaction)

Substitution of hydrogen for mercury is the most common reaction of acids with organomercury compounds which are not the addition products of mercury salts to double bonds or addition products of mercury salts to carbon monoxide.

The reaction proceeds by the following two steps:

$$R_2Hg + HX \longrightarrow RH + RHgX \qquad (1)$$

$$RHgX + HX \longrightarrow RH + HgX_2 \qquad (2)$$

At moderate temperatures, the reaction stops at the first stage if sufficiently strong concentrated mineral acids are not used. The second stage of the reaction with mineral acids occurs under more vigorous conditions. Thiophenols react with organomercury compounds in a similar manner. Under usual conditions the second reaction step does not proceed with carboxylic acids.

B. Demercuration Reaction

The reaction of carboxylic (and sometimes mineral) acids with organomercury compounds under vigorous conditions, and in some cases the reactions of compounds such as mercaptans and phenols under milder conditions, proceeds in a different way, i.e., with isolation of metallic mercury:

$$RHgR + HX \longrightarrow Hg + RH + RX \qquad (3)$$

$$RHgX + HX \longrightarrow Hg + RX \qquad (4)$$

In some cases formation of olefins as well as saturated hydrocarbons occurs.

C. Elimination Reaction, Deoxydemercuration

The effect of acids on addition products of mercury salts to olefins and to carbon monoxide is a specific reaction which proceeds with olefin or carbon monoxide regeneration:

$$HOCH_2CH_2HgX + HX \longrightarrow C_2H_4 + H_2O + HgX_2 \qquad (5)$$

$$ROCOHgX + HX \longrightarrow CO + ROH + HgX_2 \qquad (6)$$

These reactions are specific examples of reactions of β-elimination of salt or other metal derivatives from so called quasicomplex compounds by the action of various reagents (also discussed later).

Reactions of organomercury compounds with acids have been the subject of wide investigations from different points of view. The main practical use of these reactions is the decomposition of addition products of mercury salts to triple bonds which results in the isolation of aldehydes and ketones (Kucherov's reaction).

II. HYDROGEN FOR MERCURY SUBSTITUTION (PROTODEMERCURATION)

A. Reaction Mechanism

Substitution of hydrogen for mercury by the action of acids is a heterolytic reaction. Mineral acids, react with R_2Hg and $RHgX$ with second-order kinetics (S_E2 mechanism) while with excess or carboxylic acids, the kinetics are pseudo first order (S_E1 mechanism).[1-4]

The reaction mechanism apparently depends on the nature of the solvent. Thus, protolysis of β-chlorovinylmercury chloride in dioxane proceeds via an S_E2 mechanism while in dimethylsulfoxide via an S_E1 mechanism.[5] Protolysis of dibenzylmercury accompanied by elimination of one benzyl radical by HCl in a number of solvents (water, acetonitrile, butanol, tetrahydrofuran, dimethylformamide, dimethylsulfoxide) occurs through an S_E1 mechanism.[6] The solvents, as arranged above, show decreasing reaction rate. Substitution of DCl for HCl produced no isotope effect.

Taking as a basis that the reaction rate of R_2Hg is affected by the nature of the anionic part of the acid molecule (for example, the reaction of R_2Hg with HBr is faster than with HCl) and, since the reaction rate increases with a decrease in the dielectric constant of the solvent, increases with addition of an anion similar to the acid anion, and decreases upon addition of another acid (H_2SO_4 to HCl), it is suggested in compliance with Dessy[7] that the reaction occurs by an attack of an ion pair $H^+ X^-$ or by a nondissociated molecule, HCl, and not by a solvated proton. Consequently, the reaction mechanism probably includes a four-center transition state:

$$
\begin{array}{ccc}
\text{R—Hg—R} & & \text{R—Hg—R} \\
\uparrow & \rightleftharpoons & \diamondsuit \\
\text{H—Cl} & & \text{H—Cl}
\end{array} \longrightarrow \text{RH} + \text{RHgCl}
$$

It is possible that during protolysis of organomercury salts, the formation of a new C—H bond may be slower than the opening of the old C—Hg bond.[8]

Considerable changes in the energies and entropies of activation upon changing from HCl to HBr (in reaction with Ar_2Hg) proves that the anion plays a very important part in the rate-determining step of the reaction. A linear dependance of log K versus the acidity H of different solvent systems confirms the attack by ion pairs.[9] The possibility of an attack by molecular HCl is not excluded since the value of the isotope effect during substitution of DCl for HCl is insignificant and the reaction rate in HCl–dioxane system is higher than in a medium of higher dielectrical constant (dimethylsulfoxide).[7]

At the same time, if $CH_2\!\!=\!\!CHHgI$ is decomposed into $C_2H_4 + HgI^+$ by the action of acids, a considerable isotope effect is observed upon changing from H_nX^n to D_nX^n ($X = ClO_4^-$ or SO_4^{2-}). This, as believed by Kreevoy and Krechmer,[10] proves that transfer of a proton to the carbon is the rate-determining step of the reaction. The value of the entropy of activation of this reaction indicates a solvated transition state. In a similar manner during the decomposition of allylmercury iodide by aqueous $HClO_4$, the rate-determining step is proton transfer to the carbon species.[11] The acidolysis rate of allylmercury iodide by $HClO_4$ in H_2O is faster than in D_2O: K_H/K_D is 3.25.[12] The last relationship measures two different kinds of solvent isotope effects, that produced by changing the proton being transferred to carbon to a deuterium (primary solvent isotope effect) and that produced by changing the remaining protons to deuteriums (secondary solvent isotope effect).[12]

Correlation of data proves that acidolysis reactions of R_2Hg (R = aryl or alkenyl) are electrophilic in nature, however, if R = Ar, the reaction center requirements for the electrophilic effect of substituents is lower than in other common reactions of electrophilic substitution. For Ar_2Hg and $(ArC\!\!\equiv\!\!C)_2Hg$, Hammet's equation requires use of $\sigma_+\sigma^+/2$, which evidently confirms a four-centered attack by HCl or H^+Cl^- and the formation of phenonium ions,[13]

in the transition state (in case of reaction with Ar_2Hg) which requires use of both σ and σ^+ for expressing the actual and potential electronic density.[9,13] The value ρ of Hammet's equation for arylacetylene compounds is equal to -1.0 and proves that their reaction is less sensitive to the ring substituents than the reaction of Ar_2Hg compounds for which the ρ value is -2.8.[13] Energy and entropy of activation of phenylacetylene compounds in the reaction are less than those of aromatic compounds.

HCl cleaves phenyl from $C_6H_5HgC_2H_5$ more easily than from $C_6H_5HgC_6H_5$; consequently, there is a transfer of electronic effects through the mercury.[9] (Compare with Ref. 7.)

B. Effect of Acid Type

The C—Hg bond is broken more easily by HX but is more stable to attack by oxy acids. The reaction rate of organomercury compounds with HX depends on the nature of the halide, for example the rate of reactions of $(C_6H_5)_2$ Hg with HBr is higher than with HCl.[9,14]

Perchloric acid reacts in a way similar to that of hydrogen halide acids but less energetically. Its use makes it possible to exclude the effect of the acid anion.

Deutero- and tritio acids react with organomercury compounds in a way similar to that of their hydrogen analogs.[15-17] Dilute sulfuric acid substitutes hydrogen for mercury in organomercury compounds but with less energy than HX.[18] The reaction of sulfuric acid with anthraquinone-mercury derivatives results in the formation of mercury-free sulfo compounds.[19-22]

Nitric acid of 40–50% and not containing nitric oxides substitutes hydrogen for mercury in most organomercury compounds as exemplified by Eqs. (1) and (2).[23] The effect of nitric acid containing nitric oxides causes oxidizing nitration of organomercury compounds and results in cleavage of the C—Hg bond. Product formations are acid concentration dependent, diazo compound formation in dilute acid, nitrophenols, nitroso, nitro, and polynitro compounds in concentrated one.[24-30]

The reaction of phosphoric acid with bisethylmercury acetylene occurs according to Eq. (1) and results in the formation of acid mono- and diethylmercuryphosphates.[31]

Thiophenols also substitute hydrogen for mercury in aromatic and saturated organomercury salts,[32,33] however, the reaction of the thio-phenols on the organomercury compounds is sometimes accompanied by de-mercuration. The reaction of phenols with R_2Hg causes separation of one of the R groups with formation of RH. In accordance with Koton,[34-45] phenols are simultaneously mercurated and unsymmetrical organo-mercury compounds, RHgC derivatives of phenols, are formed; however, Nesmeyanov and Kravtsov have shown that the second products of the reaction of R_2Hg with phenols are phenolates of aryl and alkyl-mercury.[46]

The hydrochloride of dimethylaniline (at 130–150°C)[47] and the hydro-chloride of trimethylamine (in alcohol and benzene at 100°C)[48] react with Ar_2Hg as free hydrochloric acid, forming ArH and $HgCl_2$.

Acid amides[49] and imides[49-52] react with Ar_2Hg like acids: the hydrogen of the amides and imides adds to the radical and the RHg radical forms RHgN derivatives.

C. Effect of the Type of Organomercury Compound

The stability range of various organomercury compounds with respect to acids is very wide. Some empirical formulas pertaining to the ease of mercury separation in organomercury compounds by the action of acids are herein presented. Dessy and his co-workers[9] believe that the rates of decomposition of R_2Hg by the action of hydrochloric acid in a mixture of dimethylsulfoxide-dioxane depends on the hybridization of the attacked carbon sp, sp^2, sp^3 which are in the approximate relationship 1000:100:1. The reaction rate of aromatic mercury compounds in dimethylsulfoxide is 100–1000 times higher than Alk_2Hg.[7] Mercury is separated from mercury acetylides most easily, which agrees quite well with Dessy's observations; however, the addition products of mercury salts to triple bonds including quasicomplex compounds (addition products of mercury salts to acetylene) do not react with dilute HCl. Under the action of acids, the addition products of mercury salts to acetylenic alcohols, ketones, and acids substitute hydrogen for HgX group as well (see Sec. IV-A). Mercury is very easily eliminated from diallylmercury as well as from dicyclopropylmercury.[7] As is stated above, mercury is eliminated from aromatic mercury compounds by acids more easily than from saturated compounds. However, perfluoroarylmercury compounds are distinguished by specific stability to acids, i.e., bisperfluoroarylmercury is recrystallized from concentrated sulfuric acid[53]; the perfluoroalkylmercury compounds are stable in acids as well.[54]

The C—Hg bond in symmetrical bisphenylbarenylmercury, 2, is not broken even during 20 hr of boiling alcoholic HCl, although phenylbaren is easily separated from mercury by the action of HCl on methyl(phenylbarenyl)mercury (1).[55]

$$CH_3Hg\,C\underset{B_{10}H_{10}}{\diagdown\diagup}CC_6H_5 \qquad C_6H_5C\underset{B_{10}H_{10}}{\diagdown\diagup}C\text{---}Hg\text{---}C\underset{B_{10}H_{10}}{\diagdown\diagup}CC_6H_5$$

$$\text{(1)} \qquad\qquad\qquad\qquad \text{(2)}$$

As has been stated, the reaction of acid with organomercury compounds is an electrophilic reaction (S_E2 mechanism); therefore, electron-attracting substituents in aromatic compounds reduce the reaction rate, while electron-donating substituents accelerate it.

Deuterodemercuration of benzylmercury chloride by DCl (in dioxane at 120–140°C) is accompanied by an isotope exchange reaction to the o-position.[56]

The rate of protolysis of C_6H_5HgBr by reaction with HCl in aqueous dioxane is considerably increased by the presence of NaI, which, as the authors believe,[57] depends on the higher reactivity of the C_6H_5–HgBr–NaI complex.

Bis(p-diphenyl)mercury, (p-$C_6H_5C_6H_5$)$_2$Hg, unexpectedly does not react with hydrochloric acid even under boiling conditions, and may possibly be a polymer.[58]

Extremely high stability of the C—Hg bond to acids is observed in β- and even α-mercurybicarboxylic acids. For example, mercurydipropionic acid $Hg(CH_2CH_2COOH)_2$[59] and α-mercury-β-phenylanhydrohydroacrylic acid,[60]

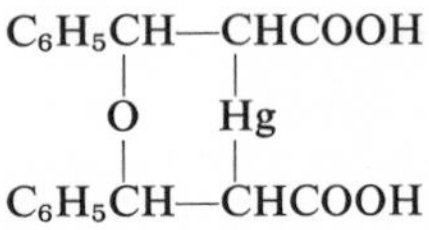

$$
\begin{array}{cc}
C_6H_5CH\!-\!CHCOOH \\
\mid \quad\;\; \mid \\
O \quad\;\; Hg \\
\mid \quad\;\; \mid \\
C_6H_5CH\!-\!CHCOOH
\end{array}
$$

are decomposed only by boiling for an extended period of time in concentrated hydrochloric, acetic, or nitric acids.

Among aliphatic mercury compounds, mercarbides are distinguished by extremely high stability to acids. The substance having, in accordance with Hofmann, the structure $(ClHg)_3C$—$C(HgCl)_3$ easily withstands boiling in concentrated HCl.[61]

Among unsaturated organomercury compounds, trans-isomers having double bonds react considerably faster than cis-isomers.[62–64] However, Beleskaya and Reutov et al.[5] find that the rate of protolysis of cis-β-chlorovinylmercury chloride by HCl or DCl in dioxane is much faster than that of the trans-isomer. Under monomolecular reaction conditions in dimethylsulfoxide, the trans-isomer reacts faster than the cis-isomer. The reaction rate of bis(cis-stilbene)mercury with HCl (in tetrahydrofuran or dioxane) is faster than that of bis(trans-stilbene)mercury.[63]

Formation of bornylchloride by the reaction of concentrated hydrochloric acid with α,α'-bis-chloromercurydibornyl ether proceeds according to the following reaction path[65]:

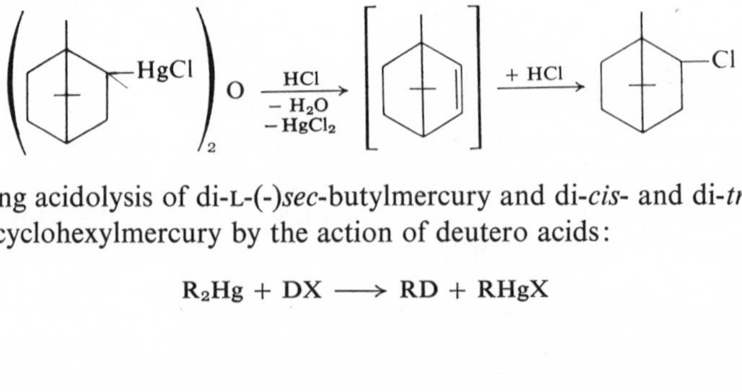

During acidolysis of di-L-(-)sec-butylmercury and di-cis- and di-trans-4-methylcyclohexylmercury by the action of deutero acids:

$$R_2Hg + DX \longrightarrow RD + RHgX$$

the configuration of R in the first case (with CH_3COOD and DCl in dioxane) remains partially unchanged, in the last two cases (with DCl in dioxane) remains almost completely unchanged, while CH_3COOD, D_2SO_4 for the last cases (in dioxane) yield RD with a *trans*-configuration.[17]

Acetylene or its homologs of mercury salts in water suspensions are decomposed by mineral acids and form acetaldehydes and ketones, respectively.[66-68] This reaction, known as Kucherov's reaction, is widely used in industry. The compound formed by acetylene, mercury(II) chloride, and water (described by Biltz and Mumm[69] as trichloromercury acetaldehyde $(ClHg)_3CCHO$), or more probably which has the structure $(ClHg)_2C=CHOHgCl$ by analogy with addition products of $HgCl_2$ with acetylenic acids,[70-72] yields aldehyde[66-69,73] and mercurt(I) and mercury (II) chlorides[73] when boiled with hydrochloric acid.

When dissolved in acid, acetone is isolated from the product produced from the mixture of methylacetylene, $HgCl_2$, and water which is described by Biltz and Mumm as tri(chloromercury)acetone, $CH_3COC-(HgCl)_3$, which probably has an enolic structure, $CH_3C(OHgCl)=C(HgCl)_2$.[67,68]

Under the action of acids, γ-mercurated alcohols and ethers, γ-$ROCH_2CH_2CH_2HgX$ (R = H or Alk, produced by the addition of mercury salts to cyclopropane compounds in water or in alcohols) form alcohols and ethers, respectively[74-84] (see Sec. IV).

1. The Kharasch Series

By the relative ease of R separation to form RH from organomercury compound, RHgR', during the following reaction,

$$RHgR' + HCl \longrightarrow RH + R'HgCl$$

Kharasch[85,91] has arranged the radicals R in the following electronegative series:

$1,3,5-(CH_3)_3C_6H_2, \alpha-C_4H_3S, p-HOC_6H_4, \alpha-C_4H_3O > CN^- > p-CH_3OC_6H_4 >$
$o-CH_3OC_6H_4 > \alpha-C_{10}H_7 > o-CH_3C_6H_4 > p-CH_3C_6H_4 > m-CH_3C_6H_4 > C_6H_5 >$
$p-ClC_6H_4 > o-ClC_6H_4 > m-ClC_6H_4 > CH_3 > C_2H_5 > C_3H_7 > n-C_4H_9 >$
$C_6H_5CH_2 > (CH_3)_3C.$

(see also Refs. 92 and 93).

The above-mentioned series was supplemented by Whitmore,[94] who has shown that highly branched aliphatic groups are located in the electronegative series as follows:

$$C_6H_5 > CH_3 > C_2H_5 > n-C_3H_7 > \left\{ \begin{array}{l} n-C_4H_9; (CH_3)_3CCH_2CH_2 \\ n-C_6H_{13}; (CH_3)_3CCH_2CH_2CH_2 \end{array} \right\}$$

$$> \underset{\underset{CH_3}{|}}{CH_3CH_2CH} > \underset{\underset{CH_3}{|}}{(CH_3)_3CCH}$$

$C_6H_5CH_2 > (CH_3)_3C, (CH_3)_3CCH_2.$

The groups enclosed in the brackets are not differentiated by this method, i.e., they produce mixtures of RHgCl and R'HgCl.

Temperature changes, solvent concentration, the nature of the hydrolytic agent and relative solubility do not affect the relative arrangement of the radicals in the Kharasch series.[88]

Kharasch's series was supplemented by many other authors with other radicals on the basis of investigation of kinetics of the reaction with hydrochloric and other acids Alk_2Hg, Ar_2Hg, symmetrical quasicomplex and other alkenyl compounds, $(XC_6H_4C{\equiv}C)_2Hg$, and some RHgX compounds. On the basis of the reaction rate of R_2Hg with HCl, R are arranged in the following series:

p-$CH_3C_6H_4$ > o-$CH_3C_6H_4$ > m-$CH_3C_6H_4$;
p-$CH_3C_6H_4$ > o-$CH_3C_6H_4$ > p-$(sec$-$C_4H_9)C_6H_4$ > α-$C_{10}H_7$ > m-$CH_3C_6H_4$ > C_6H_5 (reactions in tetrahydrofuran)[95]

$C_6H_5CH_2CH_2CH_2$ > $C_6H_5CH_2CH_2$ > $C_6H_5CH_2$ (in dioxane),[95] p-$CH_3OC_6H_4$ $\gg$ p-$C_6H_5C_6H_4$ > C_6H_5 > p-FC_6H_4 > p-ClC_6H_4 > m-O_2N C_6H_4 (in a mixture of $(CH_3)_2$ SO and dioxane).[13] The differences in the reaction rates of Alk_2Hg: C_2H_5 > i-C_3H_7 > n-C_3H_7 are not significant.[13]

The following series was found to hold in 90% water-dioxane, and some in tetrahydrofuran[62-64]:

1,3,5-$(CH_3)_3C_6H_2$ > $trans$-$CH_3CH{=}CH$ > p-$C_2H_5OC_6H_4$ > cis-$CH_3CH{=}CH$ > p-$CH_3OC_6H_4$ > $C_6H_5CH{=}CH$ > $trans$-$CH_3OCOC(CH_3){=}C(CH_3)$ > cis-$CH_3OCOCCH_3{=}C(CH_3)$ > α-C_4H_3S > o-$CH_3C_6H_4$ > p-$C_2H_5C_6H_4$ > p-$CH_3C_6H_4$ > cis-$C_6H_5CH{=}C(C_6H_5)$ > o-$CH_3OC_6H_4$ > α-$C_{10}H_7$ > m-$CH_3C_6H_4$ > $CH_2{=}CH$ > C_6H_5 > p-FC_6H_4 > m-$CH_3OC_6H_4$ > $trans$-$C_6H_5CH{=}C(C_6H_5)$ > p-ClC_6H_4 > p-BrC_6H_4 > m-FC_6H_4 > m-ClC_6H_4 > $CH_3OCOC(C_6H_5){=}C(C_6H_5)$ > o-$CH_3OCOC_6H_4$ > $trans$-$ClCH{=}CH$ > C_2H_5 > p-$CH_3OOCC_6H_4$ > cis-$ClCH{=}CH$ > o-$CH_3OOCC_6H_4$ > n-C_4H_9 > o-ClC_6H_4 > $C_6H_5CH_2$;

$CH_2{=}CH$ > $CF_2{=}CF$ > C_2H_5 (in aqueous tetrahydrofuran). But RfHgR' where Rf = $CF_2{=}CF$, R' = $CH_2{=}CH$, C_2H_5, or C_6H_5 during decomposition under the same reaction conditions, produces $CF_2{=}CFH$.[96]

Under the action of HCl (in 90% water dioxane) the rate of protodemercuration of respective R_2Hg is arranged in the series: C_5H_5Fe-C_5H_4(A) > C_6H_5 > $Mn(CO)_3C_5H_4$(B).[97]

The rate of protolysis of R_2Hg (where R = radical (A)[98] or (B)[99] is considerably higher than that of respective RHgCl.

According to the reaction rate of $RHgCHCl_2$ with HCl, R may be arranged in the series: CCl_3 > C_6H_5 > $CHCl_2$.[100]

The relative reactivity of AlkHgI with $HClO_4$ or H_2SO_4 is arranged as follows[8]:

cyclo-C_3H_5 $\gg$ CH_3 > C_2H_5 > n-C_3H_7 > i-C_3H_7 > n-C_6H_{11} > tert-C_4H_9

Kinetics of the reaction (for Alk $= i\text{-}C_3H_7$ and tert-C_4H_9) depends on the presence of oxygen.[101]

During heating in hydrochloric acid, the ease of decomposition of lower aliphatic mercury compounds, Alk_2Hg, decreases in the series[102]:

$$R_3CHg{-} > R_2CHHg{-} > RCH_2Hg{-}$$

The series obtained by Dessy et al. by the action of hydrochloric acid on R_2Hg (in a 10:1 mixture of dimethylsulfoxide and dioxan); cyclo-$C_3H_7 >$ $H_2C{=}CH > C_6H_5 \gg C_2H_5 > i\text{-}C_3H_7 > n\text{-}C_3H_7 > CH_3$ does not fully agree with the Kharasch series.

$(Z{-}C_6H_4C{\equiv}C)_2Hg$ reacts with HCl (in dimethylsulfoxide–dioxane, 10:1) faster than Ar_2Hg, however, the reaction rates of arylacetylene mercury compounds are less sensitive to ring substituents than the reactions of aromatic mercury compounds[9] (see part A above).

The rate of acetolysis of R_2Hg decreases in the series:

$$(C_6H_5)_2Hg > (sec\text{-}C_4H_9)_2Hg > (n\text{-}C_4H_9)_2Hg > [(C_6H_5)(CH_3)_2CCH_2]_2Hg^{103}$$

On the basis of the reaction with HCl (in benzene) C_6Cl_5HgR where $R = CH_3$ and C_6H_5, the following rates were observed: $C_6H_5 > C_6Cl_5 >$ CH_3,[104] which is similar to that observed for C_6F_5HgR compounds.[53]

On the basis of values of rate constants of the second-order reaction of protodemercuration of ArHgCl by HCl in a water–alcohol solution and determined with a consideration for entropy factors, Ar may be arranged in the following series[105]:

$$p\text{-}CH_3OC_6H_4 \gg p\text{-}CH_3C_6H_4 > m\text{-}CH_3C_6H_4 > C_6H_5 > m\text{-}CH_3OC_6H_4 >$$
$$p\text{-}ClC_6H_4 > m\text{-}ClC_6H_4$$

For this reaction correlation of log K with σ^+ is better than with σ. It has been proven that chloride ion participates in the rate-determining step of the reaction:

$$RHgCl + H_3O^+ \underset{slow}{\rightleftharpoons} [H{-}R{-}HgCl]^+ \xrightarrow{fast\ Cl'} RH + HgCl_2 + H_2O$$

Similar reactions for RHgCl, where $R =$ heterocyclic radicals, produced the following series: 2-furyl (27) > 2-seleniumphenyl (25) > 2-thienyl (11) > 3-furyl (1) > phenyl (0.0067).[106] The relative rate constants are enclosed in the brackets proving that the rates of the reactions of the heterocyclic compounds of mercury are considerably higher than those of aromatic (phenyl) compounds.

While $C_6H_5HgCBr_3$ reacts with HCl (in toluene) to yield C_6H_6, but not $CHBr_3$[107] and $C_6H_5HgCCl_2Br$ yields $CHCl_3$ and C_6H_6,[107] $C_6H_6HgCCl_3$ reacts with HCl (in CH_3OH) to give $CHCl_3$ and C_6H_5HgCl.[108]

III. DEMERCURATION REACTIONS

If R_2Hg or RHgX (R = aryl or alkyl) and carboxylic acids R'COOH (acetic acid in particular) are heated in a sealed tube to a temperature of 160–250°C, the organomercury compounds are decomposed with isolation of metallic mercury, hydrocarbons RH, and sometimes R'COOR and olefins R_nH_{2n}.[109,110] Analogous solvolytic type reactions may also occur at more moderate temperatures,[17,111–116] (also see Refs. 103, 117, and 118) and can be represented by the following reaction:

$$RHgX + HS \longrightarrow HX + Hg + RS + \text{olefin}$$

A. Mechanism of the Demercuration Reaction

The mechanism of demercuration at moderate temperatures was investigated for the reaction of RHgX. The rate of solvolysis (acidolysis) is proportional to the ionization of the organomercury salt, RHgX, and the relative amount of alkyl acetate and olefin formed (R = cyclo-C_6H_{11}) does not depend on the nature of anion. This suggests that a carbonium ion type mechanism is involved[115]:

$$RHgX \rightleftharpoons RHg^+ + X^-$$
$$RHg^+ \longrightarrow R^+ + Hg$$
$$R^+ + HS \longrightarrow RS + \text{olefin} + H^+ \qquad (S = \text{anion})$$

This carbonium ion mechanism is confirmed by the same rate of hydrolysis for cyclohexylmercurytosylate and perchlorate[115] and by the fact that the rate of solvolysis of alkylmercury salts including the acidolysis effected by formic and acetic acids increases with an increase of substituent branching. This is due to the fact that the branching facilitates stabilization of the carbonium ion formed.

B. Effect of the Organomercury Salt Type and Other Factors

Solvolysis reaction (in particular acidolysis) of alkylmercury salts is general for all alkyl groups studied[115] (compare Ref. 117) and is very sensitive to changes in structure[115,116] and reagent nucleophilicity.[111]

When RHgX (R = alkenyl) is heated with carboxylic acids, esters (R'COOR) are formed of the added acid R'COOH, metallic mercury is isolated, and olefins or their dimers are formed. The amount of mercury isolated decreases sharply if the double bond is not adjacent to the mercury atom and an alkenyl series may be arranged as follows[112]:

$$CH_2{=}CH > CH_2{=}CH{\cdot}CH_2 > CH_2{=}CH{\cdot}CH_2{\cdot}CH_2$$

In addition, increased temperatures are required for the decomposition of RHgX with R = alkenylradicals: from 110 to 120°C for R = vinyl and from 150 to 200°C for R = 3-butenyl.

Appropriate ethers (esters) are formed when RHgX is heated (in some cases at moderate temperatures): thiophenols form ArSR,[112,113] para-toluenesulfonic acid forms p-CH$_3$C$_6$H$_4$SO$_3$R,[113] mercaptans form AlkSR,[113] and phenols form ArOR.[113]

Decomposition of divinylmercury in p-toluenesulfonic acid has produced p-tolyl vinyl sulfone.[114] The reaction was carried out in ethanol and heating on a steam bath.

The decreasing rate constants of solvolysis of alkylmercury perchlorates by acetic acid may be arranged in the series[115]:

$$\text{tert-C}_4\text{H}_9 > \text{cyclo-C}_6\text{H}_{11} > \text{sec-C}_4\text{H}_9 > i\text{-C}_3\text{H}_7 > n\text{-C}_4\text{H}_9 > \text{C}_2\text{H}_5 > \text{CH}_3$$

If dilute perchloric acid reacts with β-oxy-(alkoxy) mercury derivatives of cyclic alkyls, the demercuration reaction may be prevalent but not deoxymercuration.

The reaction of HClO$_4$ in acetic acid with the acetates of exo-norbornyl, cyclohexyl-, and endo-norbornylmercury yield rate constants for demercuration which are in the relationship of 5000:40:1.[119]

Depending on the nature of the substituent in the β-position of ethyl-mercuryperchlorates, the rate of acidolysis of formic or acetic acids increases in the series[120]:

$$\text{H} < \text{C}_6\text{H}_5 < m\text{-CH}_3\text{C}_6\text{H}_4 < p\text{-CH}_3\text{C}_6\text{H}_4 < \text{C}_6\text{H}_5\text{C(CH}_3)_2\text{CH}_2$$

The effect of the anion in the acidolysis of crotyl- and cinnamylmercury salts has been studied and involves electrophilic (R: HgX) or nucleophilic (R :HgX) mechanisms.[121]

If X = halide, electrophilic substitution occurs, and crotylmercury bromide by the action of HCl in ethyl acetate ($S_E i$-mechanism), by the action of HClO$_4$ in acetic acid an $S_E 2$ mechanism is followed, and crotyl-mercury acetate by the action of HCl (anion exchange of RHgAc with HCl takes place) in ethyl acetate form 1-butene and mercury. The acetolysis of crotylmercury acetate by dilute perchloric acid in acetic acid involves a nucleophilic mechanism ($S_N i$) and produces butenyl acetate, mercury, and a small amount of butene:

(1) $S_E i$ (2) $S_E 2$ (3) $S_N i$

The reaction is accompanied by allyl rearrangement of the radical (1-butene is formed from crotyl radical and allylbenzene from cinnamyl radical). The rates of acidolysis of the mercury compounds are considerably higher than those of the acidolysis of the saturated organomercury compounds.

Certain organomercury esters may be demercurated by means of acids or Lewis acids. β-Arylethylmercury acetates may be demercurated with H_2SO_4, H_3PO_4, $HClO_4$, or $ArSO_3H$.[122–124] In

$$p\text{-}CH_3OC_6H_4CH_2CH_2HgOCOCH_3 \xrightarrow[40°]{HX} p\text{-}CH_3OC_6H_4OCOCH_3 + Hg$$

another instance ethyl α-2-(acetoxymercuriethyl)acetoacetate reacts with boron trifluoride etherate and acetic acid to give ethyl α(2-acetoxy-ethyl)-acetoacetate.

$$CH_3COCH(CH_2CH_2HgOCOCH_3)COOC_2H_5 \xrightarrow[(C_2H_5)_2O \cdot BF_3]{CH_3COOH}$$
$$CH_3COCH(CH_2CH_2OCOCH_3)COOC_2H_5 + Hg$$

During the reaction of diphenylmercury or phenylmercury acetate with acids [RCOOH, $(HO)_2PO(OR)$, $p\text{-}CH_3C_6H_4SO_3H$] in the presence of tributylphosphine, demercuration also takes place. However, this reaction is accompanied by the formation of the acid anhydride and tributylphosphine-oxide[125]:

$$(C_6H_5)_2Hg + (n\text{-}C_4H_9)_3P + 2RCOOH \longrightarrow$$
$$Hg + (RCO_2)_2O + (n\text{-}C_4H_9)_3PO + C_6H_6$$
$$2C_6H_5HgOCOCH_3 + 2(n\text{-}C_4H_9)_3P + 2RCOOH \longrightarrow$$
$$2Hg + (CH_3CO)_2O + (RCO)_2O + 2(n\text{-}C_4H_9)_3PO + 2C_6H_6$$

Ethyl benzoate is formed in the reaction with benzoic acid in the presence of tributylphosphine in ethanol rather than the acid anhydride, while in the presence of aniline the anilide of the acid forms rather than alcohol.[125]

IV. DEOXYMERCURATION (ELIMINATION) REACTION

A. General Survey

This specific reaction, unprecedented for other organomercury compounds, is the reaction of acids with addition products of mercury salts (in water, alcohol, and sometimes in carboxylic acid) to double bonds, i.e., β-mercurated alcohols, ethers and esters, and to carbon monoxide. The reaction is accompanied by regeneration of olefin or carbon monoxide:

$$ROCH_2CH_2HgX + HX \longrightarrow ROH + CH_2{=}CH_2 + HgX_2$$
$$(R = H, Alk, RCO)$$
$$ROCOHgX + HX \longrightarrow ROH + CO + HgX_2 \qquad (R = Alk)$$

The reaction is characteristic of these organomercury compounds and proceeds easily, even with dilute mineral acids. If the reaction is carried out with β-halogen mercuryethylamines the olefins are not isolated but ammonium salts are isolated[126]:

$$ClHgCH_2CH_2NR_2 + HCl \longrightarrow [ClHgCH_2CH_2NR_2H]^+ Cl^{-\ 126}$$

By reaction with other reagents capable of forming mercury salt complexes for example, the bulk of the symmetrization reagents such as KI,[127] KCN,[128] KCNS, etc., as well as CH_3I,[128] all types of addition products of mercury salts to olefins are decomposed with the elimination of olefin:

$$ROCH_2CH_2HgCl + 4KCN + H_2O \longrightarrow$$
$$C_2H_4 + K_2Hg(CN)_4 + ROH + KOH + KCl.$$

Except for the decomposition of mercury addition products to double bonds by the addition of acids, in other reactions (i.e. arylation[129] and some other reactions)[129–132] these compounds react in a similar manner as other organomercury compounds.

Because of the dual behavior of these compounds (on the one hand, they react similarly to common organomercury compounds; on the other hand, they easily regenerate olefin during their deoxymercuration (and easily isolate alkene by other elimination reactions)), these compounds have been named by Nesmeyanov[133] as quasicomplex ones. Quasicomplex addition products of mercury salts to acetylene, acetylenic alcohols, ketone, and acids are discussed below.

Based on the extraordinary ease of olefin elimination from alkanolmercury salts and ethers by the mineral acids (as well as other reagents)— probably the most striking example of β-elimination—a prolonged discussion was held considering their alkanol[128–131,134–138] structure, for instance, $ROCH_2CH_2HgX$ or complex[139,140] structure, such as $CH_2{=}CH_2Hg/OR/X$ or tautomerism[141,142] of the compounds shown in the following scheme:

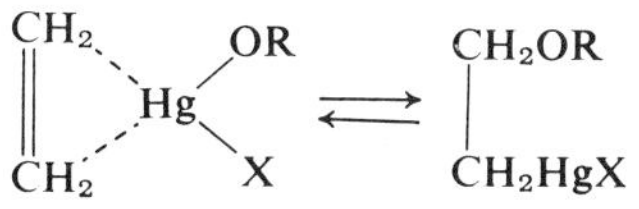

For the evidence of true organometallic structure of the addition products of mercury salts to alkenes (type $ROCR'R''CR''R'''HgX$) and to carbon monoxide (type, $ROC(O)HgX$) see the fundamental scientific work by Adams,[143,144] Nesmeyanov,[129,130] Marvel,[145,146] Pfeiffer[140] and others.[138,145–152]

The extreme ease of olefin elimination is explained by the presence of σ,σ-conjugate bond systems in the mentioned compounds[153,154]:

$$\underset{1}{\text{HO}}-\underset{2}{\text{CH}_2}-\underset{3}{\text{CH}_2}-\underset{4}{\text{HgX}}$$

These conjugate systems are characterized by the reaction proceeding not only in accordance with 1-to-2 bond, but due to displacement of reaction center[153,154] in accordance with 1-to-4 bond:

$$\text{ClHg}-\text{CH}_2-\text{CH}_2-\text{OH} \xrightarrow{\text{HCl}} \text{HgCl}_2 + \text{CH}_2{=}\text{CH}_2 + \text{H}_2\text{O}$$

The deoxymercuration reaction is related to the last reactions.

Increased polarization of Hg—C (as compared with H—C bond) and O—C bonds; for triple-bond addition products (for example, ClCH= CHHgCl) Cl—C being in favorable conjugation to a 1–3 position, is the reason for the specific properties of these compounds. Some compounds of the addition products of mercury salts to triple bonds (in the reaction with not highly concentrated acids) behave, however, similarly to common alkyl and aryl mercury compounds (see below). Actually, γ-mercurated alcohols and their ethers, $\text{ROCH}_2\text{CH}_2\text{CH}_2\text{HgX}$, discovered by Levina and her co-workers[74-84] and produced by the reaction of mercury salts with cyclopropane, compounds with XHg- and -OR are in γ-position, are not definitely quasicomplexes (see Sec. I-C).

Dual reactivity of addition products of mercury salts to carbon monoxide in alcohol, similar dual reactivity of addition products of mercury salts in water, alcohol or acetic acid to double bond of olefins is explained by σ,σ-conjugation as well:

$$\text{XHg}-\text{C}-\text{OR}$$
$$\underset{\text{O}}{\overset{\|}{}}$$

Quasicomplex properties are more vividly shown by triple bond addition products of mercury salts, i.e., the addition products of HgCl_2 to acetylene and to acetylenic alcohols, ketones, and acids.

The true organometallic structure of addition products of HgCl_2 to acetylene, for example, β-chlorovinylmercury chloride, ClCH=CHHgCl, is proved without doubt by the presence of stable *trans-* and *cis*-isomers for both types of compounds: RHgCl—ClHgCH=CHCl and respective symmetrical compounds R_2Hg, $(\text{ClCH}{=}\text{CH})_2\text{Hg}$, which can be produced by symmetrization of *trans-*[157] or *cis-*[158] ClCH=CHHgCl, respectively (see, for example, Refs. 155 and 156). Each of the stereoisomers of $(\text{ClCH}{=}\text{CH})_2\text{Hg}$ reacts with a mercury salt to produce the corresponding isomer ClCH=CHHgX.[158] No isomerization of isomers was observed and during numerous reactions of β-chlorovinyl compounds of mercury with alloys[159] or halides[160] of tin, thallium,[161,162] and boron,[163] and resulted in the formation of β-chlorovinyl compounds of the aforementioned elements and, during reverse transformations from β-chlorovinyl compounds of the said elements, to the mercury compounds.[160]

These reactions were the foundations for the retention of configuration rule during electrophilic and homolytic reactions at the olefinic carbon atom.[164,165]

β-Chlorovinylmercury salts produce 1-chloro-2-iodo-ethylene[166] by reaction with iodine. Under the action of HCl or HBr, as distinct from the addition products of mercury salts to olefins, β-chlorovinyl-mercury salts produce vinyl chloride and not acetylene.[167] On the other hand, $ClCH=CHgCl$ and $(ClCH=CH)_2Hg$ are quite susceptible to β-elimination reactions. $HgCl_2$ is easily isolated under the action of its binding agents such as KI, KCN, $Na_2S_2O_3$, H_2S, or $(C_6H_5)_3P$.[166] As distinct from addition products of mercury salts to olefins as well, both $ClCH=CHHgCl$ and $(ClCH=CH)_2Hg$ are arylated by diphenyltin(IV) chloride as free $HgCl_2$ with formation of C_6H_5HgCl in neutral medium and $(C_6H_5)_2Hg$ in alkali medium.[166] Diazomethane reacts with $ClCH=CHHgCl$ as with free $HgCl_2$, forming $ClCH_2HgCl$.[130]

Dual reactivity of addition products of mercury salts to acetylene is explained also by the fact that their molecule contains σ,σ-conjugate bond system:

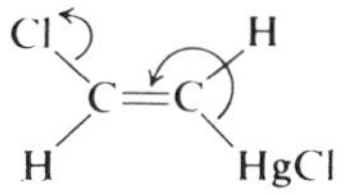

and reactions which occur are characteristic of conjugate systems not only in accordance with 1- to 2-bonds but in accordance with 1- to 4-bonds[153,154]:

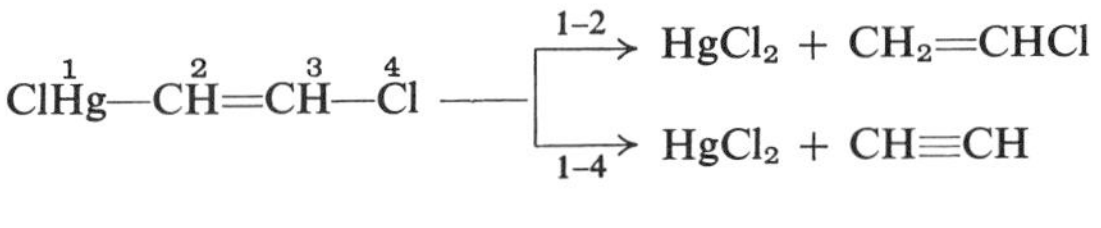

The addition products of $HgCl_2$ to acetylenic alcohols,[168] ketones,[169] acids, and esters[170] show a tendency toward β-elimination characteristic of quasicomplex compounds. For example, the addition product of $HgCl_2$ and phenylethynylmethylketone does not react with acids to form $HgCl_2$ but substitutes hydrogen for the HgCl group. However, all afore-mentioned addition products form $HgCl_2$ not only by the action of such reagents as KI, KCN, and $Na_2S_2O_3$, but also NaCl. These most vividly expressed quasi-complex properties, as compared with β-chlorovinyl compounds and the tendency toward β-elimination are explained, for the addition products of $HgCl_2$ to acetylenic ketones[169] and acids[170] by the increased mobility of the mercury atom due to simultaneous conjugation of Hg—C bond with C—Cl and C=C bonds:

$$C_6H_5\overset{3}{-}\overset{\overset{1}{HgCl}}{C}\underset{\underset{4}{Cl}}{\overset{2}{=\!\!=}}\overset{3}{C}\overset{3}{-}\underset{\underset{4}{O}}{\overset{}{C}}-CH_3 \qquad\qquad R\overset{3}{-}\overset{\overset{1}{HgCl}}{C}\underset{\underset{4}{Cl}}{\overset{2}{=\!\!=}}\overset{3}{C}\overset{3}{-}\underset{\underset{4}{O}}{\overset{}{C}}-OH$$

Thus, the deoxymercuration reaction of β-mercurated alcohols and ethers constitutes a particular case of the general process of β-elimination characteristic of quasicomplex compounds.

B. Mechanism of Deoxymercuration Reaction

Thorough investigation[171-178] of the mechanism of deoxymercuration (mainly in the case of the reaction of ROCR'R"CR"'R""HgI with, primarily, $HClO_4$) has proved that the first stage of the reaction is a fast and reversible addition of the proton to the alkoxyl atom of oxygen (7) which is followed by the rate determining step (8)—the gradual formation of the transition state (4) of a π-complex "mercurinium ion"[179] and the last rapid olefin elimination step (9):

$$\underset{\underset{OR}{|}}{\overset{|}{-C}}\!-\!\underset{\underset{HgI}{|}}{\overset{|}{C}}\!- + H_3O^+ \overset{fast}{\rightleftharpoons} \underset{\underset{\overset{+}{H}OR}{|}}{\overset{|}{-C}}\!-\!\underset{\underset{HgI}{|}}{\overset{|}{C}}\!- + H_2O \qquad (7)$$

$$(3)$$

$$\underset{\underset{\overset{+}{H}OR}{|}}{\overset{|}{-C}}\!-\!\underset{\underset{HgI}{|}}{\overset{|}{C}}\!- \overset{slow}{\longrightarrow} -\overset{|}{C}\overset{\overset{HgI}{\overset{+}{\triangle}}}{-\!-\!-\!-\!-}\overset{|}{C}- + ROH \qquad (8)$$

$$(4)$$

$$-\overset{|}{C}\overset{\overset{HgI}{\overset{+}{\triangle}}}{-\!-\!-\!-}\overset{|}{C}- + \underset{\underset{OR}{|}}{\overset{\overset{HgI}{|}}{-C}}\!-\!\overset{|}{C}\!- \overset{fast}{\longrightarrow} HgI_2 + \underset{}{\overset{}{>}C\!=\!C\overset{}{<}} + \underset{\underset{OR}{|}}{\overset{\overset{Hg^+}{|}}{-C}}\!-\!\overset{|}{C}\!- \qquad (9)$$

The reaction kinetics of the deoxymercuration of $ROCH_2CH_2HgCl$ by HCl are very similar.[172,180,181,184] (Peculiarities of the reaction kinetics in this case are discussed in Ref. 180.)

On the basis of the large isotope effect in the rate of acid ($HClO_4$) deoxymercuration[182] during substitution of deuterium for hydrogen in carbon-2 bonded to oxygen in the addition product 2-phenyl-2-methoxyethylmercury iodide, it is possible to suppose that in addition to formation of structures **3** and **4** during the transition state, structures of possible type **5** with a positive charge on the carbon atom in which oxygen separation occurs, may also take part:

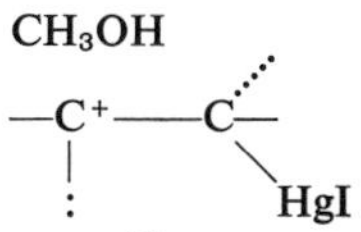

$$CH_3OH$$

(5)

The data obtained from a correlation analysis[177] for the series $R'CH(OR)$-CH_2HgI ($R = H$ or CH_3) show that the transition state is not a protonated initial state such as **3** but is the intermediate between **(3)**, **(4)** containing CH_3OH and **(5)**.

Deoxymercuration is catalysed by hydronium ions, H_3O^+, iodide ions[172,183] (or chloride ion if $ROCH_2CH_3HgCl$ is deoxymercurated by HCl[180]), with the help of the combined catalytic effects of hydronium and iodide ions. Deoxymercuration may occur without involving mineral acids, as in the case with acylated β-oxymercurated alcohols. In these cases the solvolytic reaction takes place by the action of solvents, alcohol, dioxane, acetic acid,[132] which is similarly co-catalyzed by iodide ions.[183]

Investigations of thermodynamic parameters of deoxymercuration kinetics by aqueous $HClO_4$ of the diastereomers of α- and β-2-methoxy-1-iodomercurycyclohexanes **6** and **7**[171]:

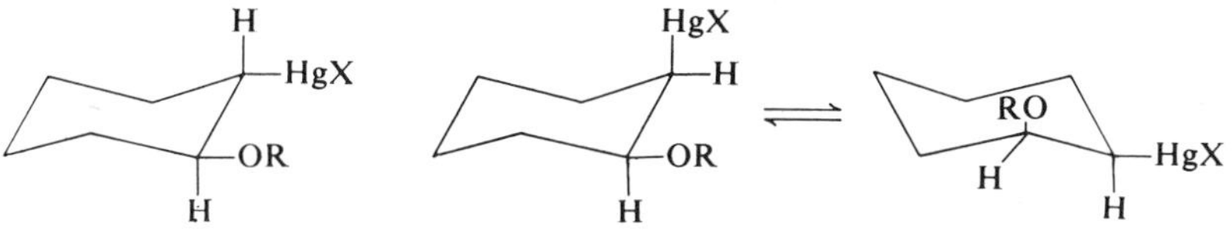

(6)   **(7)**

have made it possible to adequately solve the problem of stereochemical configuration of addition products to cycloalkenes and ascribe the structure of *trans*-isomer **(6)** to the α-product which is deoxymercurated 10^6 times faster than the β-isomer and has a lower enthalpy of activation: $\Delta H^{\ddagger} = 17.75 \pm 0.019$ kcal/mole for α-diastereomer, $\Delta H^{\ddagger} = 26 \pm 0.7$ kcal/mole for the β-diastereomer. It is due to this fact that, during synchronized isolation of ROH and formation of mercurinium ion, the required coplanarity of both carbon atoms and the groups isolated from them is obtained with less distortion of the carbocyclic bond angles when the atoms of mercury and oxygen are in a *trans*-position (in diaxial conformation).[171]

The approximate equivalence of entropy of activation values for both isomers ($\Delta S^{\ddagger} = 4.6 \pm 0.6$ cal/mole degree for α, $\Delta S^{\ddagger} = 4.5 + 2.0$ cal/mole degree for β) indicates that the solvation of both transition states is approximately equal.

The rate-determining step of the reaction in the alternate mechanism of deoxymercuration proposed by Wright[184] can be expressed by the following equation:

$$-\overset{\displaystyle |}{\underset{\displaystyle |}{C}}-\overset{\displaystyle |}{\underset{\displaystyle |}{C}}- \longrightarrow \ \ \overset{\diagdown}{\diagup}C{=}C\overset{\diagup}{\diagdown} + \underset{\displaystyle |}{RO} \longrightarrow \overset{+}{Hg}I$$
$$\underset{\displaystyle H^{+}OR\ \ HgI}{} \qquad\qquad\qquad\qquad \underset{\displaystyle H}{}$$

In this case, the higher rate of deoxymercuration of diastereomer, to which Wright ascribes a *cis*-configuration, which might be explained by higher solvation of mercury ion with alcohol rather than with water, is hardly possible. In the given case, the difference in rates of deoxymercuration of diastereomers must be dependent on the differences in entropies of activation rather than on that of enthalpies of activation.

The deoxymercuration reaction is characteristic of second-order kinetics (pseudo first order in the case of excess acid).[173,175,176,180,185–187]

C. Effect of Organomercury Compound Type

Deoxymercuration proceeds stereospecifically and results in the formation of the olefin with original configuration. The deoxymercuration reaction proceeds more easily with *trans*-isomers of addition products of mercury salts to cycloalkenes than with *cis*-isomers,[171] easier with diastereomers of lower dipole moment,[188] with addition products of mercury salts in water (β-oxyalkylmercury salts) than with addition products of mercury salts in alcohol (salts of β-alkoxyalkylmercury[172] excluding salts of tert-butoxyethylmercury whose reaction rate with aqueous $HClO_4$ is higher than that of the respective β-oxyethylmercury salt[178]) or with acylated β-mercurated alcohols.[132] The reaction rates for salts of β-alkoxyalkylmercury increases as the electron-donor properties of the alkoxy group increase and depends on steric factors and on possibility of formation of hydrogen bonding.[178,185]

Deoxymercuration of *cis-β*-mercurated alcohols is hindered (the products of oxymercuration of some substituted cycloalkenes[189]).

No deoxymercuration occurs but formation of the perchlorate RHg-ClO_4 (9) is observed when compound (8) reacts with $HClO_4$, as shown at top of following page, since the presence of electron accepting groups in position 2 and 3 hinders formation of the carbonium ion in position 5.[190] As the rate of deoxymercuration decreases (by the action of HCl in 75%

water-alcohol) involving the addition products in accordance with their double bond of ethoxy- and mercury chloride group, the olefins can be arranged in the series: isobutylene > *p*-methylstyrene > *m*-methylstyrene

(8)   **(9)**

> cyclohexene > 2-butene > styrene > *p*-chlorostyrene > *p*-bromostyrene > propylene > *m*-chlorostyrene > *m*-bromostyrene > ethylene.[180]

α-Methoxycyclohexylmercury trifluoroacetate is decomposed by trifluoroacetic acid (in methanol) slower than by hydrochloric acid,[185] although carboxylic acids usually do not decompose the addition products of mercury salts to olefins. Decomposition of addition products of mercury salts to carbon monoxide by the action of trifluoroacetic and acetic acids in dimethylsulfoxide is accelerated by the presence of a number of reaction "assistors" which bond the mercury salt in a transition state.[191] The following is a series in decreasing accelerating effect:

$$C_6H_5SH > I^- > Br^- > Cl^- > (C_6H_5)_3PO > (C_6H_5)_5P; C_6H_5OH.$$

D. Miscellaneous Observations

A mechanism of protolysis has been proposed for β-chlorovinylmercury chloride (in dimethylsulfoxide)[192,193] and for dibenzylmercury.[194,195] Catalysis of protodemercuration by means of chloride ion has been studied.[196] Protodemercuration of allylmercury compounds proceeds by attack on the double bond with over-all allylic displacement.[11] The action of acids on allylmercury iodide,[197–199] and on other allylic compounds of mercury[200] has been studied. Pernitric acid did not rupture the C—Hg bond in $(NO_2)_3C(CH_2)_2HgCl$.[201] Protolysis of C_6H_5HgBr by means of phosphoric and monochloracetic acids was reported.[202]

Reactions of ROC(O)HgX with primary and secondary amines lead to separation of Hg.[203] Mercuric derivatives of neobarenes are decomposed by acids.[204] The action of acids on β-chloromercuryethylamines has been studied[205,206] as well as the reactions of organomercury compounds with carboxylic acids,[207] and the kinetics of solvolysis.[208,209] An R group in $RHgCH_3$ containing silicon is split off more easily than CH_3 when treated with acids.[210]

Reactions of $CHCl_3$ with $C_6H_5HgCX_3$[211] and $C_6H_5HgCHX_2$[212] have also been reported.

The reaction of organic compounds of mercury with CO is analogous to the reactions of solvolysis.[213,214]

Demercuration of α-halogenmercuryoxo compounds by the action of $(RO)_3P$ has been reported.[215]

REFERENCES

1. M. M. Kreevoy, *J. Am. Chem. Soc.*, **79**, 5927 (1957).
2. O. A. Reutov, I. P. Beletskaya, and M. J. Alejnikova, *Zh. Fiz. Khim.*, **36**, 489 (1962).
3. S. Winstein and T. G. Traylor, *J. Am. Chem. Soc.*, **78**, 2597 (1956).
4. A. H. Corwin and M. A. Naylor, *J. Am. Chem. Soc.*, **69**, 1004 (1947).
5. I. P. Beletskaya, V. I. Karpov, V. A. Moskalenko, and O. A. Reutov, *Dokl. Akad. Nauk SSSR*, **162**, 86 (1965).
6. I. P. Beletskaya, L. A. Fedorov, and O. A. Reutov, *Dokl. Akad. Nauk SSSR*, **163**, 1381 (1965).
7. R. R. Dessy, G. F. Reynolds, and J.-Y. Kim, *J. Am. Chem. Soc.*, **81**, 2683 (1959).
8. M. M. Kreevoy and R. L. Hansen, *J. Am. Chem. Soc.*, **83**, 626 (1961).
9. R. E. Dessy and J.-Y. Kim, *J. Am. Chem. Soc.*, **83**, 1167 (1961).
10. M. M. Kreevoy and R. A. Kretchmer, *J. Am. Chem. Soc.*, **86**, 2435 (1964).
11. M. M. Kreevoy, P. J. Steinwand, and W. V. Kayser, *J. Am. Chem. Soc.*, **88**, 124 (1966).
12. M. M. Kreevoy, P. J. Steinwand, and W. V. Kayser, *J. Am. Chem. Soc.*, **86**, 5013 (1964).
13. R. E. Dessy and J.-Y. Kim, *J. Am. Chem. Soc.*, **82**, 686 (1960).
14. H. Zimmer and S. Makower, *Naturwissenschaften*, **41**, 551 (1954).
15. L. Steinkopf and M. Boetius, *Ann. Chem.*, **546**, 208 (1941).
16. N. R. Buu-Hoi, *Bull. Soc. Chim. France*, **1958**, 1407.
17. L. H. Gale, F. R. Jensen, and J. A. Landgrebe, *Chem. Ind.*, **1961**, 118.
18. N. N. Mel'nikov and M. S. Rokitskaya, *Zh. Obshch. Khim.*, **7**, 2518 (1937).
19. V. V. Kozlov and B. I. Belov, *Zh. Obshch. Khim.*, **25**, 410 (1955).
20. V. V. Kozlov and B. I. Belov, *Zh. Obshch. Khim.*, **25**, 565 (1955).
21. V. V. Kozlov, *Zh. Obshch. Khim.*, **18**, 2094 (1948).
22. I. S. Dokunikhin and L. A. Gaeva, *Zh. Vses. Khim. Obshchestva. im. D. I. Mendeleeva*, **6**, 112 (1961).
23. A. I. Titov and A. N. Baryshnikova, *Zh. Obshch. Khim.*, **17**, 829 (1947).
24. A. I. Titov and N. G. Laptev, *Zh. Obshch. Khim.*, **19**, 267 (1949).
25. I. Ogata and M. Tsuchida, *J. Org. Chem.*, **21**, 1065 (1956).
26. P. I. Petrovich, *Zh. Obshch. Khim.*, **30**, 2808 (1960).
27. P. I. Petrovich, *Zh. Obshch. Khim.*, **29**, 2387 (1959).
28. A. I. Titov and H. G. Laptev, *Dokl. Akad. Nauk SSSR*, **66**, 1101 (1949).
29. W. Poetke and W. Fürst, *Arch. Pharm.*, **294**, 524 (1961; *Chem. Abstr.*, **56**, 1469 (1962).
30. M. Carmack, M. M. Baiser, G. K. Handrick, L. W. Kissinger, and E. H. Specht, *J. Am. Chem. Soc.*, **69**, 785 (1947).
31. B. G. Zupancic, *Monatsh. Chem.*, **92**, 907 (1961).
32. M. M. Koton, *Zh. Obshch. Khim.*, **22**, 643 (1952).
33. G. Leandri and D. Spinelli, *Ann. Chim.* (*Roma*), **49**, 1885 (1959).
34. M. M. Koton, *Zh. Obshch. Khim.*, **22**, 1136 (1952).

35. M. M. Koton, *Zh. Obshch. Khim.*, **19**, 730 (1949).
36. M. M. Koton, *Zh. Obshch. Khim.*, **19**, 734 (1949).
37. M. M. Koton, *Zh. Obshch. Khim.*, **20**, 2096 (1950).
38. M. M. Koton and T. M. Zorina, *Zh. Obshch. Khim.*, **17**, 1220 (1947).
39. V. F. Martynova, *Zh. Obshch. Khim.*, **26**, 894 (1956).
40. M. M. Koton and I. A. Chernov, *Zh. Obshch. Khim.*, **19**, 2104 (1949).
41. M. M. Koton and V. F. Martynova, *Zh. Obshch. Khim.*, **25**, 705 (1955).
42. M. M. Koton and V. F. Martynova, *Izv. Akad. Nauk SSSR, Otd. Khim. Nauk*, **1955**, 1963.
43. V. F. Martynova, *Zh. Obshch. Khim.*, **27**, 1056 (1957).
44. M. M. Koton and V. F. Martynova, *Zh. Obshch. Khim.*, **25**, 594 (1955).
45. M. M. Koton and A. A. Bolshakova, *Zh. Obshch. Khim.*, **23**, 2023 (1953).
46. A. N. Nesmeyanov and D. N. Kravtsov, *Izv. Akad. Nauk SSSR, Otd. Khim. Nauk*, **1962**, 431.
47. M. M. Koton and V. F. Martynova, *Zh. Obshch. Khim.*, **19**, 1141 (1949).
48. M. M. Koton, *Zh. Obshch. Khim.*, **18**, 936 (1948).
49. G. A. Razuvaev and N. S. Vyazankin, *Zh. Obshch. Khim.*, **22**, 640 (1952).
50. G. A. Razuvaev, Ju. A. Ol'dekop, and N. S. Vyazankin, *Zh. Obshch. Khim.*, **21**, 1283 (1951).
51. G. A. Razuvaev and N. S. Vyazankin, Syntheses of Organic Compounds Synthesy organichesk. Soedinenii, Sbornic 2, 136 (1952) AN SSSR I. O. Kh. Moskau.
52. G. A. Razuvaev, Yu. A. Ol'dekop, and M. N. Koroleva, *Zh. Obshch. Khim.*, **21**, 650 (1951).
53. R. D. Chambers, G. E. Coates, J. G. Livingston, and W. K. Musgrave, *J. Chem. Soc.*, **1962**, 4367.
54. F. E. Aldrich, E. G. Howard, W. J. Linn, W. J. Middleton, and W. H. Sharke, *J. Org. Chem.*, **28**, 184 (1963).
55. L. I. Zakharkin, V. I. Bregadze, and O. Yu. Okhlobystin, *Zh. Obshch. Khim.*, **36**, 761 (1966).
56. Yu. G. Bundel, N. D. Antonova, and O. A. Reutov, *Dokl. Akad. Nauk SSSR*, **166**, 1103 (1966).
57. I. P. Beletskaya, A. E. Myshkin, and O. A. Reutov, *Izv. Akad. Nauk SSSR, Ser. Khim.*, **1965**, 240.
58. A. Michaelis, *Ber.*, **28**, 592 (1895).
59. E. Fischer, *Ber.*, **40**, 386 (1907).
60. W. Schoeller, W. Schrauth, and R. Struense, *Ber.*, **44**, 1057 (1911).
61. K. A. Hofmann, *Ber.*, **31**, 1904 (1898).
62. A. N. Nesmeyanov, A. E. Borisov, I. S. Savel'eva, and N. V. Kruglova, *Izv. Akad. Nauk SSSR, Otd. Khim. Nauk*, **1961**, 726.
63. A. N. Nesmeyanov, A. E. Borisov, and I. S. Savel'eva, *Izv. Akad. Nauk SSSR, Otd. Khim. Nauk*, **1961**, 2241.
64. A. N. Nesmeyanov, A. E. Borisov, and I. S. Savel'eva, *Dokl. Akad. Nauk SSSR*, **155**, 603 (1964).
65. A. N. Nesmeyanov, O. A. Reutov, A. S. Loseva, and M. Ya. Khorlina, *Izv. Akad. Nauk SSSR, Otd. Khim. Nauk*, **1959**, 50.
66. M. Kucherov, *Ber.*, **14**, 1532 (1881).
67. M. Kucherov, *Ber.*, **14**, 1540 (1881).
68. M. Kucherov, *Ber.*, **17**, 13 (1884).
69. H. Biltz and O. Mumm, *Ber.*, **37**, 4417 (1904).

70. W. W. Middleton and A. W. Barrett, *J. Am. Chem. Soc.*, **49**, 2258 (1927).
71. W. W. Middleton, R. G. Berchem, and A. W. Barrett, *J. Am. Chem. Soc.*, **49**, 2264 (1927).
72. W. W. Middleton, A. W. Barrett, and J. H. Seager, *J. Am. Chem. Soc.*, **52**, 4405 (1930).
73. K. A. Hofmann, *Ber.*, **37**, 4460 (1904).
74. R. Ya. Levina and B. M. Gladshtein, *Dokl. Akad. Nauk SSSR*, **71**, 65 (1950).
75. R. Ya. Levina and V. N. Kostin, *Zh. Obshch. Khim.*, **23**, 1055 (1953).
76. R. Ya. Levina and V. N. Kostin, *Dokl. Akad. Nauk SSSR*, **97**, 1027 (1954).
77. R. Ya. Levina, N. Mesentsova, and O. V. Lebedev, *Zh. Obshch. Khim.*, **25**, 1097 (1955).
78. R. Ya. Levina, V. N. Kostin, and V. A. Tartakovskii, *Zh. Obshch. Khim.*, **26**, 2998 (1956).
79. R. Ya. Levina, V. N. Kostin, and V. A. Tartakovskii, *Vestn. M.G.U.*, *Ser. Phys.-mat. and estestvenn. Nauk N1*, 77 (1956).
80. R. Ya. Levina, V. N. Kostin, and V. A. Tartakovskii, *Zh. Obshch. Khim.*, **27**, 881 (1957).
81. R. Ya. Levina, V. N. Kostin, and K. S. Shanasarov, *Zh. Obshch. Khim.*, **29**, 40 (1959).
82. R. Ya. Levina, V. N. Kostin, D. G. Kim, and T. K. Ustynjuk, *Zh. Obshch. Khim.*, **29**, 1956 (1959).
83. R. Ya. Levina, V. N. Kostin, P. A. Gembitskii, and A. D. Vinogradov, *Vestn. MGU N1*, 67 (1961).
84. R. Ya. Levina, V. N. Kostin, P. A. Gembitskii, S. M. Shostakovsky, and E. G. Treshchova, *Zh. Obsch. Khim.*, **31**, 1185 (1961).
85. M. S. Kharasch and M. W. Grafflin, *J. Am. Chem. Soc.*, **47**, 1948 (1925).
86. M. S. Kharasch and T. Marker, *J. Am. Chem. Soc.*, **48**, 3130 (1926).
87. M. S. Kharasch and A. L. Flenner, *J. Am. Chem. Soc.*, **54**, 674 (1932).
88. M. S. Kharasch, H. Pines, and J. H. Levine, *J. Org. Chem.*, **3**, 347 (1938).
89. M. S. Kharasch and S. Swartz, *J. Org. Chem.*, **3**, 405 (1938).
90. M. S. Kharasch, R. R. Legault, and M. R. Sprowls, *J. Org. Chem.*, **3**, 409 (1938).
91. M. S. Kharasch and S. Weinhouse, *J. Org. Chem.*, **1**, 209 (1936).
92. A. N. Nesmeyanov and K. A. Kocheshkov, *Uchen. Zap. Moskov. Gos. Univer.*, **3**, 283 (1934).
93. Ch. S. Bobashinskaya and K. A. Kocheshkov, *Zh. Obshch. Khim.*, **8**, 1850 (1938).
94. F. C. Whitmore and H. Bernstein, *J. Am. Chem. Soc.*, **60**, 2626 (1938).
95. F. Nerdel and S. Makower, *Naturwiss.*, **45**, 490 (1958).
96. R. N. Sterlin, V. G. Li, and I. L. Knunyants, *Dokl. Akad. Nauk SSSR*, **140**, 137 (1961).
97. A. N. Nesmeyanov, E. G. Perevalova, S. P. Gubin, and A. G. Kozlovsky, *J. Organometal. Chem.*, **11**, 577 (1968).
98. A. N. Nesmeyanov, A. G. Kozlovsky, S. P. Gubin, and E. G. Perevalova, *Izv. Akad. Nauk SSSR, Ser. Khim.*, **1965**, 580.
99. A. N. Nesmeyanov, A. G. Kozlovsky, and S. P. Gubin, *Izv. Akad. Nauk SSSR, Ser. Khim.*, **1966**, 387.
100. O. A. Reutov and A. N. Lovtsova, *Dokl. Akad. Nauk SSSR*, **154**, 166 (1964).
101. M. M. Kreevoy and R. L. Hansen, *J. Phys. Chem.*, **65**, 1055 (1961).
102. C. M. Marvel and H. O. Calvery, *J. Am. Chem. Soc.*, **45**, 820 (1923).

103. S. Winstein and T. Traylor, *J. Am. Chem. Soc.*, **77**, 3747, 3748 (1955).
104. F. E. Paulik, I. E. Green, and R. E. Dessy, *J. Organometal. Chem.*, **3**, 229 (1965).
105. R. D. Brown, A. S. Buchanan, and A. A. Humffray, *Australian J. Chem.*, **18**, 1507 (1965).
106. R. D. Brown, A. S. Buchanan, and A. A. Humffray, *Australian J. Chem.*, **18**, 1513 (1965).
107. D. Seyferth, J. Y.-P. Mui, and L. J. Todd, *J. Am. Chem. Soc.*, **86**, 2961 (1964).
108. A. N. Nesmeyanov, R. Kh. Freidlina, and F. K. Velichko, *Dokl. Akad. Nauk SSSR*, **114**, 557 (1957).
109. L. W. Jones and L. Werner, *J. Am. Chem. Soc.*, **40**, 1266 (1918).
110. P. Wolff, *Ber.*, **46**, 64 (1913).
111. F. R. Jensen and R. J. Ouellette, *J. Am. Chem. Soc.*, **85**, 363 (1963).
112. D. J. Foster and E. J. Tobler, *J. Org. Chem.*, **27**, 834 (1962).
113. D. J. Foster and E. J. Tobler, *J. Am. Chem. Soc.*, **83**, 851 (1961).
114. E. Tobler and D. J. Foster, *Z. Naturforsch.*, **17B**, 135 (1962).
115. F. R. Jensen and R. J. Ouellette, *J. Am. Chem. Soc.*, **83**, 4477 (1961).
116. F. R. Jensen and R. J. Ouellette, *J. Am. Chem. Soc.*, **83**, 4478 (1961).
117. J. H. Robson and G. F Wright, *Can. J. Chem.*, **38**, 21 (1960).
118. K. Ichikawa and H. Ouchi, *J. Am. Chem. Soc.*, **82**, 3876 (1960).
119. S. Winstein, E. Vogelfanger, K. C. Pande, and H. F. Ebel, *J. Am. Chem. Soc.*, **84**, 4944 (1962); *Chem. Ind.*, **49**, 2061 (1962).
120. F. R. Jensen and R. J. Ouellette, *J. Am. Chem. Soc.*, **85**, 367 (1963).
121. P. D. Sleezer, S. Winstein, and W. G. Young, *J. Am. Chem. Soc.*, **85**, 1890 (1963).
122. K. Ichikawa, S. Fukushima, H. Ouchi, and M. Tsuchida, *J. Am. Chem. Soc.*, **80**, 6005 (1958); **81**, 3401 (1959).
123. K. Ichikawa, K. Fujita, and O. Itoh, *J. Am. Chem. Soc.*, **84**, 2632 (1962).
124. K. Ichikawa and S. Fukushima, *J. Org. Chem.*, **24**, 1129 (1959).
125. T. Mikaijama, I. Kuwajima, and Z. Suzuki, *J. Org. Chem.*, **28**, 2024 (1963).
126. A. N. Nesmeyanov, N. S. Kochetkova, and R. Kh. Freidlina, *Izv. Akad. Nauk SSSR, Otd. Khim. Nauk*, **1945**, 128.
127. H. Laubie, *Bull. Soc. Pharm. Bordeaux*, **96**, 65 (1957); *Chem. Abstr.*, **51**, 15889 (1957).
128. K. A. Hofmann and J. Sand, *Ber.*, **33**, 1340 (1900).
129. A. N. Nesmeyanov and R. Kh. Freidlina, *Zh. Obshch. Khim.*, **7**, 43 (1937); *Ber.*, **69**, 1631 (1936).
130. A. N. Nesmeyanov, R. Kh. Freidlina, and F. A. Tokareva, *Zh. Obshch. Khim.*, **7**, 262 (1937); *Ber.*, **69**, 2019 (1936).
131. J. Sand, *Chem. Ber.*, **34**, 1385 (1901).
132. M. M. Kreevoy and G. B. Bodem, *J. Org. Chem.*, **27**, 4539 (1962).
133. A. N. Nesmeyanov, R. Kh. Freidlina, and A. E. Borisov, Yub. Sb., posv. 30-let Vel. Oct. soc. rev. AN, 1947, p. 658.
134. K. A. Hofmann and J. Sand, *Chem. Ber.*, **33**, 1353 (1900).
135. K. A. Hofmann and J. Sand, *Chem. Ber.*, **33**, 1358 (1900).
136. K. A. Hofmann and J. Sand, *Chem. Ber.*, **33**, 2692 (1900).
137. J. Sand, *Chem. Ber.*, **34**, 2906 (1901).
138. J. Sand and F. Singer, *Chem. Ber.*, **35**, 3170 (1902).
139. W. Manchot and A. Klug, *Ann. Chem.*, **420**, 170 (1920).
140. P. Pfeiffer, *Z. Naturwiss.*, **14**, 1100 (1926).

141. J. Sand, *Ann. Chem.*, **329**, 135 (1903).
142. J. Sand and F. Singer, *Ann. Chem.*, **329**, 166 (1903).
143. R. Adams, F. L. Roman, and W. N. Sperry, *J. Am. Chem. Soc.*, **44**, 1781 (1922).
144. L. Mills and R. Adams, *J. Am. Chem. Soc.*, **45**, 1842 (1923).
145. L. T. Sandborn and C. S. Marvel, *J. Am. Chem. Soc.*, **48**, 1409 (1926).
146. E. Griffith and C. S. Marvel, *J. Am. Chem. Soc.*, **53**, 789 (1931).
147. J. Sand and F. Breest, *J. Phys. Chem.*, **59**, 424 (1907).
148. J. Romeyn and G. F Wright, *J. Am. Chem. Soc.*, **69**, 697 (1947).
149. R. Kh. Freidlina and N. S. Kochetkova, *Izv. Akad. Nauk SSR, Otd. Khim. Nauk*, **1945**, 128.
150. A. G. Brook and G. F Wright, *Acta Cryst.*, **4**, 50 (1951).
151. A. G. Brook and G. F Wright, *J. Am. Chem. Soc.*, **72**, 3821 (1950).
152. A. G. Brook and S. Walfe, *J. Am. Chem. Soc.*, **79**, 1431 (1957).
153. A. N. Nesmeyanov, *Uchen. Zap. Moskov. Gos. Univer.*, **132**, 5 (1950).
154. A. N. Nesmeyanov and M. I. Kabachnik, *Zh. Obshch. Khim.*, **25**, 41 (1955).
155. A. N. Nesmeyanov and A. E. Borisov, *Izv. Akad. Nauk SSR, Otd. Khim. Nauk*, **1945**, 146.
156. A. N. Nesmeyanov, A. E. Borisov, and V. D. Vilchevskaya, *Izv. Akad. Nauk SSSR, Otd. Khim. Nauk*, **1949**, 578; *Uchen. Zap. Moskov. Gos. Univer.*, **132**, 33 (1950).
157. W. J. Jenkins, *J. Chem. Soc.*, **119**, 747 (1921); *Chem. Zentr.*, **1921**, III, 608.
158. A. N. Nesmeyanov, A. E. Borisov, and A. N. Gus'kova, *Izv. Akad. Nauk SSSR, Otd. Khim. Nauk*, **1945**, 639.
159. A. N. Nesmeyanov, A. E. Borisov, and A. N. Abgramova, *Izv. Akad. Nauk SSSR, Otd. Khim. Nauk*, **1949**, 570; *Uchen. Zap. Moskov. Gos. Univer.*, **132**, 73 (1950).
160. A. N. Nesmeyanov, A. E. Borisov, and A. N. Abramova, *Izv. Akad. Nauk SSSR, Otd. Khim. Nauk*, **1946**, 647.
161. A. N. Nesmeyanov, A. E. Borisov, and R. I. Shepeleva, *Izv. Akad. Nauk SSSR, Otd. Khim. Nauk*, **1949**, 582.
162. A. N. Nesmeyanov, R. Ch. Freidlina, and A. K. Kochetkov, *Izv. Akad. Nauk SSSR, Otd. Khim. Nauk*, **1948**, 445.
163. A. E. Borisov, *Izv. Akad. Nauk SSSR, Otd. Khim. Nauk*, **1951**, 402.
164. A. N. Nesmeyanov and A. E. Borisov, *Tetrahedron*, **1**, 158 (1957).
165. A. N. Nesmeyanov and A. E. Borisov, *Dokl. Akad. Nauk SSSR*, **60**, 67 (1948).
166. A. N. Nesmeyanov and R. Kh. Freidlina, *Dokl. Akad. Nauk SSSR*, **26**, 59 (1940).
167. S. L. Varshavskii, *Dokl. Akad. Nauk SSSR*, **29**, 315 (1940).
168. A. N. Nesmeyanov and N. K. Kochetkov, *Izv. Akad. Nauk SSSR, Otd. Khim. Nauk*, **1949**, 76.
169. A. N. Nesmeyanov and N. K. Kochetkov, *Izv. Akad. Nauk SSSR, Otd. Khim. Nauk*, **1949**, 305.
170. A. N. Nesmeyanov, N. K. Kochetkov, and W. M. Dashunin, *Izv. Akad. Nauk SSSR, Otd. Khim. Nauk*, **1950**, 77.
171. M. M. Kreevoy and F. R. Kowitt, *J. Am. Chem. Soc.*, **82**, 739 (1960).
172. M. M. Kreevoy, G. Stokker, R. A. Kretchmer, and A. K. Ahmed, *J. Org. Chem.*, **28**, 3184 (1963).
173. M. M. Kreevoy, *J. Am. Chem. Soc.*, **81**, 1099 (1959).
174. M. M. Kreevoy, *Bull. Soc. Chim. France*, **1963**, 2431.
175. M. M. Kreevoy, J. W. Gilje, and R. A. Kretchmer, *J. Am. Chem. Soc.*, **83**, 4205 (1961).

176. M. M. Kreevoy, J. W. Gilje, L. T. Ditsch, W. Batorewicz, and M. Turner, *J. Org. Chem.*, **27**, 726 (1962).
177. L. Schaleger, M. A. Turner, T. Chamberlin, and M. M. Kreevoy, *J. Org. Chem.*, **27**, 3421 (1962).
178. M. M. Kreevoy and M. A. Turner, *J. Org. Chem.*, **30**, 373 (1965).
179. H. J. Lucas, F. R. Hepner, and S. Winstein, *J. Am. Chem. Soc.*, **61**, 3102 (1939).
180. K. Ichikawa, K. Nishimura, and S. Takayama, *J. Org. Chem.*, **30**, 1593 (1965).
181. K. Ichikawa, H. Ouchi, and S. Araki, *J. Am. Chem. Soc.*, **82**, 3880 (1960).
182. M. M. Kreevoy and B. M. Eisen, *J. Org. Chem.*, **28**, 2104 (1963).
183. M. M. Kreevoy and M. A. Turner, *J. Org. Chem.*, **29**, 1639 (1964).
184. O. W. Berg, W. P. Lay, A. Rodgman, and G. F Wright, *Can. J. Chem.*, **36**, 358 (1958).
185. A. Rodgman, D. A. Shearer, and G. F Wright, *Can. J. Chem.*, **35**, 1377 (1957).
186. G. F. Wright, *Ann. N.Y. Acad. Sci.*, **65**, 436 (1957).
187. A. Rodgman and G. F Wright, *J. Org. Chem.*, **18**, 1617 (1953).
188. A. G. Brook, R. Donovan, and G. F Wright, *Can. J. Chem.*, **31**, 536 (1953).
189. Yu. K. Yur'ev, N. S. Zefirov, and L. P. Prikashchikova, *Zh. Obshch, Khim.*, **33**, 1793 (1963).
190. N. S. Zefirov, R. S. Filatova, and Yu. K. Yur'ev, *Zh. Obshch. Khim.*, **36**, 763 (1966).
191. R. E. Dessy and F. E. Paulik, *J. Am. Chem. Soc.*, **85**, 1812 (1963).
192. P. J. Banney, W. Kitching, and P. R. Wells, *Tetrahedron Letters*, **1968**, 27.
193. I. P. Beletzkaya, O. A. Reutov, N. S. Petrosyan, and L. V. Savinjykh, *Tetrahedron Letters*, **1969**, 485.
194. B. F. Hegarty, W. Kitching, and P. R. Wells, *J. Am. Chem. Soc.*, **89**, 4816 (1967).
195. I. P. Beletzkaya, L. A. Fedorov, and O. A. Reutov, *Zh. Organ. Khim.*, **3**, 225 (1967).
196. J. R. Coad and C. K. Ingold, *J. Chem. Soc.*, (*B*), **1968**, 1455.
197. M. M. Kreevoy, P. J. Steinwand, and T. S. Straub, *J. Org. Chem.*, **31**, 4291 (1966).
198. M. M. Kreevoy, D. J. V. Goon, and R. A. Kayser, *J. Am. Chem. Soc.*, **88**, 5529 (1966).
199. M. M. Kreevoy, T. S. Staub, W. V. Kayser, and J. L. Melquist, *J. Am. Chem. Soc.*, **89**, 1201 (1967).
200. R. B. Babb and P. D. Gardner, *Chem. Commun.*, **1968**, 1678.
201. V. A. Tartakovski, N. E. Tchlenov, and S. S. Novikov, *Izv. Akad. Nauk, Otd. Khim. Nauk*, 1842 (1966).
202. I. P. Beletzkaya, A. E. Myshskin, and O. A. Reutov, *Izv. Akad. Nauk SSSR, Ser. Khim.*, 245 (1967).
203. U. Schöllkopf and F. Gerhart, *Angew. Chem.*, **78**, 209 (1966); *ibid., Intern. Ed. Engl.*, **5**, 252 (1966).
204. L. I. Zakharkin and L. S. Podvisozkaya, *J. Organometal. Chem.*, **7**, 385 (1967).
205. J. Perie and A. Lattes, *Bull. Soc. Chim. France*, **1966**, 2153.
206. U. Schöllkopf and F. Gerhart, *Angew. Chem., Intern. Ed. Engl.* **5**, 664 (1966).
207. K. R. Brower, B. Gay, and T. L. Konkol, *J. Am. Chem. Soc.*, **88**, 1681 (1966).
208. R. G. Van Leuwen and R. Ouellette, *J. Am. Chem. Soc.*, **90**, 7056, 7061 (1968).
209. C. C. Lee and R. J. Tewari *Can. J. Chem.*, **45**, 2256 (1967); **46**, 2314 (1968).
210. M. Kumada and M. Ishikawa, *J. Organometal. Chem.*, **6**, 451 (1966).

211. D. Seyferth, J. Y.-P. Mui, L. J. Todd, and K. V. Darragh, *J. Organometal. Chem.*, **8**, 29 (1967).
212. D. Seyferth and H. D. Simmons, *J. Organometal. Chem.*, **6**, 306 (1966).
213. L. R. Barlow and J. M. Davidson, *J. Chem. Soc.*, (*A*), **1968**, 1609.
214. J. M. Davidson, *J. Chem. Soc.*, (*A*), **1969**, 193.
215. P. S. Magee, *Tetrahedron Letters*, **45**, 3995 (1965).

Chapter 10

Reactions with Alkyl and Acyl Halides

I. INTRODUCTION

Reactions with alkyl halides accompanied by formation of new C—C bonds similar to the reactions or organolithium compounds with alkyl halides are not characteristic of organomercury compounds, which is one of the basic differences in reactivity between organomercury and organolithium compounds. Under comparatively mild conditions, only those organomercury compounds in which the mobility of the mercury is enhanced by conjugation react with alkyl halides producing a new C—C bond. Interaction of other organomercury compounds with alkyl halides with the objective of forming a new C—C bond, in those few cases where it is feasible, is possible under vigorous conditions of prolonged heating at high temperatures. In most cases, however, the reaction does not go smoothly and tends not only in the desired direction but also to form RH, RX, HX, and RHgX and is accompanied by formation of appropriate decomposition products of the alkyl halide radical.

The reactivity of organomercury compounds increases considerably with respect to alkyl halides if initiating agents of radical decomposition are added: by irradiation with ultraviolet light or in the presence of peroxides. The formation of RR' compounds, which sometimes occur under these conditions, are in very small yields and are of no practical significance.

The reactions of organomercury compounds with acyl halides produce ketones, but are inferior as a synthetic method when compared to the synthesis of ketones by organocadmium or zinc compounds. Acyl compounds react specifically with α-mercurated oxocompounds and with the derivatives of monomercurated acetic acid (see below).

II. REACTIONS WITH ALKYL HALIDES

A. Reactions Forming a New C—C Bond

As is stated above, under mild conditions, α-monomercury derivatives of aldehyde and ketone compounds, where the C—Hg bond is activated by conjugation with a C=O bond;

$$ClHg—CH_2—C{=}O \qquad (R = H, CH_3 \text{ or } C_6H_5)$$
$$\diagdown R$$

react with alkyl halides producing a new C—C bond (Nesmeyanov and Lutsenko). However, these compounds react with an alkyl halide where the halide is sufficiently reactive, namely, with triphenylchloro- or triphenylbromoethane.[1-3] Interaction of chloromercury acetaldehyde, chloromercury ketone, or ω-chloromercury acetophenone with triphenylchloromethane occurs without displacement of the reaction center and produces a good yield of triphenylmethylacetaldehyde, triphenylmethylacetone, and triphenylmethylacetophenone, for example:

$$ClHgCH_2CHO + ClC(C_6H_5)_3 \longrightarrow (C_6H_5)CCH_2CHO + HgCl_2$$

$$CH_3COCH_2HgCl + ClC(C_6H_5)_3 \longrightarrow (C_6H_5)_3CCH_2COCH_3 + HgCl_2$$

These reactions take place upon heating in a water bath or even at room temperature in benzene solutions. Under similar conditions, triphenylbromoethane, as does triphenylchloromethane, reacts with chloromercuryacetaldehyde, thus producing triphenylmethylacetaldehyde[2]:

A monomercurated symmetrical derivative of acetoacetic ester is alkylated by alkyl halides[4]:

$$Hg\left(\begin{array}{c} COCH_3 \\ CH \\ COOC_2H_5 \end{array}\right)_2 + 2RI \longrightarrow 2RCH\begin{array}{c} COCH_3 \\ \\ COOC_2H_5 \end{array} + HgI_2$$

Depending on the nature of the solvent, the reaction of the ethyl ester of α-bromomercuryphenylacetic acid with triarylbromomethane and its complex with $HgBr_2$ occurs completely at the carbon atom (in dichloroethane) or practically completely at the oxygen atom, i.e., with displacement of the reaction center (in nitromethane)[5,5a]:

$$Ar_3CBr \cdot HgBr_2 + Y{-}C_6H_4CH(HgBr)COOC_2H_5 \xrightarrow{CH_2ClCH_2Cl}$$
$$C_6H_4CH(CAr_3)COOC_2H_5$$

$$(C_6H_5)_3CBr \cdot HgBr_2 + Y{-}C_6H_4CH(HgBr)COOC_2H_5 \xrightarrow{CH_3NO_2}$$

$$Y{-}C_6H_4{-}CH{=}C\begin{array}{c} OC_2H_5 \\ \\ OC(C_6H_5)_3 \end{array} \xrightarrow{H_2O} Y{-}C_6H_4CH_2COOC_2H_5 + (C_6H_5)_3COH$$

The inhibition of reactions in both solvents due to addition of $HgBr_2$, the difference in kinetics of the reaction with Ar_3CBr (first order) and with complex $Ar_3CBr \cdot HgBr_2$ (second order), and the opposite effects of substituents Y observed on the reaction rate in the first case characteristic of electrophilic reactions make it possible to assume that there are different transition states for both cases: a closed state when the reaction occurs with Ar_3CBr and an open state when the reaction occurs with the complex. In the first case, nucleophilic effect is shown by the anion of the Ar_3CBr molecule on the mercury. Apart from this effect, a positive charge on the carbon, which is the electrophilic part of the molecule that attacks the organomercury compound, evidently plays an important part in these types of reactions. $(C_6H_5)_3CBr$ alkylates sec-butylmercury bromide.[5]

Under comparatively mild conditions (4 hours boiling in benzene) chloromercuryferrocene unexpectedly reacts with triphenylchloromethane producing triphenylmethylferrocene[6] (18% yield). As was stated, there are not many other known reactions which produce new C—C bonds by reaction of alkyl halides with organomercury compounds where mercury is not activated by conjugation, and do not occur under vigorous conditions. Kekule[7] produced triphenylmethane in low yield by the reaction of diphenylmercury with benzylidene chloride at 150°C. The reactions producing diphenylalkyl(aryl)methane by the reaction of diphenylbromomethane with di-n-butyl-, diphenyl-, and di-p-tolylmercury over a temperature range from 200°C (di-p-tolylmercury) to 340°C (di-n-butylmercury)[8] belong to the same group. Others are: the reaction of p-$ClHgC_6H_4OC_6H_5$ with $C_6H_5CH_2Cl$ which produced $C_6H_5CH_2C_6H_4COC_6H_5$[9] and the reaction[10]:

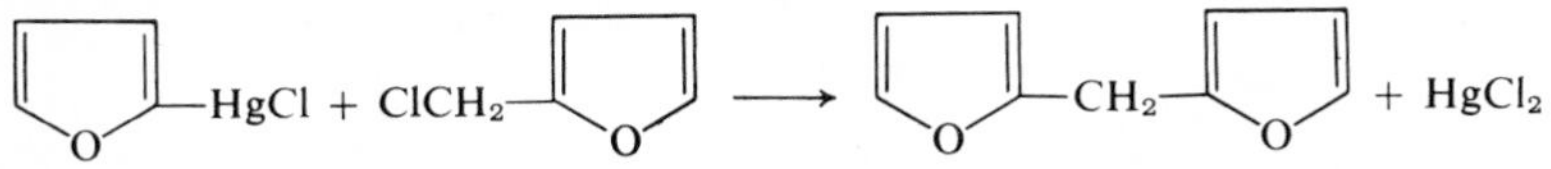

Upon heating ethylene bromide and diphenylmercury in a sealed tube from 180 to 200°C for 6–10 hours, only C_6H_5HgBr was isolated. The authors proposed that the reaction proceeds according to the following equation[11]:

$$(C_6H_5)_2Hg + CH_2BrCH_2Br \longrightarrow C_6H_5CH_2CH_2Br + C_6H_5HgBr$$

A product formed, probably diallyl, upon heating allylmercury iodide with allyl iodide.[12] It is possibly the result of dimerization of allyl radicals formed in the reaction since diallyl in addition to ethylmercury iodide and ethyl iodide is a product of the reaction of allyl iodide with diethylmercury.[13]

These reactions are all the known cases of C—C bond formation by the reaction of organomercury compounds with alkyl halides in the absence of initiating agents.

A C—C bond may be formed from organomercury compounds and carbon tetrachloride in the presence of catalytic amounts of peroxides[14] (acetyl peroxide, benzoyl peroxide). Diphenylmercury reacts with carbon tetrachloride:

$$(C_6H_5)_2Hg + CCl_4 \xrightarrow[120-125°C]{(CH_3COO)_2} C_6H_5CCl_3 + C_6H_5HgCl$$

Hexachloroethane was also produced. Other compounds are also reactive: *trans*-β-chloromercury chloride and di-*trans*-β-chlorovinylmercury. In both cases, formation of a new C—C bond occurred:

$$\text{trans-ClCH=CHHgCl + CCl}_4 \xrightarrow{\text{(CH}_3\text{COO)}_2}$$
$$\text{HgCl}_2 + \text{cis-ClHgCH=CHCl + Cl}_3\text{CCH=CHCl}$$

$$\text{trans-(ClCH=CH)}_2\text{Hg + CCl}_4 \xrightarrow{\text{(RCOO)}_2}$$
$$\text{HgCl}_2 + \text{cis—ClHgCH=CHCl + Cl}_3\text{CCH=CHCl}$$

The peroxides also caused *trans–cis* isomerization.

The reaction of *trans*-β-chlorovinylmercury chloride with CBr_4 produced an addition product of $ClCH=CH\cdot$ and $\cdot CBr_3$ radicals and was accompanied by its bromination[14]:

$$\text{ClCH=CHHgCl + CBr}_4 \xrightarrow{\text{(CH}_3\text{COO)}_2} \text{ClCH=CHCBr}_3 + \text{ClCHBrCHBrCBr}_3$$

On the basis of a similar reaction as used with nonsymmetrical organomercury compounds according to the equation:

$$\text{RHgR}' + \text{CCl}_4 \xrightarrow{\text{(C}_6\text{H}_5\text{COO)}_2} \text{RHgCl + R}'\text{CCl}_3$$

Nesmeyanov et al.[15] have shown that radicals may be arranged in the order of decreasing affinity with respect to the $\cdot CCl_3$ free radical:

$$2,4,6\text{-(CH}_3)_3\text{C}_6\text{H}_2 > \alpha\text{-C}_{10}\text{H}_7 > p\text{-CH}_3\text{C}_6\text{H}_4 > o\text{-CH}_3\text{C}_6\text{H}_4$$
$$m\text{-CH}_3\text{C}_6\text{H}_4 > \text{C}_6\text{H}_5 > \text{C}_2\text{H}_5 > \text{C}_4\text{H}_9 > \text{C}_6\text{H}_5\text{CH}_2 > \text{C}_6\text{H}_{11}$$

coinciding with the Kharasch series of radicals arranged on the basis of their decreasing affinity to protons in heterolytic reactions with hydrochloric acid.

Reaction of diphenylmercury with $CHCl_3$ and peroxide occurs as follows[14]:

$$\text{(C}_6\text{H}_5)_2\text{Hg + CHCl}_3 \xrightarrow{\text{(CH}_3\text{COO)}_2} \text{C}_6\text{H}_5\text{HgCl + C}_6\text{H}_6 + \underbrace{\text{C}_6\text{H}_5\text{CCl}_3 + \text{C}_2\text{Cl}_6}_{\text{trace}}$$

Alk_2Hg does not react by the scheme of Nesmeyanov, Borisov, et al.,[15] but undergoes a free radical reaction with β-elimination[72]:

$$\text{R}_2\text{Hg + CCl}_3 \xrightarrow{\text{(RCOO)}_2} \text{olefin + RHg}\cdot + \text{CHCl}_3$$

B. Formation of Nonsaturated Compounds

In those cases, when hydrogen halide or two atoms of halide can be isolated from alkyl halides, their reactions with organomercury compounds may occur with the formation of unsaturated compound, the following reactions take place under vigorous reaction conditions (boiling in toluene[8] for 340 hr or many hours heating at a temperature of 200°C in sealed tubes without solvents[16]) are illustrated by examples (1), (2), and (3):

$$2RCH{=}CR_2X + R_2'Hg \longrightarrow 2R_2C{=}CR_2 + HgX_2 + 2R'H \qquad (1)$$

This is the method of the reaction of tert-butylbromide and tert-amyl iodide[8] with R_2Hg ($R = C_6H_5$ and p-$CH_3C_6H_4$), and 1-bromo-1,1,2-tricarbethoxyethane[8] and neopentyl bromide and iodide[16] with (p-$CH_3C_6H_4)_2Hg$.

$$C_6H_5CHBrCHBrC_6H_5 + (p\text{-}CH_3C_6H_4)_2Hg \longrightarrow$$
$$C_6H_5CH{=}CHC_6H_5 + p\text{-}CH_3C_6H_4Br + p\text{-}CH_3C_6H_4HgBr \qquad (3)$$

C. Other Reactions with Alkyl Halides

In cases where R_2Hg reacts with R′Hal, RHgHal, RHal and hydrocarbons (including unsaturated ones) are produced.

The reaction of diethylmercury with iodoform starts at a temperature of 70°C and continues to 120°C producing ethylene, acetylene, ethyl iodide, and ethylmercury iodide.[13] Diisopropylmercury (at a temperature of 130°C for 10 hr, in the absence of oxygen) reacts with $CDCl_3$, $CHCl_3$, or CCl_4 and forms isopropylmercury chloride, propane, and other products.[17]

In the presence of $BiCl_3$ the reaction of diphenylmercury with chloroform produces C_6H_5HgCl, C_6H_6, and diphenyl.[18,19] In compliance with the other data,[20] $(C_6H_5)_2Hg$ does not react with $CHCl_3$ and $CHBr_3$; it does react with CHI_3 through the intermediate formation of $C_6H_5HgCI_3$ producing C_6H_5HgI, C_2I_4, and C_6H_6.[20] When being heated in CCl_4, differocenylmercury produces a ferrocenylmercury chloride, ferrocene, and a residue containing mercury and chlorine.[21] Diphenylmercury does not react with allyl iodide when heated in xylene for 300 hr.[8]

Reaction of o-chloromercury phenol with allyl bromide and chloride occurs extremely easily in pyridine at $-10°C$ with isolation of the complex, $[C_5H_5NC_3H_5]HgHal_4$.[22] At a temperature of 50°C the reaction of ICN with $(CH_3)_2Hg$ in ether solution produces mercury cyanide, while at a temperature of 110°C HgI_2 and methylisonitryl are produced.[23]

The reaction of alkyl halides with quasicomplex mercury compounds, which represent addition products of mercury salts to ethylene, results in the isolation of ethylene:

$$HOCH_2CH_2HgCl + CH_3I \longrightarrow CH_3OH + HgClI + CH_2=CH_2$$

This reaction is characteristic of the previously mentioned substances.[24] Similar reactions with di-β-iodomercuryethyl ether[25]

$$2CH_3I + (IHgCH_2CH_2)_2O \longrightarrow (CH_3)_2O + 2CH_2=CH_2 + 2HgI_2$$

and with addition products of mercury salts to carbon monoxide in alcohol also occur[26]:

$$ClHgCOOC_2H_5 + CH_3I \longrightarrow CH_3OC_2H_5 + CO + HgClI$$

The reaction of arylchlorosulfides with Ar_2Hg produces $ArHgCl$.[27] The effect of acyl halides of carboxylic acids on $ArHgSAr'$ produces $ArHgHal$.[27] Similar to this is the reaction of arylsulfenyl chlorides with $ArHgSAr'$[27] which occurs at room temperature:

$$ArHgSAr' + Ar''SCl \longrightarrow Ar'SSAr'' + ArHgCl$$

The effect of alkyl halides which are stronger electrophilic agents than triphenylchloromethane is similar to the effect of acyl halides; bromomercury acetaldehyde reacts with tris-(p-nitrophenyl)bromomethane and is directed not to the Hg—C bond but to the oxygen and results in the formation of the vinyl ether of tris-(p-nitrophenyl)carbinol[2]:

$$HgHal_2 + CH_2=CHOC(C_6H_4NO_2\text{-}p)_3$$

The reactions of organomercury compounds with alkyl halides are facilitated by light initiation. Radicals, formed under the effect of ultraviolet light on mercury organic compounds[10];

$$R_2Hg \xrightarrow{h\nu} RHg\cdot + R\cdot$$

react with alkyl halides so that the radical of $RHg\cdot$ abstracts the halide from the solvent and forms $RHgHal$, radical $R\cdot$ produces $RHal$ (in CCl_4[28-37]) or RH (with hydrogen-containing halo solvents such as $CHCl_3$).[38] Very often C_nH_{2n} and other unsaturated hydrocarbons are also formed. Metallic mercury and mercurous oxide salts are produced as well. Some organomercury compounds such as mercury-bis-p-benzoic acid do not react[39] with CCl_4 or $CHCl_3$ upon irradiation.

Reaction of diphenylmercury (and other R_2Hg)[28,29] with carbon tetrachloride initiated by light occurs in a different way from that of the same reaction initiated by peroxides (see above):

$$(C_6H_5)_2Hg + 2CCl_4 \xrightarrow{h\nu} C_6H_5HgCl + C_2Cl_6 + C_6H_5Cl$$

No benzotrichloride is formed under the stated conditions. In a similar way the halide is abstracted from other alkyl halides by means of the $RHg\cdot$ radical[38,40–42] with formation of RHgHal. RHgHal is not formed in the reaction with ArHal, while radical $\mathbf{R}\cdot$ enters into the aromatic nucleus, $RHg\cdot$ is further decomposed in accordance with the following reactions:

$$RHg\cdot \longrightarrow R\cdot + Hg$$
$$2RHg\cdot \longrightarrow R_2Hg + R\cdot$$

During decomposition of diphenylmercury by light in a mixture of solvents with and without halides, the radical $C_6H_5Hg\cdot$ extracts the halide from the halide-containing component of the mixture forming C_6H_5HgCl, while the phenyl radical abstracts hydrogen from the component not containing halide, producing benzene.[43] C_6H_5HgBr[44] is formed in the mixture of chloride and bromide-containing solvent. If decomposition takes place in the mixture of ArBr and AlkCl, C_6H_5HgCl[44] is formed in this case.

III. REACTIONS WITH ACID HALIDES

Organomercury compounds react with acyl halides more easily than with alkyl halides.

Organomercury compounds, derivatives of fatty aromatic, aromatic compounds, the heterocycles thiophene and furan react with acyl halides to form ketones:

$$R_2Hg + R'COCl \longrightarrow RCOR' + RHgCl$$
(Reactions are described for acyl chlorides only.)

The reactions also occur in the presence of aluminum chloride[45,46] under milder conditions. These reactions are not used for fatty organomercury compounds and were not tested for other organomercury compounds, are not carried out very easily, are often selective, and of low yield. Thus the reaction of organomercury compounds with acyl halides as a synthetic method for the preparation of ketones cannot compete with organozinc and cadmium compounds. In some cases, ketones are not produced, i.e., during boiling of phenylmercury fluoride with acetylchloride, only exchange of halides took place and C_6H_5HgCl and CH_3COF are formed.[47]

Like acyl halides, arylsulfonyl chlorides react with organomercury compounds to produce sulfones[48]:

$$R_2Hg + R'SO_2Cl \longrightarrow RSO_2R' + RHgCl$$

Organomercury compounds with a radical containing amino and oxy-groups are acylated by acid chlorides under similar conditions for acylation of the respective compounds not containing mercury (Schotten-Baumann's reaction, etc.) with formation of acylamino-[49-52] and acyloxy[52-55] compounds. Similar to this is the effect of acyl chlorides on β-oxyalkylmercury compounds which, under mild conditions, can be acylated at the oxygen without elimination of ethylene[56,57]:

$$ClHgCH_2CH_2OH + C_6H_5COCl \longrightarrow ClHgCH_2CH_2OCOC_6H_5 + HCl$$

Under conditions of the Schotten-Baumann reaction, the addition product of mercury salt to acetylenic alcohol, namely, 2-chloro-3-chloromercury-2-buten-1,4-diol is benzoylated with formation of the corresponding dibenzoate of the mercurated alcohol.[58]

α-Mercurated oxocompounds, which represent mercury derivatives of aldehydes and ketones, react with acyl halides in a different way than with alkyl halides (such as triphenylchloromethane, see above), The reaction with acyl halides of the organomercury compounds is accompanied by displacement of the reaction center due to presence of σ,π-conjugation of

$$ClHg{-}CH_2{-}C{=}O$$
$$\qquad\qquad\qquad\backslash R$$

in these compounds. The effect of acyl halides of carboxylic, sulfonic,[61] phosphoric,[62] phosphorous,[63] and phosphinic[62] acids on the compounds is directed to the oxygen and results in formation of vinyl esters of the respective acids, for example:

$$ClHg{-}CH_2CHO + RCOCl \longrightarrow CH_2{=}CHOCOR + HgCl_2$$

$$ClHg{-}CH_2COR' + RCOCl \longrightarrow RCO_2CR'{=}CH_2 + HgCl_2$$

Benzoyl chloride reacts with chloromercury acetaldehyde more slowly than acetylchloride. Consequently, by increasing the electrophilicity of the reagent (acid halides compared to alkyl halides), the displacement of the reaction center is expected for α-mercurated oxocompounds.

In accordance with this, and also during the acylation of methyl ester of mercurybisacetoacetic acid, only one product was isolated from the reaction, i.e., the ester of β-acyloxycrotonic acid which shows that the reaction occurs at the 1–4 position[4]:

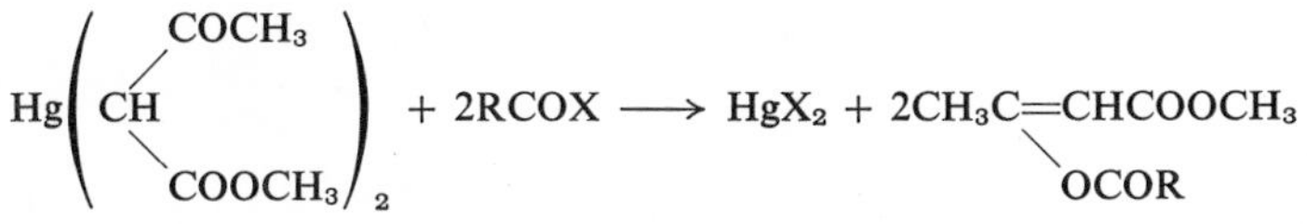

$$Hg\left(\underset{COOCH_3}{\overset{COCH_3}{CH}}\right)_2 + 2RCOX \longrightarrow HgX_2 + 2CH_3C=CHCOOCH_3 \\ \overset{|}{OCOR}$$

The reaction is carried out in benzene and xylol, and heated for several hours. The yield with C_6H_5COI is worse than with RCOCl.[4]

The reaction of acyl halides with the addition products of $HgCl_2$ and acetylenic ketone (phenylethynyl methyl ketone) occurs also at the oxygen due to the presence of conjugation between the Hg—C and C=O bonds (see Chapter 9, Sec. IV-A), and due to the reaction of acyl halides at the 1–4 bonds of the conjugate system.[64] As a result of the reaction of this compound with acetyl chloride or bromide and with benzoyl chloride (after treatment by water) one product was obtained in all cases, i.e., 1-phenyl-1-chlorobut-1-en-3-one whose formation is explained by acylation at the carboxyl group followed by successive hydrolysis of the enol acylate in the weak acid medium which is formed during treatment of the reaction mixture by water[64]:

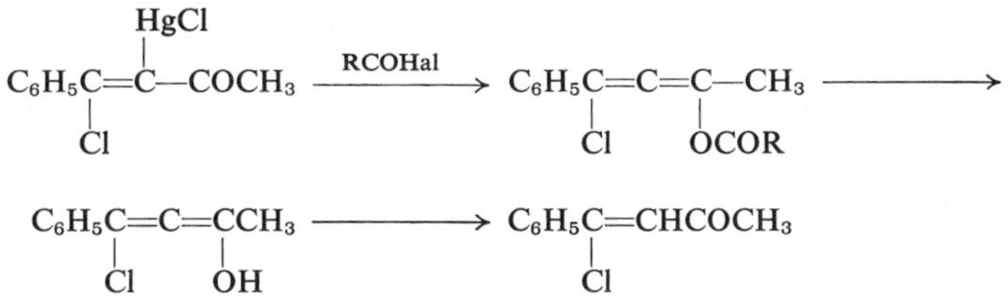

$$\underset{Cl}{\overset{HgCl}{C_6H_5C=C-COCH_3}} \xrightarrow{RCOHal} \underset{Cl\quad OCOR}{C_6H_5C=C=C-CH_3} \longrightarrow$$

$$\underset{Cl\quad OH}{C_6H_5C=C=CCH_3} \longrightarrow \underset{Cl}{C_6H_5C=CHCOCH_3}$$

Esters of mercurated acetic acid containing two conjugate systems of the following bonds: mercury-carbon σ-σ (1–2) and carbon oxygen σ-π (1–4) and σ-σ (1–4'):

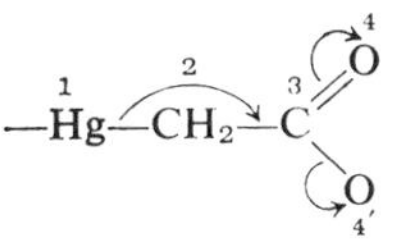

obtained by addition of mercury oxide to ketene in alcohol[65] or by the action of ketene upon other quasicomplex compounds,[66] β-mercurated oxycompounds or α-mercurated oxocompounds yield reaction products during acylation by three different reaction paths[70]:

$$\underset{1}{-}Hg\underset{2}{-}\overset{}{C}H\underset{3}{-}\overset{\overset{4}{O}}{\underset{\underset{4'}{OR}}{C}} \quad + \quad \overset{4}{R'}COCl$$

$$\longrightarrow \text{(1–2) } R'COCH_2COOR$$

$$\longrightarrow \text{(1–4) } CH_2{=}C\overset{OCOR'}{\underset{OR}{}}$$

$$\longrightarrow \text{(1–4') } R'COOR + CH_2{=}C{=}O$$

The products are formed depending on the nature of **R** and **R'**.

As a result of the reaction of acetyl chloride with the isobutyl ester of mercurybisacetic acid, isobutyl acetoacetate (reaction path 1–2) and isobutyl acetate (reaction path 1–4') were obtained. Similarly, the reaction of acetyl chloride with the methyl ester of mercurybisacetic acid and the reaction of isovaleryl chloride with the isobutyl ester of mercurybisacetic acid yielded products from reaction path 1–4[70]:

$$Hg\left(CH_2C\overset{O}{\underset{OCH_3}{}}\right)_2 + CH_3COCl \longrightarrow CH_2{=}C(OCH_3)OCOCH_3$$

$$Hg\left(CH_2C\overset{O}{\underset{OC_4H_9\text{-}i}{}}\right)_2 + i\text{-}C_4H_9COCl \longrightarrow CH_2{=}C(OC_4H_9\text{-}i)OCOC_4H_9\text{-}i$$

Quasicomplex properties of mercurated salts of phosphonium halides (type A) manifest themselves in that they are acylated by RCOCl with displacement of the reaction center[67]:

$$\left[(C_6H_5)_3P^{\oplus}\underset{(A)}{-}\overset{H}{\underset{HgCl}{C}}{-}COR\right]Cl^- + R'COCl \xrightarrow{\text{THF}}$$

$$\left[(C_6H_5)_3P^{\oplus}{-}CH{=}C\overset{R}{\underset{OCOR'}{}}\right]^+ HgCl_3^-$$

The reaction of CH_3COCl with $[R_3As^+CH(HgCl)C(O)C_6H_5]Cl^-$ **(1)** is analogous to that of phosphonium ylides, but the reaction of **1** with p-nitrobenzaldehyde (in DMFA, 100°C) gives α-(p-nitrophenyl)-β-benzoylethylene and that of phosphonium ylide gives *trans*- and *cis*-p-nitrostilbenes.[73]

TABLE VII

Reactions with Alkyl Halides[d]

R in R_2Hg or $RHgR'$ and $RHgX$	Alkyl halide	Reaction conditions	Products	Yield, %	Ref.
CH_3	ICN	Sealed tube$(C_2H_5)_2O$, 50° C, 110°C	$Hg(CN)_2$	—	23
	CHI_3	Sealed tube, 120°C, 1 day	HgI_2, CH_3NC	—	13
C_2H_5	C_3H_5I	120°C	C_2H_5I, C_2H_2, C_2H_4, $IHgC_2H_5$	—	13
C_2H_5			C_6H_{10}, C_2H_5I, $IHgC_2H_5$	—	13
n-C_4H_9	$(C_6H_5)_2CHBr$	$CH_3C_6H_5$, reflux, 340 hr	$(C_6H_5)_2CHCH(C_6H_5)_2$, $(C_6H_5)_2C(H)C_4H_9$-n	—	8
$HOCH_2CH_2HgI$[a]	CH_3I	150°C	$H_2C{=}CH_2$	—	24
			HgI_2	—	25
			CH_3OH		
$ClHgCOOC_2H_5$	n-C_3H_7I	Sealed tube, 100°C	C_3H_7-OC_2H_5	—	26
			CO		
			$HgClI$		
$(IHgCH_2CH_2)_2O$	CH_3I	Sealed tube, 110°C	$(CH_3)_2O$, C_2H_4, HgI_2	—	25
$ClHgCH_2CHO$	$(C_6H_5)_3CCl$	C_6H_6, overnight 20°C, reflux, 2 hr	$(C_6H_5)_3CCH_2CHO$	65	3
$ClHgCH_2CHO$	$(C_6H_5)_3CCl$	C_6H_6, reflux, 2 hr	$(C_6H_5)_3CCH_2CHO$	65	69
$ClHgCH_2CHO$	$(C_6H_5)_2CHCl$	C_6H_6, reflux, 2 hr	No reaction		69
$ClHgCH_2CHO$	$(C_6H_5)_2CHCl$	C_6H_5, reflux, 2 hr, $SnCl_2$	$(C_6H_5)_2CHCH_2CHO$	37	69
$ClHgCH_2CHO$	$(C_6H_5)(p$-$CH_3OC_6H_4)CHCl$	C_6H_6, reflux, 2 hr	$(C_6H_5)(p$-$CH_3OC_6H_4)CHCH_2CHO$	70	69
$ClHgCH_2CHO$	$(C_6H_5)_2C(CH_3)Cl$	C_6H_6, reflux, 2 hr	$(C_6H_5)_2C(CH_3)CH_2CHO$	7	69
$ClHgCH_2CHO$	$(C_6H_5)_2C(CH_3)Cl$	C_6H_6, reflux, 2 hr, $SnCl_2$	$(C_6H_5)_2C(CH_3)CH_2CHO$	8	69
$ClHgCH(CH_3)CHO$	$(C_6H_5)_3CCl$	C_6H_6, reflux, 2 hr	$(C_6H_5)_3CCH(CH_3)CHO$	32	69
$BrHgCH_2CHO$	$(C_6H_5)_3CBr$	C_6H_6, 50–60°C, 2 hr, reflux, 2 hr	$(C_6H_5)_3CCH_2CHO$	34	2
$BrHgCH_2CHO$	$(p$-$NO_2C_6H_4)_3CBr$	$CHCl_3$, 20°C, 2 days, heat, 3 hr	$(p$-$NO_2C_6H_4)_3COCH{=}CH_2$	40	2
$ClHgCH_2COCH_3$	$(C_6H_5)_3CCl$	C_6H_6, overnight 20°C, water bath 0.5 hr	$(C_6H_5)_3CCH_2COCH_3$	—	3

(continued)

TABLE VII (*Continued*)

R in R₂Hg of RHgR' and RHgX	Alkyl halide	Reaction conditions	Products	Yield, %	Ref.
CH(CO₂CH₃)COCH₃	CH₂=CHCH₂I	C₆H₆, 70–75°C, 58 hr	CH₂=CHCH₂CH(CO₂CH₃)COCH₃	10	4
CH(CO₂CH₃)COCH₃	C₆H₅CH₂I	C₆H₆, 70–75°C, 6 hr	(C₆H₅CH₂)₂C(CO₂CH₃)COCH₃ C₆H₅CH₂CH(CO₂CH₃)COCH₃	30 13	4
C₆H₅COCH₂HgCl	(C₆H₅)₃CCl	C₆H₆, overnight, 20°C, water bath 0.5 hr	CH₃COCH₂COOCH₃ (C₆H₅)₃CCH₂COC₆H₅	— 38	3
trans-ClHgCH=CHCl	CCl₄	(C₆H₅COO)₂ or (CH₃COO)₂, sealed tube, 60–70°C, 4 hr	Cl₃CCH=CHCl	46	14
trans-ClHgCH=CHCl	CBr₄	(CH₃COO)₂, sealed tube, 100°C, 5 hr	ClCH=CHCBr₃, ClCHBr—CHBrCBr₃	—	14
trans-ClCH=CH	CCl₄	(C₆H₅COO)₂ or (CH₃COO)₂, sealed tube, 100°C, 4 hr	Cl₃CCH=CHCl	36	14
IHgC₃H₅	C₃H₅I	—	HgI₂, C₆H₁₀	—	12
C₆H₅CH(HgBr)C(O)OC₂H₅	(C₆H₅)₃CBr	CH₃NO₂, 7 days	C₆H₅CH₂COOC₂H₅ (C₆H₅)₃COH	81 80	5a 5a
C₆H₅CH(HgBr)C(O)OC₂H₅	(C₆H₅)₃CBr	C₂H₄Cl₂, 10 days	C₆H₅CH(CO₂C₂H₅)C(C₆H₅)₃	—	5a
R—C₆H₄CH(HgBr)CO₂C₂H₅	(C₆H₅)₃CBr·HgBr₂	CH₃NO₂	R—C₆H₄COOC₂H₅ [CH₂=N(O)OC(C₆H₅)₃]	—	5, 5a
R—C₆H₄CH(HgBr)CO₂C₂H₅ R = H, C₂H₅, *i*-C₃H₇, F, Cl, Br, I, NO₂	(*p*-CH₃C₆H₄)₃CBr·HgBr₂	C₂H₄Cl₂	R—C₆H₄CH(CO₂C₂H₅)C(C₆H₄CH₃-*p*)₃	—	5, 5a
C₆H₅	CHCl₃	(CH₃COO)₂, sealed tube, 125°C, 3 hr	C₂Cl₆, C₆H₅COOH	—	14
C₆H₅	CHCl₃	BiCl₃, boiling CHCl₃, 5 hr	C₆H₆	—	19
C₆H₅	CHCl₃ or CHBr₃	Sealed tube, 130°C, 5 hr	No reaction	—	20
C₆H₅	CHI₃	Sealed tube, 130°C, 3 hr	C₆H₆, C₆H₅HgCl₃, C₆H₅HgI	—	20
C₆H₅	CHI₃	Sealed tube, 130°C, 5 hr	C₂I₄, C₆H₅HgI	29	20
C₆H₅	CCl₄	Sealed tube, >200°C	C₆H₅CCl₃, C₆H₅HgCl, Hg₂Cl₂, HgCl₂	—	32

				Yield (%)	Ref.
C_6H_5	CCl_4	$(CH_3COO)_2$ sealed tube, 120–125°C, 3 hr	C_2Cl_6, m-$NO_2C_6H_4COOH^b$	—	14
C_6H_5	CCl_4	hv, 5 hr–2 months	C_6H_5HgCl, C_6H_5Cl, C_2Cl_6	—	28
C_6H_5	CCl_4	CH_3OH or ethylcellosolve, hv, 4–5 hr	C_6H_5HgCl / C_6H_6	98 / —	43
C_6H_5	C_2H_5Br	CH_3OH, hv, 15 hr	C_6H_5HgBr, m-$(NO_2)_2C_6H_4^b$	74	40,43
C_6H_5	$C_2H_4Br_2$	Sealed tube, 180–200°C, 6—10 hr	$C_6H_5CH_2CH_2Br$	—	11
C_6H_5	$C_2H_4Br_2$ excess	Sealed tube, 180–200°C, 6—10 hr	$C_6H_5CH_2CH_2Br$ / C_6H_5HgBr	— / —	11
C_6H_5	i-C_3H_7Br	hv, 12 hr	C_6H_5HgCl, m-$(NO_2)_2C_6H_4^b$	59	40
C_6H_5	Chlorex	hv, 12 hr	C_6H_5HgCl / m-$(NO_2)_2C_6H_4^b$	71 / —	40
C_6H_5	tert-C_4H_9Br	CCl_4, reflux, 340 hr	C_6H_6, i-C_4H_8	—	8
C_6H_5	tert-$C_5H_{11}I$	$CHCl_3$, reflux, 17 hr	i-$C_5H_{10}OH$, C_6H_6	—	8
C_6H_5	C_2Cl_4	CH_3OH, hv, 18 hr	C_6H_5HgCl / Hg_2Cl_2 / C_6H_6	80 / 80 / —	43
C_6H_5	$C_6H_5CH_2Cl$	CH_3OH, hv, 80 hr	C_6H_5HgCl / $(C_6H_5CH_2)_2$ / CH_2O, C_6H_6	84 / 50 / —	42
C_6H_5	$C_6H_5CH_2I$	CH_3OH, hv, 80 hr	C_6H_5HgCl / C_6H_6, $(C_6H_5CH_2)_2$ / CH_2O	86 / 95 / —	42
C_6H_5	$C_6H_5CH_2I$	C_6H_6, hv, 80 hr	C_6H_5HgI / $(C_6H_5)_2$, $(C_6H_5CH_2)_2$	70 / —	42
C_6H_5	$C_6H_5CHCl_2$	Sealed tube, 150–155°C, 170 hr	$(C_6H_5)_3CH$, C_6H_5HgCl	90	7
C_6H_5	$(C_6H_5)_2CHBr$	Toluene, reflux, 170 hr	$(C_6H_5)_3CH$	—	8
C_6H_5	C_6H_5Cl	hv, 18 hr	Hg_2Cl_2	92	40
C_6H_5	C_6H_5Br	hv, 18 hr	C_6H_5HgBr	40	40
C_6H_5	C_6H_5Br	hv, 75 hr	p-$BrC_6H_4C_6H_5$	—	40

(continued)

313

TABLE VII (*Continued*)

R in R₂Hg of RHgR' and RHgX	Alkyl halide	Reaction conditions	Products	Yield, %	Ref.
C_6H_5	C_6H_5I	CH_3OH, hv, 2 hr	C_6H_5HgI	92	42
			C_6H_6	—	
			CH_2O	—	
C_6H_5	C_6H_5I	C_2H_5OH, hv, 24 hr	C_6H_5HgI	73	42
			C_6H_6	—	
			CH_3CHO	—	
C_6H_5	C_6H_5I	i-C_3H_7OH, hv, 24 hr	C_6H_5HgI	Quant.	42
			C_6H_6, CH_3COCH_3		
C_6H_5	C_6H_5I	C_6H_6, hv, 60 hr	C_6H_5HgI	70	42
			$(C_6H_5)_2$	20	
C_6H_5HgCl	CCl_4	$AlCl_3$, stir, 90°C, 1 hr	$C_6H_5COC_6H_5$	66	45
o-$CH_3C_6H_4$	CCl_4	hv, 5 hr–2 month	C_6H_5HgCl	—	28
			o-$CH_3C_6H_4Cl$	—	
			C_2Cl_6	—	
o-$CH_3C_6H_4$	$C_6H_5CH_2I$	CH_3OH, hv, 60 hr	o-$CH_3C_6H_4HgI$	83	42
			$CH_3C_6H_5$	—	
			$(C_6H_5CH_2)_2$	—	
o-$CH_3C_6H_4$	C_6H_5I	CH_3OH, hv, 60 hr	o-$CH_3C_6H_4HgI$	73	42
			C_6H_6, $CH_3C_6H_5$	—	
			CH_2O	—	
p-$CH_3C_6H_4$	sec-C_4H_9I	CCl_4, reflux, 380 hr	p-$CH_3C_6H_4HgI$	Trace	8
p-$CH_3C_6H_4$	tert-C_4H_9Br	CCl_4, reflux, 340 hr	i-C_4H_8	—	8
p-$CH_3C_6H_4$	tert-$C_5H_{11}I$	CCl_4, reflux, 17 hr	$CH_3C_6H_5$	Low	8
p-$CH_3C_6H_4$	$(C_2H_5COO)_2CHBr$	Toluene, reflux, 336 hr	p-$CH_3C_6H_4Br$	—	8
			$(C_2H_5COO)_2CH_2$, $[(C_2H_5COO)_2Cl]_2$	—	
p-$CH_3C_6H_4$	$(C_6H_5)_2CHBr$	Toluene, reflux, 170 hr	p-$CH_3C_6H_4(C_6H_5)_2CH$	—	8
p-$CH_3C_6H_4$	$(C_6H_5)_2CHBr$	m-xylene, reflux, 200 hr	p-$CH_3C_6H_4(C_6H_5)_2CH$	—	8
p-$CH_3C_6H_4$	1-Br-1,1,2$(C_2H_5CO_2)_3C_2H_2$	Toluene, reflux, 288 hr	$(C_2H_5CO_2)_2C{=}C(H)CO_2C_2H_5$	—	8

314

Reactant	Structure	Conditions	Product	Yield (%)	Ref.
p-CH₃C₆H₄	(bifluorenylidene structure)	Toluene, reflux, 300 hr			8

Reactant	Reagent	Conditions	Product	Yield (%)	Ref.
p-CH$_3$C$_6$H$_4$ (9-bromofluorene)	Lauryl bromide	Toluene, reflux, 340 hr	No reaction		8
p-CH$_3$C$_6$H$_4$	Ethylene bromide	Toluene, reflux, 340 hr	No reaction		8
p-CH$_3$C$_6$H$_4$	Tetrachloroethane	Toluene, reflux, 340 hr	No reaction		8
p-CH$_3$C$_6$H$_4$	Hexachloroethane	Toluene, reflux, 340 hr	No reaction		8
p-CH$_3$C$_6$H$_4$	Allyl iodide	Toluene, reflux, 340 hr	No reaction		8
p-CH$_3$C$_6$H$_4$	Benzyliodide	Toluene, reflux, 340 hr	No reaction		8
p-CH$_3$C$_6$H$_4$	Benzal chloride	Toluene, reflux, 340 hr	No reaction		8
p-CH$_3$C$_6$H$_4$	Benzotrichloride	Toluene, reflux, 340 hr	No reaction		8
p-CH$_3$C$_6$H$_4$	Cyclohexyl bromide	Toluene, reflux, 340 hr	No reaction		8
p-CH$_3$C$_6$H$_4$	β-Bromoethyl phenyl ether	Toluene, reflux, 340 hr	No reaction		8
p-CH$_3$C$_6$H$_4$	γ-Bromopropyl phenyl ether	Toluene, reflux, 340 hr	No reaction		8
p-CH$_3$C$_6$H$_4$	Ethyl dichloroacetate	Toluene, reflux, 340 hr	No reaction		8
p-CH$_3$C$_6$H$_4$	Methyl α-bromo-isobutyrate	Toluene, reflux, 340 hr	No reaction		8
p-CH$_3$C$_6$H$_4$	α-Bromocamphor	Toluene, reflux, 340 hr	No reaction		8
p-CH$_3$C$_6$H$_4$	sym-Trinitrophenyl iodide	Toluene, reflux, 340 hr	No reaction		8
p-CH$_3$C$_6$H$_4$	C$_6$H$_5$I	C$_6$H$_6$, hv, 60 hr	p-CH$_3$C$_6$H$_4$HgI	74	42
			(C$_6$H$_5$)$_2$	—	
			(CH$_3$C$_6$H$_4$)$_2$	—	
p-CH$_3$COOC$_6$H$_4$	CHCl$_3$	hv, 54 hr	p-CH$_3$COOC$_6$H$_4$HgCl	30	36
			C$_2$Cl$_6$, Hg$_2$Cl$_2$	8	
p-CH$_3$COOC$_6$H$_4$	CCl$_4$	hv, 164 hr	CH$_3$COOC$_6$H$_4$HgCl-p	51	36
			C$_2$Cl$_6$	—	
o-ClHgC$_6$H$_4$OH	CH$_2$=CH—CH$_2$Br	C$_5$H$_5$N, −10°C	p-ClC$_6$H$_4$COOH[c]	—	22
			(C$_6$H$_5$N·C$_3$H$_5$)$_2$HgBr$_4$	—	
o-ClHgC$_6$H$_4$ONa	CH$_2$=CH—CH$_2$Cl	(CH$_3$)$_2$CO, 7 hr	(C$_3$H$_5$OC$_6$H$_4$)$_2$Hg	8	22

(continued)

315

TABLE VII (Continued)

R in R₂Hg of RHgR' and RHgX	Alkyl halide	Reaction conditions	Products	Yield, %	Ref.
o-$ClHgC_6H_4ONa$	$CH_2{=}CH{-}CH_2Br$	$(CH_3)_2CO$, 7 hr	$C_3H_5OC_6H_4HgCl$	40	22
4-$ClHgC_6H_4OC_6H_5$	$C_6H_5CH_2Cl$	100°C, 6–8 hr	4-$C_6H_5CH_2C_6H_4OC_6H_5$	—	9
			$(4$-$C_6H_5CH_2C_6H_4)_2O$		
4-$N(CH_3)_2C_6H_4$	CH_3I	Stir	$I(CH_3)_3NC_6H_4 \cdot HgC_6H_4N(CH_3)_3I$	—	49
α-$C_{10}H_7$	$CHCl_3$	$h\nu$, 20 hr	α-$C_{10}H_7HgCl$	96	37
			$C_{10}H_8$	—	
			C_2Cl_6	—	
α-$C_{10}H_7$	CCl_4	$h\nu$, 20 hr	α-$C_{10}H_7HgCl$	93	37
			$C_{10}H_8$	—	
			C_2Cl_6	—	
α-$C_{10}H_7$	CCl_4	Sealed tube, 150°C	α-$C_{10}H_7CCl_3$, $C_{10}H_8$, α-$C_{10}H_7HgCl$	—	31
		Sealed tube, 160–230°C	$C_{10}H_8$, α-$C_{10}H_7HgCl$	—	
α-$C_{10}H_7HgCl$	CH_3I	$h\nu$, 24 hr	α-$C_{10}H_7HgI$	77	37
			$C_{10}H_8$	18	
			HgI_2	—	
α-$C_{10}H_7HgCl$	C_2H_5Br	$h\nu$, 24 hr	α-$C_{10}H_7HgBr$	93	37
			$C_{10}H_8$	—	
α-$C_{10}H_7HgCl$	C_2H_5I	$h\nu$, 15 hr	$C_{10}H_8$, HgI_2	—	37
α-$C_{10}H_7HgCl$	1,2-$C_2H_4Cl_2$	$h\nu$, 30 hr	α-$C_{10}H_7HgCl$	95	37
			$C_{10}H_8$	—	
α-$C_{10}H_7HgCl$	$C_2H_2Cl_4$	$h\nu$, 35 hr	α-$C_{10}H_7HgCl$	93	37
			$C_{10}H_8$	—	
2-C_4H_3OHgCl		$(C_2H_5)_2O$, stir, 24 hr		9	10
$C_5H_5FeC_5H_4$	$(C_6H_5)_3CCl$	C_6H_6, reflux, 4 hr	$C_5H_5FeC_5H_4C(C_6H_5)_3$	18	6
$C_4H_9HgC_2H_5$	CCl_4	$(C_6H_5COO)_2$, sealed tube, 100°C, 8 hr	$C_2H_5CCl_3$	20	15
			C_4H_9HgCl	70	
$C_4H_9HgCH_2C_6H_5$	CCl_4	$(C_6H_5COO)_2$, sealed tube, 100°C, 8 hr	$C_6H_5CH_2HgCl$	—	15
			$C_4H_9CCl_3$	—	

316

Starting material	Solvent	Conditions	Products	Yield (%)	Ref.
$C_4H_9HgC_6H_{11}$	CCl_4	$(C_6H_5COO)_2$, sealed tube, 100°C, 8 hr	$C_4H_9CCl_3$	60	15
			$C_6H_{11}HgCl$	84	
			$(C_6H_{11})_2Hg$	—	
$C_6H_5CH_2HgC_6H_{11}$	CCl_4	$(C_6H_5COO)_2$, sealed tube, 100°C, 8 hr	$C_6H_5CH_2COOH[c]$	42	15
			$C_6H_{11}HgCl$	60	
			$C_6H_5CH_2HgCl$	17	
$C_6H_5HgC_2H_5$	CCl_4	$(C_6H_5COO)_2$, sealed tube, 100°C, 8 hr	$C_6H_5COOH[c]$	—	15
			C_2Cl_6	86	
			C_2H_5HgCl	76	
$C_4H_9HgC_6H_5$	CCl_4	$(C_6H_5COO)_2$, sealed tube, 100°C, 8 hr	C_4H_9HgCl	—	15
			C_2Cl_6	72	
$C_6H_5CH_2HgC_6H_5$	CCl_4	$(C_6H_5COO)_2$, sealed tube, 100°C, 8 hr	$C_6H_5COOH[c]$	33	15
			$C_6H_5COOH[c]$	—	
			C_6H_5HgCl	74	
			$C_6H_5CH_2HgCl$	43	
$C_6H_5HgC_6H_{11}$	CCl_4	$(C_6H_5COO)_2$, sealed tube, 100°C, 8 hr	$C_6H_5COOH[c]$	80.7	15
			$C_6H_{11}HgCl$	—	
$(m\text{-}CH_3C_6H_4)_2Hg + (C_6H_5)_2Hg$	CCl_4	$(C_6H_5COO)_2$, sealed tube, 100°C, 8 hr	$m\text{-}CH_3C_6H_4COOH[c]$	84	15
			$m\text{-}CH_3C_6H_4HgCl$	84	
			C_6H_5HgCl	82	
$p\text{-}CH_3C_6H_4HgC_6H_5$	CCl_4	$(C_6H_5COO)_2$, sealed tube, 100°C, 8 hr	$p\text{-}CH_3C_6H_4COOH[c]$	—	15
			C_2Cl_6	89	
			C_6H_5HgCl	31	
$C_6H_5HgC_6H_2(CH_3)_3$	CCl_4	$(C_6H_5COO)_2$, sealed tube, 100°C, 8 hr	$(CH_3)_3C_6H_2COOH[c]$	94	15
			C_6H_5HgCl	62	
$o\text{-}CH_3C_6H_4HgC_6H_4CH_3\text{-}p$	CCl_4	$(C_6H_5COO)_2$, sealed tube, 100°C, 8 hr	$o\text{-}CH_3C_6H_4COOH[c]$	—	15
			$p\text{-}CH_3C_6H_4COOH$	—	
$(p\text{-}CH_3C_6H_4)_2Hg + (m\text{-}CH_3C_6H_4)_2Hg$	CCl_4	$(C_6H_5COO)_2$, sealed tube, 100°C, 8 hr	$m\text{-}CH_3C_6H_4COOH[c]$	—	15
			$m\text{-}CH_3C_6H_4HgCl$	29	
			$p\text{-}CH_3C_6H_4HgCl$	47	
			$(p\text{-}CH_3C_6H_4)_2Hg$	40	

(continued)

317

TABLE VII (Continued)

R in R₂Hg of RHgR' and RHgX	Alkyl halide	Reaction conditions	Products	Yield, %	Ref.
α-$C_{10}H_7HgC_6H_5$	CCl_4	$(C_6H_5COO)_2$, sealed tube, 100°C, 8 hr	α-$C_{10}H_7COOH$[c] C_2Cl_6	73 —	15
$(\alpha$-$C_{10}H_7)_2Hg + (p$-$CH_3C_6H_4)_2Hg$	CCl_4	$(C_6H_5COO)_2$, sealed tube, 100°C, 8 hr	α-$C_{10}H_7COOH$[c] p-$CH_3C_6H_4COOH$ α-$C_{10}H_7HgCl$ p-$CH_3C_6H_4HgCl$	— 47 36 36	15
α-$C_{10}H_7HgC_6H_2(CH_3)_3$	CCl_4	$(C_6H_5COO)_2$, sealed tube, 100°C, 8 hr	$(CH_3)_3C_6H_2COOH$[c] α-$C_{10}H_7HgCl$	36 73	15
$(o$-$CH_3C_6H_4)_2Hg + (\alpha$-$C_{10}H_7)_2Hg$	CCl_4	$(C_6H_5COO)_2$, sealed tube, 100°C, 8 hr	$C_{10}H_7COOH$[c] α-$C_{10}H_7HgCl$ o-$CH_3C_6H_4HgCl$	85 6 —	15
$(m$-$CH_3C_6H_4)_2Hg + (\alpha$-$C_{10}H_7)_2Hg$	CCl_4	$(C_6H_5COO)_2$, sealed tube, 100°C, 8 hr	α-$C_{10}H_7COOH$[c] m-$CH_3C_6H_4COOH$ α-$C_{10}H_7HgCl$ m-$CH_3C_6H_4HgCl$	— 82 30 	15

Reactions with Acyl Halides

$CH_2(OH)CH_2HgI$	C_6H_5COCl	Shake cool, 10% KOH	$CH_2(HgI)CH_2OCOC_6H_5$	—	56
$CH_2(OH)CH(OH)CH_2HgI$	C_6H_5COCl	Shake, 10% KOH	$CH_2(OCOC_6H_5)CH(OCOC_6H_5)CH_2HgI$	—	56
$ClHgCH_2CHO$	CH_3COCl	Xylene, 1 hr	$CH_3COOCH{=}CH_2$	48	1,3
$ClHgCH_2CHO$	$ClCH_2COCl$	C_6H_6, 60–70°C, 2 hr, reflux 2 hr	$ClCH_2COOCH{=}CH_2$	58	2
$ClHgCH_2CHO$	CF_3COCl	Xylene, stir, 4–5°C, 5 hr, 20°C, 1 hr	$CH_2{=}CHOOCCF_3$	32	60
$ClHgCH_2CHO$	Cl_3CCOCl	C_6H_6, 60–70°C, 2 hr, reflux, 2 hr	$Cl_3CCOOCH{=}CH_2$	53	2
$ClHgCH_2CHO$	$C_6H_5CH_2COCl$	C_6H_6, 60–70°C, 2 hr, reflux, 2 hr	$C_6H_5CH_2COOCH{=}CH_2$	50	2

				Yield (%)	Ref.
ClHgCH$_2$CHO	C=NCH$_2$COCl (phthalimide)	C$_6$H$_6$, 60–70°C, 2 hr, reflux, 2 hr	C=NCH$_2$COOCH=CH$_2$ (phthalimide)	40	2
ClHgCH$_2$CHO	C$_6$H$_5$COCl	Xylene, 50°C, 6–8 hr	C$_6$H$_5$COOCH=CH$_2$	65	1,3
ClHgCH$_2$CHO	p-CNC$_6$H$_4$COCl	—	p-CNC$_6$H$_5$COOCH=CH$_2$	50	60
ClHgCH$_2$CHO	p-NO$_2$C$_6$H$_4$COCl	C$_6$H$_6$, 60–70°C, 2 hr, reflux, 2 hr	p-NO$_2$C$_6$H$_4$COOCH=CH$_2$	55	2
ClHgCH$_2$CHO	(C$_6$H$_5$)$_3$CCOCl	C$_6$H$_6$, 60–70°C, 2 hr, reflux, 2 hr	(C$_6$H$_5$)$_3$CCOOCH=CH$_2$	48	2
ClHgCH$_2$COCH$_3$	CH$_3$COCl	Xylene, 1 hr	CH$_3$C(OOCH$_3$)=CH$_2$	66	1,3
ClHgCH$_2$COCH$_3$	C$_6$H$_5$COCl	Xylene, 50°C, 6–8 hr	C$_6$H$_5$COOC(CH$_3$)=CH$_2$	63	1,3
CH(CO$_2$CH$_3$)COCH$_3$	CH$_3$COCl	C$_6$H$_6$, 50–60°C, 5 hr	CH$_3$C(OOCH$_3$)=CHCOOCH$_3$	80	4
CH(CO$_2$CH$_3$)COCH$_3$	CH$_3$CH$_2$COCl	C$_6$H$_6$, 40–50°C, 3 hr	CH$_3$C(OCOC$_2$H$_5$)=CHCO$_2$CH$_3$	78	4
CH(CO$_2$CH$_3$)COCH$_3$	C$_6$H$_5$COCl	Xylene, 50–55°C, 10 hr	CH$_3$C(OCOC$_6$H$_5$)=CHCO$_2$CH$_3$	—	4
[(C$_6$H$_5$)$_3$P$^\oplus$CH(HgCl)COC$_6$H$_5$]Cl$^\ominus$	CH$_3$COCl	—	(C$_6$H$_5$)$_3$P$^\oplus$CH=C(OCOCH$_3$)C$_6$H$_5$	—	67
[(C$_6$H$_5$)$_3$P$^\oplus$CH(HgCl)COCH$_3$]Cl$^\ominus$	C$_6$H$_5$COCl	THF	[(C$_6$H$_5$)$_3$P$^\oplus$CH=CCH$_3$(OCOC$_6$H$_5$)]HgCl$_3$$^\ominus$	—	67
2-Chloromercurycyclohexanone	CH$_3$COCl	Xylene, 2 hr	(cyclohexene, O—COCH$_3$)	57	3
2-Chloromercurycyclohexanone	C$_6$H$_5$COCl	Xylene, water bath, 6–8 hr	(cyclohexene, OCOC$_6$H$_5$)	63	3
C$_6$H$_5$	CH$_3$COCl	—	C$_6$H$_5$COCH$_3$, C$_6$H$_5$HgCl	—	46
C$_6$H$_5$	C$_6$H$_5$COCl	Boiling toluene, 340 hr	No reaction	—	8
C$_6$H$_5$	C$_6$H$_5$COCl	AlCl$_3$, CS$_2$, reflux, 5 hr	C$_6$H$_5$COC$_6$H$_5$	47	45
C$_6$H$_5$HgF	CH$_3$COCl	Reflux, 5 hr	CH$_3$COF	90	47
C$_6$H$_5$HgCl	C$_6$H$_5$COCl	AlCl$_3$, 50°C, 2 hr	C$_6$H$_5$HgCl, C$_6$H$_5$COC$_6$H$_5$	59	45

(*continued*)

TABLE VII (Continued)

R in R_2Hg or $RHgR'$ and $RHgX$	Alkyl halide	Reaction conditions	Products	Yield, %	Ref.
$CH_3C_6H_4$	C_6H_5COCl	Boiling toluene, 340 hr	No reaction	—	8
$o\text{-}ClHgC_6H_4OH$	CH_3COCl	Heat	$o\text{-}ClHgC_6H_4OCOCH_3$	80	54
$p\text{-}ClHgC_6H_4OH$	CH_3COCl	—	$p\text{-}ClHgC_6H_4OCOCH_3$	80	54
$p\text{-}ClHgC_6H_4OC_6H_5$	C_6H_5COCl	150°C, 1 hr	$p\text{-}C_6H_5COC_6H_4OC_6H_5$	70	9
$C_6H_3(CH_3)(OH)HgCl$	C_6H_5COCl	—	$C_6H_3(CH_3)(OCOC_6H_5)HgCl$	—	53
$2\text{-}C_4H_3OHgCl$	CH_3COCl	$(CH_3)_2CO$, reflux, 0.5 hr	$2\text{-}C_4H_3OCOCH_3$	21	10
β-Chloromercurifuran	CH_3COCl	C_6H_6, reflux, 2.5 hr	2-Acetylfuran	17	68
$2\text{-}HgCl\text{-}4\text{-}BrC_4H_2O$	CH_3COCl	CH_3COCH_3 or ether	$4\text{-}BrC_4H_2OCOCH_3\text{-}2$	Low	10
$C_5H_5FeC_5H_4$	CH_3COCl	C_6H_6, reflux, 5 hr	$(C_5H_5)_2Fe$ $C_5H_5FeC_5H_4COCH_3$	94 1	6
CH_2CHO	$SOCl_2$	$i\text{-}C_5H_{12}$, stir, 2 hr	$(CH_2=CH)_2SO_3$	45	61
CH_2CHO	CH_3SO_2Cl	$i\text{-}C_5H_{12} + C_5H_5N$ stir, 2 hr	$CH_3SO_2OCH=CH_2$ $ClHgCH_2CHO$	45 —	61
CH_2CHO	$C_2H_5SO_2Cl$	$i\text{-}C_5H_{12} + C_5H_5N$ stir, 2 hr	$C_2H_5SO_2OCH=CH_2$	42	61
CH_2CHO	$C_6H_5SO_2Cl$	$C_2H_4Cl_2 + C_5H_5N$ stir, 20–30 min	$C_6H_5SO_2OCH=CH_2$ $ClHgCH_2CHO$	70 —	61
CH_2CHO	$p\text{-}CH_3C_6H_4SO_2Cl$	$C_2H_4Cl_2 + C_5H_5N$ stir, 20–30 min	$p\text{-}CH_3C_6H_4SO_2OCH=CH_2$ $ClHgCH_2CHO$	75 —	61
C_6H_5	$C_6H_5SO_2Cl$	Benzene, reflux, 8 hr	Hg (decomp)	—	48
C_6H_5	$C_6H_5SO_2Cl$	Sealed tube, 160°C, several hours	$(C_6H_5)_2SO_2$ C_6H_5HgCl	8 57	48
C_6H_5	$CH_3C_6H_4SO_2Cl$	Sealed tube, C_6H_6, 120°C, 15 hr	$(CH_3C_6H_4)_2SO_2$	75	48
CH_2CHO	$POCl_3$	C_5H_5N, stir, 0°C, 1.5 hr	Trivinyl phosphate	4	62
CH_2CHO	$ClP(OC_2H_5)_2$	$(C_2H_5)_3N + i\text{-}C_5H_{12}$, stir, 1 hr	$CH_2=CHOP(OC_2H_5)_3$ $ClHgCH_2CHO$	86 63	63
CH_2CHO	$Cl_2POC_2H_5$	$(C_2H_5)_3N + i\text{-}C_5H_{12}$, stir, 4 hr	$(CH_2=CHO)_2POC_2H_5$	60	63
$ClHgCH_2COCH_3$	$C_2H_5P(O)Cl_2$	C_5H_5N, stir, 0°C, 2 hr	Diisopropenyl ethylphosphinate	64	62

				Yield	Ref.
ClHgCH2COCH3 / α-Chloromercurydiethylketone	C2H5P(O)(OC2H5)Cl POCl3	C5H5N, stir, 0°C, 1.5 hr C5H5N + (C2H5)2O, stir, 0°C, 2 hr	Ethyl isopropenyl ethylphosphinate Tris(2-penten-3-yl) phosphate	77 76	62 62
α-Chloromercurydiethylketone	OP(OC2H5)2Cl	(C2H5)2O + C5H5N, stir, 0°C, 1.5 hr	Diethyl 2-penten-3-yl phosphate	84	62
α-Chloromercurycyclopentanone	OP(OC2H5)Cl2	(C2H5)2O + C5H5N, stir, 0°C, 1.5 hr	Ethyldicyclopentenylphosphate	74	62
CH2CO2-i-C4H9	CH3COCl	isopentan + (C2H5)2O, 12 hr, 20° O, 5 hr, 50°	CH3CO2C4H9-i CH3COCH2CO2C4H9-i	43 24	70
CH2CO2CH3	CH3COCl	CHCl3, 50–55°C, 3 hr, stand overnight	CH2=C(OCH3)OCOCH3	35	70
CH2CO2CH3 CH2CO2-i-C4H9 CH2=CHCH2HgI CH2=CHCH2HgI	C6H5COCl i-C4H9COCl (C6H5)2CHCOCl RCH2COCl R = C6H5; C6H5CH2 CH3(CH2)7(I) CH3(CH2)15(II)	CCl4, 80° (C2H5)2O, reflux, 3 hr Dioxane, reflux, 8 hr Dioxane, reflux, 8 hr	CH2=C(OCH3)OCOC6H5 CH2=C(OC4H9-i)OCOC4H9-i CH2=CHCH2C(O)CH(C6H5)2 CH3CH=CH2(A) + RCH2C(O)OC(CH2CH=CH2)=CHR(B)	47.5 42.5 73 R(A)(B) C6H5 36,46 C6H6CH2 68, 52 (I) 63, 47 (II) 51, 40	70 70 71 71 71
C5H5FeC5H4 C5H5FeC5H4	C6H5SO2J C5H5FeC5H4SO2J	C6H6, reflux, 6 hr	C5H5FeC5H4SO2C6H5 (C5H5FeC5H5)2SO2	22 27	21 21

---

[a] According to Ref. 24.

[b] After nitration.

[c] After hydrolysis. Razuvaev and Ol'dekop works given not exhausted.

[d] Some reactions detailed reviewed in the text are not included in the tables.

No acetylfuran could be isolated from the reaction of 3-chloromercury furan with acetyl chloride in acetone.[68] However, the reaction of 3-chloromercury furan with acetyl chloride in benzene yielded a 17% yield of 2-acetylfuran.[68]

Thus, the reaction of the organomercury compound with acyl chloride occurs with rearrangement.

REFERENCES

1. A. N. Nesmeyanov and I. F. Lutsenko, *Dokl. Akad. Nauk SSSR*, **59**, 707 (1948).
2. A. N. Nesmeyanov and E. G. Perevalova, *Izv. Akad. Nauk SSSR, Otd. Khim. Nauk*, **1954**, 1002.
3. A. N. Nesmeyanov, I. F. Lutsenko, and Z. M. Tumanova, *Izv. Akad. Nauk SSSR, Otd. Khim. Nauk*, **1949**, 601.
4. M. A. Kazankova, I. F. Lutsenko, and A. N. Nesmeyanov, *Zh. Obshch. Khim.*, **35**, 1447 (1965).
5. I. P. Beletskaya, O. A. Maximenko, and O. A. Reutov, *Dokl. Akad. Nauk SSSR*, **168**, 333 (1966); *Zh. Organ. Khim.*, **2**, 1129 (1966).
5a. I. P. Beletskaya, O. A. Maximenko, V. B. Volyeva, and O. A. Reutov, *Zh. Organ. Khim.*, **2**, 1132 (1966).
6. A. N. Nesmeyanov, E. G. Perevalova, and O. A. Nesmeyanova, *Dokl. Akad. Nauk SSSR*, **119**, 288 (1958).
7. A. Kekulé and A. Franchimont, *Ber.*, **5**, 907 (1872).
8. F. C. Whitmore and N. Thurman, *J. Am. Chem. Soc.*, **51**, 1491 (1929).
9. W. D. Schroeder and R. Q. Brewster, *J. Am. Chem. Soc.*, **60**, 751 (1938).
10. H. Gilman and G. F. Wright, *J. Am. Chem. Soc.*, **55**, 3302 (1933).
11. E. Dreher and R. Otto, *Ann. Chem.*, **154**, 93 (1870).
12. E. Linnemann, *Ann. Chem.*, Sp. 3, 262 (1865).
13. W. Suida, *Chem. Zentr.*, **1880**, 585.
14. A. E. Borisov, *Izv. Akad. Nauk SSSR, Otd. Khim. Nauk*, **1951**, 524.
15. A. N. Nesmeyanov, A. E. Borisov, E. I. Golubeva, and A. I. Kovredov, *Izv. Akad. Nauk SSSR, Otd. Khim. Nauk*, **1961**, 1582; *Tetrahedron*, **18**, 683 (1962).
16. F. C. Whitmore and E. Rohrmann, *J. Am. Chem. Soc.*, **61**, 1591 (1939).
17. G. A. Razuvaev, S. F. Zhil'tsov, O. N. Druschkov, and G. G. Petuchov, *Dokl. Akad. Nauk SSSR*, **156**, 393 (1964).
18. Z. M. Manulkin, A. N. Tatarenko, and F. Yu. Yusupov, *Zh. Obshch. Khim.*, sborn. II, 1308 (1953).
19. Z. M. Manulkin, *Uzbek. Khim. Zh.*, **1958**, N2, 41; *R. Zh. Khim.*, **1959**, 15244.
20. M. M. Koton, T. M. Zorina, and E. G. Osberg, *Zh. Obshch. Khim.*, **17**, 59 (1947).
21. A. N. Nesmeyanov, E. G. Perevalova, and O. A. Nesmeyanova, *Izv. Akad. Nauk SSSR, Otd. Khim. Nauk*, **1962**, 47.
22. A. N. Nesmeyanov and R. Ch. Schatzkaya, *Zh. Obshch. Khim.*, **5**, 1268 (1935).
23. G. Calmels, *Compt. Rend.*, **99**, 240 (1884).
24. K. A. Hofmann and J. Sand, *Ber.*, **33**, 1340 (1900).

25. R. Manchot and A. Klüg, *Ann. Chem.*, **420**, 170 (1920).
26. W. Manchot, *Ber.*, **53**, 984 (1920).
27. D. Spinelli and A.l Salvemini, *Ann. Chim. Ital.*, **51**, 389, 1296 (1961).
28. G. A. Razuvaev and Yy. A. Ol'dekop, *Dokl. Akad. Nauk SSSR*, **64**, 77 (1949).
29. G. A. Razuvaev and Yu. A. Ol'dekop, *Uchen. Zap. Gor'kov. Gos. Univer.*, **15**. 85 (1949).
29a. G. A. Razuvaev, Yu. A. Ol'dekop, and Z. N. Manchinova, *Zh. Obshch. Khim.*, **22**, 480 (1952).
30. G. A. Razuvaev and G. G. Petuchov, *Tr. khim. i khim. technol.* (Gor'ki) **1961**, vyp. 1, 150; *R. Zh. Khim.*, **1962**, 4, 50.
31. Yu. A. Ol'dekop, *Zh. Obshch. Khim.*, **23**, 2020 (1953).
32. G. A. Razuvaev and Yu. A. Ol'dekop, *Zh. Obshch. Khim.*, **23**, 587 (1953).
33. G. A. Razuvaev and Yu. A. Ol'dekop, *Zh. Obshch. Khim.*, **20**, 181 (1950).
34. Yu. A. Ol'dekop, *Zh. Obshch. Khim.*, **22**, 478, 2063 (1952).
35. G. A. Razuvaev, Yu. A. Ol'dekop and N. P. Nazarkova, *Uchen. Zap. Gor'kov. Gos. Univer.*, **23**, 117 (1952).
36. Yu. A. Ol'dekop and Z. B. Idel'chik, *Zh. Obshch. Khim.*, **30**, 2564 (1960).
37. G. A. Razuvaev and Yu. A. Ol'dekop, *Zh. Obshch. Khim.*, **20**, 506 (1950).
38. G. A. Razuvaev and M. A. Shubenko, *Sborn. stat. po obshch. Khim.*, **2**, 1043 (1953).
39. C. G. Krespan, *J. Org. Chem.*, **25**, 105 (1960).
40. G. A. Razuvaev and Yu. A. Ol'dekop, *Zh. Obshch. Khim.*, **21**, 1122 (1951).
41. G. A. Razuvaev and M. A. Shubenko, *Dokl. Akad. Nauk SSSR*, **67**, 1049 (1949).
42. G. A. Razuvaev and M. A. Shubenko, *Zh. Obshch. Khim.*, **20**, 175 (1950).
43. G. A. Razuvaev and Yu. A. Ol'dekop, *Zh. Obshch. Khim.*, **21**, 2197 (1951).
44. Yu. A. Ol'dekop, *Uchen. zap. Gor'kov. Gos. Univer.*, **24**, 147 (1953).
45. A. P. Skoldinov and K. A. Kocheschkov, *Zh. Obshch. Khim.*, **12**, 398 (1942).
46. M. S. Malinovski, *Trudy Gor'k. gosud.ped. instituta*, **1940**, N 5, 39; *Chem. Abstr.*, **37**, 3070 (1943).
47. G. F Wright, *J. Am. Chem. Soc.*, **58**, 2653 (1936).
48. R. Otto, *Ber.*, **18**, 246 (1885).
49. O. Dimroth, *Ber.*, **35**, 2032 (1902).
50. M. S. Kharasch, F. W. M. Lommen, and I. M. Jacobsohn, *J. Am. Chem. Soc.*, **44**, 793 (1922).
51. L. Vecchiotti and A. Michetti, *Gazz. Chim. Ital.*, **55**, 372 (1925).
52. U. H. Hodgson and D. E. Hathway, *J. Chem. Soc.*, **1945**, 123.
53. O. Dimtroth, *Ber.*, **35**, 2853 (1902).
54. F. C. Whitmore and E. Middleton, *J. Am. Chem. Soc.*, **43**, 619 (1921).
55. A. N. Nesmeyanov and E. M. Toropova, *Zh. Obshch. Khim.*, **4**, 664 (1934).
56. J. Sand, *Ber.*, **34**, 1385 (1901).
57. M. M. Kreevoy and G. B. Bodem, *J. Org. Chem.*, **27**, 4539 (1962).
58. A. N. Nesmeyanov and N. K. Kochetkov, *Izv. Akad. Nauk SSSR, Otd. Khim. Nauk*, **1949**, 76.
59. A. N. Nesmeyanov, I. F. Lutsenko, and R. M. Khomutov, *Izv. Akad. Nauk SSSR, Otd. Khim. Nauk*, **1957**, 942.
60. A. Ja. Jakubovitch, V. V. Razumovski, Z. N. Vostrukhina, and S. M. Rosenstein, *Zh. Obshch. Khim.*, **28**, 1930 (1958).
61. A. N. Nesmeyanov, I. F. Lutsenko, R. M. Khomutov, and V. A. Dubovitski, *Zh. Obshch. Khim.*, **29**, 2817 (1959).
62. I. F. Lutsenko and Z. S. Kraïtz, *Dokl. Akad. Nauk SSSR*, **135**, 860 (1960).

63. I. F. Lutsenko, Z. S. Kraïtz, and A. P. Bokovoy, *Dokl. Akad. Nauk SSSR*, **124**, 1251 (1959).

64. A. N. Nesmeyanov and N. K. Kochetkov, *Izv. Akad. Nauk SSSR, Otd. Khim. Nauk*, **1949**, 305.

65. I. F. Lutsenko, V. L. Foss, and N. L. Ivanova, *Dokl. Akad. Nauk SSSR*, **141**, 1107 (1961).

66. V. L. Foss, M. A. Zhadina, I. F. Lutsenko, and A. N. Nesmeyanov, *Zh. Obshchei Khim.*, **33**, 1927 (1963).

67. Nic. A. Nesmeyanov and V. M. Novikov, *Dokl. Akad. Nauk SSSR*, **162**, 350 (1965).

68. S. Gronowitz and G. Sörlin, *Arkiv. Kemi*, **19**, 515 (1963).

69. D. Y. Curtin and M. J. Hurwitz, *J. Am. Chem. Soc.*, **74**, 5381 (1952).

70. I. T. Lutzenko, V. L. Foss, and A. N. Nesmeyanov, *Dokl. Akad. Nauk SSSR*, **169**, 117 (1966).

71. J. Kuwajima, K. Narasaka, and T. Mukajama, *Tetrahedron Letters*, **43**, 4281 (1967).

72. F. R. Jensen and H. E. Guard, *J. Am. Chem. Soc.*, **90**, 3250 (1968).

73. Nic. A. Nesmeyanov, V. M. Novikov, and O. A. Reutov, *Zh. Organ. Khim.*, **2**, 942 (1966).

Chapter 11

Reactions with Halogens

I. INTRODUCTION

Halogens break the C—Hg bond of organomercury compounds replacing mercury with halogen. The following reaction occurs under mild conditions:

$$R_2Hg + Hal_2 \longrightarrow RHgHal + RHal$$

The further reaction with halogen leads also to the separation of the radical from RHgX and to the replacement of HgX with halogen:

$$RHgX + Hal_2 \longrightarrow RHal + HgXHal$$

Such a conversion of organomercury compound R_2Hg or RHgX into RHal, whose structure is well known, makes it possible to determine the position of the mercury. These halodemercuration reactions are of great importance from the practical viewpoint for the chemistry of organomercury compounds. Besides, some important theoretical rules are based on the investigations of these reactions.

II. REACTION STEREOCHEMISTRY, KINETICS, AND MECHANISM

A. Reaction Stereochemistry

The stereochemistry of the mercury halogenation reaction depends on such conditions of the reaction as the polarity of the reaction medium and nature of the halogenating agent (molecular halogen or halogen in the presence of halogen anion). These govern the different reaction mechanisms.

1. Compounds of Mercury at Olefinic Carbon Atoms

For the first time stereoisomeric organomercury compounds such as β-chlorovinylmercury compounds (as RHgHal and R_2Hg) were converted into RHal and RHgHal, respectively, with radical retention of configuration. The conversion took place in nonpolar solvents upon reaction with molecular halogen in the absence of halogen anion in full accord with the rule of radical retention of configuration in reactions of electrophilic and homolytic substitution of olefinic carbon atoms.[1,2] Chlorovinylmercury chloride obtained by the reaction of an equimolecular quantity of chlorine in carbon tetrachloride on *trans,trans*- and *cis,cis*-di-β-chlorovinylmercury, respectively,[3] has the same configuration as initial product R_2Hg:

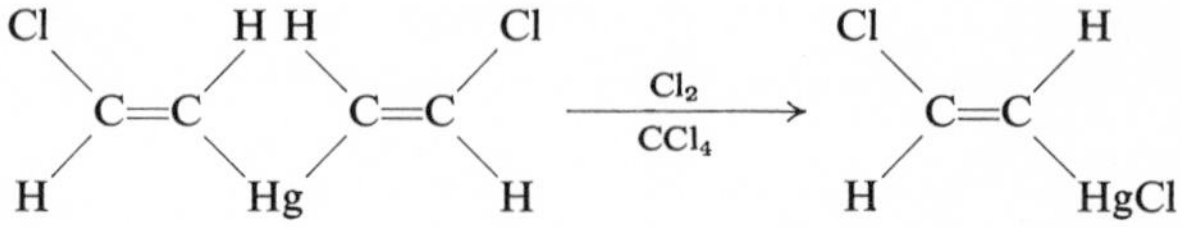

The effect of equivalent quantity of iodine upon *cis*-β-chlorovinylmercurichloride in ether solution gives only *cis*-1-chloro-2-iodoethylene[4]:

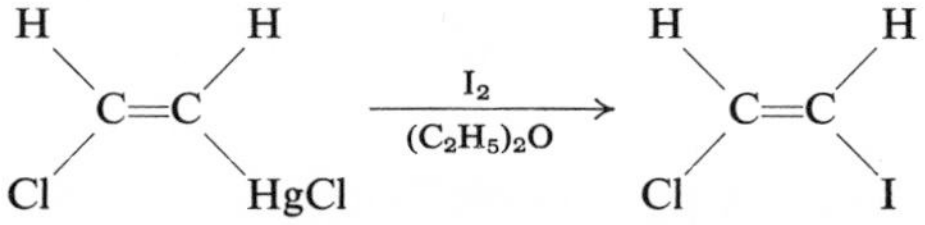

However, it is possible that under these conditions of homolytic reactions of halogenation, free radicals are not formed as kinetically independent particles. According to Reutov et al.,[6,7] retention of configuration does not occur under conditions of known free-radical reactions when photochemical halogenolysis of both isomers of β-chlorovinylmercury chloride were carried out with iodine in carbon tetrachloride or benzene[5] and when *cis* and *trans* isomers of ω-styrylmercury bromide are treated with bromine in the same solvents. The configuration is preserved in the cases of halogenolysis reactions with an S_E2 mechanism caused by use of electrophilic halogen (halogen in the presence of halogen anion) as reactions of *cis*- and *trans*-β-chlorovinylmercury chloride with iodine in the presence of cadmium iodide in methanol,[8] 85% aqueous solution of dioxane[9] and dimethylformamide[10] (see part B, below). The solvents are listed in the series of decreasing reaction rate. According to Beletskaya et al.[9] the halogenolysis reaction of *cis*-β-chlorovinylmercury chloride has a higher reaction rate than that of the *trans* isomer, which is inconsistent with the data on high reactivity of the *trans* isomer obtained by Nesmeyanov and Borisov in the course of a large number of experiments.[11-15]

Unlike S_E1 reactions in the cases of saturated atoms, S_E1 monomolecular electrophilic substitution at carbon olefinic atoms occurs with retention of configuration which is confirmed by the reaction of *trans*-β-chlorovinylmercury chloride with iodine in dimethylsulfoxide[5] and by the reaction of *trans-ω*-styrylmercury bromide with bromine in the same solvent[6]:

$$\underset{H}{\overset{Cl}{>}}C=C\underset{HgCl}{\overset{H}{<}} + I_2 \xrightarrow{(CH_3)_2SO} \underset{H}{\overset{Cl}{>}}C=C\underset{I}{\overset{H}{<}} + HgICl$$

$$\underset{H}{\overset{C_6H_5}{>}}C=C\underset{HgBr}{\overset{H}{<}} + Br_2 \xrightarrow{(CH_3)_2SO} \underset{H}{\overset{C_6H_5}{>}}C=C\underset{Br}{\overset{H}{<}} + HgBr_2$$

On the basis of these additional data, Reutov and Beletskaya[7] refined the regularities of electrophilic and homolytic substitution, first formulated by Nesmeyanov and Borisov in the following way: Both the reaction of bimolecular electrophilic substitution (S_E2 mechanism) and the reactions of monomolecular electrophilic substitution (S_E1 mechanism) proceed with retention of configuration; homolytic substitution reactions at olefinic carbon atoms, in the course of which radicals are not formed as kinetically independent particles, proceeds with retention of configuration; in free-radical substitution reactions, retention of configuration is not observed (S_R rule).

The configuration of the transferred radical is retained when RBr is formed upon the reaction of bromine in benzene solution with *trans-α-*bisstilbenylmercury. Retention of configuration also occurs if dioxane-dibromide is reacted with the above organomercury compound in dioxane and trans-α-stilbenylmercury bromide is isolated.[16] *cis-α-*Bromo-stilbene[16] is formed with *cis* isomers of R_2Hg or RHgBr, i.e.:

$$\underset{H}{\overset{C_6H_5}{>}}C=C\underset{HgBr}{\overset{C_6H_5}{<}} \xrightarrow{C_4H_8O_2 \cdot Br_2} \underset{H}{\overset{C_6H_5}{>}}C=C\underset{Br}{\overset{C_6H_5}{<}}$$

The influence of steric factors can probably not be ignored for the reactions of isomeric mercury derivatives of stilbenes.

2. Compounds of Mercury at Saturated Carbon Atoms

As early as in 1939 Whitmore et al.[17] found that under the effect of iodine in aqueous KI solution or bromine in aqueous KBr solution, a mercury in neopentylmercury chloride, may be converted into RI or RBr without rearrangement of the carbon skeleton:

$$(CH_3)_3CCH_2HgCl + Hal_2 \longrightarrow (CH_3)_3CCH_2Hal + HgClHal$$

As was previously found,[18] mercury bound to the asymmetric atom of carbon of 3-methoxy-2-bromomercuri-3-phenylpropionic acid, in various solvents and under various light conditions, gives various relative quantities of two diastereomeric 3-methoxy-2-bromo-3-phenylpropionic acids. In this work, however, the relationship between reaction conditions and stereochemistry of the reaction has not been elucidated.

Investigating the bromolysis of cis- and trans-4-methylcyclohexylmercury bromide[19,22] and sec-butylmercury bromide,[20] Jensen found that radical reactions in nonpolar solvents such as carbon disulfide, $CHCl_3$, and carbon tetrachloride are accompanied by racemization or partial preservation of the stereochemical configuration of the radical being transferred. When the bromolysis reaction of both isomers of 4-methylcyclohexylmercury bromide proceeds in a polar medium (pyridine) under conditions favorable for an ionic mechanism, it is accompanied by retention of configuration[21,22] (S_E2 mechanism).

The stereochemistry of the halogenolysis reaction may also be influenced by the presence or the absence of oxygen. In a nitrogen atmosphere halogen facilitates racemization while in air, which probably inhibits radical formation, the halogen facilitates retention of configuration.[22]

The effect of positive brominating agents such as hypobromite in a 50% aqueous solution of dioxane or the complex of bromine with zinc bromide is less favorable for a stereospecific reaction than is the case when this reaction proceeds in pyridine.[22] The radical retention of configuration occurs if sec-butylmercury bromide reacts in pyridine at $-65°C$ in the presence of oxygen.[20]

When the reaction is conducted in nonpolar solvents (CS_2, CCl_4, and $CHCl_3$) addition of alcohol increases, sometimes considerably, the retention of configuration.[20,22,23] For example, if sec-butylmercury bromide reacts with bromine in carbon tetrachloride, the addition of alcohol increases the retention of configuration from 30 to 80%.[23]

The halogenolysis reaction in nonpolar solvents was used for determining the configuration of products obtained in the oxymercuration of cycloalkene derivatives. Using the dimethyl ester of exo-4-oxy-5-chlormercury-3,6-endoxyhexahydrophthalic acid (1) and the dimethyl ester of exo-4-acetoxy-5-chlormercury-3,6-endoxyhexahydrophthalic acid (2), Yuriev and Zefirov[24,25] determined that the bromination of the oxydiester (1) in chloroform or carbon tetrachloride results in a mixture of isomers 1a and 1b with the predominance of cis-bromohydrin (1a) while the bromination in pyridine gives only one isomer, namely, *cis*-bromohydrin (**1a**). The bromination of acetoxyester (**2**) proceeds in a similar way:

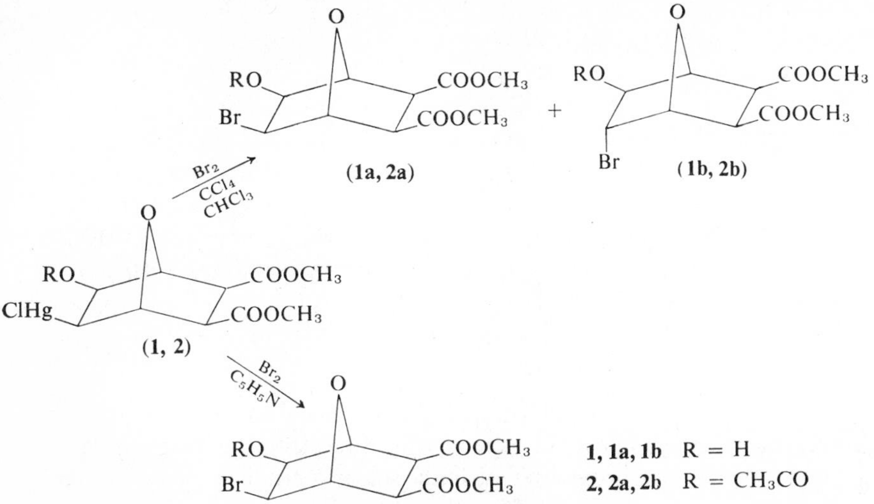

(1a, 2a) + (1b, 2b)

(1, 2)

1, 1a, 1b   R = H
2, 2a, 2b   R = CH_3CO

The reaction of the 4-oxyderivative with bromine in glacial acetic acid which proceeds with retention of configuration, contrary to previous data,[24] is not accompanied by acetylation of the hydroxyl group.[25] The chlorination of products **1** and **2** by chlorine in $CHCl_3$ or acetic acid is accompanied by retention of configuration and results in **3a** and **4a**, respectively. The reaction with product (**1**) in acetic acid is also not accompanied by the acetylation of the hydroxyl group.[25]

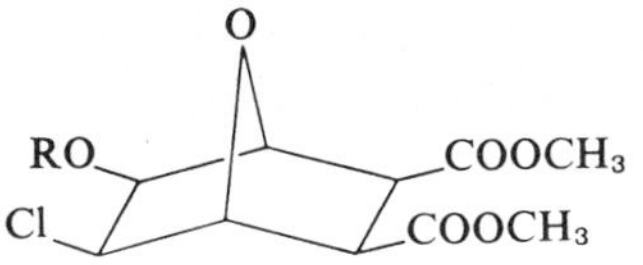

3a R = H
4a R = CH_3CO

(3a, 4a)

The bromination of the perchlorate (**5**), in acetic acid resulted in the

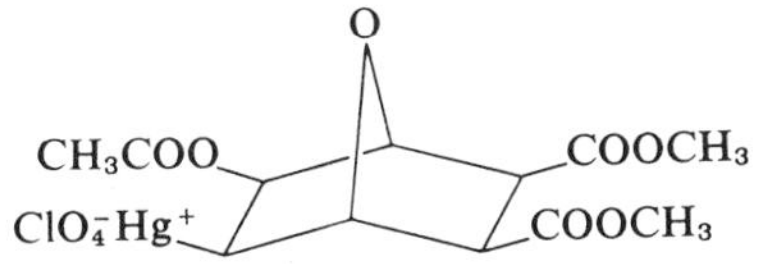

(5)

formation of *cis*-bromoacetate (**2a**).[26]

The formation of bromolactone (**6b**) with an exo-configuration of Br and produced by the reaction sequence shown below, has shown that, contrary to previous data,[27] the oxymercuration proceeds by *trans* addition[28] (with intramolecular lactonization):

(6)   (6a)   (6b)

In the case of cyclic β-mercurated alcohols, the presence of neighboring substituents may change the stereochemical results of the reaction. It was shown that in the bromolysis of both isomeric products of the oxymercuration of D-glucal,[29] configuration reversion was observed for one isomer when bromine was used in alcohol and configuration preservation was observed for another isomer when bromine was used in pyridine, methanol, and even in chloroform and carbon tetrachloride (in the latter case the yield of bromine derivative of the same configuration ranged from 60 to 85%). However, these data should be reinvestigated since some authors[30] consider that the initial products of the oxymercuration of D-glucal have another configuration.

The reaction of 4-camphenylmercury iodide with iodine in the presence of LiI in dioxane is, of necessity (reaction at a bridge head carbon) accompanied by retention of configuration.[32]

In the halogenolysis of compound 7, i.e., the product of norbornene oxymercuration, various relative quantities of *cis*- and *trans*-conformers of the respective norbornane halohydrines are obtained. These quantities depend on both the reaction medium and the kind of halogenating agent.[31] The other reaction products are also different in both cases.

The effect of bromine upon product 7 in the case of radical reaction in chloroform results in a mixture of exo- and endoisomers 7a and 7b:

(7)   (7a)   (7b)

The effect of bromine or gaseous chlorine upon the solution of product **7** in a mixture of chloroform and pyridine results in halogenhydrins and the predominant product, **7a**, and exoepoxynorbornane (**7c**)

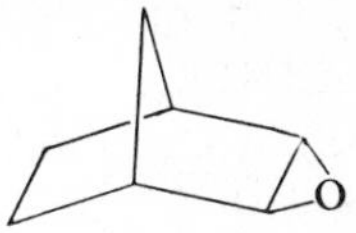

(7c)

was also produced in an insignificant quantity in the case of bromine and up to 65% in the case of chlorine. In pyridine, chlorine gave almost entirely the latter product, while bromine gave a uniform mixture of isomeric bromohydrins **7a** and **7b** and somewhat lesser amounts of epoxynorbornane (**7c**). The formation of epoxynorbornane is explained by authors as the demercuration by positive halogen:

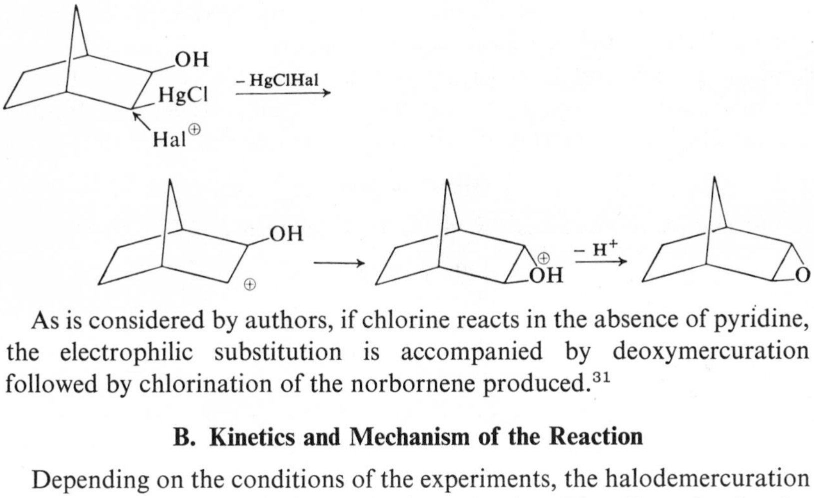

OH
HgCl   − HgClHal
Hal$^{\oplus}$ ⟶

OH
$\oplus$ ⟶ $\oplus$OH $\xrightarrow{- H^+}$ O

As is considered by authors, if chlorine reacts in the absence of pyridine, the electrophilic substitution is accompanied by deoxymercuration followed by chlorination of the norbornene produced.[31]

B. Kinetics and Mechanism of the Reaction

Depending on the conditions of the experiments, the halodemercuration reaction may have a radical or an ionic mechanism. The effect of molecular halogen in the absence of halogen anion in nonpolar media (such as CCl_4, benzene, carbon, disulfide, chloroform and, sometimes, dioxane) as well as irradiation with ultraviolet light, facilitate the reaction proceeding by a radical mechanism. Under such conditions the radical mechanism of the reaction is supported by kinetic data. These reactions may be of second order in case of some organomercury compounds (reactions of camphenyl-mercury iodide[32] and n-C_4H_9HgI[32] with iodine in dioxane) and of first order in the case of other organomercury compounds (reactions of sec-C_4H_9HgBr,[23] $C_6H_5CH_2HgCl$ with bromine[33] or iodine[34] in CCl_4, $C_6H_5CH_2HgBr$ with I_2 in dioxane,[35] trans- and cis-β-chlorovinylmercury bromide with iodine in carbon tetrachloride[5] or benzene,[7] cis- and *trans*-ω-styrylmercury bromide with bromine in the same solvents,[7] ester of α-bromomercuryphenylacetic acid with iodine in dioxane,[35] C_6H_5HgBr with I_2 in C_6H_6[36] and correspondingly they may have second order (for $C_{10}H_{17}HgI$,[32] n-C_4H_9HgI[32]), first order (for $C_6H_5CH(HgBr)CO_2C_2H_5$ with I_2 dioxane,[35] $ClCH{=}CHHgCl$ with I_2 in C_6H_6[5] and CCl_4,[7] $C_6H_5CH{=}CHHgBr$[7] with Br_2 in C_6H_6 and CCl_4, $C_6H_5CH_2HgCl$ with I_2 in CCl_4,[34] $C_6H_5CH_2HgBr$ with Br_2 and I_2 in CCl_4[34]) or zero order (for sec-butylmercury bromide[23] with Br_2 in CCl_4) relative to halogen. However, all these reactions are of zero order relative to the organo-mercury compound. The majority of these reactions are of a photo-chemical nature. The following chain mechanism was suggested by Winstein and Traylor for radical reactions:

$$I\cdot + RHgI \longrightarrow R\cdot + HgI_2$$
$$R\cdot + I_2 \longrightarrow RI + I\cdot$$

Since the rate constants of radical reactions of the same halogen with various RHgX have the same order or differ only slightly from one another, Reutov and Beletskaya[34] consider that these reactions have a common reaction governing the reaction rate. They consider that this step is photochemical or thermal dissociation of the halogen:

$$Hal_2 \xrightarrow{h\nu} 2Hal\cdot$$

(or formation of photoexcited molecule of halogen

$$Hal_2 \longrightarrow Hal_2^*)$$

followed by a fast step:

$$RHgX + 2Hal\cdot \longrightarrow RHal + HgXHal$$

The halodemercuration reaction may proceed by an ionic mechanism (S_E mechanism) in those cases in which the reaction is conducted in the presence of halide (I_2 in the presence of LiI[32] or CdI_2[9,35] and Br_2 in the presence of NH_4Br) or in polar solvents (alcohol, dimethylformamide, aqueous dioxane, acetic acid, and especially pyridine). When the reaction of RHgX with halogen (tests were carried out with the use of bromine) is conducted in a nonpolar solvent, the addition of oxygeneous substances (such as water, alcohol, or ether) may change the radical mechanism into an ionic one[34,38]; the reaction rate in this case does not depend on light.

In polar solvents or in the presence of halide, the halogen molecule undoubtedly polarizes, loses its ability to dissociate homolytically, and becomes a source of electrophilic halogen. This may be clearly illustrated by the following scheme:

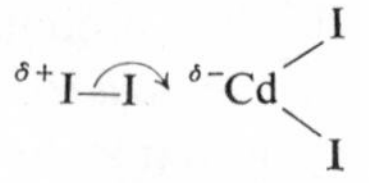

In addition, the polar solvent probably polarizes the C—Hg bond and inhibits radical formation.

Not only the polarity of medium, but also its ability to form complexes as well, probably plays a role in these reactions. For instance, solvents, according to their effect upon the rate of reaction of benzylmercury chloride with iodine are arranged in order of their solvating ability in the following way: methanol > aqueous dioxane (30% water) > ethanol > dimethylformamide, while the order in which they should be arranged, according to the magnitudes of their dielectric constant, is as follows: dimethylformamide > methanol > ethanol > 70% aqueous dioxane.[39]

The formation of a complex of solvent B with a molecule of initial

$$\underset{\diagup}{\overset{\text{X}}{}}$$

organomercury compound, type R—Hg···B and final product HgXHal or the solvation at the border of the organomercury compound complex or transitory state results in a weakening of the C—Hg bond; in case of formation of the solvent complex with a molecule of halogen, the latter loses its ability to react homolytically.[39]

In the majority of cases, the halogenolysis reactions conducted in polar media or under the effect of halogen in the presence of halide are bimolecular (S_E2 mechanism). For instance, reactions of RHgX with halogen in the presence of halogen anion [such reactions as benzylmercury bromide[37] and phenylmercury bromide[36] with bromine and NH_4Br in dimethylformamide, methyl alcohol, aqueous dioxane; benzylmercury chloride in aqueous dioxane[40]; phenylmercury bromide[36] cis-[8,10] and trans-[8,10] β-chlorovinylmercury chloride in aqueous dioxane,[9,36] methanol,[8,36] dimethylformamide[10,36] with iodine and CdI_2 and 4-camphenylmercury iodide with I_2 and LiI in aqueous dioxane[32] have second and first orders (for three latter RHgX and for $C_6H_5CH_2HgBr$[37])] relative to RHgX and halogen (S_E2 mechanism).

Jensen and Gale[22] have considered it probable that the polar reaction may proceed by an S_E2 mechanism (effect of positive halogen upon C—Hg bond or carbon atom), by an S_Ei mechanism (effect of halogen upon mercury atom and following rearrangements) or by passing through a four-center transitory state (simultaneous effect of Hal—Hal upon both C and Hg, the version which was suggested by Winstein and Traylor[32] and accepted by Reutov and Beletskaya)[36,41]:

$$
\begin{array}{c}
\text{R} \text{-----} \text{Hg—X} \\
\text{Hal} \text{----} \text{Hal}
\end{array}
$$

The reaction of β-chlorovinylmercury chloride with iodine in dimethylsulfoxide having first order with respect to the organomercury compound and zero order with respect to iodine, at the same time has an over-all first-order reaction rate. The reaction in this case proceeds by an S_E1 mechanism which has, as a limiting step, the ionization of RHgX with formation of an ion pair[7]

$$ClCH=CHHgCl \overset{slow}{\rightleftharpoons} ClCH=\overset{\ominus}{CH}:\overset{\oplus}{HgCl}$$

$$ClCH=\overset{\ominus}{CH}:\overset{\oplus}{HgCl} + I_2 \overset{fast}{\longrightarrow} ClCH=CHI + HgClI$$

The reaction of ω-styrylmercury bromide with bromine in the same solution also proceeds by an S_E1 mechanism.[6,7]

However, both reaction mechanisms (ionic or radical) and reaction rate may also be determined by the structure of the organomercury compound. According to Razuvaev and Savitsky,[42] for instance, the reaction

$$R_2Hg + I_2 \longrightarrow RHgI + RI,$$

depending on the electronegativity of the R radical, may proceed in a nonpolar solvent (carbon tetrachloride) by an ionic mechanism (in case of more electronegative radicals) or by a radical mechanism (in case of less electronegative radicals).

The rate of radical (photochemical) reaction of benzylmercury bromide with iodine is considerably higher than that of the same reaction proceeding in the presence of cadmium iodide.[35] At the same time, the rate of the heterolytic reaction of $C_6H_5CH(HgBr)COOR$ (in the presence of CdI_2) is higher than that of the radical reaction.[35] The high rate of the electrophilic reaction of the latter compound, as compared with benzylmercury bromide, is probably explained by the fact that the conjugation of the Hg—C bond with the C=O group facilitates heterolytic breakage of this bond:

$$C_6H_5CH\underset{HgBr}{\overset{}{\longrightarrow}}C\overset{O}{\underset{OR}{\diagdown}}$$

Different reactions of the hypobromites,[43] N-bromoamides,[44] N-bromoimides[43,45-47] of acids with the organomercury compounds in the presence or in the absence of ultraviolet light irradiation (see Sec. III-B) are explained by different reaction mechanisms in these cases. The reaction without irradiation proceeds by an ionic or latent ionic mechanism in which the electrophilic bromine atom serves as a reacting agent. This is facilitated by the distribution of electronic charge:

$$R—C\overset{O}{\underset{O \leftarrow Br}{\diagdown}} \qquad \overset{\delta\oplus}{Br} \rightarrow \overset{\ominus}{N} \overset{}{\underset{R}{\diagup}} C \overset{O}{\diagdown}$$

The ionic mechanism of the reaction is supported by the reaction of a bromosuccinimide with $RHgC_6H_5$, which, depending on the nature of R, proceeds in two directions[45]:

$$RHgC_6H_5 + BrNC_4H_4O_2 \longrightarrow \begin{cases} C_6H_5Br + RHgNC_4H_4O_2 \\ RBr + C_6H_5HgNC_4H_4O_2 \end{cases}$$

The reaction proceeds in the second direction if more electronegative radicals (p-$(CH_3)_2NC_6H_4$, p-$CH_3OC_6H_4$, and $(CH_3)_3C_6H_2$) are used, and in the first direction when more positive radicals (p-$NO_2C_6H_4$, p-BrC_6H_4, C_2H_5, and $C_6H_5CH_2$) are used.

In the presence of ultraviolet light, these reactions proceed by a radical mechanism and, as a rule, the bromine attacks the mercury.[46]

III. REACTION CONDITIONS

A. Effect of Halogen Type

Among the gaseous halogens, chlorine reacts quite vigorously with organomercury compounds (with Alk_2Hg with combustion). Alk_2Hg and other organomercury compounds react with bromine and iodine without combustion. The comparative effect of various halogens upon other organomercury compounds has not been widely investigated.

In solvents, the effect of halogens upon organomercury compounds is less active and therefore the halo-demercuration reaction is usually conducted in solutions. The reaction may be carried out in such solvents as alcohol, chloroform, carbon tetrachloride, carbon disulfide, acetic acid, pyridine, and dioxane. The substitution reactions with bromine and iodine are also conducted in aqueous KBr and KI.

As was shown in the reaction mechanism, solvent type affects the rate and mechanism of the reaction and plays an important role in the stereochemistry of the reaction for halogenating the HgX group.

By comparing rate constants[36] of reactions of various halogens with the same organomercury compounds, it is possible to conclude that bromine is more reactive than iodine. The difference in effects of chlorine and bromine have already been shown in the example of the products obtained as a result of adding the mercury salt to norbornene.[31] see II-A-2.

For easy stepwise substitution of mercury in R_2Hg (or RHgX) by bromine, it is suitable to use a mild brominating agent such as dioxanedibromide.[16] Pyridinebromide perbromide was used for obtaining bromoferrocene from the chloromercuriferrocene.[48]

Chlorine may be generated in situ. For this purpose, RHgX is treated with a mixture of hydrochloric acid with potassium chlorate.[49]

It might be well to point out that in reactions of RHgCl with iodine, and especially when the reaction is conducted for extended times, the ·HgCl radical may substitute not only by iodine, but also by chlorine due to the following exchange reaction of halogens between the initially formed RI and mercury salt containing chloride ion.[50,51]

B. Related Reactions

Similarly to halogen molecules, iodine chloride reacts with organomercury compounds, but always with the formation of RI[52]:

$$R_2Hg + 2ICl \longrightarrow 2RI + HgCl_2$$

$$RHgX + ICl \longrightarrow RI + HgXCl$$

Chapter 1 discusses the reactions of organomercury compounds with iodine trichloride which result in the substitution of mercury not for halogen, but for ICl— or ICl_2—groups; i.e., lead to the synthesis of iodinium salts and iodide chlorides.

Free thiocyanogen[53] also reacts with organomercury compounds:

$$RHgX + (CNS)_2 \longrightarrow RCNS + HgXCNS$$

After treatment with sodium thiosulfate, the reaction of diferrocenylmercury with thiocyanogen results in the substitution of mercury by sulfur.[54] The mechanism of this reaction is not clear.

Thiocyanogen- and 1,1'-dithiocyanogenferrocene was obtained by heating 1,1'-dichlorferrocenylmercury with $Cu(SCN)_2$ in benzene.[55] Covalent selenocyanates react with organomercury compounds in a somewhat different way than thiocyanogen and halogens. $Se(SeCN)_2$ reacts as follows[56]:

$$R_2Hg + Se(SeCN)_2 \longrightarrow RHgSeCN + Se + RSeCN$$
$$(R = CH_3, C_6H_5)$$

$Se(CN)_2$ with $(C_6H_5)_2Hg$ reacts without separation of selenium:

$$(C_6H_5)_2Hg + Se(CN)_2 \longrightarrow C_6H_5HgCN + C_6H_5SeCN$$

The first stage of the reactions of compounds containing mobile halogens such as $ArICl_2$, PCl_3, etc., with R_2Hg is often the converting the latter into RHgCl compounds.

N-Bromoamides, N-bromoimides and hypobromites of acids react differently with R_2Hg depending on the presence or the absence of ultraviolet light (see Sec II-B).

A suitable method of haloferrocene synthesis is based on the reaction of organomercury compounds with the positive halogen of N-halosuccinimides. For instance, a 57% yield of bromoferrocene was obtained by the reaction of N-bromosuccinimide in dimethylformamide with chloro-mercuryferrocene. Under similar conditions, 1,1'-dibromoferrocene was obtained from 1,1'-dichloromercuryferrocene. Comparable yields of bromoferrocene were obtained as a result of the reactions of the mercury derivatives of ferrocene with N-bromosuccinimide in dichloromethane and with bromoacetamide in dimethylformamide, but these reactions are less suitable as they result in simultaneous formation of ferrocene.[48] Iodo-ferrocene was obtained in the same way by the reaction of N-iodosuccini-mide with chloromercuryferrocene in CH_2Cl_2 (yield about 85%[48]). N-Iodosuccinimide in dimethylformamide converts the less soluble 1,1'-dichloromercuryferrocene to the di-iodo derivative with a yield of 42%.[48]

Haloferrocenes were obtained as a result of an exchange reaction between the HgX residue and halide from $CuHal_2$; this reaction is un-precedented for organomercury compounds of other classes. The reaction proceeds under conditions consisting of a boiling, aqueous acetone sus-pension under a heptane layer. The following compounds were obtained in this way: chloro- (58%), bromo- (69%), 1,1'-dichloro- (17%) and 1,1'-dibromo- (30%) ferrocene.[55] (Yields in parentheses.)

In some cases the effect of nitrosyl chloride upon $(C_nF_{2n+1})_2Hg$ com-pounds is accompanied by the substitution of mercury by chlorine,[57] while the ordinary reaction of organomercury compounds R_2Hg, in-cluding $(C_nF_{2n+1})_2Hg$ compounds, results in the formation of nitroso compounds.

C. Effect of Organomercury Compound Type

Though liquid diethyl- and dimethylmercury are more reactive with gaseous chlorine in the absence of solvent than aromatic mercury com-pounds, the latter, as usual, are more active than saturated aliphatic compounds when the reaction is conducted in solution. As a result of increasing activation energy of the reaction:

$$R_2Hg + I_2 \xrightarrow{CCl_4} RHgI + RI$$

the radicals may be arranged in the following series[42]:

p-$CH_3OC_6H_4$ < o-$CH_3OC_6H_4$ < β-$C_{10}H_7$ < α-$C_{10}H_7$ < p-$CH_3C_6H_4$ <
o-$CH_3C_6H_4$ < C_6H_5 < p-$C_2H_5O_2CC_6H_4$ < p-FC_6H_4 < α-C_4H_3S < p-ClC_6H_4 <
CH_3 < α-$C_2H_5O_2CC_2H_4$ < C_2H_5 < n-C_4H_9 < n-C_6H_{13} < i-C_5H_{11} < i-C_3H_7 <
$C_6H_5CH_2$

The series agrees largely with the Kharasch series obtained by the reaction of acids upon organomercury compound in aqueous solution.

The reaction of halogens with quasicomplex compounds and other compounds with mercury at double bonds of noncyclic alkenes results in normal substitution of mercury for halogen and does not lead to the elimination of a nonsaturated compound. The halogenolysis of products of mercury salt addition to double[58] or triple bonds proceeds under more mild conditions than halogenolysis reactions of mercury derivatives RHgX and R_2Hg containing radicals on the left side of the above series. In addition to the series, C_6H_5Br and $Cl_2CHHgBr$[59] are obtained as a result of the reaction of bromine with $C_6H_5HgCHCl_2$ in $CHCl_3$; the reaction of bromine with $C_6H_5HgC\equiv C-CH=CH_2$ and $C_6H_5HgCH_2CH=CH_2$ in CCl_4 even at $-15°C$ results in breaking the bond between unsaturated moiety and mercury with the formation of C_6H_5HgCl.[60]

Thus, the reactions of halogens with organomercury compounds, containing Hg with multiple bond, including the reaction with mercury acetylide, breaks, first of all, the C—Hg bond. The addition of halogen to the multiple bond may be avoided by not using an excess of halogen.[3,4,61] However, the reaction of $(CFI=CH)_2Hg$ with bromine under mild conditions (in carbon tetrachloride) results in the formation of $(CFIBr-CHBr)_2Hg$ without demercuration.[62] In the reaction of dicyclopentadienylmercury with even insufficient bromine at $-20°C$, the substitution of mercury by bromine is accompanied by bromine addition to the double bond and leads to the formation of tribromocyclopentane.[63] At temperatures from -2 to $-5°C$, use of excessive bromine results in a mixture of pentabromocyclopentanes. When bromination is carried out with the use of potassium hypobromite at temperatures from 0 to 2°C, the substitution of mercury for bromine is accompanied by the substitution of hydrogen by bromine, and the addition of bromine to the double bond does not occur. The reaction results in the formation of hexabromocyclopentadiene.

Similarly to β-mercurated alcohols and their derivatives, α-mercurated oxo-compounds (monomercury derivatives of aldehydes, ketones and acids) normally substitute mercury by halogen by reaction with halogen, but this occurs only in the absence of iodide ion.[64] The effect of iodine in a KI solution results in the decomposition of these compounds and does not lead to the substitution of mercury by iodine (see Part D, below).

In monomercury derivatives of superaromatic heterocyclic compounds, the substitution of mercury by halogen occurs as easily as in the case of the mercurated derivatives of benzene and the aromatic compounds containing electron-donating substituents. It is more difficult to substitute mercury by halogen in mercurated pyridine and similar compounds.

The following reaction takes place under mild conditions (boiling chloroform)[65]:

$$ClHgC_5H_4Mn(CO)_3 + I_2 \longrightarrow IC_5H_4Mn(CO)_3 + HgClI$$

Haloferrocenes[66,67] were obtained for the first time by the substitution of the XHg radical by halogen in mercurated ferrocene. The halogenolysis of mercurated haloferrocenes was conducted under the same conditions[66,67] (as discussed previously).

Under strong irradiation, bis-perfluoroethylmercury[68] and bis-perchlorovinylmercury[69] substitute mercury by chlorine; the reactions of bisperchlorovinylmercury[70] with bromine (in CCl_4) and iodine (in xylene) did not stop at RHgX even in the case when equimolar quantities of halogen were used. This may be explained by the instability of these RHgHal or by the higher reaction rate of RHgHal with halogen as compared with R_2Hg. However, the substitution of mercury by halogen in secondary saturated and aromatic perfluorinated mercury compounds requires very severe conditions: for the secondary perfluoralkylmercury compounds, contact with bromine or iodine in an autoclave at 200°C for many hours[71] is required while $(C_6F_5)_2Hg$ requires heating with iodine for 15 hr in a sealed tube at temperatures ranging from 150 to 155°C, but even under these conditions only small quantities of pentafluoroiodobenzene were obtained.[72] The reaction of bis(perfluoro-tert-butyl)mercury with bromine resulted in the respective RBr, while the reaction of bis-(perfluoro-tert-butyl)mercury with iodine is more capricious. Only one experiment, which could not be repeated, resulted in the formation of a product which probably corresponds to RI.[71] (Compare, however, with Ref. 73.)

$$\text{Bis(phenylbarenyl)mercury, } C_6H_5-\underset{\underset{B_{10}H_{10}}{\diagdown O \diagup}}{C-C}-Hg-\underset{\underset{B_{10}H_{10}}{\diagdown O \diagup}}{C-C}-C_6H_5$$

is especially stable to the action of bromine and does not react upon boiling with bromine in methylene chloride or bromobenzene.[74]

The substitution of several mercury atoms for halogens in polymercurated compounds requires more severe conditions than the halogenolysis of monomercurated compounds of the same class. Mercurated polystyrene (in p-position) was converted into poly(p-iodostyrene) under comparatively mild conditions (three hours of boiling with iodine in chloroform).[75] The conversion was quantitative.

The effect of iodine and bromine upon the product obtained as a result of mercury salt addition to acetylenic alcohol such as 2-chloro-3-chloromercury-2-butene-1,4-diol leads to the substitution of HgX by halogen[76]:

$$HOCH_2C(HgCl)=C(Cl)CH_2OH + Br_2 \longrightarrow$$
$$HOCH_2C(Br)=C(Cl)CH_2OH + HgClBr$$

However, the effect of bromine is probably complicated by the fact that bromine also reacts with the product, since as a result of the bromination only a 10% yield of the respective 2-chloro-3-bromo-2-butene-1,4-diol was obtained. Other reaction products were a mixture of several substances whose structures are not known.

D. Secondary Reactions

Besides the normal reaction of the substitution of mercury by halogen which may be called the halodemercuration reaction, halogen, and especially when it is in excess, may halogenate the organic part of the organomercury compounds. In the case of halodemercuration of mercury derivatives of unsaturated compounds, excessive halogen may add to the multiple bond. Iodine in KI solution cannot be used for the substitution of mercury for iodine in the following compounds: organomercury compounds which may be easily symmetrized by the effect of potassium iodide (for which respective symmetrical compounds R_2Hg are less soluble than initial $RHgX$[77]), mercurated phenols with free hydroxyl group which decompose under the effect of potassium iodide according to the following scheme[78]:

$$HOC_6H_4HgI + 3KI + H_2O \longrightarrow C_6H_5OH + K_2HgI_4 + KOH$$

quasicomplex compounds which eliminate unsaturated compounds under the effect of potassium iodide, for example[79]:

$$ROCH_2CH_2HgCl + 4KI + H_2O \longrightarrow K_2HgI_4 + C_2H_4 + ROH + KOH + KCl$$
$$ClCH{=}CHHgCl + 4KI \longrightarrow CH{\equiv}CH + K_2HgI_4 + 2KCl$$

and α-mercurated oxo-compounds and acids which also decompose under the effect of potassium iodide in protonic solvents in accordance with the following scheme[80]:

$$ClHgCH_2CHO + 4KI + ROH \longrightarrow K_2HgI_4 + CH_3CHO + KCl + KOR$$

The substitution of mercury by hydrogen under the effect of aqueous, alcoholic, or acetone solutions of potassium iodide is observed usually for other compounds containing HgX groups at the α-carbon atom of aldehydes, ketones, or acids.

In the compounds listed above, mercury should be substituted by iodine in organic solvents without potassium iodide.

The products obtained by addition of mercury salts to acetylenic alcohols,[76] ketones,[81] and acids[82] are also decomposed by potassium iodide (and in these cases also by bromides and even chlorides of alkali metals). However, the adducts of mercury(II) chloride and esters of acetylenic acids are more stable[82] than adducts of acetylenic alcohols, ketones, and acids as they decompose more slowly. Only 70–80% decomposition is observed in a saturated solution of potassium iodide and they are not decomposed by a saturated solution of sodium chloride, while the adducts of acetylenic alcohols, acids, and especially ketones are easily decomposed in these cases. (See also Chapters 7 and 15.)

In addition to products of normal decomposition, the ester of iodo-fumaric acid[82] was also unexpectedly obtained from the reaction of potassium iodide solution with the adduct of mercury(II) chloride and the ester of acetylenedicarboxylic acid. The only explanation of this result may be that the atoms of chlorine in the adduct molecule were first substituted by iodine and the then unstable intermediate product, on one hand, decomposes with separation of mercury(II) iodide and the ester of acetyl-enedicarboxylic acid and, on the other hand, substitutes HgI group by hydrogen, retaining the second halogen atom:

$(CH_3OOC)ClC=CHgCl(COOCH_3) \longrightarrow$

$$[(CH_3O_2C)IC=CHgI(CO_2CH_3)] \longrightarrow \begin{cases} CH_3O_2CC\equiv CCO_2CH_3 \\ CH_3O_2CIC=CHCO_2CH_3 \end{cases}$$

This is supported by immediate quantitative separation of sodium chloride when the reaction was conducted in acetone.

In the case of products obtained as a result of adding mercury(II) salts to alkenes, halogens may cause the substitution of the mercury by various groups: hydroxyl (in water), by alkoxyl (in alcohols), etc.; this reaction takes place instead of the halodemercuration reaction.[83]

The analogous formation of ROH instead of RHal during the reaction of halogens with RHgX may occur when an RHal, which is easily hydro-lyzed, is formed. For instance, the reaction of bromine or iodine with 2-phenyl-5-acetoxymercury-1,3,4-oxadiazole (8) formed the 2-phenyl-5-oxy-1,3,4-oxadiazole (8b) rather than the 5-halogen-derivative (8a)[84]:

(8)

(8a)   (8b)

IV. MISCELLANEOUS OBSERVATIONS

The action of iodine on β-chlorovinylmercuric chloride has been reported.[91] The mechanism of bromodemercuration of γ-dinitromercuric compounds catalyzed by peroxides is analogous to that of Winstein.[92] The rate of halodemercuration depends on polarity of the medium and its ability to undergo complexation.[93] Kinetics of halodemercuration in various solvents have been discussed.[94] A four-center mechanism[95] and an S_E1 mechanism of halodemercuration of Ar_2Hg and $ArHgBr$ have been proposed.[96,97] Fluorine, when diluted fivefold with nitrogen, substitutes F for Hg in R_2Hg and $RHgBr$ (R = 1,1,1-trinitropropyl).[98] Reactions also proceed in CCl_4 at room temperature for R_2Hg, and at 40–50°C for $RHgBr$ in glass apparatus.[98]

Halogenolysis of alkene aminomercuration products are discussed.[99] $(o\text{-}BrC_6F_4)_2Hg$ substitutes halogens for mercury under the action of bromine at 115° and of iodine at 230° in vacuum.[100] Mercury in bis-methoxy mercurated products of allene is substituted by iodine under mild conditions: at $-6°$ in benzene during 2 hr.[101] The removal of mercury from $(C_6F_5)_2Hg$ by use of iodide ion has been discussed.[102]

V. PREPARATIVE SYNTHESES

1. 2-iodo-n-butane from Di-sec-butylmercury[85]

An ether solution of di-sec-butylmercury absorbed 2 atoms of iodine with instantaneous decoloration. 2-Iodobutane with m.p. 118–119°C (743 mm) was obtained after separation of $n\text{-}C_4H_9HgI$, removal of ether, and vacuum distillation of the residue.

2. trans-β-Chlorovinylmercury Chloride[3]

Chlorine was gently passed through a solution of 0.2 g of trans-di-(β-chlorovinyl)mercury in 5 ml of carbon tetrachloride at room temperature until a precipitate appeared. After 5 min of stirring, the residue was filtered and dried to give 0.14 g (76.5%) of product, m.p. 124–124.5°C. The melting point of the product recrystallized from benzene and petroleum ether (3:1) did not change.

3. cis-α-Bromostilbene[16]

A solution of 0.0144 mole of dioxane-dibromide in 50 ml of dioxane was added to a solution of 0.0072 mole of cis-bis(α-stilbene)mercury in 140 ml of dioxane. The mixture was allowed to stand for 1.5 hr. The solvent was removed under vacuum and the residue was treated with 3% HCl and ether. After ether distillation, the residue was extracted with petroleum ether and was then recrystallized from CH_3OH upon cooling with solid CO_2 affording 3.45 g of cis-α-bromostilbene, m.p. 19°C.

4. o-Iodophenol[86]

The technique described in Organic Syntheses[86] gives an example of the standard method for substituting mercury by halogen in aromatic compounds with active substituents in the nucleus; see also Ref. 87.

5. Cleavage of Bis(perchlorovinyl)mercury with Bromine[70]

a. 1:2 Molar Ratio. To 15.0 g (0.033 mole) of bis(perchlorovinyl)-mercury in 50 ml of CCl_4 a solution of 10.6 g (0.066 mole) of bromine was added with stirring over 2 hr. The mixture was stirred for 6 hr. The white precipitate which formed soon after the start of the bromine addition was filtered off giving crude mercury(II) bromide (10.6 g, m.p. 227–234°C) which was recrystallized from methanol (10.4 g, 87%, m.p. 237°C). The filtrate was shaken with 20 ml of 10% aqueous sodium thiosulfate, dried over anhydrous magnesium sulfate, and distilled to give 8.7 g (63%) of bromotrichloroethylene, b.p. 52°C (27 mm) n_D^{23} 1.5394.

b. 1:1 Molar Ratio. Bromine (5.3 g) in 30 ml of CCl_4 was added to 15.0 g of R_2Hg in 50 ml of CCl_4 for 1 min. Mercury(II) bromide [5.9 g of crude, 5.6 g (93%) of recrystallized, m.p. 235°C] was filtered off. The filtrate was evaporated to dryness, leaving a white solid (7.4 g, m.p. 68–72°C), which was recrystallized from pentane to give 7.2 g (96%) of R_2Hg, m.p. 73°C.

6. Bromination of mercury acetylides[61]

A solution of bromine (1.81 g) in benzene (20 cc) was added dropwise to a stirred solution of 13.4 g of the mercury derivative of 1-*trans*-2'-bromovinyl-ethynylbenzene in benzene (150 cc). A red color persisted after 80% of the bromine had been added. When addition was complete, the mixture was stirred for 10 min and the solvent was then removed by vacuum at 60°C. The residue was extracted several times with chilled, light petroleum ether and the extract freed from dissolved mercury(II) bromide by passage through a short column of charcoal (5 g) containing one part in ten of 10% palladium-charcoal. Evaporation of two eluates gave 1-*trans*-2'-bromovinyl-2,2'-bromoethynylbenzene as a pale yellow oil (2.45 g, 79%), m.p. −10°C, b.p. (bath) 60°C (10^{-4} mm). 1-Bromo-oct-1-yne and bromophenylacetylene were similarly obtained in 75 and 87% yields, respectively; CCl_4 was used as a solvent instead of benzene.

7. Bromovinylacetylene[88]

Freshly prepared di(vinylethynyl)mercury reacts with bromine or iodine to give the corresponding bromo- and iodo -vinyl acetylenes. To 23.5 g of freshly prepared di(vinylethynyl)mercury in 177 cc of $CHCl_3$, a 10% solution of Br_2 in $CHCl_3$ was slowly added. Br_2 (22.64 g) was absorbed before decoloration ceased. The chloroform was removed, and distillation of the residue (11 g) then gave 3 g of 1-bromo-2-vinylacetylene, b.p. 50–52°C (210 mm).

8. 2-Bromoheptafluoropropane[71]

Bis(perfluoroisopropyl)mercury (162 g, 0.30 mole) and 96 g (0.60 mole) of bromine were heated in a bomb at 200°C for 4 hr. The bomb was cooled to about 50°C and the gases were condensed in a Dry Ice–acetone trap. Distillation gave 102 g (68%) of 2-bromo-heptafluoropropane, b.p. 14–18°C.

9. 2-Chlorodiphenylene[49]

2-Acetoxymercurydiphenylene (180 mg) was stirred with concentrated hydrochloric acid (50 ml) and $CHCl_3$ (25 ml) while a saturated aqueous solution of $KClO_3$ (20 mg) was added dropwise. After half an hour a further portion of $KClO_3$ (5 mg) was similarly added and stirred. The stirring was continued for 4 hr. The $CHCl_3$ layer was washed with aqueous Na_2CO_3 and the sublimed product (crude material, 56 mg, m.p. 56–60°C) was converted into the 2,4,7-trinitrofluorenone complex (scarlet needles, m.p. 133–134°C). Decomposition of this complex (199 mg) by passing its benzene solution through alumina (20 × 2 cm), elution with the same solvent, and sublimation at 110–120°C (12 mm) gave very pale yellow plates, m.p. 67.5–68.5°C.

10. Dimethyl Ester of Exo-cis-4-oxy-5-Crom-3,6-endoxohexahydrophthalic Acid[24]

Bromine (1.5 ml) is added dropwise with stirring and cooling to 20 ml of pyridine. The obtained mixture is added to 5 g of the dimethyl ester of exo-cis-4-oxy-5-chloromercury-3,6-endoxohexahydrophthalic acid in 20 ml of pyridine, stirred for another 7 hr, and left overnight. After the vacuum distillation of pyridine, chloroform is added to the residue, the mixture is stirred, and the solvent is separated from residue. The chloroform solution is chromatographed on Al_2O_3, the solvent distilled off under vacuum, and the 0.7 g of bromohydrin with m.p. 162° are recrystallized from methanol.

The bromination of 2 g of exo-cis-mercury derivatives (by 0.64 g of bromine in 80 ml of CCl_4 with further stirring for 25 hr) gave 0.7 g of exo-cis-bromohydrin and 0.4 g of oil whose chromatography on aluminum oxide in a thin, unfixed layer in the solvent system of 4% methanol in chloroform showed the presence of cis- and trans-bromohydrins.

11. 2,4,5-Triiodo-3-nitrothiophene[89].

2,4,5-Tri(acetoxymercury)-3-nitrothiophene (5g) with 9 g of I_2 and 15 g of KI in 30 ml of H_2O were shaken for 3 hr with heating on a steam bath. After cooling, the filtered green residue is extracted by ether and the ether evaporated. The residue was recrystallized from benzene, m.p. 169.5–170.5°C.

12. 2,2″-Dibromo-o-terphenyl[90]

Dimeric o-terphenylenemercury (0.5 mole) and 2.5 mole of bromine in 3 ml of glacial acetic acid are left for 16 hr. Excessive bromine is removed by boiling; the solution is diluted with 100 ml of water and the suspension obtained is extracted by ether. The ether residue is treated with petroleum ether. A small quantity of $HgBr_2$ in this case remains undissolved. After distillation of the solvent, the residue is dissolved in a small amount of methanol. Cooling to $-60°C$ gives 140 mg of 2,2″-dibromo-o-terphenyl in the form of colorless crystals with a melting point of 74–83°C which remains unchanged after further recrystallizations. The yield is 72%.

13. Bromoferrocene[48]

A solution of N-bromosuccinimide (1.15 g, 6.4 mole) in 100 ml of dry nitrogen-flushed dimethylformamide was added dropwise to a cold, stirred solution of chloromercuryferrocene (2.10 g, 5 moles) in 50 ml of the same solvent. The reaction was conducted at 0°C in a nitrogen atmosphere for 3 hr, after which time 200 ml of 10% sodium thiosulfate solution was added, and the resulting dark solution was poured into 2 liters of cold water. The aqueous solution was extracted four times with 100-ml portions of petroleum ether, and the combined organic extract was washed with water and dried over magnesium sulfate. Evaporation of the solvent left an orange oil which was taken up in Cellosolve B and chromatographed on an alumina column (Fisher, activity 1). Evaporation of solvent from the single yellow band, which developed on elution with Cellosolve B, gave 0.81 g of an orange oil, which solidified on cooling. Recrystallization from cold ethanol gave 0.75 g (57%) of bromoferrocene as yellow plates, m.p. 31–32°C.

REFERENCES

1. A. N. Nesmeyanov and A. E. Borisov, *Dokl. Akad. Nauk SSSR*, **60**, 67 (1948).
2. A. N. Nesmeyanov and A. E. Borisov, *Tetrahedron*, **1**, 158 (1957).
3. A. N. Nesmeyanov, A. E. Borisov, and A. N. Gus'kova, *Izv. Akad. Nauk SSSR, Otd. Khim. Nauk*, **1945**, 639.
4. A. N. Nesmeyanov and A. E. Borisov, *Izv. Akad. Nauk SSSR, Otd. Khim. Nauk*, **1945**, 146.
5. I. P. Beletskaya, I. V. Karpov, and O. A. Reutov, Tr. Konf. po problem. primenenya correlationn. uravnenji v organ. khim., t.I, Tartu. **1962**, 1355; *R. Zh. Khim.*, **1963**, 22Ж, 24.

6. I. P. Beletzkaya, I. V. Karpov, and O. A. Reutov, *Izv. Akad. Nauk SSSR, Otd. Khim. Nauk*, **1964**, 1707.

7. I. P. Beletzkaya, V. I. Karpov, and O. A. Reutov, *Dokl. Akad. Nauk SSSR*, **161**, 586 (1965).

8. I. P. Beletzkaya, O. A. Reutov, and V. I. Karpov, *Izv. Akad. Nauk SSSR, Otd. Khim. Nauk*, **1961**, 2125.

9. I. P. Beletzkaya, O. A. Reutov, and V. I. Karpov, *Izv. Akad. Nauk SSSR, Otd. Khim. Nauk*, **1961**, 1961.

10. I. P. Beletzkaya, O. A. Reutov, and V. I. Karpov, *Izv. Akad. Nauk SSSR, Otd. Khim. Nauk*, **1961**, 2129.

11. A. N. Nesmeyanov, A. E. Borisov, and R. I. Shepeleva, *Izv. Akad. Nauk SSSR, Otd. Khim. Nauk*, **1949**, 582.

12. A. E. Borisov, *Izv. Akad. Nauk SSSR, Otd. Khim. Nauk*, **1961**, 1036.

13. A. N. Nesmeyanov, A. E. Borisov, I. S. Savel'eva, and N. V. Kruglova, *Izv. Akad. Nauk SSSR, Otd. Khim. Nauk*, **1961**, 726.

14. A. N. Nesmeyanov, A. E. Borisov, and I. S. Savel'eva, *Izv. Akad. Nauk SSSR, Otd. Khim. Nauk*, **1961**, 2241.

15. A. N. Nesmeyanov, A. E. Borisov, and I. S. Savel'eva, *Dokl. Akad. Nauk SSSR*, **155**, 603 (1964).

16. A. N. Nesmeyanov, A. E. Borisov, and N. A. Vol'kenau, *Izv. Akad. Nauk SSSR. Otd. Khim. Nauk*, **1956**, 162.

17. F. C. Whitmore, E. L. Wittle, and B. R. Harriman, *J. Am. Chem. Soc.*, **61**, 1585 (1939).

18. E. J. Van Loon and H. E. Carter, *J. Am. Chem. Soc.*, **59**, 2555 (1937).

19. F. R. Jensen and L. H. Gale, *J. Am. Chem. Soc.*, **82**, 145 (1960).

20. F. R. Jensen, L. D. Whipple, D. K. Wedegaertner, and J. A. Landgrebe, *J. Am. Chem. Soc.*, **82**, 2466 (1960).

21. F. R. Jensen and L. H. Gale, *J. Am. Chem. Soc.*, **81**, 1261 (1959).

22. F. R. Jensen and L. H. Gale, *J. Am. Chem. Soc.*, **82**, 148 (1960).

23. O. A. Reutov, E. V. Uglova, I. P. Beletzkaya, and T. B. Svetlanova, *Izv. Akad. Nauk SSSR, Otd. Khim. Nauk.*, **1964**, 1383.

24. Yu. K. Yur'ev, N. S. Zefirov, and L. P. Prikashchikova, *Zh. Obshch. Khim.*, **33**, 1793 (1963).

25. N. S. Zefirov, P. P. Kazjausnas, Yu. K. Yur'ev, and V. N. Bazanova, *Zh. Obshch. Khim.*, **35**, 1499 (1965).

26. N. S. Zefirov, R. S. Filatova, and Yu. K. Yur'ev, *Zh. Obshch. Khim.*, **36**, 763 (1966).

27. N. S. Zefirov, L. P. Prikashchikova, M. A. Bondareva, and Yu. K. Yur'ev, *Zh. Obshch. Khim.*, **33**, 4026 (1963).

28. N. S. Zefirov, L. P. Prikashchikova, and Yu. K. Yur'ev, *Zh. Obshch. Khim.*, **85**, 822 (1965).

29. P. M. Manolopoulos, M. Mednik, and N. N. Lichtin, *J. Am. Chem. Soc.*, **84**, 2203 (1962).

30. G. Inglis and J. Schwarz, *J. Chem. Soc.*, **1962**, 1014.

31. E. Tobler and D. J. Foster, *Helv. Chim. Acta*, **48**, 366 (1965).

32. S. Winstein and T. G. Traylor, *J. Am. Chem. Soc.*, **78**, 2597 (1956).

33. I. P. Beletskaya, T. A. Azizyan, and O. A. Reutov, *Izv. Akad. Nauk SSSR, Otd. Khim. Nauk*, **1962**, 223.

34. I. P. Beletskaya, O. A. Reutov, and T. P. Gur'anova, *Izv. Akad. Nauk SSSR, Otd. Khim. Nauk*, **1961**, 2178.

35. O. A. Reutov and I. P. Beletskaya, *Izv. Akad. Nauk SSSR, Otd. Khim. Nauk*, **1960**, 1716.
36. I. P. Beletskaya, L. V. Ermanson, and O. A. Reutov, *Izv. Akad. Nauk SSSR, Ser. Khim.*, **1965**, 231.
37. I. P. Beletskaya, T. A. Azizyan, and O. A. Reutov, *Izv. Akad. Nauk SSSR, Otd. Khim. Nauk*, **1962**, 424.
38. I. P. Beletskaya, T. A. Azizyan, and O. A. Reutov, *Izv. Akad. Nauk SSSR, Otd. Khim. Nauk*, **1963**, 1332.
39. I. P. Beletskaya, O. A. Reutov, and T. P. Gur'anova, *Izv. Akad. Nauk SSSR, Otd. Khim. Nauk*, **1961**, 1997.
40. I. P. Beletskaya, O. A. Reutov, and T. P. Gur'anova, *Izv. Akad. Nauk SSSR, Otd. Khim. Nauk*, **1961**, 1589.
41. I. P. Beletskaya, T. P. Fetisova, and O. A. Reutov, *Dokl. Akad. Nauk SSSR*, **166**, 861 (1966).
42. G. A. Razuvaev, and A. V. Savitsky, *Dokl. Akad. Nauk SSSR*, **85**, 575 (1952).
43. G. A. Razuvaev and N. S. Vyazankin, *Zh. Obshch. Khim.*, **22**, 640 (1952).
44. G. A. Razuvaev and N. S. Vasilejskaya, *Dokl. Akad. Nauk SSSR*, **74**, 279 (1950).
45. G. A. Razuvaev and V. S. Vasilejskaya, *Dokl. Akad. Nauk SSSR*, **67**, 851 (1949).
46. G. A. Razuvaev and Z. I. Bugaeva, *Uchen. Zap. Gor'kov. Gos. Univer.*, **24**, 143 (1953).
47. E. Tobler and D. J. Foster, *Z. Naturforsch.*, **17B**, 135 (1962).
48. R. W. Fish and M. Rosenblum, *J. Org. Chem.*, **30**, 1253 (1965).
49. W. Baker, J. W. Barton, and J. F. W. McOmie, *J. Chem. Soc.*, **1958**, 2666.
50. R. K. Summerbell and S. R. Forrester, *J. Org. Chem.*, **26**, 4834 (1961).
51. A. N. Nesmeyanov and I. F. Lutsenko, *Izv. Akad. Nauk SSSR, Otd. Khim. Nauk*, **1942**, 366.
52. F. C. Whitmore and M. A. Thorpe, *J. Am. Chem. Soc.*, **55**, 782 (1933).
53. E. Söderbäck, *Ann. Chem.*, **419**, 266 (1919).
54. A. N. Nesmeyanov, E. G. Perevalova, and O. A. Nesmeyanova, *Dokl. Akad. Nauk SSSR*, **119**, 288 (1958).
55. V. A. Nefedov and M. N. Nefedova, *Zh. Obshch. Khim.*, **36**, 122 (1966).
56. E. E. Aynsley, N. N. Greenwood, and M. J. Spragna, *J. Chem. Soc.*, **1965**, 2395.
57. P. Tarrant and D. E. O'Connor, *J. Org. Chem.*, **29**, 2012 (1964).
58. J. Sand, *Ber.*, **34**, 1385 (1901).
59. O. A. Reutov and A. N. Lovtsova, *Dokl. Akad. Nauk SSSR*, **154**, 166 (1964).
60. K. B. Rall and A. A. Petrov, *Zh. Obshch. Khim.*, **32**, 1095 (1962).
61. G. Eglinton and W. McCrae, *J. Chem. Soc.*, **1963**, 2295.
62. K. M. Smirnov, V. A. Ginsburg, and A. Ya. Yakubovich, *Zh. Vses. Khim. Obshch. imv. Mendeleeva*, **8**, (2), 231 (1963).
63. A. N. Nesmeyanov, N. S. Kochetkova, and R. B. Materikova, *Dokl. Akad. Nauk SSSR*, **147**, 113 (1962).
64. A. N. Nesmeyanov, I. F. Lutsenko, and R. N. Khomutov, *Izv. Akad. Nauk SSSR, Otd. Khim. Nauk*, **1957**, 942.
65. A. N. Nesmeyanov, K. N. Anisimov, and Z. P. Valujeva, *Izv. Akad, Nauk SSSR, Otd. Khim. Nauk*, **1962**, 1683.
66. A. N. Nesmeyanov, E. G. Perevalova, and O. A. Nesmeyanova, *Dokl. Akad. Nauk SSSR*, **100**, 1099 (1955).
67. A. N. Nesmeyanov, V. A. Sazonova, V. N. Drozd, and L. A. Nikonova, *Dokl. Akad. Nauk SSSR*, **131**, 1088 (1960).

68. J. Banus, H. J. Emeléus, and R. N. Haszeldine, *J. Chem. Soc.*, **1950**, 3041.
69. K. A. Hofmann and H. Kirmreuther, *Ber.*, **42**, 4232 (1909).
70. D. Seyferth and R. H. Towe, *Inorg. Chem.*, **1**, 185 (1962).
71. P. E. Aldrich, E. G. Howard, W. J. Linn, W. J. Middleton, and W. H. Sharkey, *J. Org. Chem.*, **28**, 184 (1963).
72. P. L. Coe, R. Stephens, and J. C. Tatlow, *J. Chem. Soc.*, **1962**, 3227.
73. W. T. Miller and M. B. Freedman, *J. Am. Chem. Soc.*, **85**, 180 (1963).
74. L. I. Zakharkin, V. I. Bregadze, and O. Yu. Okhlobystin, *Zh. Obshch. Khim.*, **36**, 761 (1966).
75. R. Okawara, Y. Tanaka, and E. Imoto, *Kogyo Kagaku Zasshi*, **64**, 235 (1961); *Chem. Abstr.*, **57**, 4854 (1962).
76. A. N. Nesmeyanov and N. K. Kochetkov, *Izv. Akad. Nauk SSSR, Otd. Khim. Nauk*, **1949**, 76,
77. F. C. Whitmore and R. J. Sobatzki, *J. Am. Chem. Soc.*, **55**, 1128 (1933).
78. F. C. Whitmore and E. Middleton, *J. Am. Chem. Soc.*, **45**, 753 (1923).
79. H. Laubie, *Bull. Soc. Pharm. Bordeaux*, **96**, 65 (1957); *Chem. Abstr.*, **51**, 15889 (1957).
80. A. N. Nesmeyanov and I. F. Lutsenko, *Dokl. Akad. Nauk SSSR*, **59**, 707 (1948).
81. A. N. Nesmeyanov and N. K. Kochetkov, *Izv. Akad, Nauk SSSR, Otd. Khim. Nauk*, **1949**, 305.
82. A. N. Nesmeyanov, N. K. Kochetkov, and V. M. Daschunin, *Izv. Akad. Nauk SSSR, Otd. Khim. Nauk*, **1950**, 77.
83. L. Berman, R. H. Hall, R. G. Pyke, and G. F Wright, *Can. J. Chem.*, **30**, 541 (1952).
84. O. P. Schwayka and G. P. Klimischa, *Zh. Obshch. Khim.*, **35**, 290 (1965).
85. J. Tafel, *Ber.*, **39**, 3626 (1906).
86. R. Adams, *Organic Syntheses*, Vol. 4, 1921, p. 37.
87. O. Dimroth, *Ber.*, **35**, 2853 (1902).
88. W. H. Carothers, R. A. Jacobson, and G. J. Berchet, *J. Am. Chem. Soc.*, **55**, 4665 (1933).
89. W. Steinkopf, *Ann. Chem.*, **545**, 38 (1940).
90. G. Wittig, E. Hahn, and W. Tochtermann, *Chem. Ber.*, **95**, 431 (1962).
91. P. J. Banney, W. Kitching, and P. R. Wells, *Tetrahedron Letters*, **1968**, 27.
92. V. A. Tartakovski, B. G. Gribov, and S. S. Novikov, *Izv. Akad. Nauk SSSR, Ser. Khim.*, **1965**, 1074.
93. A. Lord and H. O. Pritchard, *J. Phys. Chem.*, **70**, 1689 (1966).
94. I. P. Beletzkaya, T. P. Fetisova, and O. A. Reutov, *Izv. Akad. Nauk. Ser. Khim.*, **1967**, 990.
95. F. Gerart, U. Schöllkopf, and H. Schumacher, *Angew. Chem.*, **79**, 50 (1967).
96. I. P. Beletzkaya, N. K. Genkina, A. L. Kurz, and O. A. Reutov, *Zh. Organ. Khim.*, **4**, 1117, 1120 (1968).
97. O. Itoh, H. Taniguchi, A. Kawabe, and K. Ichikawa, *Chem. Abstr.*, **65**, 19951 (1966).
98. L. T. Eremenko, F. Ja. Nazibullin, and G. N. Nesterenko, *Izv. Akad. Nauk SSSR, Ser. Khim.*, **1968**, 1360.
99. J. Perie and A. Lattes, *Bull. Soc. Chim. France*, **1966**, 2153.
100. S. C. Cohen and A. G. Massey, *J. Organometal. Chem.*, **12**, 341 (1968).
101. W. L. Waters and E. F. Kiefer, *J. Am. Chem. Soc.*, **89**, 6261 (1967).
102. G. B. Deacon, *J. Organometal. Chem.*, **9**, 1 (1967).

Author Index

Numbers in parentheses are reference numbers and show that an author's work is referred to although his name is not mentioned in the text. Numbers in square brackets indicate the pages on which the full references appear.

Abercrombie, M. J., 254(33), 261(33), 262(33), [268]
Abramova, A. N., 190(1), 200(66), 207(66), 221(1,66), [224, 225], 229(6), 232(6), 235(6), [239], 289(159,160), [298]
Acree, S. F., 213(85), [226]
Adams, R., 287, [298], 342(86), [348]
Adcock, W., 41(207), [53]
Ahmed, A. K., 290-292(172), [298]
Aishima, I., 41(204), [53]
Akermark, B., 104, 113(114), [116, 117]
Aldrich, P. E., 255(43), [269], 279(54), [295], 339(71), 344(71), [348]
Alei, M., Jr., 197(32), 215(32), [224]
Alejnikova, M. J., 276(2), [294]
Allegra, G., 7(28), [48]
Allen, A. D., 44, [54], 88, 89(39,40), 90-93, 95, 103(40), 105, 106(40), [115]
Allpress, C. F., 144(113), 169(113), [176]
Allred, A. L., 27(127), [51], 197(42), 205(42), 216(42), 218(42), [225]
Amberger, E., 29(145), [51]
Ambrecht, F. M., Jr., 186(17,18), [189]
Amma, E. L., 6(25), [48]
Anderson, R. C., 32(165), [52]
Anderson, W. R., Jr., 16(77), [49]
Anisimov, K. N., 338(65), [347]
Ansaloni, A., 132(21), [173]
Antonova, N. D., 279(56), [295]

Apperson, L. D., 37, [53]
Araki, M., 44(235), [54], 71(89), 95(89), 107, 108(106), [116, 117]
Araki, S., 290(181), [299]
Aresta, M., 110, [116]
Armor, J. N., 90(100), 111, [116]
Aronheim, B., 140(73), 161(73), [175]
Ashby, E. C., 31(158), [52]
Aynsley, E. E., 336(56), [347]
Azizyan, T. A., 331(33), 332(38), 333(37), [346, 347]

Babb, R. B., 293(200), [299]
Bachmann, W. E., 214(87), [226]
Badische-Anilin & Soda Fabrik, 16(75), [49]
Baine, O., 195(23), 196(23), 204(23), 211(23), 213-215(23), [224]
Baird, W. C., Jr., [270]
Baiser, M. M., 278(30), [294]
Baker, A. W., 255(37), 261(37), [268]
Baker, W., 335(49), 344(49), [347]
Bamford, C. H., 198(46), 218(46), [225]
Banney, P. J., 293(192), [299], 341(91), [348]
Banus, J., 339(68), [348]
Barbaras, G. D., 28(139), [51]
Barlow, L. R., 294(213), [300]
Barrett, A. W., 281(70-72), [296]
Bartocha, B., 23(110), 29(110, 141, 142), [50, 51] 136(37), 142(37, 85a), 155(37), 164(37,85a), [174,

176], 194-196(12), 203(12), 209(12), 213(12), 215(12), 219 (110), [224, 226], 241(1,2), 243(1, 2), [246]

Barton, D. H. R., 241(5), 243(5), [247], 253(15), 262(15), [268]

Barton, J. W., 335(49), 344(49), [347]

Baryshnikova, A. N., 278(23), [294]

Basolo, F., 112, [116]

Bassi, I. W., 42(210), [53]

Batorewicz, W., 290(176), 292(176), [299]

Baukov, Yu. I., 139(65-67), 140(65,72), 150(66), 159(65-67), 160(72,143, 144), 161(65,144), [175, 177], 231 (9), 234(9), 239(9), [240], 242, 244 (7,8,11), 245(8,11), 246(7,8), [247], 253(28), 254(28,31), [268]

Bazanova, V. N., 328(25), 329(25), [346]

Beatty, H. A., 3, 8, [47-49]

Beck, W. W., 144(106), 167(106), 168 (106), [176]

Becker, P., 136(40), 142(40), 155(40, 140), 166(40), [174, 177]

Becker, W. E., 24(116), [50]

Beermann, C., 42(214,215), [53], 142 (90), 153(90), 165(90), [176]

Belavin, I. Yu., 160(143), [177], 244 (11), 245(11), [247], 253(28), 254 (28), [268]

Beletskaya, I. P., 132(23,25-29), 134(26, 27,34), [174], 276(2,5,6), 280, 293 (193,195,202), [294, 299], 302(5,5a), 303(5), 312(5,5a), [322], 326, 327, 328(23), 331(5,7,23,33-35), 332, 333, 334(6,7,35), 342(94,96), [345-348]

Belii, A. A., 58(8), 59(8), 65(8), 66(8), [114]

Belmondi, G., 132(21), [173]

Belov, B. I., 278(19,20), [294]

Benedikt, G., 28(134), [51]

Benz, E., 209(80), 213(80), [226]

Berchem, R. G., 281(71), [296]

Berchet, G. J., 343(88), [348]

Berg, O. W., 290(184), 292(184), [299]

Bergstrom, J. W., 146(125), 170(125), [177]

Berman, L., 257(87), [270], 341(83), [348]

Bernstein, H., 281(94), [296]

Bestian, H., 42(214), [53]

Beveridge, W. D., 99, 100, [115]

Bharucha, K. R., 254(33), 261(33), 262 (33), [268]

Bibler, J. P., 248(13), [250]

Bichin, L. P., 58(6), 59(6), [114]

Bickelhaupt, F., 194(17,18), 204(17), 208(17), 209(17,18), 211(17), [224], 272(3), [274]

Biilman, E., 255(45,46), 256(60), [269]

Bilbo, A. J., 23(110), 29(110,141,142), [50, 51], 219(110), [226], 241(1, 2), 243(1,2), [246]

Biltz, H., 281, [295]

Binger, P., 3(7), 7(7), 18, 27(131), 28(7, 131,136), 39(131), [47, 51]

Bisnette, M. B., 83(32,33), [114]

Blasi, N., 143(105), 166(105), 167(105), [176]

Bleschinsky, S. V., 145(116), [176], 255, [268]

Blicke, F. F., 166(167), 167(167), [178]

Blitzer, S. M., 8(44), 30(44,149,150), 32 (150,167,168), 37(167), [48, 51, 52]

Bobashinskaya, Ch. S., 281(93), [296]

Boche, G., 59(28), 67, 72, 73(28), 95 (60), 113(114), [114, 115, 117]

Bodem, G. B., 287(132), 291(132), 292 (132), [297], 308(57), [323]

Boetius, M., 278(15), [294]

Bogdanov, I. F., 141(80), 163(80), [175]

Bokovoĭ, A. P., 164(150), [177], 320 (63), [324]

Bolshakova, A. A., 278(45), [295]

Bondareva, M. A., 329(27), [346]

Borchert, A. E., 10(57), 11(57), [49]

Bordwell, F. C., 263-265(95), [270]

Borisov, A. E., 130(1-5), 136(2), 137(1, 49-55), 138(1,50,53,58), 148(2), 155(2), 157(1,49,50,52-55,141), 158(58,141), 159(58), [173-175, 177], 190, 191, 194(13), 200(66, 70), 207(66), 208(13), 221(1,66,70), [224, 225], 229(6), 230(7), 232(6, 7), 234(7,10,11), 235(6,7,11,13), 236(7,11), 237(7), 238(7), [239, 240], 256(55,56), [269], 280(62-64), 282(62-64), 287(133), 288(155,

156,158), 289(159-161,163-165), [295, 297, 298], 303(14), 304, 312 (14), 313(14), 316-318(15), [322], 326, 327, 335(16), 338(3,4), 342(3, 16), [345, 346]

Borod'ko, Yu. G., 71(48,49), 88(38), 89, 90, 91(49), 92, 93(49), 100(43), 105, 112, [115-117]

Bottomley, F., 44, [54], 89, 92(40), 95, 103(40), 104, 105(40), 106(40), [115, 116]

Braack, A., 201(75), 223(75), [225]

Brainina, E. M., 147(133), 171(133), [177]

Brandt, J., 15(68), 17, [49]

Bratzev, V. A., 139(64), 159(64), [175]

Breest, F., 287(147), [298]

Bregadze, V. I., 265(98), [270], 279 (55), [295], 339(74), [348]

Breil, H., 32(170), 43(170), [52]

Breindel, A., 32(172), [52]

Breitbeil, F. W., 186(40), [189]

Breitinger, D., 253(18), 263(18), [268]

Breslow, D. S., 41(209), [53], 78(29), [114]

Breuer, F., 195(27), 196(27), 211(27), 213(27), 214(27), [224]

Brewster, R. Q., 303(9), 316(9), 320(9), [322]

Brinckman, F. E., 136(37), 139(70), 142 (37), 155(37), 161(70), 164(37), [174, 175]

Brindley, W. H., 165(157), [178]

Brintzinger, H., 70(26), 72, [114]

Brockway, L. O., 6(19), 7(19), [48]

Brodersen, K., 131, [173], 253(18), 263 (18), [268]

Brook, A. G., 287(150-152), 292(188), [298, 299]

Brower, K. R., 293(207), [299]

Brown, H. C., 5-7(15), [48], 208(78), [226, 270]

Brown, R. D., 283(105,106), [297]

Brown, R. E., 216(92,95), [226]

Brown, T. L., 10, 23(112), 27(59), [49, 50]

Bruno, G., 27(125), 28(125), [51]

Bryce-Smith, D., 267(77,78), [270]

Bublitz, D. E., 29(141), [51], 219(110), [226], 241(1), 243(1), [246]

Bubnov, Ju. I., 131(12), 132(12), [173]

Buchanan, A. S., 283(105,106), [297]

Buckton, G. B., 144(107), 161(107, 146), 168(107), [176, 177], 209 (81), 217(103), 218(105), [226, 227]

Bugaeva, Z. I., 334(46), 335(46), [347]

Bukreev, V. S., 89, 100(43), [115]

Bundel, Yu. G., 279(56), [295]

Burdon, J., 200(69), 222(69), [225], 249(8), [250], 253(17), 255(17), 256(17), 262(17), [268]

Burg, A. B., 215(89), [226]

Burlatchenko, G. S., 139(66,67), 150 (66), 159(66,67), [175]

Burlitch, J. M., 179(1), 180(4,5), 181 (5), 182(1, 8-11), 183(11), 184(4, 11), 185(15), 186(10,29,30), 187 (11), 188(11), [188, 189]

Bush, G. C., 249(9), [250]

Buu-Hoi, N. R., 278(16), [294]

Buza, M., [270]

Cabassi, F., 17(83), 18(83), 25(83), [49]

Cahours, A., 164(148), 165(148), [177], 215(90), 216(90,96,97,99), 217(90), 218(90,96,97,99), [226]

Calingaert, G., 3, 8, [47-49]

Calmels, G., 305(23), 311(23), [322]

Calvery, H. O., 283(102), [296]

Campbell, C. H., 109, [116]

Campbell, I. G. M., 146(129), 171(129), [177]

Campbell, T. W., 146(126), 170(126), [177]

Carey, M. A. D., 248(12), [250]

Carley, D. R., 30(150), 32(150), [51]

Carmack, M., 278(30), [294]

Carothers, W. H., 210(82), [226], 343 (88), [348]

Carrick, W. L., 144(115), 169(115), [176]

Carter, H. E., 328(18), [346]

Casper, J., 142(92), 153(92), 165(92), [176]

Cava, M. P., 273(7), [274]

Cerfontain, H., 132(15), [173]

Cermeno, F. A., 219(109), [226]

Challenger, F., 144(113), 169(113), [176]

Chamberlin, T., 290(177), 291(177), [299]

Chambers, R. D., 34(182,182a,183), [52], 279(53), 283(53), [295]

Chamot, E. M., 199(61), 221(61), [225]

Chang, G., 43(225), [54]

Chapovskaya, N. K., 58(10), 59(10), 61 (10), [114]

Chapovski, Yu. A., 148(139), 172(139), [177]

Charman, H. B., 133(32), [174]

Chatt, J., 89, 91, 104, 105, 111, [115, 116]

Chernikov, S. S., 112, [116]

Chernov, I. A., 278(40), [295]

Chivers, T., 34(183), [52]

Chow, D., 253(21), 260(21), [268]

Clarke, H. C., 248(12), [250]

Clarke, H. T., 214(87), [226]

Clement, R., 137(44), 156(44), [174]

Coad, J. R., 293(196), [299]

Coates, G. E., 5(13), 6(13,18), 27(126), [48, 51], 199(55), 220(55), [255], 279(53), 283(53), [295]

Coe, G. R., 197(37), 215(37), [224]

Coe, P. L., 200(69), 222(69), [225], 249(8), [250], 253(17), 255(17), 256(17), 262(17), [268], 339(72), [348]

Coffmann, D. D., 194(11), 210(82), [224, 226]

Cohen, H. M., 186(24), [189]

Cohen, S. S., 248(11), [250], 342(100), [348]

Collman, J. P. 91, 106, 109, [115, 116]

Comyns, A. E., 147(131), 171(131), [177], 201(74), 223(74), [225]

Cook, S. E., 30(151), 37, [51]

Cope, A. C., 265(97), [270]

Corradini, P., 42(210), [53]

Corwin, A. H., 276(4), [294]

Cowan, D. O., 197(40), 215(40), 216 (40), [225]

Coyne, D. M., 16(76a), [49]

Crane, H. I., 199(63), 221(63), [225]

Crawford, J. V., 256(58), 266(58), [269]

Criegee, R., 141(78), 161-163(78), [175]

Croatto, U., 132(21), [173]

Cross, J. M., 132(20), [173]

Cross, R. J., 186(17,19,31), [189]

Csakvari, B., 35(184), [52]

Cubberly, B., 108, [117]

Cunningham, D., 34(182), [52]

Cunningham, J. A., 34(182a), [52]

Curtin, D. Y., 311(69), [324]

Czempik, H., 142(90), 153(90), 165 (90), [176]

Dahlig, W., 30(149a), [51]

Damrauer, R., 185(41), 186(21,23,36, 41), [189]

Darragh, K. V., 186(16,26,27), [189], 293(211), [300]

Das, P. K., 106(81), 111, [116]

Dashunin, W. M., 289(170), 290(170), [298], 340(82), 341(82), [348]

Davidson, J. M., 294(213,214), [300]

Davidson, N., 5(15), 6(15,19), 7(15,19), [48]

Davies, W. C., 165(158), 167(171), [178]

Davis, B. R., 44(234), [54], 97, [116]

Deacon, G. D., 342(102), [348]

Dehmlow, E. V., 186(22), [189]

De Luca, D., 18(89), [50]

Dennerlein, D. T., 186(40), [189]

Dennis, L. M., 140(71), 161(71), [175], 199(56,61), 206(56), 220(56), 221 (61), [225]

Derbyshire, D. N., 267(75), [270]

Dertouzos, H., 185(15), [189]

Dessy, R. E., 131, 133, [173], 197(34, 37), 215(34,37), [224], 276, 277(7, 9,13), 278(7,9), 279, 282(13), 283 (9,104), 293(191), [294, 297, 299]

De Vries, H., 10(58,61), 42(216), [49, 53]

Dewhurst, K. C., 42(217), [53]

Dias, A. R., 109, [116]

Dickson, R. S., 36(188), 40(188), 41 (188), [53]

Dierks, H., 32(171), [52]

Dietrich, H., 32(171), [52]

Dillard, G., 28(139), [51]

Dimroth, O., 308(49,53), 320(53), [323], 342(87), [348]

Dimroth, P., 141(78), 161-163(78), [175]

Di Pietro, J., 18(87), [50]
Ditsch, L. T., 290(176), 292(176), [299]
Dittmar, P., 199(52), 219(52), [225]
Dobratz, E. H., 38(192), [53]
Dokunikhin, I. S., 278(22), [294]
Donovan, R., 292(188), [299]
Douglas, C. M., 142(85a), 164(85a), [176], 194-196(12), 203(12), 209 (12), 213(12), 215(12), [224]
Douglass, M. L., 263-265(95), [270]
Dow Chemical Co., 42(213), [53]
Dowd, S. R., 182(10,11), 183(11), 184 (11), 186(10), 187(11), 188(11), [188]
Drake, L. R., 142(84), 151(84), 164 (84), [176]
Dreher, E., 164(152), [178], 213(86), 217(86), 223(86), [226], 249(5,6), [250], 303(11), 313(11), [322]
Drozd, V. N., 339(67), [347]
Druschkov, O. N., 305(17), [322]
Dubovitski, V. A., 320(61), [323]
Dunn, J. H., 42(222), [54]
Duppa, B. F., 201(72), 215(91), 216 (72), 217(72), [225, 226]
Dvorjantseva, G. G., 139(63), 158(63), [175]

Ebel, H. F., 267(80), [270], 285(119), [297]
Eden, C., 42(220), [53]
Eglinton, G., 338(61), 343(61), [347]
Egmont, J. G. van, 36, [53]
Eisch, J. J., 34, 35(179), [52]
Eischens, R. P., 87, 88, 112, [114]
Eisen, B. M., 290(182), [299]
Eisenbach, W., 23(111), 27(111), [50]
Eisert, M. A., 185(12), [188]
Ela, S. W., 59(28), 72, 73(28), [114]
Ellermann, J., 44(240), [54]
Emeléus, H. J., 255(42), [269], 339 (68), [348]
Emelyanova, L. I., 161(145), [177], 227 (2,3), 231(3), 233(3), 234(3), [239]
Emerson, M. T., 199(58), 220(58), [225]
Enemark, J. H., 44(234), [54], 97, [116]
Eremenko, L. T., 342(98), [348]

Ermanson, L. V., 332(36), 333(36), [347]
Eskin, I. T., 236(14), 238(14), [240], 256(54), [269]
Ethyl Corp., 39(201), [53]
Evans, D. F., 25(118), [50]

Fazakerly, V., 25(118), [50]
Fechter, R. B., 59(28), 72, 73(28), 113 (114), [114, 117]
Federov, L. A., 276(6), 293(195), [294, 299]
Fedotov, M. S., 267(85,86), [270]
Feigel, H., 145(118), 169(118), [177]
Feilchenfield, H., 42(220), [53]
Fergusson, J. E., 89(80), 91, 105, [115, 116]
Fernandez, V. P., 6(22), [48]
Fetisova, T. P., 333(41), 342(94), [347, 348]
Fetter, N. R., 33(176), [52]
Fiebig, A. E., 186(40), [189]
Fienner, A. L., 281(87), [296]
Filatova, R. S., 292(190), [299], 329 (26), [346]
Finhold, A. E., 28(139), [51]
Finzi, C., 168(176,177), [178]
Fischer, E., 280(59), [295]
Fischer, E. O., 74(63,64), [115]
Fish, R. W., 335(48), 337(48), 345(48), [347]
Fleck, H. P., 216(94), [226]
Florinsky, F. S., 252(9), [268]
Forrester, S. R., 336(50), [347]
Foss, V. L., 166(179), [178], 309(65, 66), 310(70), 321(70), [324]
Foster, D. J., 284(112-114), 285(112-114), [297], 330(31), 331(31), 334 (47), 335(31), [346, 347]
Franchimont, A., 303(7), 313(7), [322]
Frankenburg, W. G., 56, [113]
Frankland, E., 201(72), 215(91), 216 (72), 217(72), [225, 226]
Fray, G. I., 199(50), 219(50), [225]
Freedman, M. B., 339(73), [348]
Freidlina, R. Kh., 137(48), 141(75), 147 (132-134), 150(75), 154(132), 157 (48), 162(75), 171(132-134), 172 (132), [174, 175, 177], 253(14), 254(34), [268], 283(108), 287

(126,129,130,133,149), 289(130, 162,166), [297, 298]

Frey, F. W., 30(151), 37, 38(193), [51, 53]

Freyer, W., 164(149), [177]

Friedel, C., 219(112), [226]

Frolov, N. Ja., 132(17), [177]

Fujita, K., 286(123), [297]

Fukushima, S., 286(122,124), [297]

Fulton, M., 200(69), 222(69), [225], 249(8), [250], 253(17), 255(17), 256(17), 262(17), [268]

Fürst, W., 278(29), [294]

Fyodorov, L. A., 69(36), 83(36), [114]

Gaeva, L. A., 278(22), [294]

Gale, L. H., 278(17), 281(17), 284(17), [294], 328(19,21,22), 333, [346]

Gardner, P. D., 266(102), [270], 293 (200), [299]

Garner, C. S., 131(8), [173]

Garze, G., 39(196), [53]

Gatti, G., 15(67), 16(67), 17(83), 18(67, 83,87,89), 25(67,83), [49, 50]

Gavrilenko, V. V., 8(42), 22(106), 23 (109), 24(106,109,115,117), 42 (223), [48, 50, 54]

Gay, B., 293(207), [299]

Gellert, H. G., 22(101-103), 23(101), 24 (101), [50]

Gembitskii, P. A., 281(83,84), 288(83, 84), [296]

Genkina, N. K., 342(96), [348]

Gensler, W. J., 194(14), 209(14), [224]

Geoghegan, P., Jr., [270]

George, M. V., 160(142), [177], 242-244(6), [247]

George, T. A., 32(169), [52]

Gerard, W., 137(45), 156(45), [174]

Gerart, F., 342(95), [348]

Gerhart, F., 293(203,206), [299]

Giannini, U., 42(211,212), [53]

Gibson, C. S., 142(91), 165(91), [176]

Gilje, J. W., 290(175,176), 292(175, 176), [298, 299]

Gilliam, W. F., 6(23), 7(23), 11(23), [48], 218(104), 220(104), [226]

Gilman, H., 37, [53], 137(43), 160(142), [174, 177], 195(23,27), 196(23,27), 197(33,38), 199(51,60,64), 204(23,

38,77), 206(60), 208(77,79), 210 (79), 211(23,27,77,83), 212(84), 213(23,27), 214(23,27,77,79,83), 215(23,33,38), 216(92,95), 219(51, 79), 220(60), 221(64), [224-226], 242-244(6), [247], 303(10), 306 (10), 316(10), 320(10), [322]

Ginsberg, H., 274(11), [274]

Ginsburg, V. A., 338(62), [347]

Gladshtein, B. M., 281(74), 288(74), [296]

Gladyshev, E. N., 223(116), [226], 253 (25,27), [268]

Glemser, O., 81, [114]

Glocking, F., 38, [53]

Gluschkova, V. P., 138(60-62), 149(60), 150(60), 158(60-62), 159(62), [175]

Godchaux, E., 145(120), 169(120), [177]

Goddard, A. E., 137(46,47), 138(59), 156(46,47), 158(47,59), 159(47), [174, 175], 199(65), 219(65), 221 (65), [225]

Goddard, D., 137(46), 156(46), [174], 199(65), 219(65), 221(65), [225]

Golubeva, E. I., 137(54), 157(54), [175], 194(13), 208(13), [224], 304(15), 316-318(15), [322]

Goon, D. V. J., 293(198), [299]

Gordon, M. E., 180(4), 184(4), 186(16, 21,29), [188, 189]

Graeff, F., 164(153), [178]

Graf, H., 6(16), [48]

Grafflin, M. W., 281(85), [296]

Gray, M. I., 29(141), [51], 142(85a), 164(85a), [176], 194-196(12), 203 (12), 209(12), 213(12), 215(12), 219(110), [224, 226], 241(1), 243 (1), [246]

Grebenshchikov, Yu. B., 78(111), [117]

Greeley, R. G., 67, 95(60), 113(114), [115, 117]

Green, I. E., 283(104), [297]

Green, M. L. H., 109, [116]

Greenwood, N. N., 336(56), [347]

Griasnow, G., 28(136), [51]

Gribov, B. G., 341(92), [348]

Griffith, E., 287(146), [298]

Grignard, V., 17(80), [49]

Grim, S. O., 186(28), [189]

Groenewege, M. P., 10(58,61), [49]
Gronowitz, S., 320(68), 322(68), [324]
Grosse, A. von, 9, 13, 15(54), 17(54), [49]
Gruber, W. H., 44(240), [54]
Gruescu, C., 142(95), 143(95), 166(95), 167(95), [176]
Grüttner, G., 5(14), [48], 198(45), 201 (45), 209(45), 213(45), 215-218(45), 219(111), 222(45), [225, 226]
Guard, H. E., 304(72), [324]
Gubin, S. P., 282(97-99), [296]
Guichard, F., 142(83), 164(83), [176]
Güntner, A., 144(108), 168(108), [176]
Gur'anova, T. P., 331(34), 332(34), 333 (39,40), [346, 347]
Gus'kova, A. N., 288(158), [298], 326 (3), 338(3), 342(3), [345]
Gutowski, H. S., 7(29), 11(29), 17(29), [48], 218(106), [226]

Haage, K., 15(69), [49]
Haber, 57
Hafner, W., 74(63,64), [115]
Hager, F. D., 194(10,11), 201(10), 208 (10), [224]
Hahn, E., 194(20), 209(20), [224], 272 (5), [274], 345(90), [348]
Haight, G. P., 69, [114]
Hall, R. H., 257(87), [270], 341(83), [348]
Ham, N. S., 3(8), 10(8), 19(96), 25(121, 122), [48, 50, 51]
Hamilton, C. S., 144(106), 167(106), 168(106,173-175), [176, 178]
Hammar, W. J., [270]
Handrick, G. K., 278(30), [294]
Hansen, R. L., 276(8), 282(8), 283 (101), [294, 296]
Hanson, E. M., 186(37), [189]
Harriman, B. R., 327(17), [346]
Harris, R. O., 89(40), 92(40), 103(40), 105(40), 106(40), [115]
Harrison, D. F., 93, 94, 104(69), [115]
Hart, C. R., 135(36), [174]
Hartman, H., 142(90), 153(90), 165(90), [176]
Hartmann, H., 8, [48]
Hasegawa, I., 10(57), 11(57), [49]
Hasenbäumer, I., 144(109), 168(109),

[176]
Haszeldine, R. N., 255(42), [269], 339 (68), [348]
Hata, G., 43(232), [54]
Hatada, K., 16(78), 17(78), [49]
Hathway, D. E., 308(52), [323]
Hayamizu, K., 11(63), 12, [49]
Hechenbleikner, I., 196(29), 211(29), 212(29), [224]
Heeren, J. K., 182(8), 186(28), [188, 189]
Hegarty, B. F., 293(194), [299]
Hein, F., 70(19,20), [114], 208(76), 217(76), 218(76), [226]
Heitsch, C. W., 6(17), [48]
Henbest, H. H., 253(19), 255(36), 257 (19), 258(36), 259(19,36), 260-262 (19), [268]
Henke, H., 39(198), [53]
Henrici-Olivé, G., 59(15), 66, 67(16), 70, 104, [114, 116]
Hepner, F. R., 290(179), [299]
Herman, D. F., 42(221), [54]
Herrmann, G., 43(228), 44(242), 45 (228), [54]
Herwig, W., 194(19), 209(19), [224], 272(4), [274]
Heumann, K., 145(119), 169(119), [177]
Hidai, M., 44(235), [54], 71(89), 95 (89), 107, 108(90,106,109), 109, [116, 117]
Hillers, S. A., 253(26), 262(26), [268]
Hilpert, S., 5(14), [48], 198(45), 201 (45), 209(45), 213(45), 215-218 (45), 219(111), 222(45), [225, 226]
Hirota, K., 20(98), [50]
Hnizda, V., 17(81), [49]
Hodgson, U. H., 308(52), [323]
Hoffmann, E. G., 6(20), 7(20), 10(56), 11, 15(68), 17, 22(20), [48, 49]
Hofmann, K. A., 145, 169(118), [177], 255, 256(49, 50), 257(38), [268, 269], 280(61), 281(73), 287(128, 134-136), [295-297], 306(24), 311 (24), 321(24), [322], 339(69), [348]
Holley, Ch. E., Jr., 197(32), 215(32), [224]
Holtz, J., 194(9), 195(9), 208-213(9),

[224]
Honeycutt, J. B., 24(116), [50]
Hooton, K. A., 38, [53]
Horder, J. R., 36(189), 39(189), [53]
Höring, M., 142(92), 153(92), 165(92), [176]
Hornberger, H., 197(35), 215(35), [224]
Hosking, J. W., 106(85), 109(85), [116]
Howard, E. G., 248(2), 249(2), [250], 255(43), [269], 279(54), [295], 339(71), 344(71), [348]
Howard, M., 137(45), 156(45), [174]
Huether, E., 34(180), [52]
Huggins, M. L., 16(77), [49]
Hughes, E. D., 131(13), 132(16,24), 133 (16,32), [173, 174]
Hughes, W. B., 186(28), [189]
Humffray, A. A., 283(105,106), [297]
Humpfner, K., 199(59), 219-221(59), [225]
Hümphner, K., 7(37), [48]
Hurwitz, M. J., 311(69), [324]

Iatsimirskii, K. B., 69(18), 70, [114]
Ibers, J. A., 44(234), [54], 97, 100(59), 104, [115, 116]
Ichikawa, K., 284(118), 286(122-124), 290(180,181), 291-293(180), [297, 299], 342(97), [348]
Idelchik, Z. B., 267(66), [269], 306(36), 315(36), [323]
Ikeda, S., 43(227,229-231), 44(233,236, 241), [54], 71, 79, 95, 96(52,53), 98, 99(73), 108(91), 109, 110(73), [115, 116]
Ikegami, Sh., [270]
Ilatovskaya, M. A., 58(4,7), 59(4,7,61, 62), 74(61,62), 76(62), 77(61,62), 78(7), 80(7), 81, 113(4,7), [113-115]
Immirzi, A., 7(28), [48]
Imoto, E., 339(75), [348]
Inglis, G., 330(30), [346]
Ingold, C. K., 5(12), [48], 132(16,24), 133, 134, 135(36), [173, 174], 293 (196), [299]
Inoue, I., 261(89), 263(89), [270]
Isacescu, D. A., 142(95), 143(95), 166 (95), 167(95), [176]
Ishikawa, M., 293(210), [299]

Issleib, K., 201(75), 223(75), [225]
Itoh, O., 286(123), [297], 342(97), [348]
Itzkovitch, I. J., 105, 106(78), [116]
Ivanov, L. L., 8(42), 24(117), [48, 50]
Ivanova, N. L., 309(65), [324]
Iyoda, J., 31(155), [52]

Jacknow, J., 87, 88, 112, [114]
Jacobs, G., 227(1), 233(1), [239]
Jacobsohn, I. M., 308(50), [323]
Jacobson, R. A., 343(88), [348]
Jakubovich, A. Y., 39(200), [53], 318 (60), 319(60), [323]
Janssen, M. J., 36(187a), [53]
Jaworski, R., 197(31), 215(31), [224]
Jeffery, E. A., 4(10), 6(21), 7(10,30,31, 35), 9(30,31), 10(59a), 15(10), 16 (10,21), 18(30,31), 19(31,35,96), 20(10), 21(10), 25(119-122), [48-51]
Jenei, S., 39(196), [53]
Jenkins, R., 17(80), [49]
Jenkins, W. J., 288(157), [298]
Jenkner, H., 16(71), 26, 31(123), 32, 34 (123), 35(123), 39(71,123), 40(123), 41, [49, 51, 52]
Jensen, F. R., 278(17), 281(17), 284(17, 111,115,116), 285(115,120), [294, 297], 304(72), [324], 328, 333, [346]
Jerchel, D., 164(151), [177]
Johannsen, T., 199(54), 220(54), [225]
Johnson, G. L., 99, 100, [115]
Johnson, M. R., [270]
Johnston, W. K., 36, [53]
Jolly, P. W., 44(243), [54], 111, [116]
Jonas, K., 44(243), [54], 111, [116]
Jones, L. W., 284(109), [297]
Jones, R. G., 199(60,64), 206(60), 220 (60), 221(64), [225]
Jones, W. E., 143(99), 166(99), [176]
Joo, W. C., 39(198), [53]
Jula, T. F., 185(41), 186(41), [189]

Kabachnik, M. I., 288(154), 289(154), [298]
Kaesz, H. D., 136(37), 142(37,85), 152 (85), 155(37,85), 164(37,85), [174, 176]

Kali-Chemie, 16(72), 31(159,160), 32 (162,163), [49, 52]
Kalil, E. O., 17(79), [49]
Kalinina, R. V., 252(11), [268]
Kalyavin, V. A., 131(14), 132(14,30), 134(14,30,33), 135(33,35), [173, 174]
Kane-Maguire, L. A. P., 112, [116]
Kang, J. W., 91, 106, 109(83), [115, 116]
Kaplan, F., 197(37), 215(37), [224]
Kaplin, Ju. A., 267(76), [270]
Karpov, V. I., 276(5), 280(5), [294], 326(5-10), 327(5-7), 331(5,7), 332 (5,7,9), 333(8-10), 334(6,7), [345, 346]
Kawabe, A., 342(97), [348]
Kawai, M., 20(98), [50]
Kawakami, J. H., [270]
Kayser, R. A., 293(198), [299]
Kayser, W. V., 277(11,12), 293(11,199), [294, 299]
Kazankova, M. A., 302(4), 309(4), 312 (4), 319(4), [322]
Kazjausnas, P. P., 328(25), 329(25), [346]
Keim, W., 42(217,218), [53]
Kekulé, A., 303, 313(7), [322]
Kelbe, W., 165(163,164), 167(164), [178]
Keough, A. H., 186(24), [189]
Kerk, G. J. M. van der, 35, 36(187a), 37, [52, 53]
Khalkin, V. A., 148(137), 172(137), [177]
Kharasch, M. S., 281, 282, [296], 308 (50), [323]
Khasapov, B. N., 139(67), 159(67), 160 (144), 161(144), [175, 177], 231(9), 234(9), 239(9), [240], 242(8), 244 (8), 246(8), [247], 254(31), [268]
Khasskina, E. E., 145(121), 169(121), [177]
Khidekel', M. L., 58(7), 59(7), 61(7), 78 (7,111), 80(7), 81, 89, 92, 100(43), [114, 115]
Khomutov, R. M., 148(138), 172(138), [177], 195(21), 196(21), 208(21), 213(21), [224], 320(61), [323], 338(64), [347]

Khorlina, I. M., 17(84,85), 23(84,108, 109), 24(109), 29(140), [50, 51]
Khorlina, M. Ya., 280(65), [295]
Khu, K. V., 131(9), [173]
Kiefer, E. F., 342(101), [348]
Kiljakova, G. A., 141(81), 163(81), [176]
Kim, D. G., 281(82), 288(82), [296]
Kim, J.-Y., 131(11), [173], 276(7), 277 (7,9,13), 278(7,9), 279(7,9), 282 (13), 283(9), [294]
King, R. B., 83(32,33), [114]
Kirby, G. W., 266(100), [270]
Kirby, R. H., 208(79), 210(79), 211 (83), 214(79,83), 219(79), [226]
Kirmreuther, H., 255(49,50), 256(49, 50), [269], 339(69), [348]
Kissinger, L. W., 278(30), [294]
Kitazume, S., 44(233,236,241), [54], 71, 79, 95, 96(52,53), 98, 99(73), 110(73), [115, 116]
Kitching, W., 293(192,194), [299], 341 (91), [348]
Klein, H. F., 39(197), [53]
Klimischa, G. P., 341(84), [348]
Klug, A., 287(139), [297], 306(25), 311 (25), [323]
Knoll, P. G., 130(6), 132(6,31), [173, 174], 191(2,4), [224]
Knotes, W. H., 110, [116]
Knoth, W. H., 44(241a), [54]
Knunyants, I. L., 136(39), 142(98), 153 (98), 166(98), [174, 176], 282(96), [296]
Kobetz, P., 24(116), 38(193), [50, 53]
Kobs, H. D., 5, 18(11), [48]
Kocheshkov, K. A., 138(60-62), 140 (74), 141(76,77,79), 144(114), 149 (60), 150(60), 151(74,79), 158(60-62), 159(62), 161(74,114), 162(76, 114), 163(74,77,79,114), 169(114), [175, 176], 194(15), 197(43), 200 (68), 201(68,73), 205(43), 207(68), 209(15), 217(43), 221(68), 222(68), 223(73), [224, 225], 228, 229(4), 231(8), 232(5), 234(4), 235(4), 236 (4,8,14), 237(4,5,8), 238(4,8,14), 239(8), [239, 240], 256(52-54), [269], 281(92,93), [296], 307(45), 314(45), 319(45), [323]

Kochetkov, A. K., 137(48), 141(75), 150(75), 157(48), 162(75), [174, 175], 289(162), [298]
Kochetkov, N. K., 256(51), [269], 289 (168-170), 290(169,170), [298], 308(58), 309(64), [323, 324], 339 (76), 340(76,81,82), 341(82), [348]
Kochetkova, N. S., 139(63), 158(63), [175], 254(34), [268], 287(126, 149), [297, 298], 338(63), [347]
Köchlin, P., 145(119), 169(119), [177]
Kolk, E., 41(208), [53]
Kondo, H., 43(232), [54]
Konkol, T. L., 293(207), [299]
Kooyman, E. C., 248(10), [250]
Korabliova, L. G., 81, [114]
Korneva, S. P., 223(116), [226], 242 (9,10), 244(9,10), 245(10), [247], 254(29,30), [268]
Koroleva, M. N., 267(72), [269], 279 (52), [295]
Kost, A. N., 145(121), 169(121), [177]
Köster, R., 3(7), 7(7), 16(73), 18, 27 (125,131), 28(7,125,131-136), 31 (153,154), 34(153), 35(153,186), 39(131), [47, 49-52]
Kostin, V. N., 252(12), [268], 281(75, 76,78-84), 288(75,76,78-84), [296]
Kosyakova, L. V., 59(61,62), 74(61,62), 76(62), 77(61,62), [115]
Koton, M. M., 248(1), 249(1), [250], 251(1,3), 252(1,5-10), [267, 268], 271(1,2), [274], 278, [294, 295], 305(20), 312(20), [322]
Kovredov, A. I., 304(15), 316-318(15), [322]
Kowitt, F. R., 290-292(171), [298]
Kozlov, V. V., 278(19-21), [294]
Kozlovsky, A. G., 282(97-99), [296]
Kozminskaya, T. K., 144(114), 161-163 (114), 169(114), [176]
Kozub, G. I., 89, 100(43), [115]
Krafft, F., 146(128), 171(128), [177], 249(7), [250]
Kraffts, H., 219(112), [226]
Kraïtz, Z. S., 164(150), [177], 320(62, 63), 321(62), [323, 324]
Kramer, L., 17(79), [49]
Kraus, C. A., 17(81), [49]
Krause, E., 199(52), 219(52), [225]

Kravtsov, D. N., 278, [295]
Kreevoy, M. M., 276(1,8), 277, 282(8), 283(101), 287(132), 290(171-178, 182), 291(132,171,172,177,183), 292(132,171-173,175,176,178), 293 (11,197-199), [294, 296-299], 308 (57), [323]
Krespan, C. G., 306(39), [323]
Kretchmer, R. A., 277, 290(172,175), 291(172), 292(172,175), [294, 298]
Kroll, W. R., 14(66), 18(91), 22(100), 26(100), [49, 50]
Kröner, M., 43(224), [54]
Kruglova, N. V., 280(62), 282(62), [295], 326(13), [346]
Kubota, M., 91(46), 106, 109(83,85), [115, 116]
Kucherov, M., 281, [295]
Kudinova, V. V., 166(179), [178]
Kudryavtsev, L. F., 267(76), [270]
Kudryavtsev, R. V., 82(112), [117]
Kudryavtseva, T. A., 137(56), 157(56), [175], 235(12), [240]
Kula, M. R., 29(145), [51]
Kumada, M., 293(210), [299]
Kurras, E., 70(21), [114]
Kurz, A. L., 342(96), [348]
Kuse, T., 107(90), 108(90,109), 109, [116, 117]
Kuwajima, I., 286(125), [297]
Kuwajima, J., 321(71), [324]
Kuz'min, S. G., 112, [116]
Kuznucki, R. E., 186(40), [189]
Kvashina, E. F., 113(115), [117]

La Coste, W., 142(96), 143(102,103), 154(103), 165(96,156), 166(102, 103), 167(102,168), [176, 178]
Ladenburg, A., 139(68,69), 159(68), 160(69), [175]
Lagowski, J. J., 251(4), [267]
Landgrebe, J. A., 186(34), [189], 278 (17), 281(17), 284(17), [294], 328 (20), [346]
Lane, J., 195(24), 212(24), [224]
Lange, G., 194(16), 197(16), 209(16), 215(16), [224]
LaPlaca, S. T., 100(59), [115]
Lappert, M. F., 32(169), 36(189), 39 (189), [52, 53]

Laptev, N. G., 278(24,28), [294]
Larbig, W., 22(100), 26(100), 28(136), [50, 51]
Larikov, E. I., 58(7), 59(7), 61(7), 78(7), 80(7), [114]
La Roy, J. M., 248(10), [250]
Latjaeva, V. N., 58(9), 69(36), 80(9), 81 (9), 83(36), [114], 141(81,82), 163 (81), 164(82), [176], 267(69), [269]
Lattes, A., 266(99), [270], 293(205), [299], 342(99), [348]
Laubengayer, A. W., 6(23), 7(23), 11 (23), 15(70), [48, 49], 218(104), 220(104), [226]
Laubie, H., 287(127), [297], 340(79), [348]
Lauter, W., 142(92), 153(92), 165(92), [176]
Lavrov, V., 215(88), [226]
Lay, W. P., 290(184), 292(184), [299]
Leandri, G., 278(33), [294]
Lebedev, O. V., 281(77), 288(77), [296]
Lee, C. C., 293(209), [299]
Lee, I. K., 131(11), [173]
Lee, J. B., 108, [117]
Legault, R. R., 281(90), [296]
Lehmkuhl, H., 5, 18, 22(102,103), 23, 24(93), 27(93,111), [48, 50]
Leicester, H. M., 145(123), 146(123-125), 170(123-125), 171(124), [177]
Leigh, G. I., 111, [115, 116]
Lengnick, G. F., 15(70), [49]
Lengyel, B., 35(184), 39(195,196), [52, 53]
Levesque, G., 186(35), [189]
Levi, D. L., 198(46), 218(46), [225]
Levin, J. H., 281(88), 282(88), [296]
Levina, R. Ya., 252(12), [268], 281(74-84), 288, [296]
Lewis, P. H., 6(24), [48]
Li, V. H., 136(39), 142(98), 153(98), 166(98), [174, 176], 282(96), [296]
Lichtenstadt, L., 142(92), 153(92), 165 (92), [176]
Lichtenwalter, G. D., 160(142), [177], 242-244(6), [247]
Lichtin, N. N., 330(29), [346]
Light, J. R. C., 38(194), [53]

Linder, J., 165(165), [178]
Link, A., 142(88), 165(88), 166(88), [176]
Linn, W. J., 255(43), [269], 279(54), [295], 339(71), 344(71), [348]
Linnemann, E., 303(12), 312(12), [322]
Little, W. F., 106(87), [116]
Livingston, J. G., 279(53), 283(53), [295]
Lockhart, J. C., 3, 8, [47]
Lodochnikova, E. I., 141(76,77,79), 151 (79), 162(76), 163(77,79), [176]
Loeffler, P. K., 216(93), [226]
Löhr, P., 198(44), 215-217(44), [225]
Lokshin, B. V., 148(139), 172(139), [177]
Lommen, F. W. M., 308(50), [323]
Long, L. H., 201(71), 222(71), [225]
Long, W. P., 41(209), [53]
Lorberth, J., 29(145), [51], 108, [117]
Lord, A., 342(93), [348]
Loseva, A. S., 280(65), [295]
Louise, E., 198(47), 218(47), [225]
Love, J. L., 89(80), 105, [116]
Lovtsova, A. N., 282(100), [296], 338 (59), [347]
Lowe, W. G., 168(173,174), [178]
Lucas, H. J., 290(179), [299]
Luijten, J. G. A., 35, 36(187a), [52, 53]
Lukevics, Z., 253(26), 262(26), [268]
Lundeen, A. J., 16(76a), [49]
Lutsenko, I. F., 139(64-67), 140(65,72), 148(138), 150(66), 159(64-67), 160 (72,143,144), 161(65,144), 164 (150), 166(179), 172(138), [175, 177, 178], 195(21), 196(21), 208 (21), 213(21), [224], 231(9), 234 (9), 239(9), [240], 242, 244(7,8,11), 245(8,11), 246(8), [247], 253(13, 28), 254(28,31), 255(44), [268, 269], 302, 309(4,65,66), 310(70), 311(3), 312(3,4), 318(1,3), 319(1,3, 4), 320(61-63), 321(62,70), [322-324], 336(51), 338(64), 340(80), [347, 348]
Lyons, R. E., 146(128), 171(128), [177], 249(7,9), [250]

McCloskey, A. L., 31(161), [52]
McCoy, C. R., 27(127), [51], 197(42),

205(42), 216(42), 218(42), [225]
McCrae, W., 338(61), 343(61), [347]
McCullough, J. D., 146(126), 170(126), [177]
McDonald, B. J., 267(77), [270]
McElhinney, R. S., 255(36), 258(36), 259(36), [268]
McGinnety, J. A., 44(234), [54], 97, [116]
McKay, W. T., 30(150), 32(150), [51]
McKervey, M. A., 265(97), [270]
McOmie, J. F. W., 335(49), 344(49), [347]
Magee, P. S., 294(215), [300]
Magnuson, V. R., 20, [50]
Mahadevan, A. P., 194(14), 209(14), [224]
Mai, V. A., 186(26), [189]
Maier, L., 40(203), [53]
Makarova, L. G., 119, 148(139), 161 (145), 172(139), [177], 227(2,3), 231(3), 233(3), 234(3), [239], 272 (6), [274]
Makower, S., 278(14), 288(95), [294, 296]
Malaiyandi, M., 255(41), 261(41), 262 (41), [269]
Malinovski, M. S., 307(46), 319(46), [323]
Malinowski, S., 112(104), [117]
Malyscheva, A. V., 141(81), 163(81), [176]
Manchinova, Z. N., 167(68), [269], 306 (29a), [323]
Manchot, W., 287(139), [297], 306(25, 26), 311(25,26), [323]
Mangham, J. R., 30(152), 32(152), [51]
Mann, F. G., 165(158), [178]
Manolopoulos, P. M., 330(29), [346]
Manulkin, Z. M., 144(111,112), 169 (111,112), [176], 305(18,19), 312 (19), [322]
Marinangelli, A., 15(67), 16(67), 17(83), 18(67,83), 25(67,83), [49]
Marker, T., 281(86), [296]
Marmor, R. S., 186(39), [189]
Marple, K., 199(51), 219(51), [225]
Marquardt, A., 217(100), [226]
Marsel, C. J., 17(79), [49]
Martin, H., 18(86), 22-24(101), [50],

76, [115]
Martynova, V. F., 278(39,41-44,47), [295]
Marvel, C. S., 142(84), 151(84), 164 (84), [176], 194(10,11), 201(10), 208(10), [224], 283(102), 287, [296, 298]
Maskill, R., 71, [114]
Massey, A. G., 248(11), [250], 342 (100), [348]
Materikova, R. B., 139(63), 158(63), [175], 338(63), [347]
Mathis, R. D., 186(34), [189]
Matlack, A. S., 78(29), [114]
Matsumiya, K., 167(170), [178]
Matteson, D. S., 265(96), [270]
Mavity, J. M., 9, 13, 15(54), 17(54), [49]
Maximenko, O. A., 302(5,5a), 303(5), 312(5,5a), [322]
Mazzanti, G., 42(211,212), [53], 76 (65), [115]
Mednik, M., 330(29), [346]
Meisenheimer, J., 142(92), 165(92), [176]
Mellet, R., 197(31), 215(31), [224]
Melnikov, N. N., 138(57), 146(127), 154 (127), 158(57), 170(127), [175, 177], 278(18), [294]
Melquist, J. L., 293(199), [299]
Menzel, W., 219(113), [226]
Mesentsova, N., 281(77), 288(77), [296]
Me-Tchy, T., 132(22), [174]
Meyer, F. J., 194(16), 197(16), 209(16), 215(16), [224]
Michaelis, A., 136, 137(41,42), 142(40, 86-88,93), 143(100-104), 144(108), 145(120), 149(41,42), 154(103, 104), 155(40-42,140), 156(41), 164 (86,147,153), 165(87,88,93,154- 156,159,161,162,166), 166(40,88, 100-104), 167(102,104,168,169), 168(108), 169(120), [174, 176- 178], 280(58), [295]
Michetti, A., 308(51), [323]
Middleton, E., 255(48), [269], 308(54), 320(54), [323], 340(78), [348]
Middleton, W. J., 248(2), 249(2), [250], 255(43), [269], 279(54), [295],

339(71), 344(71), [348]
Middleton, W. W., 281(70-72), [296]
Mieg, W., 142(97), 153(97), 165(97), [176]
Mikaijama, T., 286(125), [297]
Miklukhin, G. P., 267(61), [269]
Miller, D. B., 22, [50]
Miller, W. T., 339(73), [348]
Mills, J. S., [270]
Mills, L., 287(144), [298]
Miltenberger, K., 7(37), [48]
Minasz, K. J., 182(10,11), 183(11), 184 (11), 186(10), 187(11), 188(11), [188]
Mingos, D. M. P., 111, [116]
Misono, A., 43(227,229-231), 44(235, 239), [54], 71(56,89), 95(89), 96, 97, 107, 108(90,106,109), 109, [115-117]
Mitrofanova, E. V., 219(114), [226]
Miyake, A., 43(232), [54]
Moedritzer, K., 3, 8, [47]
Mole, T., 1, 3(8,9), 4(10), 6(21), 7(10, 31-33,35), 8(43,45,50-52), 9(30,31, 50,52), 10(8,59a), 11(43), 12(43, 64), 13(43,64,65), 14(64,65), 15(9, 10), 16(10,21), 18(30,31,50), 19(31-33,35,50,52,96), 20(10), 21(10,43), 23(43,50,51,65), 24(9,43,50,51), 25 (119,121,122), 26(45), 27(43), 30 (43), 31(43), 46(43), [48-51]
Montecatini, 33(175), [52]
Moon, S., [270]
Mooney, E. F., 137(45), 156(45), [174]
Moore, L., 137(43), [174]
Morifuji, K., 43(227,229-231), [54]
Morton, A. A., 196(29), 211(29), 212 (29), [224]
Mosher, H. S., 197(40), 215(40), 216 (40,93), [225, 226]
Moskalenko, V. A., 276(5), 280(5), [294]
Motserev, G. V., 39(200), [53]
Moy, D., 199(58), 220(58), [225]
Mui, J. Y.-P., 179(3), 180(4), 181(3), 184(4), 185(13,41), 186(25-27,29, 30,36,41), 187(11), 188(11), [188, 189], 283(107), 293(211), [297, 300]
Mukajama, T., 321(71), [324]

Müller, A., 218(107), [226]
Müller, H., 19(95), [50]
Muller, N., 7(38), 10(55), [48, 49]
Mumm, O., 281, [295]
Murahashi, S., 218(108), [226]
Murray, R., 112(103), [117]
Musgrave, W. K., 279(53), 283(53), [295]
Muxart, R., 197(31), 215(31), [224]
Myshkin, A. E., 280(57), 293(202), [295, 299]

Nadj, M. M., 140(74), 144(114), 151 (74), 161(74,114), 162(114), 163 (74,114), 169(114), [175, 176], 200 (68), 201(68), 207(68), 221(68), 222(68), [225]
Nagel, K., 22-24(101), 26(124), [50, 51]
Nakagawa, T., 261(89), 263(89), [270]
Narasaka, K., 321(71), [324]
Natta, G., 18(89), 42(210-212), [50, 53], 76, [115]
Naylor, M. A., 276(4), [294]
Nazarkova, N. P., 267(67), [269], 306 (35), [323]
Nazarova, L. M., 27(128,129), [51]
Nazibullin, J. Ja., 342(98), [348]
Neal, H. R., 8(47), [48]
Nechiporenko, G. N., 72, 81, [114]
Nefedov, V. A., 336(55), 337(55), [347]
Nefedov, V. D., 132(17,18), 148(137), 172(137), [173, 177]
Nefedova, M. N., 336(55), 337(55), [347]
Neller, R. N., 31(156), [52]
Nelson, W. K., 42(221), [54]
Nenitzescu, C. D., 142(95), 143(95), 166(95), 167(95), [176]
Nerdel, F., 282(95), [296]
Nesmeyanov, A. N., 130(1,3-6), 131, 132(6), 137(1,48,49,51-54,56), 138 (1,53), 139(63,64), 141(75), 145 (122), 146(122), 147(132-134), 148 (138,139), 150(75), 154(132), 157 (1,48,49,52-54,56,141), 158(63, 141), 159(64), 161(145), 162(75), 164(150), 169(122), 171(122,132-134), 172(132,138,139), [173-175,

177], 190, 191, 194(13), 195(21), 196(21), 197(43), 199(53), 200(66, 70), 205(43,53), 207(66), 208(13, 21), 213(21), 217(43), 219(53), 220 (53), 221(1,66,70), [224, 225], 227 (3), 228, 229(4,6), 230(7), 231(3,8), 232(5-7), 233(3), 234(3,4,7,10,11), 235(4,6,11-13), 236(4,7,8,11,14), 237(4,5,7,8), 238(4,7,8,14), [239, 240], 253(14), 255(44), 256(51-57), [268, 269], 272(6), [274], 278, 280 (62-65), 281(92), 282(62-64,97-99), 283(108), 287, 288(153-156,158), 289(130,153,154,159-162,164-166, 168-170), 290(169,170), [295-298], 302, 303(6), 304, 305(21,22), 306 (2), 308(55,58), 309(4,64,66), 310 (67,70,73), 311(2,3), 312(3,4), 315 (22), 316(6,15,22), 317(15), 318(1- 3,15), 319(1-4,67), 320(6,61), 321 (70), [322-324], 326, 327, 335(16), 336(51,54), 338(3,4,63-65), 339(66, 67,76), 340(76,80-82), 341(82), 342 (3,16), [345-348]

Nesmeyanova, O. A., 145(122), 146 (122), 169(122), 171(122), [177], 256(57), [269], 273(9), [274], 303 (6), 305(21), 316(6), 320(6), [322], 336(54), 339(66), [347]

Nesterenko, G. N., 342(98), [348]

Neumann, W. P., 28(138), 29(138,143, 144), 30(138,147), 32(166), 35(166, 186), 36, 47(166), [51, 52]

Newitt, D. M., 198(46), 218(46), [225]

Niermann, H., 29(143,144), [51]

Nikolls, R., 253(19), 257(19), 259-262 (19), [268]

Nikolsky, A. B., 104, 105(77), [116]

Nikonova, L. A., 339(67), [347]

Nishimura, K., 290-293(180), [299]

Nordman, C. E., 6(17), [48]

Norseev, Yu. V., 148(137), 172(137), [177]

Nöth, H., 22, [50], 108, [117]

Novatorov, N. F., 267(73,74), [269]

Novikov, S. S., 293(201), [299], 341 (92), [348]

Novikov, V. M., 310(67,73), 319(67), [324]

Novikova, N. V., 137(49,51,52,55), 157

(49,52,55), [174, 175], 199(53), 205(53), 219(53), 220(53), [225], 230(7), 232(7), 234(7,10,11), 235 (7,11), 236(7,11), 237(7), 238(7), [239, 240], 256(55,56), [269]

Nozakura, S., 218(108), [226]

Nyborg, S. C., 89, 104, [115, 116]

Obreimova, L. I., 145(121), 169(121), [177]

O'Brien, J. F., 10(57), 11(57), [49]

O'Connor, D. E., 337(57), [347]

O'Connor, G. N., 256(58), 266(58), [269]

Oda, J., 261(89), 263(89), [270]

Odling, W., 218(105), [226]

Ofele, K., 74(64), [115]

Ogata, I., 278(25), [294]

Ogawa, T., 20(98), [50]

Okawara, R., 339(75), [348]

Okhlobystin, O. Yu., 9(53), 27(53,130), 30(148), 31(157), 34(148,177,181), 35, 36(148,177,187), 38(177,187), 40(148,181,202), 41(148), 42(223), 47(148), [49, 51-54], 198(49), 218 (49), [225], 265(98), [270], 279 (55), [295], 339(74), [348]

Ol'dekop, Ju. A., 267(62,64-72,79,81- 84), [269, 270], 279(50,52), [295], 306(28,29,29a,31-37), 307(28,29, 40,43,44), 312(32), 313(28,40,43), 314(28), 315(36), 316(31,37), 321, [323]

Olivé, S., 59(15), 66, 67(16), 70, 104, [114, 116]

Oliver, J. P., 7(34,36), 11(61a), [48, 49], 199(54a,57,58), 220(54a,57, 58), [225]

Oppermann, G., 41(208), [53]

Optiz, G., 253(18), 263(18), [268]

Orndorff, W. K., 140(71), 161(71), [175]

Osberg, E. G., 305(20), 312(20), [322]

Osipova, M. A., 138(58), 158(58), 159 (58), [175], 230(7), 232(7), 234- 238(7), [239], 256(55), [269]

Ostapchuk, G. M., 192(5,7), 193(7), [224]

Otermat, A. L., 7(38), [48]

Othen, C. W., 167(171), [178]

Otto, J., 70(21), [114]
Otto, R., 164(152), [178], 213(86), 217
  (86), 223(86), [226], 249(5,6),
  [250], 303(11), 308(48), 313(11),
  320(48), [322, 323]
Ouchi, H., 284(118), 286(122), 290
  (181), [297, 299]
Ouellette, R. J., 284(111,115,116), 285
  (115,120), 293(208), [297, 299]

Pacevitz, H. A., 195(23), 196(23), 204
  (23), 211(23), 213-215(23), [224]
Paetsch, J. D. H., 186(39), [189]
Page, J. A., 105, 106(78), [116]
Pande, K. C., 285(119), [297]
Paneck, Cl., 165(159), [178]
Panov, E. M., 141(76,77,79), 151(79),
  162(76), 163(77,79), [175]
Papa, D., 274(11), [274]
Parham, W. E., 179(2), 186(20), [188,
  189]
Park, W. R. R., 255(35,40), 257(40,88),
  258(35), [268, 270]
Parry, R. W., 6(17), [48]
Parshall, G. W., 83(34,35), 98, [114,
  116]
Paske, R. J., 111(97,98), [115, 116]
Pasquon, I., 17(83), 18(83,89), 25(83),
  [49, 50]
Pasynkiewicz, S., 30(149a), [51]
Patnode, W., 199(56), 206(56), 220
  (56), [225]
Paulik, F. E., 283(104), 293(191), [297,
  299]
Pavlova, V. K., 69(18), 70, [114]
Pearson, R. G., 112, [116]
Pearson, T. H., 8(44), 30(44,149,150),
  32(150,167,168), 37(167), [48, 51,
  52]
Pengilly, B. W., 267(77), [270]
Pennella, F., 144(115), 169(115), [176]
Perego, G., 7(28), [48]
Perevalova, E. G., 145(122), 146(122),
  169(122), 171(122), [177], 256
  (57), [269], 273(9), [274], 282(97-
  99), [296], 302(2), 303(6), 305(21),
  306(2), 311(2), 316(6), 318(2), 319
  (2), 320(6), [322], 339(66), [347]
Perie, J., 266(99), [270], 293(205),
  [299], 342(99), [348]

Perrine, J. C., 31(156), [52]
Peters, F. M., 23(110), 29(110), 33
  (176), [50, 52]
Petree, H. E., 22(104), [50]
Petrosyan, N. S., 293(193), [299]
Petrov, A. A., 338(60), [347]
Petrovich, P. I., 278(26,27), [294]
Petterson, L., 255(47), [269]
Petukhov, G. G., 219(114), [226], 252
  (11), 267(61,63,76), [268-270], 305
  (17), 306(30), [322, 323]
Petzchner, E., 208(76), 217(76), 218
  (76), [226]
Pfeiffer, P., 287, [297]
Pfohl, W., 22(102,103), 26(124), [50,
  51]
Pinajian, J. J., 132(20), [173]
Pinazzi, C., 186(35), [189]
Pines, H., 281(88), 282(88), [296]
Pinkerton, R. C., 24(116), [50]
Pino, P., 42(211,212), [53]
Pitzer, K. S., 7(29), 11(29), 17(29),
  [48], 218(106), [226]
Podall, H., 42(222), [54]
Podall, H. E., 22(104), [50]
Podvisozkaya, L. S., 293(204), [299]
Poetke, W., 278(29), [294]
Polak, L. S., 57(2), [113]
Pollard, D. R., 192(6), [224]
Pollick, P. Y., 248(13), [250]
Polovyanyuk, I. V., 148(139), 172(139),
  [177]
Polster, R., 37(190), [53]
Ponzio, G., 217(101), [226]
Pope, W., 142(91), 165(91), [176]
Postnikova, G. V., 253(13), [268]
Potoski, J. R., 186(20), [189]
Potrosov, V. I., 197(43), 205(43), 217
  (43), [225]
Praisnar, B., 134(34), [174]
Pratt, D. E., 137(45), 156(45), [174]
Pratt, J. M., 71, 106(81), 111, [114,
  116]
Pregaglia, G., 76(65), [115]
Prikashchikova, L. P., 292(189), [299],
  328(24), 329(24,27), 330(28), 344
  (24), [346]
Prince, M. I., 16(74), 42(219), [49, 53]
Pritchard, D. E., 10(55), [49]
Pritchard, H. O., 342(93), [348]

Prodayko, L. A., 82(112), [117]
Prokai, B., 186(17,19,31-33), [189]
Pu, L. S., 44(233,236), [54], 71, 79, 95, 96(52,53), 108(91), 109, [115, 116]
Puzyreva, V. P., 231(8), 236-239(8), [239]
Pyke, R. G., 257(87), [270], 341(83), [348]

Rabb, R. B., 266(102), [270]
Rabideau, Sh. W., 197(32), 215(32), [224]
Radcliffe, L. G., 165(157), [178]
Rall, K. B., 338(60), [347]
Ramey, K. C., 10(57), 11(57), [49]
Rausch, M. D., 273(8), 274(10), [274]
Ray, R. L., 30(150), 32(150), [51]
Razumovski, V. V., 318(60), [323]
Razuvaev, G. A., 141(80-82), 163(80, 81), 164(82), [175, 176], 186(38), [189], 200(67), 201(67), 219(114), 221(67), 222(67), 223(116), [225, 226], 242(9,10), 244(9,10), 245 (10), [247], 248(1), 249(1), [250], 251, 252(1,5,6,11), 253(25,27), 254 (29,30), 267(61-65,67-74,76,79,81, 82,86), [267-270], 271(1,2), [274], 279(49,52), [295], 305(17), 306(28, 29,29a,30,32,33,35,37,38), 307(28, 29,38,40-43), 312(32), 313(28,40, 42,43), 314(28,42), 315(42), 316 (37), 319(60), 321, [322, 323], 334, 335(45,46), 337(42), [347]
Reiche, W. T., 144(115), 169(115), [176]
Reidlinger, A., 17(79), [49]
Reilly, C. A., 42(217), [53]
Reinäcker, R., 18(86), [50]
Reinheckel, H., 15(69), [49]
Reinsalu, V. P., 89(40), 92(40), 103(40), 105(40), 106(40), [115]
Rekasheva, A. F., 267(61), [269]
Reutov, O. A., 130(6), 131, 132(6,12, 14,22,23,25-31), 133, 134(14,26,27, 30,33,34), 135(33,35), [173, 174], 191(2-4), 192(3,5,7), 193, [224], 253(23), 260(23), 263(90,91), [268, 270], 276(2,5,6), 279(56), 280, 282 (100), 293(193,195,202), [294-296, 299], 302(5,5a), 303(5), 310

(73), 312(5,5a), [322, 324], 326, 327, 328(23), 331(5,7,23,33-35), 332, 333, 334(6,7,35), 338(59), 342 94,96), [345-348]
Reynolds, G. F., 276(7), 277(7), 278(7, 9), 279(7,9), [294]
Rheinboldt, H., 147(130), 171(130), [177]
Richards, R. L., 89(80), 104, 105, [116]
Rickborn, B., [270]
Rinze, P. V., 108, [117]
Roberts, R. M., 132(24), [174]
Robinson, G. C., 38(193), [53]
Robinson, R., 199(50), 219(50), [225]
Robson, J. H., 253(21,24), 257(24), 260 (21), [268], 284(117), [297]
Rochow, E. G., 199(61), 221(61), [225]
Rodgman, A., 254(33), 261(33), 262 (33), [268], 290(184), 292(184, 185,187), 293(185), [299]
Rodina, N. B., 253(23), 260(23), 263 (90), [268, 270]
Roeder, G., 143(105), 166(105), 167 (105), [176]
Rohler, H., 165(154), [178]
Rohrmann, E., 305(16), [322]
Rokitskaya, M. S., 138(57), 146(127), 154(127), 158(57), 170(127), [175, 177], 278(18), [294]
Roman, F. L., 287(143), [298]
Romeyn, J., 287(148), [298]
Rosenblum, M., 335(48), 337(48), 345 (48), [347]
Rosenfelder, W. J., 241(5), 243(5), [247], 253(15), 262(15), [268]
Rosenstein, S. M., 318(60), 319(60), [323]
Rosser, R. J., 143(99), 166(99), [176]
Rossi, M., 44(237,238), [54], 79, 96, 97, 107, 108, [115, 116]
Roux, E., 198(47), 218(47), [225]
Rozantsev, E. G., 138(62a), [175]
Ruff, J. K., 8, 29(39), [48]
Rundle, R. E., 6(24), [48]
Russell, G. A., 39(199), [53]

Sacco, A., 44(237,238), [54], 79, 96, 97, 107, 108, 110, [115, 116]
Sackman, J. F., 201(71), 222(71), [225]
Saito, T., 43(227,229-231), 44(239),

[54], 71(56), 96, 97, 108(106), 109, [115-117]
Sakurada, Y., 16(77), [49]
Salinger, R. M., 197(37), 215(37), [224]
Salvemini, A., 169(178), [178], 306 (27), [323]
Samuel, W., 142(92), 153(92), 165(92), [176]
Sand, J., 254(32), 257(32), [268], 287 (131,134-138,141,142,147), [297, 298], 306(24), 308(56), 311(24), 318(56), 321(24), [322, 323], 338 (58), [347]
Sandborn, L. T., 287(145), [298]
Sanders, D. A., 11(61a), [49]
Sanders, J. R., 104, 105(77), [116]
Sarry, B., 70(22), [114]
Sauermann, G., 223(117), [226]
Sauerwald, A., 218(107), [226]
Saunders, J. K., 3(10), 7(10,30,31), 9(30, 31), 15(10), 16(10), 18(30,31), 19 (31), 20(10), 21(10), 25(121,122), [48, 51]
Savel'eva, I. S., 137(54), 157(54), [175], 194(13), 208(13), [224], 280(62-64), 282(62-64), [295], 326 (13-15), [346]
Savinjykh, L. V., 293(193), [299]
Savitsky, A. V., 334, 337(42), [347]
Savlevitch, Kh., 148(137), 172(137), [177]
Sazonova, V. A., 339(67), [347]
Schäfer, R., 18(88,94), [50]
Schaleger, L., 290(177), 291(177), [299]
Schapiro, A. P., 138(62a), [175]
Schatzkaya, R. Ch., 305(22), 315(22), 316(22), [322]
Scheidegger, H. A., 90(100), 111, [116]
Scheinker, Yu. N., 139(63), 158(63), [175]
Schempf, R., 141(78), 161-163(78), [175]
Schenk, A., 165(161,162), [178]
Scherer, H., 16(76), [49]
Schick, R., 35(186), [52]
Schlenk, W., 194(9), 195(9), 197(41), 205(41), 208-213(9), 216(41), [224, 225]
Schlenker, U., 131(7), [173]

Schlesinger, H. I., 7(26), 22(26), 28 (139), [48, 51], 198(48), 208(78), 215(89), 218(48), [225, 226]
Schmidbauer, H., 5, 39(197), [48, 53]
Schmidt, M., 6(16), [48]
Schneider, B., 28(138), 29(138,143), 30 (138), [51]
Schneider, J., 22-24(101), [50]
Schneller, S. W., 113(114), [117]
Schoeller, W., 280(60), [295]
Schöllkopf, U., 293(203,206), [299], 342(95), [348]
Schorygin, P. P., 195(25,26,28), 196(25, 26,30), 209(28), 210(25,26,28,30), 212(28,30), [224]
Schott, H., 45(245), [54]
Schrauth, W., 280(60), [295]
Schroeder, W. D., 303(9), 316(9), 320 (9), [322]
Schulte, C., 165(166), [178]
Schulze, F., 197(33,38), 204(38), 215 (33,38), [224]
Schumacher, H., 342(95), [348]
Schumb, W. C., 199(63), 221(63), [225]
Schwarz, J., 330(30), [346]
Schwayka, O. P., 341(84), [348]
Schwenke, E., 274(11), [274]
Schwirten, K., 5(11a), [48]
Scott, R., 69, [114]
Seager, J. H., 281(72), [296]
Segitz, F. A., 208(76), 217(76), 218 (76), [226]
Segre, A. L., 15(67), 16(67), 18(67), 25 (67), [49]
Seifert, F., 7(37), [48]
Senoff, C. V., 88, 89(39,40), 91, 92(40), 93, 103(40), 105(39,40), 106(40), [115]
Seydel, G., 16(76), [49]
Seyferth, D., 164(149), [177], 179, 180 (4-6), 181(3,5,7), 182(1,6-11), 183 (11), 184(4,11), 185(6,12-15,41), 186(10,16-19,21,23,25-33,36,37,39, 41), 187(11), 188(11), [188, 189], 223(115), [226], 283(107), 293 (211,212), [297, 300], 339(70), 343 (70), [348]
Shah, S. W., 266(100), [270]
Shanasarov, K. S., 281(81), 288(81), [296]

Shapiro, H., 42(222), [54]
Sharkey, W. H., 248(2), 249(2), [250],
  255(43), [269], 279(54), [295],
  339(71), 344(71), [348]
Shcherbakov, A., 216(98), [226]
Shearer, D. A., 292(185), 293(185),
  [299]
Shepeleva, R. I., 130(1), 137(1), 157(1),
  [173], 289(161), [298], 318(1),
  326(11), [346]
Sheridan, P. S., 112, [116]
Shetsov, Yu. A., 58(7), 59(7), 61(7), 78
  (7), 80(7), [114]
Sheverdina, N. I., 201(73), 223(73),
  [225]
Shiihara, J., 31(155), [52]
Shilov, A. E., 69, 71(48,49), 72, 81, 88
  (38), 89, 90(105), 91(49), 92, 93
  (49), 100(43), 105, 106, 112(105),
  113, [114-117]
Shilova, A. K., 69, 71(48,49), 72, 89(49,
  79), 91(49), 92, 93(49), 105, 106,
  113(115), [114-117]
Shindler, A., 62(14), [114]
Shiner, V. J., 6(22), [48]
Shostakovsky, S. M., 281(84), 288(84),
  [296]
Shteinmen, A. A., 88(38), [115]
Shubenko, M. A., 306(38), 307(38,41,
  42), 313-315(42), [323]
Shulgaitser, L. A., 58(9), 80(9), 81(9),
  [114]
Shur, V. B., 55, 58(3-13), 59(3-8,10-13,
  61,62), 60(3), 61(3-5,7,10,11,13),
  62(4), 63(4), 64(3,4), 65(3,4,8), 66
  (8), 68(4), 69(13,36), 72(3,4), 74
  (61,62), 76(62), 77(61,62), 78(4,7,
  11), 80(7,9), 81(9), 82(12,112), 83
  (36), 87(13), 100(13), 106(13),
  [113-115, 117]
Shurov, A. E., 200(67), 201(67), 221
  (67), 222(67), [225]
Shustorovich, E. M., [117]
Shvetsov, Yu. A., 81, 92, [114, 115]
Siemens-Schuckertwerke, 35(185), [52]
Simmons, H. D., Jr., 180(6), 182(6,11),
  183(11), 184(11), 185(6,15), 187
  (11), 188(11), [188, 189], 293
  (212), [300]
Sims, L. L., 30(150), 32(150), [51]

Singer, F., 254(32), 257(32), [268], 287
  (138,142), [297]
Singh, G., 180(6), 182(6), 185(6,13),
  186(28), [188, 189]
Sinn, H., 41(208), [53]
Sinotova, E. N., 132(17-19), 133(19),
  148(137), 172(137), [173, 177]
Sistrunk, J. O., 38(193), [53]
Skoldinov, A. P., 307(45), 314(45), 319
  (45), [323]
Sleddon, G. J., 32(165), [52]
Sleezer, P. D., 285(121), [297]
Sloan, M. F., 78(29), [114]
Smidt, J., 10(58,61), [49]
Smirnov, K. M., 338(62), [347]
Smith, D. C., 112(103), [117]
Smith, F. D., 166(167), 167(167),
  [178]
Smith, J. J., 144(115), 169(115), [176]
Smith, R. G., 106(81), 111, [116]
Smolina, T. A., 131(9,12,14), 132(12,
  14,22,30), 134(14,30,33), 135(33,
  35), [173, 174]
Sobatzki, R. J., 340(77), [348]
Societa Edison, 31(160a), [52]
Söderbäck, E., 336(53), [347]
Sörlin, G., 320(68), 322(68), [324]
Sokolov, E., 217(102), [226]
Sokolov, V. I., 132(23,25-28), 134(26,
  27,34), [174], 253(23), 260(23),
  263(90,91), [268, 270]
Sommer, R., 30(147), [51]
Song, K. M., 96, [115]
Soroos, H., 8(48,49), [49]
Specht, E. H., 278(30), [294]
Sperry, W. N., 287(143), [298]
Spinelli, D., 169(178), [178], 278(33),
  [294], 306(27), [323]
Spragna, M. J., 336(56), [347]
Sprowls, M. R., 281(90), [296]
Stamm, W., 32(172), [52]
Staub, T. S., 293(197,199), [299]
Stauffer Chemical Co., 32(173), [52]
Steacie, E. W. R., 267(75), [270]
Stear, A. N., 251(4), [267]
Stecher, O., 199(54), 220(54), [225]
Steinkopf, L., 278(15), [294]
Steinkopf, W., 142(97), 153(97), 165
  (97), 167(172), [176, 178], 344
  (89), [348]

Steinwand, P. J., 277(11,12), 293(11, 297), [294, 299]

Stephens, R., 339(72), [348]

Sterlin, R. N., 136(39), 142(98), 153 (98), 166(98), [174, 176], 282(96), [296]

Steudel, O. W., 22(100), 26(100), [50]

Stevens, J. R., 90, [115]

Stevens, L. G., 7(34), [48], 199(54a, 57), 220(54a,57), [225]

Stokker, G., 290-292(172), [298]

Stone, F. G. A., 136(37), 139(70), 142 (37,85), 152(85), 155(37,85), 161 (70), 164(37,85), [174-176]

Strecker, M., 165(165), [178]

Strohmeier, W., 7(37), [48], 199(59), 219-221(59), [225]

Struense, R., 280(60), [295]

Strunin, B. N., 36(187), 38(187), [52]

Stucker, J. F., 273(7), [274]

Stucky, G. D., 20, [50]

Studiengesellschaft-Kohle, 28(137), 29 (146), 43(226), 45(244), [51, 54]

Subbotina, A. I., 251(2), [267]

Suida, W., 303(13), 305(13), 311(13), [322]

Sullivan, M. F., 106(87), [116]

Summerbell, R. K., 336(15), [346]

Sun, J. Y., 106, 109(83), [116]

Sun, T.-Y., 9(46), [115]

Sundermeyer, W., 33, [52]

Surtees, J. R., 7(32,33), 8(50), 9(50), 18(50), 19(32,33,50), 22(105), 24 (50), 25(50), 41(206), [48, 49, 53]

Suzuki, Z., 286(125), [297]

Svetlanova, T. B., 328(23), 331(23), 332 (23), [346]

Swanwick, M. G., 109, [116]

Swartz, S., 281(89), [296]

Swinden, G., 106(81), 111, [116]

Szekely, T., 39(195,196), [53]

Tabern, D. L., 140(71), 161(71), [175]

Tabrina, G. M., 72, [114]

Tafel, J., 342(85), [348]

Takashi, Y., 41(204,205), [53]

Takayama, S., 290-293(180), [299]

Takeda, S., 18(90), [50]

Takeuchi, S., 218(108), [226]

Talalaeva, T. V., 194(15), 209(15), [224]

Talbot, M. L., 265(96), [270]

Tamelen, E. E. van, 59(28), 67, 72, 73, 95, 104, 113, [114-117]

Tanaka, Y., 339(75), [348]

Taniguchi, H., 342(97), [348]

Tarao, R., 18(90), [50]

Tarrant, P., 337(57), [347]

Tartakovskii, V. A., 252(12), [268], 281 (78-80), 288(78-80), 293(201), [296, 299], 341(92), [348]

Tatarenko, A. N., 144(111), 169(111), [176], 305(18), [322]

Tatlow, J. C., 253(17), 255(17), 256 (17), 262(17), [268], 339(72), [348]

Taube, H., 90(100), 93, 94, 104(69), 111, [115, 116]

Tchlenov, N. E., 293(201), [299]

Terentjev, A. P., 145(121), [177]

Tewari, R. J., 293(209), [299]

Thonstad, J., 77(67), [115]

Thorpe, F. G., 132(16), 133(16), [173]

Thorpe, M. A., 336(52), [347]

Thurman, N., 303(8), 305(8), 311(8), 313-315(8), 319(8), 320(8), [322]

Tille, D., 70(20), [114]

Titov, A. I., 278(23,24,28), [294]

Tobler, E. J., 284(112-114), 285(112-114), [297], 330(31), 331(31), 334 (47), 335(31), [346, 347]

Tochtermann, W., 194(20), 209(20), [224], 272(5), [274], 345(90), [348]

Todd, L. J., 179(3), 181(3,7), 182(7), 185(12), 186(27), [188, 189], 283 (107), 293(211), 297, [300]

Tokareva, F. A., 287(130), 289(130), [297]

Toropova, E. M., 308(55), [323]

Toropova, M. A., 148(137), 172(137), [177]

Towe, R. H., 339(70), 343(70), [348]

Traylor, T. G., 131(8), [173], 241(3,4), 243(3,4), [246, 247], 253(16,20), 255(37), 261(16,37), 262(20), [268, 270], 276(3), 283(103), 284(103), [294, 297], 330(32), 331(32), [332], 333, [346]

Treshchova, E. G., 281(84), 288(84),

[296]
Trieber, A. J.-H., 182(10,11), 183(11), 184(11), 186(10), 187(11), 188(11), [188]
Troitzkaya, L. L., 263(91), [270]
Tscherbakov, V. I., 186(38), [189]
Tsuchida, M., 278(25), 286(122), [294, 297]
Tsutsui, M., 43(225), [54]
Tumanova, Z. M., 302(3), 311(3), 312 (3), 318(3), 319(3), [322]
Turner, E. E., 146(129), 171(129), [177]
Turner, M. A., 290(176-178), 291(177, 183), 292(176,178), [299]
Twelves, R. R., 179(2), [188]
Tyrlik, S., 112(104), [117]

U, Y. T., 131(12), 132(12), [173], 191 (3,4), 192(3), [224]
Uchida, Y., 43(227,229-231), 44(235, 239), [54], 71(56,89), 95(89), 96, 97, 107, 108(90,106,109), 109, [115-117]
Uglova, E. V., 328(23), 331(23), 332 (23), [346]
Ukhin, L. Yu., 92, [115]
Ulrich, S. E., 195(24), 212(24), [224]
Uson, R., 6(16), [48]
Ustynjuk, T. K., 281(82), 288(82), [296]
Usubakunov, M., 145(116), [176], 255, [268]

Valujeva, Z. P., 338(65), [347]
Van Akern, G. M. F., 132(15), [173]
Van der Kelen, G. P., 199(62), 221(62), [225]
Van Leuwen, R. G., 293(208), [299]
Van Loon, E. J., 328(18), [346]
Van Wazer, J. R., 3, 8, [47]
Varshavskii, S. L., 289(167), [298]
Vasilejskaya, N. S., 334(44,45), 335 (45), [347]
Vaska, L., 100, [115]
Vastine, F., 91(46), 106, 109(83), [115, 116]
Vaughan, L. S., 223(115), [226]
Vecchiotti, L., 308(51), [323]
Velichko, F. K., 283(108), [297]

Verbeek, W., 33, [52]
Vicentini, G., 147(130), 171(130), [177]
Vilchevskaya, V. D., 130(3), 137(53), 138(53), 157(53), [173, 175], 288 (156), [298]
Vinogradov, A. D., 281(83), 288(83), [296]
Vinogradova, V. N., 161(145), [177], 227(3), 231(3), 233(3), 234(3), [239], 272(6), [274]
Visser, H. D., 7(36), [48]
Vogelfanger, E., 285(119), [297]
Vohwinkel, F., 76, [115]
Volger, H. C., 131(13), 132(16), 133(16, 32), [173, 174]
Vol'kenau, N. A., 157(141), 158(141), [177], 200(70), 221(70), [225], 235(13), [240], 327(16), 335(16), 342(16), [346]
Vol'pin, M. E., 55, 58(3-13), 59(3-8,10-13,61,62), 60(3), 61(3-5,7,10,11, 13), 62(4), 63(4), 64(3,4), 65(3,4, 8), 66(8), 68(4), 69(13,36), 72(3,4), 74(61,62), 76(62), 77(61,62), 78(4, 7,11), 80(7,9), 81(9), 82(12,112), 83(36), 87(13), 100(13), 106(13), [113-115, 117]
Volyeva, V. B., 302(5a), 312(5a), [322]
Voorhoeve, R. J. H., 38, [53]
Vostroknutova, Z. N., 106, [116]
Vostrukhina, Z. N., 318(60), 319(60), [323]
Vranka, R. G., 6(25), [48]
Vyazankin, N. S., 223(116), [226], 242 (9,10), 244(9,10), 245(10), [247], 253(25,27), 254(29,30), [268], 279 (49-51), [295], 334(43), [347]
Vyshinskaya, L. I., 58(9), 69(36), 80(9), 81(9), 83(36), [114], 141(82), 164 (82), [176]

Wade, K., 5(13), 6(13,18), 27(126), [48, 51]
Waga, F., 197(39), 216(39), [225]
Wagler, K., 208(76), 217(76), 218(76), [226]
Walfe, S., 287(152), [298]
Wang, C.-H., 256(58,59), 266(58), [269]

Warburne, S. S., 185(14), [189]
Wartik, T., 7(26), 22(26), 28(139), [48, 51], 198(48), 218(48), [225]
Waters, W. L., 342(101), [348]
Waxmann, B. H., [270]
Wazynski, K., 30(149a), [51]
Wedegaertner, D. K., 328(20), [346]
Wedekind, E., 142(94), 153(94), 165 (94), [176]
Weinberg, N. L., 253(22), 260(22), 262 (22), [268]
Weinhouse, S., 281(91), [296]
Weinshenker, N. M., 265(97), [270]
Weiss, A., 145(117), [176]
Weiss, E., 223(117), [226]
Weiss, K., 16(74), 42(219), [49, 53]
Weiss, R., 70(19), [114]
Weissberger, E., 94, 104(69), [115]
Weitkamp, A. W., 168(175), [178]
Weller, J., 165(160), [178]
Wells, P. R., 293(192,194), [299], 341 (91), [348]
Werner, L., 284(109), [297]
West, B. O., 36(188), 40(188), 41(188), [53]
Westwood, J. V., 192(6), [224]
Wheland, G. W., 8(41), [48]
Whipple, L. D., 328(20), [346]
Whitman, B., 274(11), [274]
Whitmore, F. C., 195(22), 211(22), [224], 255(48), [269], 281, [296], 303(8), 305(8,16), 308(54), 311(8), 313-315(8), 319(8), 320(8,54), [322, 323], 327, 336(52), 340(77, 78), [346-348]
Whittaker, D., 6(22), [48]
Wiberg, E., 6(16), 22, [48, 50], 199(54), 220(54), [225]
Wiberg, v. N., 39(198), [53]
Wilke, G., 19(95), 32(170), 43(170,224, 228), 44(242), 45(228,245), [50, 52, 54]
Willemsens, L. C., 37, [53]
Willgerodt, C., 147, 171(136), 172(135, 136), 173(135,136), [177]
Williams, K. C., 10, 23(112), 27(59), [49, 50]
Wilzbach, K. E., 28(139), [51]
Winstein, S., 131(8), [173], 241(3), 243 (3), [246], 253(16), 261(16), [268],

276(3), 283(103), 284(103), 285 (119,121), 290(179), [294, 297, 299], 330(32), 331(32), 332, 333, 341, [346]
Witt, J., 255(46), [269]
Wittenberg, D., 17(82), 36(82), 40(82), [49], 197(36), 204(36), 215(36), [224]
Wittig, G., 194(16-20), 197(16,35,36), 204(17,36), 208(17), 209(16-20,80), 211(17), 213(80), 215(16,35,36), [224, 226], 267(80), [270], 271, 272(3-5), [274], 327(17), 345(90), [346, 348]
Wojicicki, A., 248(13), [250]
Wolff, P., 284(110), [297]
Wolfsberger, W., 5(11a), [48]
Woods, L. W., 144(110), 168(110), [176]
Woods, W. G., 31(161), [52]
Woodward, F. N., 143(99), 166(99), [176]
Woolcock, W. J. V., 106(81), 111, [116]
Work, R. W., 199(61), 221(61), [225]
Wright, G. F., 253(21,22,24), 254(33), 255(35,40,41), 257(24,40,87,88), 258(35), 260(21,22), 261(33,41), 262(22,33,41), [268-270], 284 (117), 287(148,150,151), 290(184), 292, 293(185), [297-299], 303(10), 306(10), 308(47), 316(10), 319(47), 320(10), [322, 323], 341(83), [348]

Yagupolsky, L. M., 142(89), 143(89), 152(89), 165(89), [176]
Yakubovich, A. Ya., 338(62), [347]
Yamamoto, A., 10(60), 43(227,229-231), 44(233,236,241), [49, 54], 71, 79, 95, 96, 98, 99(73), 108(91), 109, 110, [115, 116]
Yamamoto, O., 7(27), 11(63), 12, [48, 49]
Young, R., 204(77), 208(77), 211(77), 212(84), 214(77), [226]
Young, W. G., 285(121), [297]
Yufa, P. A., 142(89), 143(89), 152(89), 165(89), [176]
Yuki, H., 16(78), 17(78), [49]
Yur'ev, Yu. K., 292(189,190), [299],

328, 329(24-27), 330(28), 344(24), [346]

Yurkevitch, A. M., 145(121), 169(121), [177]

Yusupov, F., 144(111), 169(111), [176], 305(18), [322]

Zakharkin, L. I., 8(42), 9(53), 17(84, 85), 22(106), 23(84,108,109), 24 (106,109,115,117), 27(53,130), 29 (140), 30(148), 31(157), 34(148, 177,181), 35, 36, 38(177,187), 40 (148,181,202), 41(148), 42(223), 47(148), [48-54], 198(49), 218(49), [225], 279(55), 293(204), [295, 299], 339(74), [348]

Zambelli, A., 15(67), 16(67), 17, 18(67, 87,89), 25(67,83), [49, 50]

Zefirov, N. S., 292(189,190), [299],

328, 329(24-27), 330(28), 344(24), [346]

Zeiser, H., 248-250(3), [250]

Zhadina, M. A., 309(66), [324]

Zhil'tsov, S. F., 305(17), [322]

Zhilzov, S. T., 186(38), [189]

Zhitkova, L. A., 201(73), 223(73), [225]

Ziegler, K., 3(6), 14(66), 16(73), 18, 22 (100-103), 23, 24(101,113,114), 26 (100,124), 34, 35(178), [47, 49-52]

Zietz, J. R., 22(104), 30(150), 32(150), [50, 51]

Zimmer, H., 278(14), [294]

Zook, H. D., 195(22), 211(22), [224]

Zorina, T. M., 278(38), [295], 305(20), 312(20), [322]

Zosel, K., 22(102,103), [50]

Zupancic, B. G., 278(31), [294]

Subject Index

Acid, crotonic, dichloromethyl ester-
    trans, 188
    trans, 188
Allen-Senoff complex, 93, 94
Allyl rearrangement, radicals, 286
Aluminum, bis(pentafluorophenyl
    bomide, 34
    bistrimethylamine, 6
    dialkyl hydrides, 7
    diethyl azide, 16
    diethyl bromide, 23
    diethyl t-butoxide, 12, 13
    diethyl chloride, 16, 18, 22, 23, 33,
      36, 41
    diethyl deuteride, 30
    diethyl diethylamide, 14
    diethyl ethoxide, 13, 36, 43, 95, 96
    diethyl ethoxide dimer, 13
    diethyl hydride, 22, 23, 28, 29
    diethyl isopropoxide, 12, 13
    diethyl methoxide trimer, 13
    diethyl phenoxide, 24
    diethylphenyl, 18, 19, 24
    diethyl(phenylethynyl), 8
    diisobutyl hydride, 30, 36
    diisobutylphenyl, 19
    dimethyl, dimethylamide, 7, 32
      ethoxide, 96
    dimethyl azide, 39
    dimethyl bromide, 14, 16, 25
      dimer, 4, 6, 16
    dimethyl t-butoxide dimer, 12
    dimethyl chloride, 3, 7, 14, 16, 24
    dimethyl diphenylamide dimer, 20
    dimethylethyl, 25

dimethyl hydride, 22, 28
dimethyl iodide, 39
dimethyl isopropoxide, 12, 25
dimethyl isopropoxide dimer, 4, 6, 12,
    21
dimethyl methoxide, 7, 13, 25
dimethyl methoxide trimer, 6
dimethyl(α-naphthylethynyl), 19
dimethyl phenoxide, 6
dimethylphenyl, dimer, 7, 9
dimethylphenyl etherate, 19
dimethyl(phenylethynyl), 7, 19
dimethyl thiomethoxide, 7
dimethyl-p-tolyl, 18
μ-diphenylamino-μ-methyl-tetramethyl
    di, 20
diphenyl chloride, 9, 15
diphenyl hydride, 22
diphenyl lithium dihydride, 241
ethoxide, 14
ethyl dibromide, 23, 24
ethyl dichloride, 18, 39
ethyl diethoxide, 13
ethyl dihydride, 23
ethyl sesquibromide, 38
isopropoxide, 13, 14
lithium Tetravinyl, 241
methyl chloride etherate, 24
methyl diazide, 39
methyl dibromide, 16
methyl dichloride, 3, 7, 25
methyl dichloride etherate, 24
methyldiisobutyl, 11
methyldiisobutyl etherate, 11
methyl dimethoxide, 13

methyl diphenyl, 9
methyl diphenyl etherate, 25
methyl iodideazide, 39
methyl sesquibromide, 16, 17
methyl sesquichloride, 15, 17, 38, 39
methyl sesquiiodide, 17
mixed bridges organo compounds, 20
organo compounds, redistribution re-
    actions, 24
  synthesis of, 241
organo halides, halogen bridged
    dimers, 14
organo hydrides, 22
pentafluorophenyl dibromide, 34
phenyl dichloride, 9
phenyl sesquichloride, 36, 40
tetraphenyl lithium, 241
tetravinyl lithium, 243
tribenzyl, 27
tributyl, 10, 27, 28, 29
tri-(n-butyl), dimer, 218
tricyclopropyl, 11
triethyl, 10, 11, 12, 13, 16, 17, 18, 22,
    24, 25, 26, 27, 28, 29, 31, 32, 33,
    34, 38, 42, 97, 218
triethyl etherate, 31
triethyl trimethylamine, 23
trihydride trimethylaminate, 8
triisobutyl, 7, 9, 11, 12, 22, 24, 25,
    26, 36, 38, 46, 78, 80, 81, 96, 97,
    102, 108, 109
tri-(isopentyl), 218, 219
triisopropyl, 42, 218
trimethyl, 3, 8, 9, 10, 11, 12, 13, 14,
    16, 17, 18, 22, 23, 24, 26, 28, 31,
    34, 39, 40, 41, 42, 45, 46, 198,
    218
  dimer, 6
  preparation of, 205
trimethylamino trihydride, 241
trimethyl etherate, 25
tri-(-2-methylpentyl), 219
trimethyl trimethylamine, 23
tri-m-tolyl, 199, 219
tri-o-anisyl, 199
tri-o-tolyl, 199, 219
tris-(o-anisole), 219
triorgano, reactions with halides, 15
tri-p-chlorophenyl, 199, 219
tri-p-fluorophenyl, 199, 219

triperfluorovinyl, trimethylamine,
    241, 243
triphenyl, 9, 18, 22, 24, 25, 26, 28,
    30, 40, 119, 219
triphenyl, preparation of, 205
tripropyl, 10, 30, 218
tri-pseudocumene, 199
tri-(p-tolyl) ethyl etherate, 219
tri-(see butyl), 218
tris-(biphenyl), 219
tris-(α-naphthyl), 220
tris-(tri-methylphenyl), 219
trivinyl, 29, 219
trivinyl trimethylamine, 241, 243
unsaturated organo compounds, 18
vinyl trimethylamino dihydride, 241
Amines, dialkyltrichlorovinyl, 186
  1,1-dichloro-2,2-bis, 186
  trichlorovinyl, 186
Antimony, diphenyl oxide, 168
  diphenyl trichloride, 144, 168
  methylmercuriochlorodimethyl di-
      chloride, 144
  tetraethyl salts, 41
  triethyl, 201
  triethyl dichloride, 41
  trimethyl, 201
  trimethyl dichloride, 144
  triphenyl dichloride, 144
Arsenic, p-anisole dichloride, 167
  p-chlorophenyl dichloride, 167
  2-chlorovinyl dichloride, 166
  diacetonylphenyl, 166
  diethylmethylketo phenyl, 166
  difuranyl chloride, 168
  m-dimethylphenyl dichloride, 167
  diphenyl chloride, 166
  dithiophene chloride, 167
  ethyl 4-dibromoarsenic-s-bromo-2-
      furoate, 167
  ethyl dichloride, 142, 153, 165
  furanyl dichloride, 168
  methyl 4-dibromoarsenic-5-bromo-2-
      furoate, 167
  α-naphthyl dichloride, 167
  α-naphthyl oxide, 167
  organo compounds, preparation of,
      153
  phenyl benzene, di, 253
  phenyl dichloride, 166

phenyl oxide, 166
α-thiophene dichloride, 167
O-tolyl dichloride, 167
p-tolyl dichloride, 167
p-tolylphenyl chloride, 167
trifluorovinyl dichloride, 153, 166
trioxo diphenyletherate, 167
Arsine, aryl dichloride, 143
3 benzothienone-2-dichloro, 168
5-bromothiophene-2-dichloro, 168
β-chlorovinyl dichloride, 143
diaryl chloride, 143
diethyl phenyl, 165
diisobutylchloro, 40
di-(α-naphthyl) chloride, 167
diphenyl chloride, 143, 154
di-p-tolyl chloride, 154
isobutyldichloro, 40
α-naphthyl dichloride, 153
perfluorovinyl dichloride, 142
phenyl, 253
phenyl dichloride, 154
phenyl-p-tolyl chloride, 154
trialkyl, 32
tri-2-chlorovinyl, 166
triethyl, 40
trifuranyl, 168
trifuryl, 143
tri-α-naphthyl, 167
triphenyl, 143, 154, 166, 222
tri-n-propyl, 165
trithiophene, 167
Astatine, diphenyliodo chloride, 172
organo, 148

Benzanilide, 103
Benzene, O-dilithium, 194
Benznorborneol, 261
3,4-Benzodiphenylene, 273
Beryllium, Bis(dimethylamino), 33
diaryl, 197
Diethyl, 215
Dimethyl, 2, 8, 197, 215
Diphenyl, 215
preparation, 204
Di-n-propyl, 215
Di-p-tolyl, 204, 215
Ethyl chloride, 33
Methyl hydride, 28
Biferrocenyl, 274

Bismuth, diphenyl, 169
(diphenyl)α-naphthyl, 169
α-naphthyl dibromide, 144
α-naphthyl diphenyl, 144
O-[tri-ethyl benzoic acidester], 169
trialkyl, 41
triethyl, preparation of, 47
trimethyl, 222
tri-α-naphthyl dibromide, 144
triphenyl, 144, 201, 222, 223
Bond, carbon-mercury, conjugation of, 290
Boric acid, β-chlorovinyl, 148
di-β-chlorovinyl, 148
Bornyl chloride, 280
Boron, α-naphthyl dichloro, 156
aryl di halide, 136
β-naphthyl dichloro, 156
di-β-chlorovinyl chloride, 148
m-dimethylphenyl dichloro, 155
O-dimethylphenyl dichloro, 155
p-dimethylphenyl dichloro, 155
diphenyl bromide, 149, 155
diphenyl chloro, 149, 155
diphenyl chloro dimer, 156
diphenyl hydroxide, 149
O-ethoxyphenyl dihydroxide, 156
p-ethoxyphenyl dichloro, 156
ethyl dichloride, 35
o-methoxyphenyl dichloro, 155, 156
methyl dichloride, 35
organo compounds, preparation of, 148
phenyl bromochloro, 156
phenyl dibromo, 149, 155
phenyl dichloride, 137, 155
phenyl dihydroxide, 155
O-tolyl dichloro, 155
p-tolyl dibromo, 155
trans di(2-chlorovinyl)chloro, 155
trans di(2-chlorovinyl)hydroxide, 155
tribenzyl, 27
tributyl, 27, 28
triethyl, 27, 31
trifluorovinyl dichloro, 155
triisobutyl, 27
trimethyl, 27, 33
1,2,4-trimethylphenyl dibromo, 155
tri-n-butoxyboron, 137
triphenyl, 27

vinyldifluoro, 155
Boroxine, tributyl, 31
Bromotrichloroethylene, 343

Cadmium, bis(phenyl ethylnyl)-, 19
    diethyl, 28
    diisobutyl, 30
        preparation of, 46
    dimethyl, 8, 27, 30
        preparation of, 46
    diphenyl, 198, 218
    ethyl, 217
Camphane, 80, 243, 261
Carbene, bromochloro, 182
    carbon hydrogen bond, insertion of,
        179
    dichloro, 182
    dihalo, 180
        generation of, 179
    insertion, reaction mechanism, 179
    monohalo, generation of, 179
    oxygen-hydrogen bond, insertion of,
        179
Carbide, hexa(chloromercury), 280
Carboranes, alkenyl, 186
    gem-dichlorocyclopropyl, 186
Carvomenthen, 259
Chromium, di-hexamethylbenzene ion,
        43
    dimesitylene, ion, 43
Cis-1-chloro-2-iodoethylene, 326
Cis-hex-3-enol, 257
Cobalt, acetylacetonate, 43
    1,2-bis(Diphenylphosphino)ethane,
        hydrido, complex, 43
    di[di(diphenylphosphine)-σ-ethylene]
        hydride, 97, 107
    diethylbis-(2′,2′-dipyridyl), 43
    [ethyl, tri(triphenylphosphine)] , 96
    [methyl, tri(triphenylphosphine)] , 96
    poly[tris(triphenylphosphine)carbon-
        yl] , complex, 109
    tetrakis(diphenyl-phosphinomethyl)
        methane, mixed hydridoethyl,
        complex, 44
    (triacetylacetonate), 71, 95, 96, 97,
        107, 108
    trihydridephosphine complex, 96
    triphenylphosphine hydride complex,
        with molecular nitrogen, 96

[tri(triphenylphosphine)chloride] , 107
[tri(triphenylphosphine)cyano methyl
    hydride] , 109
[tri(triphenylphosphine)dihydride] ,
    80, 96
[tri(triphenylphosphine)dinitrogen
    complex, 95, 96, 97, 98, 105, 107
[tri(triphenylphosphine)ethylene] , 96
[tri(tri-phenylphosphine)formate] ,
    109
[tri(triphenylphosphine)hydra, nitro-
    gen, complex, 80
tri(triphenylphosphine)nitride, 71
tri(triphenylphosphine)nitrogen com-
    plex, 80, 107
[tri(triphenylphosphine)nitrogen,
    hydride] , 44, 108, 109
tri(triphenylphosphine)trihydride, 80
Complexes, π-cyclopentadienyl, 22
Coupling reactions, organomercury com-
    pounds, 271
Cyclohexan-1β-4α-diox, 259
Cyclohex-3-encarboxylic acid, 259, 260
Cyclohexene, [(2,2-dichloro)cyclo-
    propyl]-4-, 183
Cycoohexyl, ethyl methyl amine, 266
Cyclopentyl(1-phenylethyl) amine, 266
Cyclopentyl phenyl methyl amine, 266
Cyclopropanes, alkenyl, 183
    1,1 dibromo, 187
    1,1 dichloro, 187
    1,1-dichloro-2-(chlorodimethylsilyl)-,
        182
    1,1-dichloro-2-vinyl, 187
    1,1-dihalo-2-tri-methylsilyl, 182
    gem-dihalo, 182
        preparation of, 186
    hexachloro, 182
        preparation of, 187
    mono and gemdihalo, synthesis of, 181

Dehydrobenzene, 84, 86
Demercuration reactions, effect of
    organomercury salt type, 284
    mechanism, 284
Deoxy mercuration, β-elimination, 287
    in quasicomplex compounds, 289
    protolysis, mechanism, 293
    rate, 292
    reaction assistors, 293

Deoxy mercuration reactions, effect of
    organomercury compounds, 292
Devarda's alloy, 89
Diethyl phenyl amine, 266
Dioxane-di-bromide, 335
α,α-Diphenyl-γ-phenylallyl alcohol, 208
Diphenyl, 169, 273
    O,O'-dilithium, 194
    O-lithium, 194
Diphenylene, 272
2,2-Diphenylethyl phenyl ketone, 208
Disproportionation, organoaluminum
    compounds, 17

Electrophilic acidolysis, 285
Elimination reactions, deoxydemercura-
    tion, 275
    deoxymercuration, 286
Entropy activation, protolysis, organo-
    mercury salts, 277
Ether, bis(4-benzylphenyl), 316
    bis-(iodomercuri-2-ethyl), 311
    di-β-iodomercuryethyl, 306
Ethylene, tetrachloro, 182
Ethyl methyl phenyl amine, 266
Ethyl phenyl amine, 266
Exchange reactions, alkali metals with
    tetraorgano aluminates, 23
    alkenyl and aryl mercury derivatives
        with boron halides, 136
    organoaluminums and other metal
        compounds, 26
    organomercury compounds, to acid,
        275
    with metal halides, 135
    trialkylaluminums, and aluminum
        alkoxides, 13
    and dialkylaluminum alkoxides, 12
Exchange reactivity, monomeric tri-
    methylaluminum, 10
Exoepoxynorbornane, 331

Ferrocene, 190, 201, 223, 274
    acetyl, 320
    2-biphenyl, 273
    3-biphenyl, 273
    4-biphenyl, 273
    bromo, 335, 336, 337, 345
    chloromercury, 303, 335, 345
    1,1'-dibromo, 337

1,1-dichloromercury, 274
1,1'-dithiocyanogen, 336
    halo, 339
    iodo, 337
    monsulfo, dimer, 169
    phenyl, 273
    triphenylmethyl, 303, 316
Ferrocenylene, polymercuri, 274

Gallium, cis-tri(1-Propene), 220
    diethyl, 199
    dimethyl(octynyl), 7
    dimethyl(phenylethynyl), 7, 19
    dimethylvinyl, 7
    dipropenyl, 199
    divinyl, 199
    trialkyl, 35
    triethyl, preparation of, 206
    tri(iso-propyl), 220
    trimethyl, 7, 10, 22, 27, 220
    tri(n-propyl), 220
    triphenyl, 7, 199, 220
        preparation of, 206
    tris(1-methylethene), 220
    trivinyl, 220
Germane, dibutyl, 242, 246
    dibutyl acetic acid, 241
    dibutyl acetic acid methyl ester, 246
    dipropyl acetic acid methyl ester, 246
    hexaethyl, di, 38
    isobutyl, poly, 38
    tetraethyl, 38
    tetraisobutyl, 38
Germanium, alkyl, carbene insertion,
    179
    aryl trichloride, 140
    dibenzyloxo di, oxide, 161
    di-β-naphthyl oxide, 234
    dibutyl bis acetic acid methyl ester,
        241, 246
    dibutyl di(-α-acetic acid methyl ester),
        244
    dibutyl iodide, dimer, 233
    diethyl, 227
    di-m-chlorophenyl diiodide, 233
    di-m-tolyl diiodide, 233
    di-(n-butyl)bisacetic acid methyl ester,
        hydride, 242
    di(n-butyl) chloro α acetic methyl
        ester, 253

di-(*n*-butyl) chloro hydride, 253
di(*n*-butyl) methyl acetic acid ester
 hydride, 160
di-*n*-propyl α-acetic acid methyl ester,
 chloride, 244, 253
di(*n*-propyl) chloro hydride, 253
di(*n*-propyl) di(methylacetic acid
 ester), 160
di(*n*-propyl) methylacetic acid ester
 hydride, 160, 242, 244
di-*o*-bromophenyl diiodide, 233
di-*o*-ethoxyphenyl diiodide, 234
di-*o*-tolyl diiodide, 233
di-*p*-anisyl iodide, 234
di-*p*-anisyl oxide, 234
di-*p*-bromo diiodide, 234
di-*p*-bromophenyl dioxide, 234
di-*p*-bromophenyl oxide, 233
di(*p*-bromophenyl oxo) oxide, 233
di-*p*-chlorophenyl diiodide, 233
di-(*p*-tolyl metal oxo) oxide, 231
(di-*p*-tolyl) oxide, dimer, 231, 233
di-*p*-tolyloxo dioxide, 161
(di-*p*-tolyl oxo) oxide, 233
diphenyl diiodide, 227, 228, 233
diphenyloxo di, oxide, 161
dipropyl diacetic acid, 140
dipropyl iodide, 140
m-tolyl triiodide, 233
m-tritolyl, di oxide, 161
organo compounds, synthesis of, 227,
 231, 242
p-bromophenyl triiodide, 233
p-chlorophenyl triiodide, 233
p-ethoxyphenyl diiodide, 234
phenyl diiodide, 228
trialkyl acetic acid esters, 140
trialkyl acetic acid methyl ester, 242
trialkyl iodide, 140
tri-(β-naphthyl) iodide, 228, 234
triethyl, 30
triethyl hydride, 253
tri-*m*-tolyl, iodide, 233
tri-*n*-butyl α-acetic acid propyl ester,
 244
tri-(*n*-butyl) ethyl acetic acid ester,
 161
tri(*n*-butyl) methyl acetic acid ester,
 160
tri(*n*-butyl) propyl acetic acid

 ester, 161
tri(*n*-propyl) ethyl acetic acid ester,
 160
tri-*m*-tolyl, iodide, 228
tri(*o*-anisyl) iodide, 228
tri(*o*-bromophenyl) iodide, 228
tri-*o*-tolyl, iodide, 228
tri-*p*-bromo iodide, 234
tri-(*p*-bromophenyl) iodide, 228, 233,
 234
tri-*p*-chlorophenyl iodide, 228, 233
tri-*p*-ethoxyphenyl iodide, 234
tri(*p*-ethylphenyl), iodide, 228
triphenyl dichloromethyl, 181
triphenyl iodide, 233
tripropyl (α-acetic acid propyl ester),
 244
tri-*p*-tolyl, iodide, 228
vinyl trichloride, 139, 161

Hammet's equation, rho(ℓ) value, 277
 sigma value, 277
Heptane, 7-bromo-7-chlorobicyclo-
 (4,1,0)- 179, 184
 1,1-dichlorobicyclo-(4,1,0)-, 183
 7-7-dichlorobicyclo-(4,1,0)-, 184
Hexabromocyclopentadiene, 338
α-Hydrogen sulfite acetic acid, 169
Hydrogen bonding, α-mercurated oxo
 compounds, 132
6-Hydroxbicyclo[2,2,1] heptane-2-
 carboxylic acid lactone, 261
5-Hydroxy-1,4-methylene cyclohexane-
 2-carboxylic acid, 262
5 Hydroxy-1,4-methylene cyclohexane-
 2-3, dicarboxylic acid, 262
6-Hydroxy-1,4-methylenecyclohexane-
 2-carboxylic acid γ-lactone, 261,
 262

Indium, dimethyl(phenylethynyl), 19
 diphenyl chloride, 137, 156
 phenyloxo diindiumtrioxide, 221
 trialkyl, 35
 trimethyl, 10, 27, 199, 221
 triphenyl, 199, 221
Insertion reaction, O-H bond, dichloromethylene of, 183
 organo mercury compounds, methylene of, 179

titanium-hydrogen bond, nitrogen of, 81
titanium-phenyl bond, nitrogen of, 82
Iodine, β-chlorovinyl chloride, 147
2-chlorovinyl dichloro, 171
β-naphthyl phenyl, chloride, 172
O-tolyl phenyl chloride, 172
phenyl oxide, 172
p-tolyl phenyl chloride, 172
Iodo-fumaric acid ester, 341
Iodonium, di-β-chlorovinyl chloride, 147
diphenyl chloride, 154, 172, 173
phenyl chloride, 173
Ion, phenonium, 277
π-complex"mercurinium," 290
Iridium, [bis(triphenylphosphine) chloro, nitrogen], 109
[di(trialkylphosphine) carbonyl] complex, 106
di(triphenylphosphine) carbonyl chloride, 91
[di(triphenylphosphine) dinitrogen chloride], 105
[di(triphenylphosphine), dinitrogen] complex, 106
Iron, acetylacetonate, 43
1,2-bis(diphenylphosphine)ethane, -π-ethylene complex, 43
π-cyclopentadienyl dicarbonyl aryl-σ, 148
π-cyclopentadienyl dicarbonyl iodide, 148
π-cyclopentadienyl σ-ethyl benzoic acid ester dicarbonyl, 172
π-cyclopentadienyl σ-phenyl dicarbonyl, 172
diethyl bis(2,2'-dipyridyl), 43
hydroxychloride vinylate, 148
1-oxovinyl hydroxy chloride, 172
triacetylacetonate, 109
[tris(diphenylethylphosphine) carbonyl dihydride], 110
[tris(trialkylphosphine) nitrogen, dihydride], 110
Isotope, primary solvent effect, 277

Kharasch series, radicals in electronegative series, 281

Lead, α-naphthyl triacetate, 151, 163

α-naphthyl triisobutyrate, 151
α-naphthyl tripropionate, 163
aryl tetraacylate, 141
di-α-naphtyyl diacetate, 151
diaryl diacetate, 140
di-β-chlorovinyl diacetate, 150, 162
di-β-naphthyl diacetate, 163
diethyl diacetate, 161
di(ethylphenylether) diacetate, 163
dimethyldiethyl, 8
di-o-anisyl diacetate, 151
di(o-methoxyphenyl) diacetate, 163
di(p-ethylbenzoic acid ester) diacetate, 163
diphenyl diacetate, 162
diphenyl dichloride, 37
di-(trans-β-chloro-vinyl), 141
methyltriethyl, 8
phenyl triacetate, 162
phenyl triisobutyrate, 162
p-iosophenyl triacetate, 163
p-iodophenyl tribenzoate, 163
p-iodophenyl triisobutyrate, 163
p-iodophenyl tripropionate, 163
p-methoxyphenyl triacetate, 163
p-tolyl triacetate, 162
tetracarboxylates, 30, 37
tetraethyl, 8, 30
tetraisobutyrate, 151
tetramethyl, 8
tetraphenyl, 28, 37
trimethylethyl, 8
triphenyl chloride, 37
Lithium, alkyls, preparation of, 201
aluminum hydride, 28
benzyl, 209
2-butylacetylene, 209
ethyl, 202, 208, 223
isoamyl, 202
iso-pentyl, 208
methyl, 208
m-toly, 84, 85
napthalene, 70, 104
n-butyl, 202, 208
n-heptyl, 202, 208
n,n-dimethylaniline, 209
n-propyl, 208
o-diphenyl, 86, 209
o-phenyl, di, 209
organo compounds, carbanion

donors, 22
o-tolyl, 84, 85
p-tolyl, 84, 85, 209
   preparation of, 204
phenyl, 69, 102, 209
tetraethyl aluminate, 40
tetraethynylaluminate, 29
tetramethylaluminate, 23
tetra(m-tolyl), boron, 209
tetra-(o-tolyl), boron, 209
tetra(p-tolyl), boron, 209
tolyl, 86
vinyl, 195, 203
vinyloxy, 208
Lithium sodium dialkyl, complex, 196

Magnesium, -benzyl bromide, 24
-dialkyl ethyletherate, 197
-diethyl, 29, 197, 215, 216
   preparation, 204
-di(iso-phenyl), 216
-di(Isopropyl), 216
-dimethyl, 215
-di(n-butyl), 216
-di(n-propyl), 216
-diphenyl, 197, 216
-ethyl bromide, 22, 71, 78, 93, 101
-ethyl chloride, 72
-methyl iodide, 23
-phenyl bromide, 24, 93
Manganese, -(chloromercurophenyl) tricarbonyl, 338
-N,N-salicyladehyde-1,3-propane di-
   imino, nitrogen complex, 99
-N,N-salicylaldehyde-1,3-propane di-
   imino, 99
Mechanism, -SE[1], 276
   -SE[2], 276
Mercurated Polystyrene, 339
α-Mercurybromo-α Phenylacetic acid
   ethyl ester, 332
l-Mercurichloro-1-methyl carboxylate-2-
   acetic acid ethylene, 341
3-Mercurichloro-3-triphenyl phosphonium acetophenone chloride, 319
1-Mercuriodo 2,3, Dibenzoic acid n-
   propyl ester, 318
β-Mercuriodoethylbenzoic acid ester,
   318
Mercuri-β-Phenylanhydrohydroacrylic

acid, 280
Mercury, -1-acetic acid methyl ester
   iodide, 159
-2-acetoxy diphenylene, 344
-acid ester halide, 275
-alkenyl halide, 284
-4-alkoxy-n-butyl halide, 281
-alkyl amino derivative, 279
-alkyl perchlorates, 285
-allyl iodide, 277, 293, 303, 312, 321
-α,α'-bis-chloro dibornyl ether, 280
-α-bromo-cyclohexanone, 131
-α-bromophenyl acetic acid, 302
-α-bromo phenylacetic acid ester, 332
-α-bromo-phenylacetic acid ethyl ester,
   131
-α-bromophenyl acetic acid, 1-methyl
   ether, 131, 191
-α-chlorocyclopentanone, 321
-α-chlorodiethyl ketone, 321
-α-dinaphthyl, 316
-α-furyl salt, 143
-α-halogen oxo, 294
-α-methoxycyclohexyl trifluoroacetate, 293
-α-2-methoxy-1-iodo cyclohexanes,
   291
-α-methylvinyl bromide, 235
-α-naphthyl bromide, 316
-α-naphtyl chloride, 159, 316, 318
-α-naphthyl iodid, 316
-α-oxo-compounds of, 131, 338
-α-phenyl-α-bromomercuriacetic acid
   ethyl ester, 312, 334
-α-thienyl, 143
-α-thienyl chloride, 159, 192
-ammonim salts, 287
-anthraquinone derivatives, 278
-arylsulfenyl chloride, 306
-benzene(trichloromethyl), 283
-3,4-benzodiphenylene, 273
-benzyl acetate, 162
-benzyl bromide, 131, 236, 331, 332,
   334
-benzyl butyl, 316
-benzyl chloride, 158, 235, 279, 316,
   317, 331, 333
-benzyl cyclohexyl, 317
-benzyl phenyl, 317
-4β-acetoxy-2β-acetoxymetalo-1-α-

methoxcyclohexane, 259
-β-alkoxyethyl chloride, 290
-β-alkoxyalkyl salts, 292
-β-alkoxyethyl salt, 286
-β-arylethyl acetates, 286
-β-benzoateethyl chloride, 308
-4β-benzyloxy-2β-chlorometalo-1-α-
    methoxycyclohexane, 259
-2β-chlorometalo-4-βcyan-1-α-
    methoxycyclohexane, 259
-β-chloro metal ethylamines, 293
-β-chloro metalo furan, 320
-3β-chlorometal-4α-methoxycyclo-
    hexan-1β-ol, 257
-β-chlorovinyl, 136, 200, 289
-β-chlorovinyl acetate, 151, 162
-β-chlorovinyl chloride, 276, 288, 293,
    332, 333, 341
-β-chlorovinyl halides, 131
-β-chlorovinyl iodide, 143
-β-ethanol chloride, 254
-β-hydroxyethyl chloride, 288, 306,
    308
-(β-hydroxyethyl) iodide, 311, 318
-β-2-methoxy-1-iodocyclohexanes, 291
-β-naphthyl acetate, 163
-β-naphthyl bromide, 170
-β-naphthyl propionate, 163
-β-oxyalkyl salts, 292
-β-phenylvinyl bromide, 332
-3-bromo-3-benzylcamphor, 131
-bisacetaldehyde, 195, 196
-bis-α-acetic acid methyl ester, 231,
    242, 246, 253
-bisacetic acid isobutyl ester, 310
-bisacetic acid propyl ester, 246
-bisacetoacetic acid, 309
-bis (α-naphthyl), 256
-bis (α-stibene), 342
-bis arylacetylene, 283
-bis (β-naphthyl), 256
-bis (bromoacetylenyl), 256
-2,5 bis(bromometalomethyl)tetra-
    hydrofuran, 265
-bis-(carbomethoxymethyl), 150
-bis (1-carboxylicstyrene), 257
-bis(cis-stibene), 280
-bis(cis-α-stibenyl), 229, 256
-bis-[2-(2,3-dihydrobenzofuranyl)
    methyl], 265

-bis[(diisopropyl)metal] oxide, 255
-bis(hydrazinetetrafluorophenyl), 255
-bis(m-tolyl), 317
-bis(o-Bromotetrabromophenyl), 342
-bis(o-iodophenyl), 256
-bis-o-oxyphenyl, 256
-bis(o-tolyl), 256
-bis-p-benzoic acid, 306
-bis(p-diphenyl), 280
-bis(p-phenylacetic acid ester), 315
-bis(p-tetrafluorophenol), 256
-bis(p-tolyl), 256
-bis(pentafluorophenyl), 255, 339
-bis perfluoroaryl, 279
-bis(phenoxyallyl), 315
-bis(phenylbarenyl), 279, 339
-bis-perchlorovinyl, 339, 343
-bis-perfluoroethyl, 339
-bis(perfluoroisopropyl), 344
-bis(perfluoro-tert-butyl), 339
-bis(trans-α-chlorostilbenyl), 230
-bis(trans-stibene), 280
-bromo acetaldehyde, 311
-3-bromo-camphor, 131
-3-butenyl halide, 284
-butyl chloride, 158, 316, 317
-butyl ethyl, 316
-4-camphenyl iodide, 330, 331, 333
-3-camphor bromide, 192
-4-camphyl chloride, 241
-carbides, 145
-2-carbmethoxy-1-butene-3-chloride,
    157
-2-carbmethoxy-1-methyl-1-propene
    chloride, 157
-carbomethoxyphenyl, 137
-carbmethoxyphenyl bromide, 158
-chloro acetaldehyde, 145, 148, 302, 308,
    311, 317, 318, 320, 340
-chloro acetone, 311
-chloro acetophenone, 312
-chloro β-chlorovinyl, 312
-2-chloro-3-chloro-metalo-2-butene-1,4
    diol, 329
-2-chloro cyclohexanone, 319
-3 chloro-2-ethyl tetrahydrofuran, 262
-chloroferrocene, 337
-chloro formic acid ethyl ester, 306
-α-chloro furan, 262
-3-chloro furan, 322

-chloro ketone, 302
-chlorometalmethyl-1-methoxycyclo-
    hexan, 257
-chloromethyl chloride, 289
-chloro propionate, 311
-2-chloro thiophen, 262
-2-chlorovinyl chloride, 157, 325
-cinnamyl salt, 285
-*cis*-β-chlorovinyl, 191
-*cis*-β-chlorovinyl bromide, 331
-*cis*-β chlorovinyl chloride, 280
-*cis*-β-stilbene chloride, 200
-*cis,cis*-di-β-chlorovinyl, 130, 326
-*cis*,1,2-diphenylvinyl chloride, 157,
    158
-*cis*-(exo-3-hydroxy-exo-2-
    norcamphanyl), 261
-*cis*-2-methoxycyclohexylneophenyl,
    131
-*cis*-1-methyl-2-acetoxy-1-propene-1-
    yl, 138
-*cis*-propene chloride, 157
-*cis*-dipropenyl, 138
-crotyl salt, 285
-cyclic β- mercurated alcohols, 330
-cyclohexyl butyl, 317
-cyclohexyl chloride, 317
-cyclohexyl tosylate, 284
-cyclopentadienyl chloride, 201
-decalin iodide, 332
-diacetic acid methyl ester, 140
-diacetoxy -2D-indole, 266
-dialkyl, reaction of, 195
-diallyl, 279
-di -α-furyl, 262
-di-(α-methylvinyl), 230
-di-(α-naphthyl), 151, 153, 169, 229,
    230
-diaryl, 136, 145
-dibenzyl, 198, 248, 251, 253, 271,
    276, 293
-di-(β-chlorovinyl), 138, 229, 289, 325
-di(β-fluoroiodovinyl), 338
-di-β-naphthyl diacetate, 163
-di-β-phenylethyl, 251
-di-butyl, 8; 135
-dichloro ferrocene, 336, 337
-(dichloromethyl)bromide, 338
-di(2-chlorovinyl)iodochloro,
    dichloride, 171

-di-(*cis*-β-chlorovinyl), 232
-dicyano, 311
-di(cyclohexyl), 317
-di-cyclopentadienyl, 139, 141, 201,
    338
-dicyclopropyl, 279
-diethyl, 27, 29, 34, 142, 153, 154,
    202, 206, 227, 229, 230, 232, 278,
    305, 337
-di(ethylmethylene ketone), 253
-diethyl phosphates, 278
-diferrocenyl, 145, 146, 256, 273, 305,
    336
-di(2,2,2-fluoroiodo bromo-2-
    bromoethyl), 338
-difuryl, 253
-di(heptafluoroisopropyl), 255
-di(3-hyroxy-3-phenylpropyl), 252
-2,3-dihydroxypropane iodide, 318
-diisobutyl, 199
-diisopropenyl, 230
-diisopropyl, 198, 305
-di-*l*-(-)sec-butyl, 280
-dimeric *o*-terphenylene, 345
-di(*m*-chlorophenyl), 152
-dimethyl, 10, 28, 33, 144, 197, 205,
    305, 337
-di(2-methylacetylene), 8
-di(2-methylpentyl), 199
-di-naphthyl, 267
-di-(*n*-butyl), 151, 227, 303
-di-*n*-propyl, 137
-di(*N,N*-dimethylaniline), 316
-di-*O*-acetoxyphenyl, 138
-di-*O*-anisyl, 151
-di[*O*-bromotetrafluorophenyl], 248
-di-*O*-carboxyphenyl, 144
-di-*O*-tolyl, 76, 248, 250
-di-*p*-aminophenyl, 256
-di-*p*-anisyl, 267
-di-*p*-bromophenyl, 198
-di-*p*-carbethoxyphenyl, 198
-di-*p*-chlorophenyl, 205, 232
-di-*p*-iodophenyl, 198
-di-*p*-tolyl, 154, 204, 231, 265, 303,
    305, 314, 315, 317
-di-(pentalfluorophenyl), 200, 262,
    342
-di-pentyl, 196
-diperfluoroisoproyl, 248

-diperfluorovinyl, 241
-diphenothio, 249
-diphenyl, 8, 27, 28, 136, 137, 140,
  142, 149, 154, 173, 185, 186, 200,
  201, 205, 213, 218, 222, 223, 230,
  239, 251, 252, 253, 265, 278, 286,
  303, 304, 305, 307, 312
-di-phenylacetylene, 277
-diphenyl bromide, 170
-diphenyl etherate, preparation, 205
-diphenyliodo dichloride, 173
-diphenyl iodochloro dichloride, 171
-dipropenyl, 230
-dipropionic acid, 280
-dipropyl, 30
-di-sec-butyl, 342
-di(1,2,2,2-Tetrafluoroethyl), 251,
  252
-di-(tetrafluorophenyl hydrazine), 262
-ditolyl, 196
-di(trans-β-chlorovinyl), 150, 304
-di-trans-4-methyl cyclohexyl, 280
-di(triethylgermanium), 253
-di[trimethyl(dichlorosilyl)], 186
-divinyl, 8, 136, 139, 140, 142, 152,
  196, 203, 241, 285
-di(vinylethynyl), 343
-endo-norbornyl, 285
-ethanehexamercarbide, 257
-ethaneoxyhexamercarbide, 255
-ethyl acetate, 161
-ethyl α-2-(acetoxymethalethyl)
  acetaacetate, 286
-ethyl bromide, 234
-ethyl chloride, 153, 158, 161, 166,
  168, 171, 229, 234, 254, 317
-ethyl fluoride, 171
-ethyl iodide, 311
-ethyl perchlorate, 285
-exo-4-acetoxy-5-chlormercury-3,6
  endoxy-hexahydrophthalic acid
  dimethylester, 328
-ethyl(triethylsilyl), 253
-exo-cis-4-oxy-5-chloromercury-3,6
  endoxohexa-hydrophthalic acid
  dimethyl ester, 344
-exo-4-oxy-5-chlor-metal-3,6
  endoxyhexahydrophthalic acid
  dimethylester, 328
-ferrocenyl chloride, 192

-2-furyl chloride, 253, 283
-3-furyl chloride, 143, 283
-γ-alkoxy-n-propyl halide, 288
-hexachloro ethane, 145
-2-hydroxy-5-isopropyl, chloride, 258
-2-hydroxy-5-methyl chloride, 258
-iodo-(carbomethoxymethyl), 150
-isopentyl chloride, 156
-isopropyl chloride, 305
-mercarbides, 280
-m-tolyl chloride, 318
-3-methoxy-2-bromomercuri-3-phenyl
  propionic acid, 328
-M-phenylene, polymer, 272
-M-tolyl iodide, 161
-methyl acetate, 254
-methyl chloride, 144, 161, 164, 230
-methylester of bis acetic acid, 239
-methyl(phenylbarenyl), 279
-methylpentafluorophenyl, 34
-2,2-methylphenylpropyl acetate, 162
-1-methylvinyl bromide, 157
-neopentyl chloride, 328
-norbornene oxymercuration, 330
-n-butyl acetate, 161
-n-butyl chloride, 152
-n-butyl iodide, 331, 332
-N-cyclohexyl N-methyl N-2-
  ethylamino chloride, 266
-ω-chlorometal acetophenone, 302
-ω-styryl bromide, 326, 334
-N-phenyl N-ethyl N-2 ethylamino
  chloride, 266
-organo compounds, reaction of, 194
  exchange reaction mechanism, 190
  exchange reactions with metals and
    metal alloys, 190
  exchange reactions stereochemistry,
    190
  methylene insertion, 179
  reaction mechanism, 130
-organo halides, preparative syntheses,
  342
-O-diphenylene, 272
-O-diphenylene tetramer, 194
-O-iodophenyl iodide, 267
-O-methoxyphenyl diacetate, 163
-O-nitrophenyl, 143
-(O-phenol) chloride, 320
-(O-phenoxysodium) chloride, 315, 316

-*O*-phenylene, 137
  hexamer, 194
-*O*-polyphenylenes, 271
-*O*-terphenylene, 272
-*O*-terphenylene dimer, 194
-*O*-tolyl chloride, 318
-*O*-tolyl iodide, 314
-(*p*-anisylethyl) acetic acid ethylester,
  286
-*p*-anisyl phenyl, 335
-(*p*-bromophenyl) phenyl, 335
-*p*-chloro benzoic acid, 137
-*p*-chloro phenol, 137
-(*p*-*N,N*-dimethylaniline) phenyl, 335
-*p*-ethybenzoic acid ester acetate, 163
-*p*-ethylbenzoic acid ester, bromide, 238
-*p*-methoxy acetate, 163
-*p*-nitrobenzyl bromide, 134, 135
-(*p*-nitrophenyl) phenyl, 335
-*p*-phenoxyphenyl chloride, 303, 320
-(*p*-phenylacetic acid ester) chloride,
  315
-*p*-tolyl acetate, 162
-*p*-tolyl bromide, 170, 314
-*p*-tolyl chloride, 170, 207, 230, 317
-*p*-tolyl iodide, 314
-*p*-tolyl O-tolyl, 317
-pentachlorophenyl methyl, 282
-pentachlorophenyl phenyl, 283
-pentyl acetate, 161
-perfluoroalkyl compounds, 339
-(phenoxyallyl) chloride, 315
-phenyl acetate, 162, 274, 286
-phenyl chloride, 141
-phenyl hydroxide, 267
-2-phenyl-5-acetoxymercury-1,3,4
  oxadiazole, 341
-phenyl[alkyl(bromohalomethyl)] ,
  185
-phenyl allyl, 338
phenyl bromide, 144, 169, 170, 181,
  182, 187, 188, 192, 280, 293,
  303, 313, 332, 333
-phenyl bromo sodium iodide,
  complex, 280
-phenyl chloride, 149, 154, 162, 163,
  164, 166, 169, 172, 173, 186, 201,
  239, 243, 253, 283, 289, 305, 313,
  317, 319, 338
-phenyl cyanide, 336

-phenyl (dibromochloromethyl), 181,
  182
-phenyl (dibromomethyl), 180, 182
-phenyl (dichlorobromomethyl), 179,
  181, 182, 183, 186, 187, 188, 283
-phenyl (dichloromethyl), 338
-phenyl dihalomethyl, 293
-phenyl ethyl, 278
-2-phenyl ethyl acetate, 162
-phenyl fluoride, 308, 319
-phenyl iodide, 228, 233, 305, 313,
  314
-phenyl isobutyrate, 162
-2-phenyl-2-methoxyethyl iodide, 290
-phenyl(tribromomethyl), 181, 186,
  187, 283
-phenyl(trichloromethyl), 181, 185,
  186
-phenyl(trihalomethyl), 179, 180, 293
-phenyl triiodomethyl, 305
-2-phenylvinyl bromide, 157
-sec-butyl bromide, 303, 328, 331
-piperidine-*N*-ethyl-2 chloride, 254
-2-seleniumphenyl chloride, 283
-tetracetate pyrrole, 256, 266
-2-thienyl chloride, 253, 283
-tolyl chloride, 192, 320
-*trans*-α-Bisstibenyl, 327
-*trans*-α-stilbenyl bromide, 327
-*trans*-β-chlorovinyl, 191, 207
-*trans*-β-chlorovinyl bromide, 331
-*trans*-β-chlorovinyl chloride, 342
-*trans*-β-stilbene chloride, 200
-1-*trans*-2'-bromovinylbenzene
  derivative, 343
-*trans*-dipropenyl, 138
-*trans*-1-methyl-2-acetoxy-1-propen-1-
  yl, 138
-*trans*-4-methylcyclohexyl bromide,
  328
-*trans*-ω-styryl bromide, 327, 332
-*trans, trans*-β-Chlorovinyl, 326
-*trans*-*trans*-Di-β-chlorovinyl, 148
-2,4,5-tri(acetoxymercury)-3-
  nitrothiophene, 344
-trichloro acetaldehyde, 281
-tri(chloromet-al) acetone, 281
-trifluoromethyl iodide, 255
-trifluorovinyl chloride, 166
-trifluorovinyl ethyl, 153

-trifluorovinyl vinyl, 153
-trimethylphenyl chloride, 159
-(trimethylphenyl) phenyl, 317, 335
-1,1,1-trinitropropyl bromide, 342
-4-trinitro-N-butyl chloride, 293
-vinyl chloride, 153, 157, 164, 166
-vinyl iodide, 277
-vinyl halide, 284
Metadibromobenzene, 272
Metalhalogenides, 22
Metanoylphenyl, 273
Metaterphenyl, 273
3-Methoxy-2-Bromo-3-phenylpropionic
    acid, 328
4-α-Methoxycyclohexanecarboxylic acid,
    260
4-α-Methoxycyclohexane-1βcarboxylic
    acid, 259
4-α-Methoxycyclohexan-1β-ol, 259
4-α-Methoxycyclohexyl-1βcarbinol, 259
4-α-Methoxy-1β-cyclohexyl methanol,
    259
4-Methoxyhexan-1-ol, 257
2-Methoxy-1,4-methylencyclohexane,
    261
4-Methyl[2,2,2] Bicyclo-8-oxaoctane, 262
Methylene, -carbon-hydrogen bond in-
    sertion of, 180
-dichloro, carbon-hydrogen bond in-
    sertion of, 185
-dihalo, germanium hydrogen bond
    insertion of, 185
    nitrogen hydrogen bond insertion
    of, 186
    silicon hydrogen bond insertion of,
    186
    germanium, hydrogen bond, insertion
    of, 181
-olefinic bond additon, 181
-silicon, hydrogen bond, insertion of,
    180
-1,4methylenecyclohexane-2-
    carboxylic acid, 261
Methyl vinyl sulfoxide, 320
Molybdenum, -cyclopentadienyl di-
    carbonyl phenylazo derivative, 83
-cyclopentadienyl tricarbonyl anion,
    83

Nickel, acetylacetonate, 43, 45

bis(1,5-cyclooctadiene), 45
bis[di(trialkylphosphine)nickel]
    nitrogen, 111
bis(tricyclohexylphosphine)(Nitrogen),
    44
bis(tri-O-phenylphenoxyphosphine)-π-
    ethylene, 44
bistrityl, 45
diacetylacetonate, 111
diethyl(2,2'-dipyridyl)-, 43
dimethyl, 45
[di(trihexylphosphine)nitrogen],
    complex, 111
Nitrogen, activation by transition metal
    compounds, 57
bond energies, 56
chemical activation, 57
chemical fixation, 55
chemisorption, 87
complexes, π-back bonding, 113
fixation, 85
    activity, solvent effect, 64, 68
    catalytic, 73
    effect of Grignard reagent, 62, 63
    effect of reagents, 24
    experimental procedures, 101
    hydrogen effect, 78, 79, 80
    inhibitors effect, 65
    mechanism of reaction, 70
    systems, 59, 60, 61
    titanium derivatives and organo-
    metallic compounds, 61
hydrogenation, activation energy of,
    56
N-N stretching vibration, 87
physical activation, 57
reduction, 58, 69
surface complex with metals, 88
p-Nitrostilbenes, 310
Norborneol, 261, 265, 285
Norcarane, 185
    dichloro, 184, 186
    7,7-dichloro, preparation of, 188
Nucleophilic Acidolysis, 285

O-aminodiphenyl, 103
O-benzoylaminodiphenyl, 103
Olefins, 1-halo, preparation of, 185
Osmium, complex with molecular
    nitrogen, 89

[nitrogen pentaamino silver] [3+], 111
pentaamino dinitrogen borofluoride, 90
pentaamino dinitrogen chlorate, 90
[pentaamino dinitrogen] complex, 105, 106
pentaamino dinitrogen dihalide, 90
pentaamino nitrogen complex, 111
[tetraamino tetranitrogen] complex, 111
[tris(Di-n-butylphenylphosphine) dinitrogen dichloride], 111
6-Oxa-tricyclo[3.2.1.1.]-nonane, 261

Pentabromocyclopentane, 338
Pentafluoroiodobenzene, 389
1-Phenyl-1-chloro-2-mercurichlorbuten-3-one, 309
Phenylethnylmethylketone, 289
Phenylmercuri-1-yne-3-butene, 338
Phenylvinylsulfoxide, 320
Phosphines, alkydichloro, 40, 141
  alkyldichlorosulfide, 40
  anisolephenylchloro, 165
  benzyldichloro, 164
  β-naphthyldichloride, 165
  dialkylchlorosulfide, 40
  [2,4-dimethylphenyl] dichloro, 165
  di(N,N-dimethyl)chlorosulfide, 40
  di(N,N-dimethyl)ethylsulfide, 40
  diphenylchloride, 142, 153, 165
  ethyldichloro, 164
  isopentyldichloro, 164
  isopropyldichloro, 164
  m-chlorophenyldichloride, 152, 165
  m-tolyldichloro, 165
  naphthyldichloro, 165
  n-butyldichloride, 151, 164
  n-propyldichloro, 164
  o-Tolydichloro, 165
  p-bromophenylphenylchloro, 165
  phenyl, 253
  phenyldibromo, 165
  phenyldichloride, 142, 153, 164
  phenyltolylchloro, 165
  p-N,N-dimethylanilinedichloro, 165
  p-tolyldichloro, 165
  p-tolylphenylchloride, 153
  tributyl, 286
  tributyloxide, 286

  trimethylphenyldichloro, 165
  trimethylsilanemethylenedichloro, 164
  trimethylsulfide, 40
  tri-n-propyl, 164
  triphenyl, 40, 71, 164, 185
  triphenylsulfide, 40
  trivinyloxo, 164
  vinyldibromo, 164
  vinyldichloride, 152, 164
  vinyldihalide, 142
Phosphite, triethyl, 185
  trimethyl, 185
Phosphorous, naphthyloxyacid, 165
  organo compounds, preparation of, 151
  tetraphenyl, 253
Photodecomposition, organic mercury compounds, 271
Platinum, di(triethylphosphine) chlorohydride, 83
  di(triethylphosphine)chlorophenyl-azoderivative, 83
Polyferrocenylene, 274
Potassium, benzyl, 195
  methyl, 223
  vinyl, 196, 203
Proportionation reactions, 15
Protodemercuration, acid type effect, 278
  allyl mercury compounds, 293
  effect of organomercury compound type, 279
  rate, alkyl series, 282
  reaction mechanism, 276
Pyridine bromide, 335

Radicals, activation energy reaction, 337
  phenylmercury, 307
Reactions, amineformation by nitrogen, 81
  demercuration, 275, 284
  deoxymercurations, mechanism, 290
  dibutylgermane with mercury bisacetic acid methyl ester, 242
  halodemercuration, kinetics and mechanism of, 331
    stereochemistry of, 325
  mechanism, carbonium ion type, 284
  new C-C bond formation, 301
  Organomercury, with elements of the

sulfur group, 248
  with organo tin compounds, 231
Organomercury compounds with
    acylhalides, 307
  with alkyl and acid halides, 301
  with arylhydrides of metals, 241
  with halogens, 325
  with hydrides, 240
  with elements of group VIII, 201
  with low valence compounds of
    metals, 227
  with metal alloys of group V, 201
  with metal alryls, 241
  with reducing agents, 251
  protodemercuration, 275
  redistribution, mechanism, 4
  trans-β-chlorovinyl mercury chloride
    with tin sodium alloy, 207
Rhodium, [di(trialkylphosphine)nitro-
    gen] complex, 106
  di(triphenylphosphine)carbonyl halide,
    106
  di(triphenylphosphine)dinitrogen
    chloride, 92
  hydridotetrakis(triphenylphosphine),
    44
  tetra(triphenylphosphine)mono
    hydride, 99
  triacetylacetonate, 99
Ruthenium, ammonium hexachloro, 103
  [aqua Di(ethyleneamine)nitrogen]
    [Di(tetraborophenylate)], 112
  [aqua pentaamino] di-cation, 95
  aquo Pentaamine complex, 94
  aquopentaamine methansulfonate, 103
  [aquopentachloro] di-cation, 88
  binuclear complex, 95
  diammonium hexachloride, 88
  [diaqua dichloro nitrogen]-tetra-
    hydrofuran complex, 106
  [diethyleneammine aquo dinitrogen]
    complex, 112
  [diethyleneamine, azo nitrogen] com-
    plex, 112
  dihydridotetrakis(triphenylphosphine),
    44
  hexaamino dichloride, 91
  hydroxytrichloride, 92
  nitrogen borofluoride complex, 89
  nitrogen hexafluorophosphate

    complex, 89
  nitrogenpentaamine complexes, syn-
    thesis of, 103
  [nitrogenpentaamino Di(borotetra-
    fluoride)], 104, 105
  [nitrogenpentaamine] dichloride, 91
  nitrogenpentaamine salts, 103
  pentaamino aquo cation, 94
  [pentaamino aquo] tri(methyl sulfo-
    trioxide) complex, 88
  [pentaamino chloro] dichloride, 95,
    105
  [pentaamino dinitrogen complex] 105,
    111
  [pentaamino dinitrogen] dianion, 88,
    89
  pentaamino dinitrogen dichloride, 104
  [pentaamino dinitrogen] dihalide, 92
  pentaamino dinitrogen tetra(borotetra-
    fluoride), 94
  [pentaamino hydrazine] complex,
    105, 106
  [pentaamino nitrito]$^{2+}$ complex, 111
  8-phenylmonohydride, 99
  potassium pentachloroaquo, 103
  tetra(triphenylphosphine) dihydride,
    99
  triacetylacetonate, 98
  [tris(triphenylphosphine) dihydride,
    nitrogen], 110
  tris(triphenylphosphine) tetrahydride,
    110
  tri(trialkylphosphine) trichloride, 88

Sacco-Rossi complex, 98
S_E^1 Mechanism, organomercury com-
    pounds, 133
S_E^2 Reactions, four-centered transition
    state, 133
Selenium, chloromercurio di(phenyl) p-
    tolyl chloride, 171
  dialkyl, 146
  di(α-naphthyl), 170, 249
  diaryl, 145
  di(β-naphthyl), 170
  di(biphenyl), 170
  dichloromercurio, di(p-tolyl) phenyl
    chloride, 171
  tri(m-tolyl) chloride, 170
  tri(p-tolyl), chloride, 171

diethyl, 154, 170
diferrocenyl, 146, 171
dihalodiaryl, 146
di[m-tolyl]phenyl chloride, 170
di-O-tolyl, 249
    preparation of, 250
di[O-tolyl]phenyl chloride, 170
di-p-tolyl, 170, 249
di-pentafluoro, 249
diphenyl, 170, 249
diphenyl p-tolyl chloride, 170
di-selenocyano, 336
mercuriodichloro tri(p-tolyl) chloride,
    170
octafluoroselenanthrene, 248
organo compounds, preparation of,
    154
phenyl, Di[p-tolyl] chloride, 170
phenyl cyanide, 336
tri(m-tolyl)chloride, 170
triphenyl chloride, 170
tri(p-tolyl)chloride, 170
Selenocyanates, 336
Silane, alkyl, carbene insertion, 179
alkoxy, 139
α-triethyl oxyvinylmenthyl ether, 139
β-methylvinyl trimethyl, 223
diaryl dihydride, 180
diethyl dibutoxy, 32
diethyl divinyloxo, 159
dimethylvinyl chloride, 182
ethyllithium, 223
ethyl tributoxy, 32
hexaphenyl, di, 160
1-methoxyvinyloxodiethyl, 159
monoaryltrichloro, 139
phenyl trichloro, 159
p-tolyl trichloro, 160
tetramethyl, 33, 39
tetra[1-Methylvinyloxo], 159
triethyl 1-acetic acid methyl ester, 159
trimethylfluoro, 39
trimethylvinyloxo, 159
triphenyl hydroxy, 160
triphenyl p-tolyl, 160
tetravinyloxy, 139, 159
tetraphenyl, 160, 243, 244
triethyl, 30, 253
triethyl butoxy, 32
triethyl-(carbomethoxymethyl), 150

triethyl iodide, 150
triethyllithium, 223
trimethyl bromo, 39
trimethylethyl, 39
trimethyl(trichloromethyl), 186
trimethylvinyl, 182
triphenyl, dimer, 242, 243, 244
triphenyl aryl, 241
triphenyl(bromomethyl), 180
triphenyl hydride, 180
triphenyl hydroxide, 243, 244
triphenyl lithium, 241, 242
triphenyl p-tolyl, 244
trivinyloxophenyl, 159
vinyl, 182
vinyloxyalkylaryl, 139
vinyltrichloro, 139, 159
Silicon, organo compounds, preparation
    of, 150, 241
tetraalkyl, 185
tetraphenyl, 28
trimethyl azide, 39
trimethyl isothiocyanate, 39
triphenyl azide, 39
Sodium, aluminum methyl trichloride,
    33
benzyl, 195, 213
    preparation of, 204
bis(2-butylethnyl)dihydridealuminate,
    24
bis(2-phenylethnyl)diethylaluminate,
    24
ethyl, 196, 210, 211
    preparation of, 204
isobutyltrihydridealuminate, 24
lithiumdiphenyl, 213
methyl, 209
naphthalene, 67, 72
n-propyl, 211
octyl, 212
petene, Di, 196
phenyl, 195, 213
tetraethylaluminate, 23, 24, 30, 38, 39
tetramethylaluminate, 23
tetraphenylaluminate, 23
tolyl, 195
vinyl, 195, 196, 203
vinyloxo, 213
Spiropentanes, 183
Stereochemical isomers, entropy

activation, 292
Stilbene, cis-α-bromo, 342
  triethyl, 40, 41, 168
  trimethyl, 222
  triphenyl, 222
Sulfide, dialkyl di, 248
  di-O-tolyl, 248, 249
  diphenyl, 249
  pentafluoroethyl, dimer, 249
Sulfur, diferrocenyl, 145
  hexafluorothioacetone, 249
  O-nitrophenyl, O-tolyl, 169
  p-Nitrophenyl M-tolyl, 169
  p-Nitrophenyl-p-chlorophenyl, 169
  p-tolyl O-nitrophenyl, 169
  p-tolyl phenyl, 169
  perfluorothioacetone, 248
  trifluorothioacetyl fluoride, 249
  tri[N,N-dimethylaniline] chloride, 169
Symmetrizators, substitution hydrogen,
  by, 254

Tellurium, 4-anisole α-naphthyl, 171
  α-naphthyl β-naphthyl, 171
  aryl trichloride, 146
  di-α-naphthyl, 249
  di-O-tolyl, 249
  diphenyl, 249
  di-p-tolyl, 249
  p-phenoxyphenyl β-naphthyl, 171
  4-phenoxyphenyl α-naphthyl, 171
  phenyl α-naphthyl, 171
  phenyl anisole, 171
  phenyl β-naphthyl, 171
  phenyl diphenyletherate, 171
  trichloro, 4 carboxyphenyl phenyl-
    etherate, 171
  trichloro tolylphenyletherate, 171
Tetra fluoro ethane, 251
2,2,5,5-tetramethyl-1,4 dioxan, 264
Thallium, α-di-naphthyl chloride, 159
  α-naphthyl Di(isobutyrate), 158
  α-thienyl Di-isobutyrate, 159
  β naphthyl Di(isobutyrate), 159
  cis,cis-Di-β-chlorovinyl, 130
  cis, Di[2-Chlorovinyl] chloride, 157
    Dipropene bromide, 157
  cyclopentadienyl, 139
  dialkyl chloride, 35
  di(α-naphthyl)chloride, 159

di-β-chlorovinyl, 138
di 2-carbmethoxy-1-butene 3-chloride,
  157
di[2-carbmethoxy-1-methyl-1-
  propene] chloride, 157
diethyl bromide, 135
di-1-methyl-2-acetoxy-1-propene-1-yl
  dichloride, 138
di(1-methylvinyl)bromide, 157
di-O-acetoxyphenyl bromide, 138
di(O-anisole)chloride, 158, 159
di[p-anisole]isobutyrate, 158
diphenyl chloride, 158, 159
diphenyl isobutyrate, 150, 158
di[2-phenylvinyl]bromide, 157
dipropyl chloride, 156
di p-tolyl isobutyrate, 158
di-trans-propenyl chloride, 138
divinyl chloride, 157
isobutyrate, 138
monoaryl carboboxylate salt, 138
n-Propyl chloride, 137
O-carbmethoxyphenyl chloride, 158
organo compounds, preparation of,
  149
p-anisole di isobutyrate, 158
p-chlorophenyl Di isobutyrate, 158
phenyl dichloride, 159
phenyl Di isobutyrate, 149, 158
p-tolyl Di-isobutyrate, 158
thienyl derivatives, 138
trans dipropene chloride, 157
tricarboxylate salt, 138
8-tri(isobutyloxy)-2,2,4,6-tetramethyl-
  1,2,3,4-tetrahydroquinoline-
  nitrogenoxide, 159
Thiophenol, 248, 285
Tin, alkyl trichloride, 229
  α-naphthyl tribromide, 230
  β-chlorovinyl, 191
  β-chlorovinyl trichloride, 200, 221
  bis(α-naphthyl)dichloride, 256
  bis(β-naphthyl)dichloride, 256
  bis(β-tolyl)dichloride, 256
  bis(o-iodophenyl)dichloride, 256
  bis(o-tolyl)dichloride, 256
  cis-di(β-chlorovinyl)dichloride, 235
  cis-(di-1-propene), dibromide, 234
  cis-[distibene] oxide, 235
  (cis-stibene)oxo hydroxide, 235

cis-stilbene trichloride, 235
dialkyl dichloride, 229
di-(α-methylvinyl)dibromide, 230, 234, 235
di-(α-naphthyl)dibromide, 230, 238
di-(α-naphthyl)dichloride, 238
diaryl dihalides, 228
dibenzyl diacetate, 30
dibenzyl dibromide, 236
dibenzyl dichloride, 235
di-(β-chlorovinyl)dichloride, 221
di-cis-β-chlorovinyl chloride, 232
diethoxy dihalide, 228
diethyl, 231, 239
diethyl(α-acetic acid methyl ester), chloride, 245
diethyldibenzyl, 30
diethyl dibromide, 234
diethyl dichloride, 35, 36, 47, 229, 230, 232, 234
  preparation of, 232
diisobutyl dichloride, 36
di(isobutyl)oxide, 36
diisopropenyl dibromide, 230
di-n-butyl(α-acetic acid methyl ester) chloride, 245
di-o-tolyl, dichloride, 230, 237
  oxide, 237
di(p-bromophenyl)dibromide, 237
di-(p-bromo phenyl)dichloride, 237
di-(p-chlorophenyl)dibromide, 237
di-p-chlorophenyl dichloride, 232, 237
di(p-ethyl benzoic acid ester), dibromide, 238
  dichloride, 238
di(p-iodophenyl)dichloride, 237
diphenyl chloride, 140, 289
diphenyl dichloride, 236
diphenyl diethyl, 236
diphenyldiethyl, preparation of, 239
diphenyl oxide, 236
dipropenyl dibromide, 230
dipropyl(α-acetic acid methyl ester) chloride, 245
dipropylchloro hydride, 254
di(p-tolyl)dichloride, 230
di(trans-β-chlorovinyl), dichloride, 200
hexaaryl, di, 201
hexaethyl di, 231, 239
hexaethylditin, 239

n-tributyl bromide, 181
organo compounds, synthesis of, 228, 242
O-oxyphenyl trichloride, 256
p-aminophenyl, trichloride, 256
p-n,n-dimethylaniline triethyl, 238
p-tolyl oxo hydroxide, 237
phenyl, oxo hydroxide, 235
phenyl trichloride, 36
tetra-(α-methylvinyl), 230, 234, 235
tetrabutyl, 36, 47
tetraethyl, 28, 32, 36, 47
tetraisobutyl, 36
tetramethyl, 33
tetraoctyl, 36
tetra(p-chlorophenyl), 222
tetra(p-tolyl), 222
tetra-p-tolyl, preparation of, 207
tetra(pentafluorophenyl), 200, 222
tetraphenyl, 28, 36, 200, 221, 222
trans-di-(β-chlorovinyl), dichloride, 229, 235
trans-(di-1-propene)dibromide, 234
tri-[benzoic acid ethyl ester], chloride, 222
tribenzyl chloride, 200, 221
tri-(β-chlorovinyl)chloride, 221
tributyl acetate, 30
tributyl hydride, 29, 181
triethyl, dimer, 242, 245, 254
  preparation of, 239
triethyl acetate, 244, 254
triethyl acetic acid methyl ester, 231, 234, 239, 241, 242, 245
triethyl acetic acid propyl ester, 246
triethyl bromide, 234
triethylchloride, 35, 36, 47, 161, 229, 239, 244, 245
triethyl deuteride, 30
triethyl hydride, 242, 246
triethyl p-tolyl, 242, 245, 254
triethyl phenol, 237
triethyl phenyl, 236, 239, 242, 245, 254
trimethyl, dimethylamide, 32
trimethyl ethyl, 36
tri-(p-chlorophenyl), dimer, 222
tri-(p-tolyl), chloride, 222
tri(p-tolyl), dimer, 22
triphenyl, dimer, 222

triphenyl bromide, 230, 236
triphenyl chloride, 207, 222, 235
triphenyl hydroxide, 236
tris-β-chlorovinyl, chloride, 200
tri[trans-β-chlorovinyl], chloride, 200,
207
vinyl trichloride, 140, 161
Tin-sodium alloy, 191
Titanium, aluminum halide nitrogen
complexes, 76
bis(cyclooctatetraene), 32
dichlorotitanocene diethyl alumino,
complex, 42
dicyclopentadienic, 71
di(cyclopentadienyl)chloride, 41
dicyclopentadienyl dichloride, 141
(dicyclopentadienyl dihydride) anion,
72
dicyclopentadienyl hydride), dimer, 72
di(hexyl-oxy), dichloride, 72
dimethyl dichloride, 42
di(methyloxy)dichloride, 72
dimethyl-titanocene, 81
di(phenyloxy)dichloride, 72
diphenyltitanocene, 81
di(ter-butyl-oxy), dichloride, 72
ethylchlorotitanocene, 41
ethyl trichloride, 42
hexa(cyclopentadienyl)nitrogen, 71
isobutyl trichloride, 42
methyl trichloride, 42
nitrogen complex, 77
o-tolyl, 86
phenylazo derivative, 83
tetrabutoxide, 32
tetraethoxy, 80, 81, 102
tetraisopropoxide, 67, 68, 104
titanocene, diphenyl, 68, 69, 82, 86
titanocene dichloride, 41, 62, 65, 68,
69, 71, 72, 73, 78, 82, 84, 85, 101,
102, 104
tri(ethoxy)cyclopentadienyl, 78
tris(cycooooctatetraene)di, 32
Titanocene, 81
anionic hydride titanium complex, 104

p-Toluene sulfonic acid, 285
p-Tolyl vinyl sulfone, 285
p-Tolyl vinyl sulfoxide, 320
Trans-hex-3-enol, 262
Triaryl bromomethane, 302
Tri-bromocyclopentane, 338
Triethoxy vinyl phosphate, 320
Trimethylamine, 278
2,4,7 Trinitro fluorenone, 344
Triphenyl chloromethane, 302, 303
Triphenylene, 83, 272
Triphenylmethylacetaldehyde, 302
Triphenylmethylacetophenone, 302
Triphenylmethyl bromide, 303
Tris(2-penten-3-yl)phosphate, 321
Trivinyl trisulfoxide, 320

Vanadium, phenyl dichloroxo, 144
phenyl trichloride, 144
Vinyl alkalimetal compounds, preparation of, 203

Zinc, bis(phenylethynyl), 19
dibenzyl, 217
di-(β-naphthyl), 217
dibutyl, 27
diethyl, 8, 27, 28, 33, 46, 216
diisobutyl, 30, 317
di(iso-pentyl), 217
dimethyl, 9, 10, 27, 30, 31, 34, 216
preparation of 46, 245
di-(n-butyl), 216, 217
di-(n-propyl), 216
di-o-tolyl, 198, 217
di-p-chlorophenyl, 198, 217
preparation of, 205
di-p-dimethylaminophenyl, 198
di-p-fluorophenyl, 198, 217
di[p-n,n-dimethylaniline], 217
diphenyl, 198, 217
ethoxide, 31, 46
Zirconium, Di(chlorozirconocene)
oxide, 41
methylchlorozirconocene, 41
zirconocene dichloride, 41